Estuarine and Wetland Processes

WITH EMPHASIS ON MODELING

MARINE SCIENCE

Coordinating Editor: Ronald J. Gibbs, *University of Delaware*

Volume 1 – *Physics of Sound in Marine Sediments*
Edited by Loyd Hampton

An Office of Naval Research symposium
Consulting Editors: Alexander Malahoff and Donald Heinrichs
Department of the Navy

Volume 2 – *Deep-Sea Sediments: Physical and Mechanical Properties*
Edited by Anton L. Inderbitzen

An Office of Naval Research symposium
Consulting Editors: Alexander Malahoff and Donald Heinrichs
Department of the Navy

Volume 3 – *Natural Gases in Marine Sediments*
Edited by Isaac R. Kaplan

An Office of Naval Research symposium
Consulting Editors: Alexander Malahoff and Donald Heinrichs
Department of the Navy

Volume 4 – *Suspended Solids in Water*
Edited by Ronald J. Gibbs

An Office of Naval Research symposium
Consulting Editors: Alexander Malahoff and Donald Heinrichs
Department of the Navy

Volume 5 – *Oceanic Sound Scattering Prediction*
Edited by Neil R. Andersen and Bernard J. Zahuranec

An Office of Naval Research symposium

Volume 6 – *The Fate of Fossil Fuel CO_2 in the Oceans*
Edited by Neil R. Andersen and Alexander Malahoff

An Office of Naval Research symposium

Volume 7 – *Transport Processes in Lakes and Oceans*
Edited by Ronald J. Gibbs

Consulting Editor: Richard P. Shaw

Volume 8 – *Ocean Wave Climate*
Edited by Marshall D. Earle and Alexander Malahoff

Volume 9 – *Marine Geology and Oceanography of the Pacific Manganese Nodule Province*
Edited by James L. Bischoff and David Z. Piper

Volume 10 – *Ecological Processes in Coastal and Marine Systems*
Edited by Robert J. Livingston

Volume 11 – *Estuarine and Wetland Processes: With Emphasis on Modeling*
Edited by Peter Hamilton and Keith B. Macdonald

A Continuation Order Plan is available for this series. A continuation order will bring delivery of each new volume immediately upon publication. Volumes are billed only upon actual shipment. For further information please contact the publisher.

Estuarine and Wetland Processes

WITH EMPHASIS ON MODELING

Edited by

Peter Hamilton

Science Applications, Inc.
Raleigh, North Carolina

and

Keith B. Macdonald

Woodward–Clyde Consultants
San Diego, California

Springer Science+Business Media, LLC

Library of Congress Cataloging in Publication Data

Symposium on Estuarine and Wetland Processes and Water Quality Modeling, New Orleans, 1979.
Estuarine and wetland processes, with emphasis on modeling.

(Marine science series; 11)
Sponsored by the U.S. Army Engineer Waterways Experiment Station, Vicksburg, Miss.
Includes index.
1. Estuarine oceanography–Mathematical models–Congresses. 2. Estuarine oceanography–United States–Congresses. 3. Wetlands–Mathematical models–Congresses. 4. Wetlands–United States–Congresses. I. Hamilton, Peter, 1949- II. Macdonald, Keith B. III. U. S. Army Engineer Waterways Experiment Station. IV. Title.
GC96.5.S95 1979 551.46'09 80-14721

ISBN 978-1-4757-5179-6 ISBN 978-1-4757-5177-2 (eBook)
DOI 10.1007/978-1-4757-5177-2

Proceedings of the Workshop on Estuarine and Wetland Processes and Water Quality Modeling, held in New Orleans, Louisiana, June 18–20, 1979, and sponsored by the U.S. Army Engineer Waterways Experiment Station, Vicksburg, Mississippi.

Originally published by Plenum Press, New York in 1980.

Softcover reprint of the hardcover 1st edition 1980

PREFACE

Estuaries and Wetlands are important coastal resources which are subject to a great deal of environmental stress. Dredging, construction, creation of intertidal wetlands, regulation of freshwater flow, and pollution are just a few of the activities which affect these coastal systems. The need to predict the effects of these perturbations upon ecosystem dynamics, particularly estuarine fisheries, as well as on physical effects, such as sedimentation and salt intrusion, is of paramount importance. Prediction requires the use of models, but no model is likely to be satisfactory unless fundamental physical, chemical, sedimentological, and biological processes are quantitatively understood, and the appropriate time and space scales known.

With these considerations in mind, the Environmental Laboratory, U. S. Army Engineer Waterways Experiment Station,* Vicksburg, Mississippi, sponsored a workshop on "Estuarine and Wetland Processes and Water Quality Modeling" held in New Orleans, June 1979. The contents of this volume have been selected from the workshop papers. The resulting book, perhaps more than any other symposium proceedings on estuaries and wetlands, attempts to review important processes and place them in a modeling context. There is also a distinct applied tinge to a number of the contributions since some of the research studies were motivated by environmental assessments. The difference in title between this volume and the workshop reflects more accurately the contents of the published papers.

The editors have attempted to make the contributions accessible to a wide range of coastal marine scientists. Therefore, this volume consists of a heterogeneous collection of papers ranging from an almost monograph length review to short research notes covering a variety of disciplines involved in research and environmental assessments. Since understanding complex interactions occurring in these environments requires knowledge of a number of disciplines, the editors hope that biologists will read the front half of the book (despite the equations) as well as the wetland sections and vice-versa for the physical oceanographers and engineers (despite the Latin names).

Much of the book is devoted to reviews of processes and studies of large scale systems. Reviews of turbulence, numerical hydrodynamics, sediment transport modeling, salt marsh modeling, and various aspects of marsh-estuarine interactions are in the former category, while reviews of circulation in Chesapeake Bay, engineering applications of two dimensional numerical tidal models, hydrography of Gulf coast estuaries, and sediment and nutrient cycling in salt marshes are in the latter. Complementing these reviews are papers reporting results of smaller studies.

The first half of the volume emphasizes estuarine physical processes. In addition to the topics mentioned above, a number of contributions are devoted to circulation modeling with emphasis on engineering and environmental applications. A majority of these papers report the application of one or two dimensional versions of the vertically integrated equations of motion. However, one paper critically examines the use of steady-state box models to calculate the distribution of dissolved substances in an estuary. These kinds of models are often used (or sometimes abused) as the basis for water quality studies. Another contribution presents a simple model used to evaluate the dispersion of a turbidity plume from an operating clam shell dredge.

The second half of the volume brings together ecologically oriented contributions. Papers describing the flux of ATP, suspended particulates and coliform bacteria in wetlands reflect an increasing appreciation for the complexity and variability of such systems. Contributions on the role of sedimentation in salt marsh nutrient flux, seepage in salt marsh soils, and the use of remote sensing to measure emergent vegetation biomass introduce important topics that warrant more attention in future research. A seminal review discusses twenty years of speculation and research on the role of salt marshes in estuarine productivity and water chemistry. Progress reports on salt marsh and estuarine ecosystem modeling contrast sharply in scale and approach with a paper that describes a detailed model of small-scale movements of zooplankton populations in a tidal lagoon. The volume closes with two papers outlining field experiments designed to assess the effects of estuarine modifications upon fisheries resources and chronic impact detection among marine organisms.

The editors wish to thank contributing authors, workshop participants, and numerous outside reviewers of the papers. Don Robey and Ross Hall provided valuable assistance in planning the workshop. Assistance of the staff members from the Raleigh, North Carolina, Boulder, Colorado, and La Jolla, California offices of Science Applications, Inc., particularly Ken Fucik, Joanne Brown, Martin Miller, Paul Debrule, and Ivan Show, in running the workshop is gratefully acknowledged. We are particularly grateful to

Paula Bonaminio for her excellent preparation and typing of the manuscripts.

Peter Hamilton, Raleigh, North Carolina
Keith B. Macdonald, San Diego, California
February 1980

*The findings in these papers are not to be construed as an official Department of the Army position unless so designated by other authorized documents.

CONTENTS

TURBULENT PROCESSES IN ESTUARIES

George B. Gardner, Arthur R. M. Nowell and
J. Dungan Smith

INTRODUCTION

Estuaries usually are regions in which mixing of miscible fluids with differing densities is occurring; thus they are regions in which turbulent processes are of particular importance. Moreover, the nature of these turbulent processes, although highly variable, has a profound effect on the circulation. As a first approximation, theoreticians often write the diffusion of mass and momentum in terms of a tensor eddy coefficient having spatially varying components. In some cases this mathematical procedure closely approximates the physics of the problem whereas in others, it is reasonable neither from an oceanographic nor mathematical point of view. Without a physical understanding of the basic processes that ultimately result in mixing in systems of various types, it is impossible to judge the validity of such simplifications for specific embayments. Nevertheless, surprisingly little basic research has been done on the small scale processes that are responsible for turbulence production in such environments.

In many estuarine situations, time dependent processes associated, in particular, with tidal currents are very important in regard to the mixing. In these cases a host of nonlinearities are included in a tidal average. Not only does tidal averaging hide the basic physical processes responsible for mass and momentum transfer in estuaries, but, worse yet, it makes the momentum diffusion impossible to specify. Therefore it becomes a free coefficient that must be adjusted for each specific estuary. Studies that utilize a momentum diffusion coefficient for turbulent mixing in estuaries have been reviewed recently by Bowden (1977); rather than summarize these works and the early experimental observations, we will focus here on the processes responsible for turbulent mixing in well-mixed and highly stratified estuaries.

Proper classification of estuaries, with end members of the well-mixed and highly stratified types, requires a description of the scales and intensities of the turbulent flow field. The turbulent processes responsible for mixing in estuarine waters may be examined through assessment of the role of gravity in driving the flow and in inhibiting or enhancing entrainment and mixing. Well mixed estuaries can be studied using boundary layer models because the dominant source of turbulent energy occurs just above the lower boundary. However, in highly stratified estuaries turbulent transport across the pycnocline is controlled by local processes on the pycnocline. These estuaries may be modeled only when the mixing processes at the free shear layer including their interaction with the wall-bounded turbulent boundary layers, are understood. Much experimental work on free shear layer mixing in the presence of stratification has been carried out in the laboratory but few experiments in the natural environment have measured the turbulent field with sufficient spatial coverage to describe adequately the nature of the mixing processes.

The present survey reviews some of the ideas and results that have been published recently in the fluid dynamics literature on the mechanics of mixing in boundary layers and in free shear layers, then applies these concepts along with recent estuarine measurements to an assessment of the present understanding of estuarine dynamics with particular emphasis on those estuaries with relatively large tidal ranges, where tidal averages are less useful.

GENERAL CHARACTERISTICS OF TURBULENCE IN ESTUARIES

Turbulence is best defined by a syndrome of characteristics and it is described using statistical methods; its physics is conceptualized using comprehensive flow measurements and visual tracers. In laboratory experiments, hydrogen bubbles have been used extensively to examine the generation of Reynolds stresses, while in estuaries high frequency echo sounders can be used to reveal the details of mixing events. A limited use of echo sounding for studying the pycnocline was described by Edgerton (1966), and a fairly extensive acoustic survey is presented by Fukashima et al. (1966). Recent work with this technique is described later in this paper (Figures 11-14 and 16-18). The effects of turbulence are predictable in a gross sense, utilizing closure theories of various types. These theories range from the simplest eddy viscosity models to complicated postulates such as the third order model prescribed by Long (1977). In estuaries a chief symptom of interest is the hierarchy of scales associated with the different mechanisms of turbulent production. In well mixed estuaries the outer length scale is proportional to the distance from the bed and following Kolmogorov's notions, to the wavelength that corresponds to the peak of the energy spectrum. In stratified flows, this macro-scale is proportional to the shear layer width. Other traditionally mentioned turbulence characteristics (Tennekes and Lumley, 1972; Lumley and Panofsky, 1964; Bradshaw, 1972) are three dimensionality and high rates of energy dissipation. Although equally applicable to estuarine flows, these features do not comprise a particularly useful basis for classifying the turbulent effects in such embayments.

One important characteristic of turbulence that has received much emphasis in recent laboratory studies is intermittency. That turbulent kinetic energy dissipation is intermittent is well documented (Townsend, 1948; Boeing Symposium, 1970; that stress production is intermittent in smooth wall boundary layers and free shear layers and is characterizable by identifiable coherent motions is well established. However, it is likely that the prevalent fetish about bursting (Liepmann, 1979) will not have global applicability in depth-limited estuarine boundary layers; the absence of an outer intermittent region into which irrotational flow is entrained and the absence of a continuous viscous sublayer removes two of the elements necessary in the feedback loop for bursting (Lighthill, 1970). Recent work by Chen and Blackwelder (1978) indicates a strong coupling, across the buffer layer, between the outer region coherent eddies and the bursting at a smooth wall. A consistent model of the bursting process, based on flow visualization results, requires that the bursting be instigated by events originating in the outer intermittent region and that the inflows (sweeps) be caused by these regions of irrotational flow. Such a region cannot exist in many estuarine, depth-limited boundary layers.

Estuarine turbulence is strongly anisotropic, especially at the scales which control the rate of mixing and entrainment; the redistribution term in the Reynolds stress transport equation, therefore must play a dominant role in the mechanics of mixing. The modeling of turbulence takes as its starting point the Reynolds stress transport equation,

Production Dissipation

$$\frac{\partial \overline{u_i u_j}}{\partial t} + \overline{u_k}\,\frac{\partial \overline{u_i u_j}}{\partial x_k} = -\overline{u_i u_k}\,\frac{\partial \overline{u_k}}{\partial x_j} - \overline{u_k u_j}\,\frac{\partial \overline{u_k}}{\partial x_i} - 2\nu\left(\overline{\frac{\partial u_i}{\partial x_j}\,\frac{\partial u_j}{\partial x_i}}\right)$$

$$+ \underbrace{\frac{p}{\rho}\left(\frac{\partial u_i}{\partial x_j} + \frac{\partial u_j}{\partial x_i}\right)}_{\text{Redistribution}} - \underbrace{\{\frac{\partial}{\partial x_k}(\overline{u_i u_j u_k} - \nu \frac{\partial \overline{u_i u_j}}{\partial x_k} + \frac{\overline{p}}{\rho}(\delta_{jk} u_i + \delta_{ik} u_j))\}}_{\text{Diffusion}}$$

$$+ \frac{1}{\rho}\underbrace{(g_i \overline{u_j \rho'} + g_j \overline{u_i \rho'})}_{\text{Buoyancy}} \tag{1}$$

Equations for the density-velocity covariance and density variance complete the set. In many boundary layers the last three terms in (1) may be ignored yielding a simple balance between production and dissipation. Such is the case in models for well mixed estuaries where only small corrections need be made for the last term to account for the buoyant production or destruction of turbulence. These estuaries also can be modeled successfully using eddy viscosity models which give good predictions at a low cost (e.g., Long and Smith, 1979). However, when strong stratification enters the problem it is obvious that the redistribution and buoyancy terms become very significant. The body force enters explicitly in both terms and it can be shown that the redistribution term behaves differently in free shear layers and in wall-bounded shear flows. Rotta (1951) indicated that in neutral flows, the two sources of contribution to the redistribution term arise, one from the mean shear and the other due to turbulence interactions. In stratified flows there is a contribution due to the interaction of the density fluctuations and velocity fluctuations in the gravitational term. An extensive review of this single term in the Reynolds stress transport equation is given by Launder (1975, 1976) and Gibson and Launder (1976).

Two important points emerge whatever type of turbulence model is attempted. First the equations represent an average condition, the theoretical ensemble average, and averaging requires that we know what the significant mixing scales are. Approximating the ensemble average by using a time average requires that we know where the spectral gap occurs. Few detailed time series over a sufficient period have been obtained that document this required characteristic, though McLean and Smith (in press) present spectra from a natural flow indicating a flattening of the energy spectrum at low frequency. Tidal averaging and depth averaging the equations of motion and the Reynolds stress equation hides the physics of the mixing. Because the time scales of mixing will be proportional to the time scales of decay of the large eddies, it would seem realistic to average on this dynamic scale. Second, if intermittency is important in stress generation, then the average form of the

Reynolds stress equation will likely reproduce poorly the details of the physics, unless the mean of the generating process is a good measure of the magnitude and importance of the process, or unless it scales all moments of the resulting frequency distribution. That most turbulence models, even the simplest eddy viscosity type, work well for smooth-wall boundary layers where intermittency is most readily observed (Willmarth, 1975), indicates the details of intermittency may not be as important as implied by Gordon and Witting (1977) in modeling the effects of turbulent mixing.

Well Mixed Estuaries

Turbulence production in weakly stratified or unstratified embayments is associated primarily with the bottom boundary layer. Therefore, in order to develop an understanding of the turbulence fields in such estuaries it is necessary (1) to examine bottom boundary layer data procured under relatively simple conditions in order to ascertain whether the knowledge gained over many decades from laboratory and theoretical studies is directly applicable to the field situation and (2) to investigate the important and complicated effects that arise due to non-planar bottom topography. The second category further can be divided into two-dimensional or quasi two-dimensional effects and effects due to lateral gradients and channel curvature. Classification of such natural turbulence production and near bottom flow problems in regard to topographic effects provides a framework within which theoretical and experimental research efforts can be focused. This approach provides an alternative to the weakly founded, but often used set of generalizations that typically obscures the physics of estuarine circulation and often leads both modeling groups and field workers astray.

Numerous velocity field measurements have been made near the bottom in various types of estuaries, and several stress field measurements in this environment are available. Nevertheless, few experiments have been carried out in which velocity, stress and density fields have been measured simultaneously, and in none of these has the local- and regional-scale bottom topography been determined accurately. Topographic effects can induce distinct variation in near-bottom velocity and stress data and it is the obligation of the experimentalist to identify the source of this variability. In those cases for which the beds can be presumed to have been relatively smooth on all but the smallest scales and for which velocity components were measured from stable, bottom mounted frames, the mean flow data indicate a velocity profile of the form $u = (u_*/k) \ln (z/z_o)$, as expected. Here u_* is the shear velocity, k is von Karman's constant, z is the distance from the bed and z_o is a constant of integration called the roughness parameter. As no estuarine experiments have provided accurate estimates of von Karman's constant most bottom boundary layer researchers presume that is has a value of 0.40 in clear water.

In contrast to the apparently well understood situation in regard to velocity profile shape, an understanding of what causes effective changes in the roughness parameter and a knowledge of how to predict this variable is lacking. The roughness parameter, like von Karman's constant, is affected by erroneous interpretation of measured velocity profiles such as occur when topographic and stratification effects are present but ignored. Even accounting for this difficulty, typical values of z_o in natural flows are from one to two orders of magnitude higher than those found for pipe flow by Nikuradse (1933). Moreover, this parameter varies with boundary shear stress in a manner substantially different from that predicted using Nikuradse's work. There are many causes for this situation: one is the sediment transport process; a second is pressure drag on large scale bed topography. A third effect arises due to the nonlinear interaction between large-amplitude flows of greatly differing frequencies. For example, using a theory analagous to that derived by Smith (1977), Grant and Madsen (1979) show that high frequency gravity waves (e.g., wind waves) cause a substantial increase in z_o.

Due to a paucity of complete data sets for the boundary layers in well mixed estuaries, much reliance will be placed in this paper on a comprehensive experiment carried out in the Columbia River by Smith and his co-workers. In this study, measurements of velocity, Reynolds stress, sediment concentration, water temperature and bottom topography were procured at a large number of locations over a field of 2 m high and 100 m long sand waves. Bottom topography was determined over a large region at the beginning of the investigation and again every few days in the immediate neighborhood of the experimental site. The relatively complete nature of the data set from this study permitted Smith and McLean (1977a, b) to obtain an accurately spatially averaged flow field; hence to examine what might be considered the zero order velocity profile in a simple, natural flow that includes topographic and sediment transport effects. Here it should be noted that although important in many boundary layer problems, downstream variation in the velocity field over a topographic feature is significantly smaller than the spatially averaged effect. These nonuniform flow results are believed to be of general applicability and they certainly are relevant to the well mixed estuary situations in which large scale bed forms are a common feature.

From a careful, fluid mechanically based analysis of the Columbia River data, Smith and McLean obtained a von Karman's constant of 0.38 and an expression for the variation of z_o with boundary shear stress, under sediment transporting conditions, of the form

$$z_o = \alpha_o \frac{\tau_b - \tau_c}{(\rho_s - \rho)g} + (z_o)_N. \quad (2)$$

Here $\tau_b = \rho u_*^2$ is the shear stress on the boundary averaged over an area of order 100 D, D is the diameter of the sediment grains, τ_c is the boundary shear stress at which the sedimentary particles first move, ρ_s is the sediment density, ρ is the fluid density, $(z_o)_N$ is the value of z_o given by the Nikuradse results and α_o is an empirical coefficient found to be about 26.

The spatially averaged velocity data obtained by Smith and McLean are presented in Figure 1 along with the semi-empirical profile that they fit to it. The pronounced profile curvature in the upper part of the flow is due in part to the decrease in the eddy coefficient as the upper boundary is approached but also in large measure to form drag on the 2 m high sand waves. The flow depth at this location was approximately 15 m at the time the measurements were made. Some stable stratification resulted from the suspended load being transported by the river, and the spatially averaged profile of suspended sediment concentration is also given in the figure.

Figure 2 presents the measured and calculated Reynolds stress fields. In the outer part of the flow where the streamlines are nearly parallel to the mean riverbed, all of the stress on the boundary is associated with the Reynolds shear stress. However, in the immediate vicinity of the boundary, the Reynolds stress is associated only with what might be called skin friction on the actual river bottom at the location immediately below, the rest of the stress being transmitted to the riverbed through streamline distortion and the pressure distribution on the uneven surface. Referring back to Figure 1, we see that both regions yield nearly logarithmic profiles but that the slopes are considerably different. The value of u_* associated with the slope of the velocity profile in the outer region is much higher as it includes the stress ultimately exerted on the boundary through the pressure distribution.

In physical oceanographic investigations, it is the effect of the roughness parameter associated with the outer logarithmic region that is of importance whereas in sediment transport investigations, it is the shear velocity associated with the inner logarithmic region that must be determined; that is, the major features of the circulation in a given estuary are associated with the overall drag on the boundary, whereas pressure forces exerted on a sediment bed

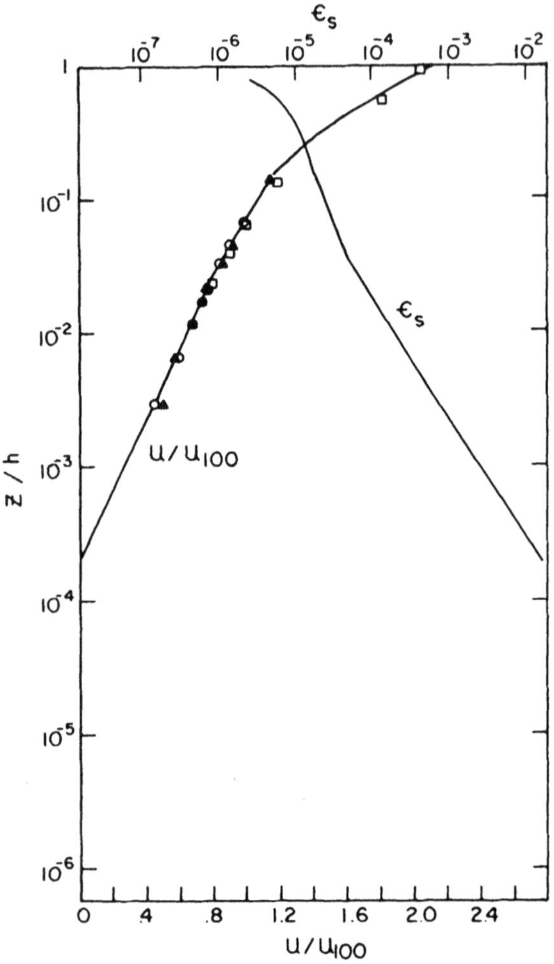

Figure 1. Spatially averaged velocity and suspended sediment (ϵ_s) profiles for flow over a sand wave field in the Columbia River. The symbols on the velocity profile are data from three separate experiments and the smooth curve is the prediction of a model by Smith and McLean (after Smith and McLean, 1977a).

at a scale larger than a few tens of grain diameters have little effect on the erosion and deposition of sedimentary materials. This is because the lift and drag forces on particles in the immediate vicinity of the seabed are scaled by the effective shear velocity below that site. At any given location, the velocity field varies from this spatial average due to the effects of convective accelerations, with the greatest variation being in

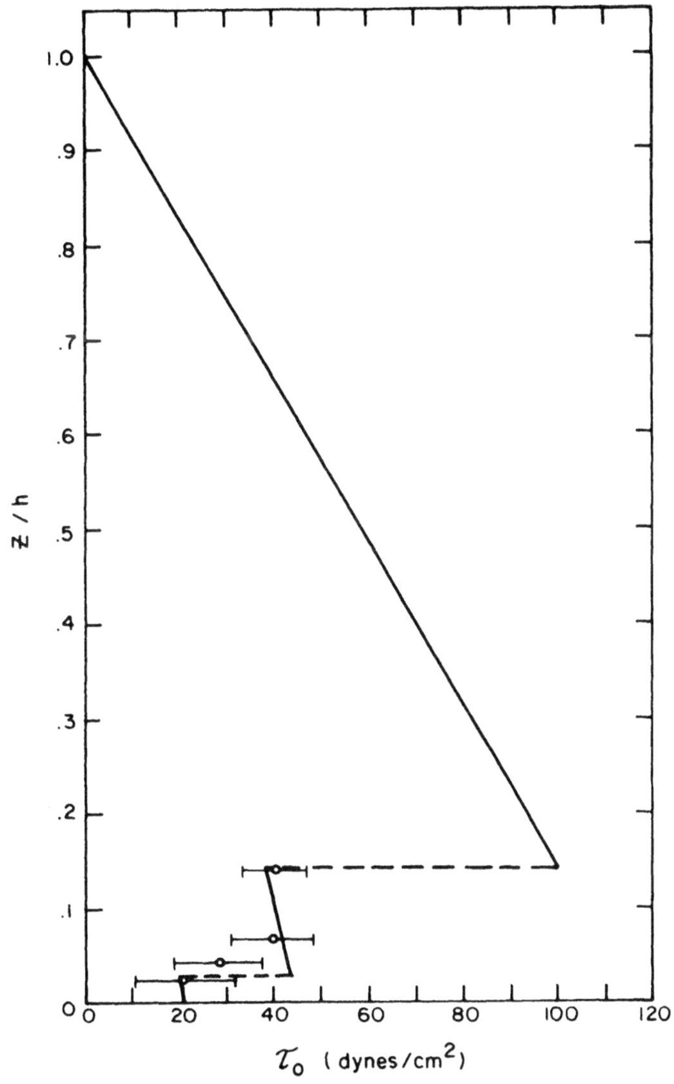

Figure 2. Experimental and theoretical shear stress profiles. The theoretical profile is related to the curves shown in Figure 1 (from Smith and McLean, 1977a).

the immediate neighborhood of the seabed and the least being at the top of the profile near the sea surface. Nevertheless, the general structure of the profile is preserved and can be anticipated once the fluid mechanics of the situation are understood. Likewise, the Reynolds stress field varies from position to position over a given topographic feature, but its general structure is dominated by the spatially averaged profile. That this is the case can be seen by reference to Figure 3, which displays a set of typical Reynolds shear stress profiles taken from various locations over the experimental sand wave field.

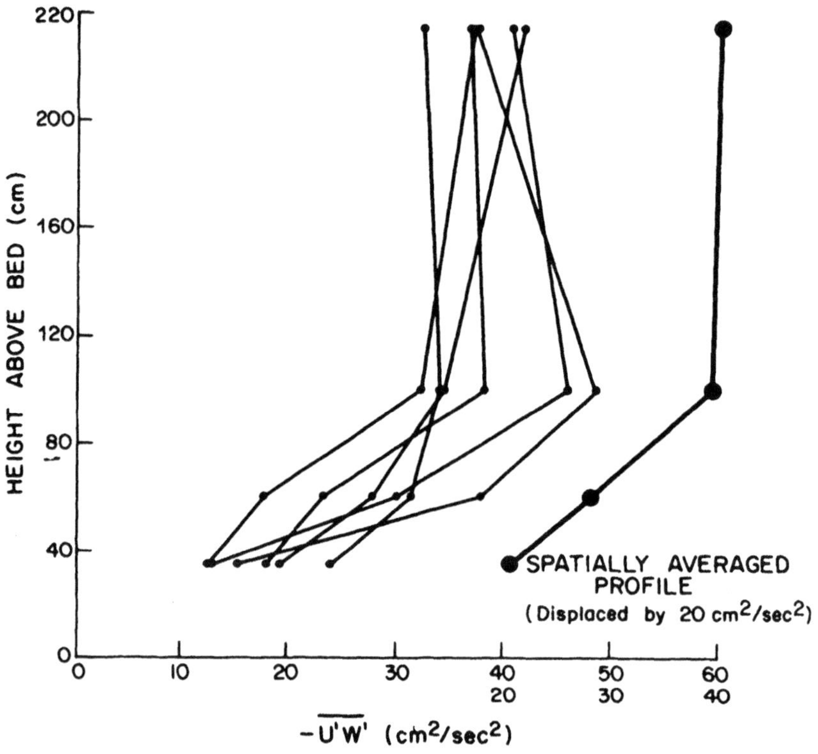

Figure 3. Selected profiles of Reynolds stress on the upstream side of a sand wave. Note that the basic structure of the stress profiles does not greatly differ from that of the spatially averaged version (from Smith, 1978).

The only estuarine related investigations known to the authors, in which more than two mean velocity and Reynolds stress measurements were made in a profile, is the study of Tochko and Williams (Tochko, 1978). In this investigation, four acoustic current meters were deployed in a section of Nantucket Sound near Woods Hole, Massachusetts. The velocity and stress profiles were relatively complicated and unfortunately the local topography was not determined very well during the experiment. Nevertheless, the Tochko and Williams results confirm the importance of making more than one or two Reynolds stress measurements if useful data are to be obtained and they demonstrate the need for making accurate maps of bed geometry in the neighborhood of the deployment site.

For the estuarine modeler, two points are worth extracting from this discussion. First, even in complicated situations, spatially averaged velocity profiles are logarithmic in segments, and within these regions, the mean velocity and Reynolds shear stress fields are quite accurately represented by classical turbulent boundary layer theory. Second, what the physical oceanographer considers boundary shear stress is not the same thing that is called boundary shear stress in sediment transport calculations. Therefore, the relatively complicated near bottom flow processes, which can be neglected in many general circulation studies, have a profound effect on sediment transport patterns in estuaries. Whether or not these complications can be modeled successfully, they must be appreciated if the goal of a given program is to predict erosion and deposition. A third point also might be made here, not in regard to the modeling of the estuarine situations but rather with respect to the procurement of boundary layer data. Clearly the stress field as presented in Figure 3 is sensitive to the local topography. Indeed, it is much more sensitive than the velocity field. Therefore, making Reynolds stress measurements at only one or two levels may be of little value when estimates of boundary shear stress are desired, especially if the bottom topography is not determined accurately at the experimental site. Use of an interior flow model in conjunction with a set of accurate velocity profiles may indeed provide a better estimate of total boundary shear stress and use of a Preston tube will provide a better estimate of local skin friction.

The rate of turbulent mixing in boundary layers is governed by the interior velocity, the boundary shear stress, the boundary morphology and the channel topography. In unstratified flow over hydrodynamically smooth beds there is considerable evidence that the mixing is achieved by the interaction of large scales of motion at the bottom of the logarithmic region with small scale viscous instabilities and that this interaction results in short, identifiable bursts of momentum transfer (Willmarth, 1975); in the carefully controlled laboratory experiments the bursts have emerged only rarely beyond the logarithmic layer. Few laboratory experiments have examined the coherent burst-sweep cycle in sediment transporting, rough-wall flows such as occur in nature, the only exceptions being measurements in a flume over sand and gravel (Grass, 1971) and work on rough wall pipe flow (Sabot et al., 1977). The latter authors conclude that the bursts are still the dominant events controlling the shear stress but the scaling for their frequency of occurrence is different for the rough wall case. They point out that the common scaling for bursting, in terms of free stream velocity and flow depth, does not collapse the data for the rough wall case. In rough wall flows, where there is no continuous viscous sublayer, the frequency of bursts will be likely influenced by the flow instabilities associated with vortex shedding by the particles on the boundary and by small scale bed topography. As these small scale instabilities are produced frequently (being related to the Strouhal frequency) the background level of fluctuating vorticity will be high. Thus the identification of intermittent bursts of vorticity become of little practical value.

The specific character of the velocity fluctuation field in the vicinity of the seabed has been the focus of several investigators, the most notable of these are the studies by Bowden (1963), Gordon and Dohne (1973), Heathershaw (1974) and McLean and Smith (in press). Of these, only the investigations of Bowden and Gordon and Dohne were carried out under truly estuarine circumstances, that of Heathershaw being in the Irish Sea and that of McLean and Smith being in the Columbia River upstream of the Bonneville Dam. Nevertheless, from a fluid mechanical point of view, the environmental differences are less critical than the experimental ones. Only in the McLean and Smith case was velocity fluctuation information collected at more than two levels and only in that case were fluctuations with frequencies up to 3 Hz measured. Figure 4 shows their spectra of $\overline{u^2}$, $\overline{v^2}$ and $\overline{w^2}$ at a height of 2.15 m. Note the anisotropy at low frequency, related to turbulence production by the vertical shear in u. At high frequencies the three spectra are similar and approximately follow a -5/3 decay. Cross-spectra of $\overline{uw}$ are shown for four levels in Figure 5. At high frequency the cross-spectrum is highest for the lower positions while at lower frequency the reverse is true. This is consistent with the expected increasing characteristic length scale of the turbulence with increasing distance from the boundary. A significant feature of the co- and cross-spectra is the important contribution at periods of a few seconds and less. This provides a guide for the sampling design of measurement programs and a warning concerning the use of slow response instruments.

Time series of velocity correlations have been studied by Heathershaw (1974), Gordon (1974) and McLean and Smith (in press). The Heathershaw and McLean and Smith data sets do not differ qualitatively and display features that might be expected on theoretical grounds. In contrast, the Gordon and Dohne data as interpreted by Gordon and Witting (1977) show a marked intermittency. Unfortunately neither the topography in the immediate vicinity of the measurement site nor the density field during the experiment were measured in the latter case, and the sampling

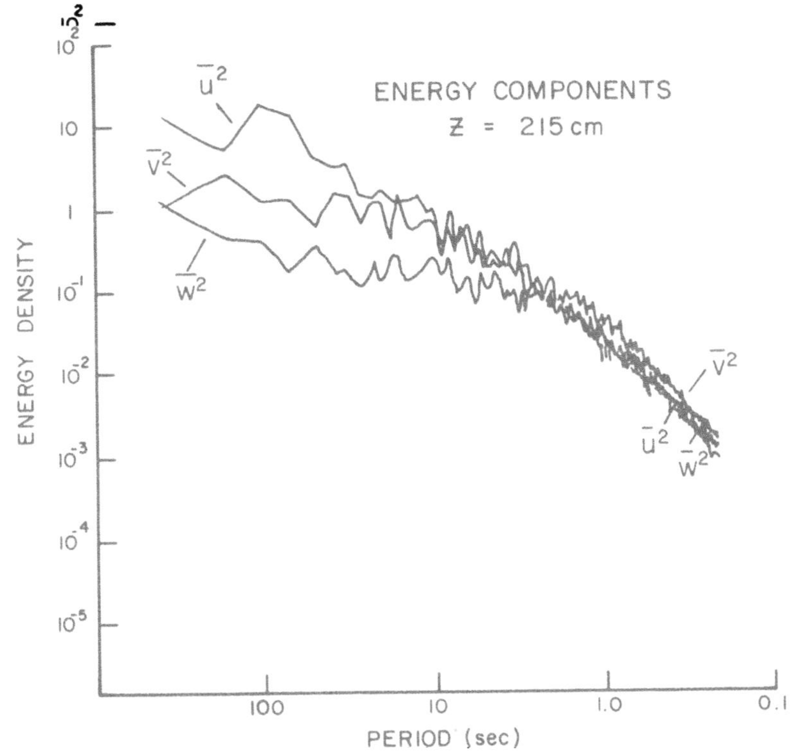

Figure 4. Spectra of $\bar{u}^2$, $\bar{v}^2$, and $\bar{w}^2$ at 215 cm showing increasing isotropy at higher frequencies (after McLean, 1976).

interval at which the data were recorded was only 2.6 seconds. Therefore it is not possible to determine whether the reason for these differences is environmental or experimental. McLean and Smith (in press) have calculated Reynolds stress by averaging the velocity data for 0.8 sec and 2.2 sec before computing u'w'. The results are shown in Figure 6 together with the u'w' calculated without averaging. Note the apparent increase in significant intermittent events with increasing sample interval. The mean stress was 24, 21 and 16 dynes/cm^2 for the unaveraged (0.2 sec), 0.8 sec averaged, and 2.2 sec averaged records respectively. This provides further support for the importance of sampling at high frequency and millitates against uncritical acceptance of all boundary layer measurements.

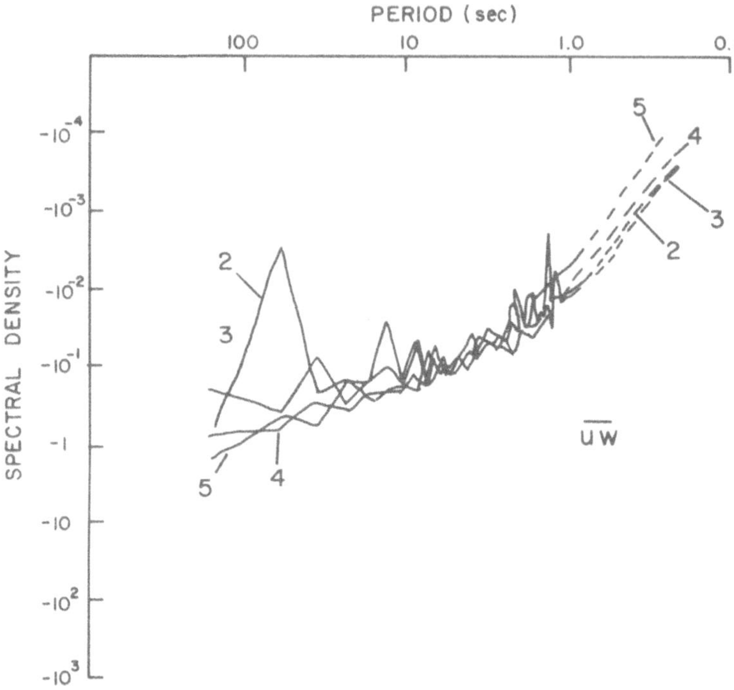

Figure 5. Cross spectra of $\overline{uw}$ at 35 cm, 60 cm, 100 cm and 215 cm (2, 3, 4, 5 respectively). Note that the spectral density is highest for the lower current meters at high frequency and for the higher meters at low frequency.

HIGHLY STRATIFIED ESTUARIES

Relevant Laboratory Results

In highly stratified estuaries, most transport of mass and momentum depends on processes at the pycnocline. Often this feature can be regarded as a depth and stratification limited, free shear layer. Many laboratory investigations of free shear layer flows have been carried out, and a review with consideration given to oceanographic applications is presented by Maxworthy and Browand (1975). These experiments, however, have all related the freely growing shear layer, which is an initial value phenomenon, rather than the depth and stratification limited case, which is best formulated as a boundary value problem. For this reason the laboratory work is for the most part not directly transferable to

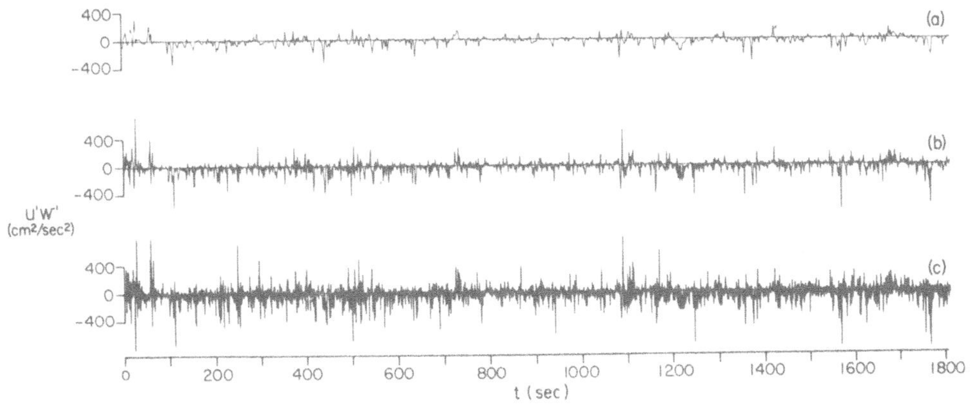

Figure 6. Time series of u'w' at 35 cm using a velocity sample interval of 2.2 sec (a), 0.8 sec (b) and 0.2 sec (c) (from McLean and Smith, in press).

the overall problem of pycnocline mixing. It is, however, generally relevant. As will be described in some detail later, recent field studies have indicated that a large percentage of cross-pycnocline transport occurs in relatively brief periods related to internal hydraulic and internal wave phenomena.

The growth and structure of incompressible, unstratified, freely growing shear layers is determined by the flow geometry, the starting conditions, and the Reynold's number. Due to Reynold's number similarity, for any given flow geometry, the turbulence characteristics are a function of the initial conditions. Many experiments have shown that the initial growth of a free shear layer is dominated by vortex pairing (c.f. Demotakis and Brown, 1967). The evolution of the flow has been visualized as sequential pairing of coherent vortices; hence since each interaction depends on the scale of the previous pairing, the initial conditions retain a strong influence over the turbulent structure. The question arises as to what heppens in depth limited and stratification limited cases.

Remarkably little is known about the structure of stratified shear flows and the variation of the mixing processes in stably stratified free shear layers. Laboratory experiments, such as those of Kato and Phillips (1969) have measured mixed layer growth but few measurements of turbulent structure and even fewer visualization studies that characterize the mechanism of mixing have been carried out. Measurements by Koop and Browand (1979) show that the transition occurs quite rapidly from a neutral free shear layer to a stratification bounded case. Because turbulent energy is lost in working against buoyancy forces, it was originally conjectured that when this loss (in addition to viscous dissipation) exceeds the production of turbulence through mean shear, then the turbulent field would decay (Townsend, 1957). Often it does not do so. Kelvin-Helmholz billows and lateral intrusions have been postulated to be responsible for the mixing across sharp interfaces; however, Sherman et al. (1978), following Turner (1973) indicate the ineffectiveness of such mechanisms. Detailed visual laboratory studies of entrainment and mixing at a free shear layer in the presence of a density gradient were initiated by Brown and Roshko (1974) who examined non-Boussinesq density effects on the large scale structures responsible for free shear layer spreading. They suggested that large changes in the ratio of the densities had a relatively small effect on the spreading angle or on the thickness of the mixed layer, but recent work by Koop (1976) shows that with decreasing Richardson number the spatial growth indeed is affected and that the growth is inhibited by the stabilizing influence of gravitation on the largest scales of motion. Detailed measurements of the influence of buoyancy on these largest scales indicate that the turbulence becomes more anisotropic with increasing Richardson number (Hopfinger, 1978), the vertical buoyancy flux decreases relative to the horizontal component and that the buoyancy forces affect the vortex pairing at Richardson numbers equal to one-third, after which turbulence decays in the order of one eddy lifetime. Hopfinger defines the Richardson number as $R_i = g'2\delta/(\Delta u)^2$ where $g' = g\,\rho'/\bar{\rho}$, δ is the half-width of the free shear layer and Δu is the initial velocity difference across the layer. He also shows that the degree of anisotropy depends on the ratio of the eddy lifetime (l/u) to the buoyancy time scale $(l/g')^{1/2}$ where l is the integral (large eddy) length scale, u is the RMS turbulent velocity and ρ is the density. This ratio has the dimensional structure of a Froude number.

Because so little quantitative work has been carried out on the mechanism of mixing in bounded shear layers, we must speculate on the exact processes that are responsible for the mixing at the pycnocline. That the internal Froude number exerts a dominant control on mixing has been demonstrated by Partch and Smith (1978) and Gardner and Smith (1978).

The initial conditions in many free shear layers exert a dominant effect on the turbulence characteristics. Experiments by Oster et al. (1978) show that the rate of spreading is influenced by small upstream disturbances. In estuarine situations, bed morphology and channel topography would be expected to influence strongly the incidence of mixing events. Topographic controls in Knight Inlet allow a free shear layer to develop and vertical mixing processes similar to those described earlier may well be expected. However, the continual input of energy at the shear layer interface suggests that pycnocline processes must best be regarded as boundary value problems. Vortex pairing may well occur in some cases, but this clean structure is likely to be subsumed in the finer scale random turbulence.

Available Theories

Most existing models of circulation in highly stratified estuaries are tidally averaged, the mixing processes that produce fluxes across the pycnocline being parameterized in a variety of ways. For example, Rattray (1967) presents a similarity solution for the steady state circulation in fjords using an eddy coefficient to relate mixing to the mean flow. This model is based on the assumption that the frictional stress and turbulent mixing depend on the mean circulation rather than on the tidal currents. Similarly Rattray and Mitsuda (1974) present an analysis of salt wedge circulation which assumes a laminar pycnocline separating turbulent surface and bottom layers. They use constant eddy coefficients in the interior of the two layers and match the resulting velocity fields to logarithmic profiles at the bottom and at the pycnocline. Long (1975a) describes a two-layer, steady state model of circulation in fjords which assumes no flow in the bottom layer and uses an entrainment velocity based on his laboratory results (Long, 1975b). The model does not accurately predict pycnocline depth, probably because of the direct extrapolation of laboratory results to complicated field situations. Careful scaling is required to ensure the applicability of such approaches. Pearson and Winter (1978) use a two-layer, steady state model to calculate the pycnocline depth and mean velocities in fjords. They model mixing processes at the pycnocline with upward and downward mass fluxes which are derived from the densities in the two layers. The results of the model are applied to Knight Inlet, British Columbia and compare favorably with available measurements.

All of these models assume steady state, which implies not only tidal averaging, but asserts that any short term processes can be adequately modeled on the mean (gravitational) circulation. Field observations by Partch and Smith (1978) and Gardner and Smith (1978) have shown the importance of shorter term processes in salt wedge estuaries. Also, Farmer and Smith (1978) describe time dependent mixing processes in Knight Inlet, a fjord type estuary. In view of these results, steady state models will inevitably mask interesting and important physics. If steady state models must be used in highly time dependent systems, it would seem that the technique of Pearson and Winter is most appropriate. In this case, the mean density field, which is used to set the vertical fluxes, contains at least some information on the mean transport across the pycnocline, guaranteeing that the vertical mixing is reasonably well modeled.

Field Studies

The Duwamish River and Knight Inlet investigations of Smith and his co-workers will be described in some detail here to clarify the nature and importance of time dependent mixing processes. The Duwamish River enters Puget Sound through a small estuary in Seattle, Washington (Figure 7). Under most conditions of freshwater flow it is of the salt wedge type. The small size of the Duwamish, as well as its proximity to the University of Washington, makes it an ideal location to study the physical processes active in a salt wedge estuary.

For the past several years, a research program has been conducted in the Duwamish River by the University of Washington Department of Oceanography. The early work, described by Partch and Smith (1978) consisted of velocity and density measurements made from an anchored research vessel. The original expectation was that maximum turbulence production and resulting vertical salt flux would occur during the flood, when shear is maximum. It was found, however, that maximum turbulence was produced on the ebb during a period of minimum mean shear. In fact, up to 79% of the total salt flux occurred during a period of about four hours around maximum ebb (Figure 8). The ebb mixing events are evident as large peaks in the turbulent kinetic energy time series. As the internal Froude number was near critical at the time of maximum turbulence production, Partch and Smith postulated that the mixing was due to breaking internal waves.

To investigate further the mechanism of mixing in the Duwamish, a new program was instigated in 1976. It was based on the development of a system that allowed continuous measurement of velocity, temperature and salinity at fixed depths from a moving research vessel. In addition, vertical profiles of temperature and salinity were obtained with a CTD. An electromagnetic current meter attached to the CTD also was used to give an indication of any sharp shear layers which might be missed by the fixed current meters and permitted gradient Richardson number profiles to be calculated. This system was described by Gardner and Smith (1978), who also presented evidence that considerable mixing occurs near the upper limit of the dredged channel. The mixing continues as the patch of high salinity surface water advects downstream. Further analysis of these data has confirmed this general picture and provided additional insight into the mechanisms which generate the turbulence. Data from the profiling CTD-current meter system demonstrate that the shear is strongest during the flood and relatively small during

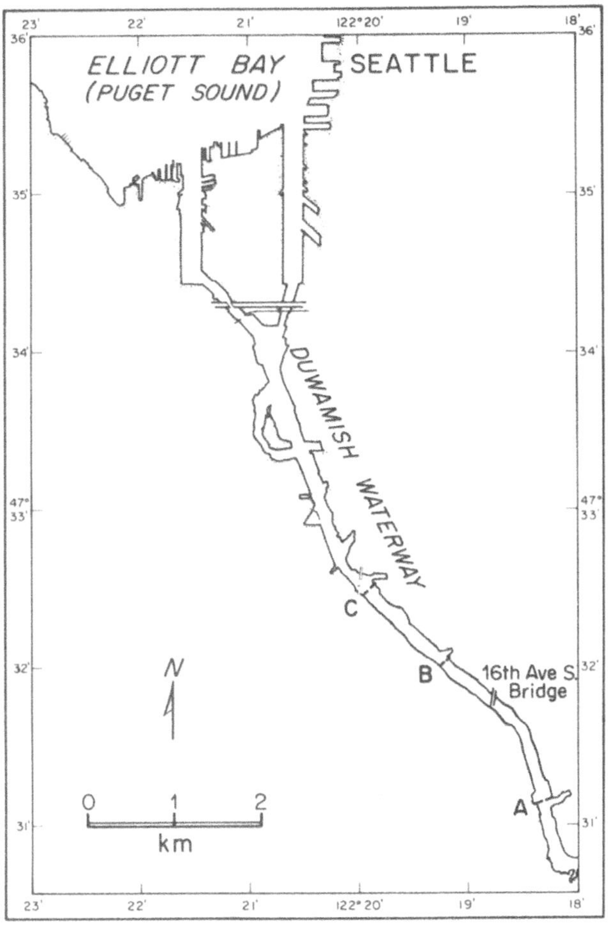

Figure 7. Map of the Duwamish estuary. Velocity and salinity data were obtained in the section from A to C, while echo sounder records were made from A to B. The large internal lee-wave described in the text was observed at the 16th Ave. S. Bridge (after Gardner and Smith, 1978).

the ebb. Moreover, the gradient Richardson number (R_i) at the pycnocline is significantly less during the flood, even though little mixing due to shear instabilities occurs at this time. Figure 9 shows plots of σ_t and velocity for mid and late flood. On the ebb, the Richardson number is normally very large, indicating virtually no mixing due to the mean shear. However, there frequently are periods when R_i is less than unity, indicating

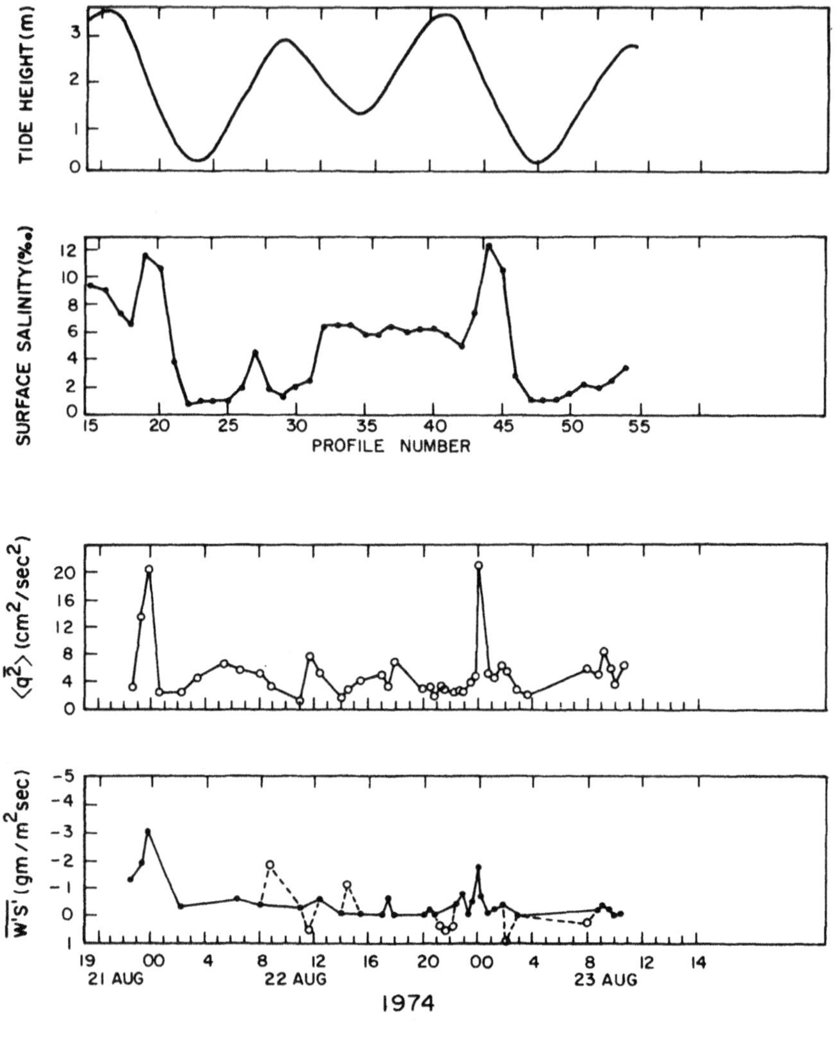

Figure 8. Time series of tide height, surface salinity, turbulent kinetic energy, and salt flux across the pycnocline at an anchor station in the Duwamish. Note the peaks in the latter three quantities during the strong ebbs (after Partch and Smith, 1978).

significant internal wave shear-produced turbulence (Figure 10). Strong, small scale shear layers are particularly evident during the mid ebb and result in locally unstable flow in the pycnocline. The surface density increases from mid to late ebb (Figure 10) indicating the importance of the ebb mixing events; thus, it appears that under the near critical internal Froude number of the

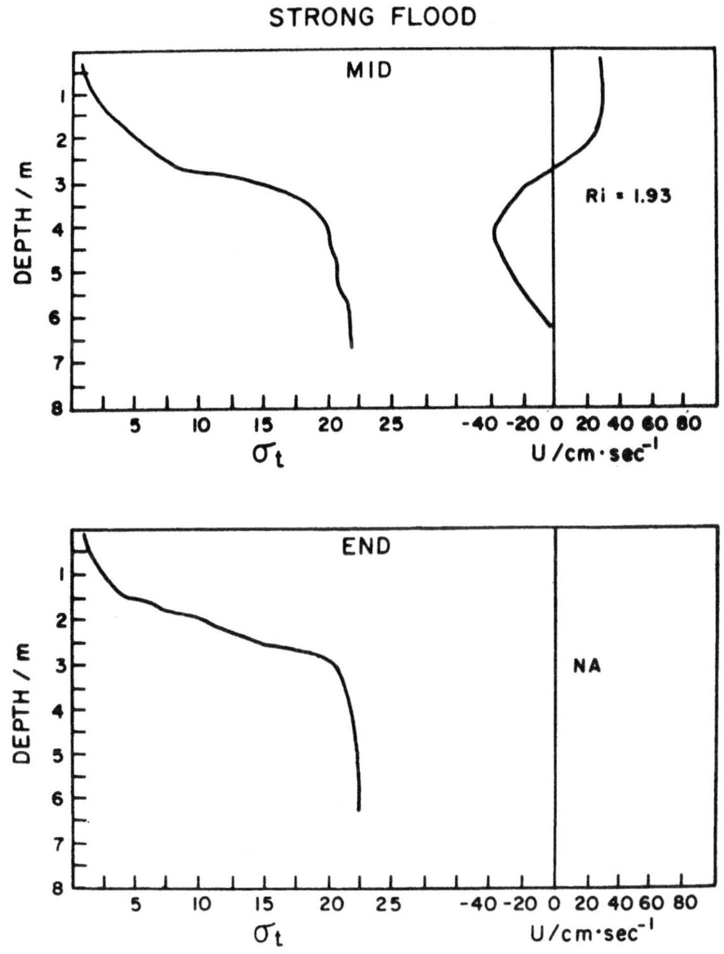

Figure 9. Profiles of σ_t and velocity made in the Duwamish during a strong flood, indicating a Richardson number of 1.93 during mid-flood.

ebb, internal waves are generated with sufficient amplitude to produce regions of instability.

The processes responsible for the generation of these internal waves are still being investigated. It is likely, as was suggested by Gardner and Smith (1978) that considerable energy is put into the internal wave field through a sharp change in bottom topography at the upper end of the dredged region. Continued mixing downstream probably is related to internal waves forced by flow over local bottom topography.

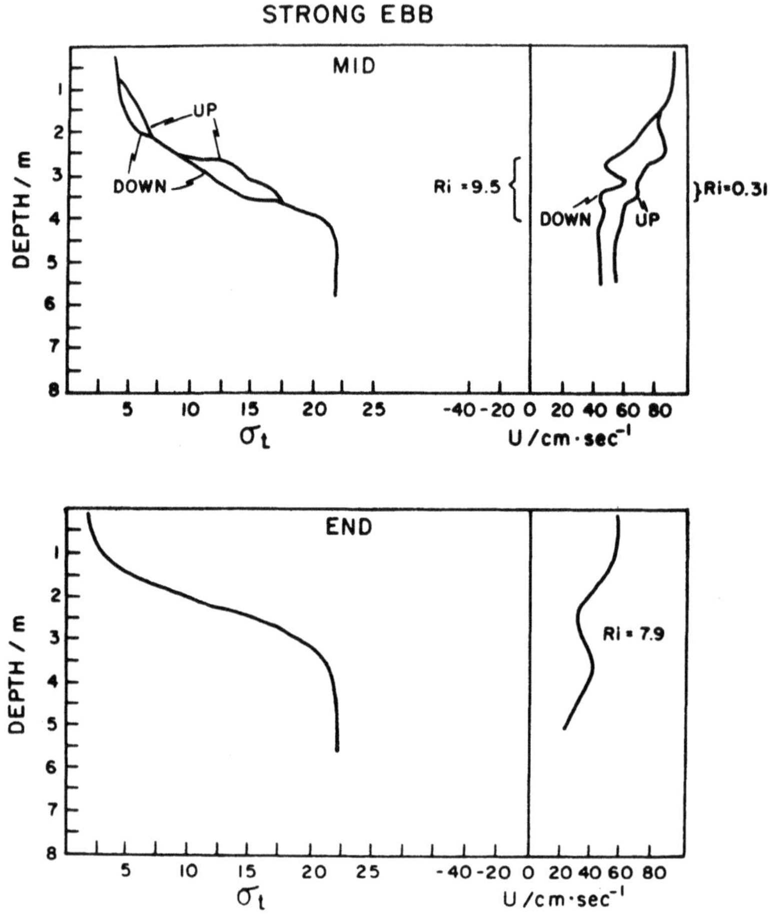

Figure 10. Profiles of σ_t and velocity made in the Duwamish during a strong ebb. The mean shear is much lower than during the flood, but small scale shear layers at the pycnocline produce a local Richardson number of 0.31. Also note that the surface salinity is higher than on the flood.

Recently it has been found that an echo-sounder can be used to visualize the pycnocline in the Duwamish and in October, 1978, this technique was used on a short cruise. Figures 11 and 12, taken from records procured during this cruise, show the development of the internal wave field on a strong ebb. The pycnocline is observed as the base of the dark region near the surface. Note that as the ebb approaches its maximum, very large amplitude waves develop throughout the reach. While additional data analysis is required to describe fully the processes, it seems reasonable that

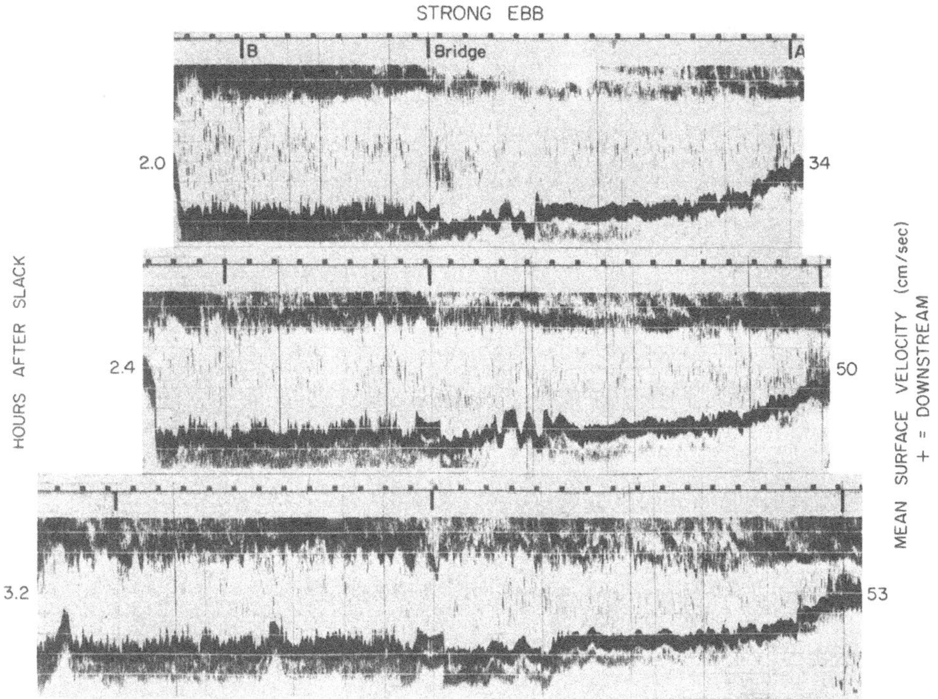

Figure 11. Echo-sounder records made in the Duwamish during the first half of a strong ebb. The base of the dark region near the river surface at top of record represents the pycnocline. Note the increasing internal wave activity as the flow accelerates, and the large lee wave which develops at the Bridge. Upstream is to the right in this figure and in the next three figures.

significant production of turbulence and resulting mixing is induced by these waves. The very large feature that develops during maximum ebb near the center of the section is a large amplitude lee wave generated by flow through a constriction at a bridge. This feature was found in salinity and velocity records on an earlier cruise, and was described in some detail by Gardner and Smith (1978) who showed that considerable mixing is caused by turbulence produced in this lee wave. There is substantially less internal wave activity on the flood tide (Figures 13 and 14). The sloping features in the upper layer near the upstream (right) side of the records seem to be weak fronts associated with small lateral embayments in the river. That such fronts can persist is indicative of the low turbulent intensity on the flood.

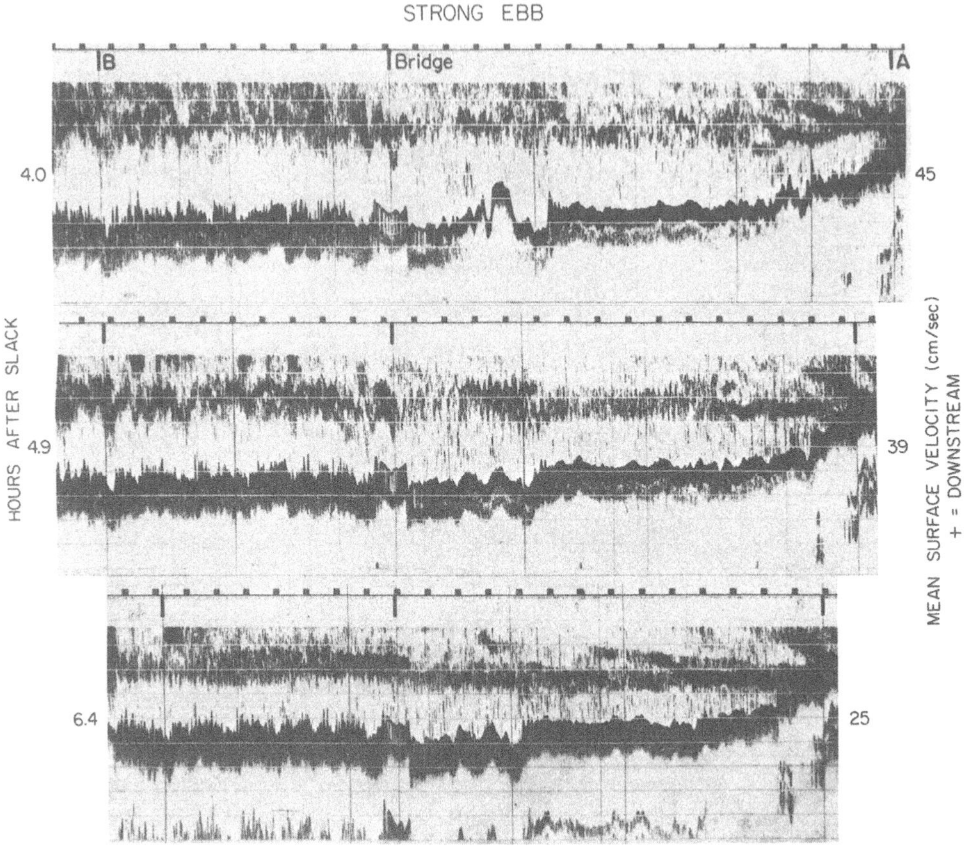

Figure 12. Echo sounder records made during the last half of a strong ebb.

Another example of the importance of time dependent processes related to internal hydraulics has been found in Knight Inlet, a fjord type estuary in British Columbia, Canada (Figure 15). Farmer and Smith (1978) describe the propagation of large amplitude internal waves up-inlet from the sill, and show that significant mixing is generated as the waves propagate. Subsequent work has shown that these waves are generated by the breakdown of an internal hydraulic jump which forms at the sill on the ebb. As the tide turns, some of the energy associated with flow over the sill propagates up-inlet as non-linear internal waves. In addition, they have shown that considerable mixing is caused by flow over this feature and that free shear layers much as described by Koop and Browand (1979) are produced in the water of the sill.

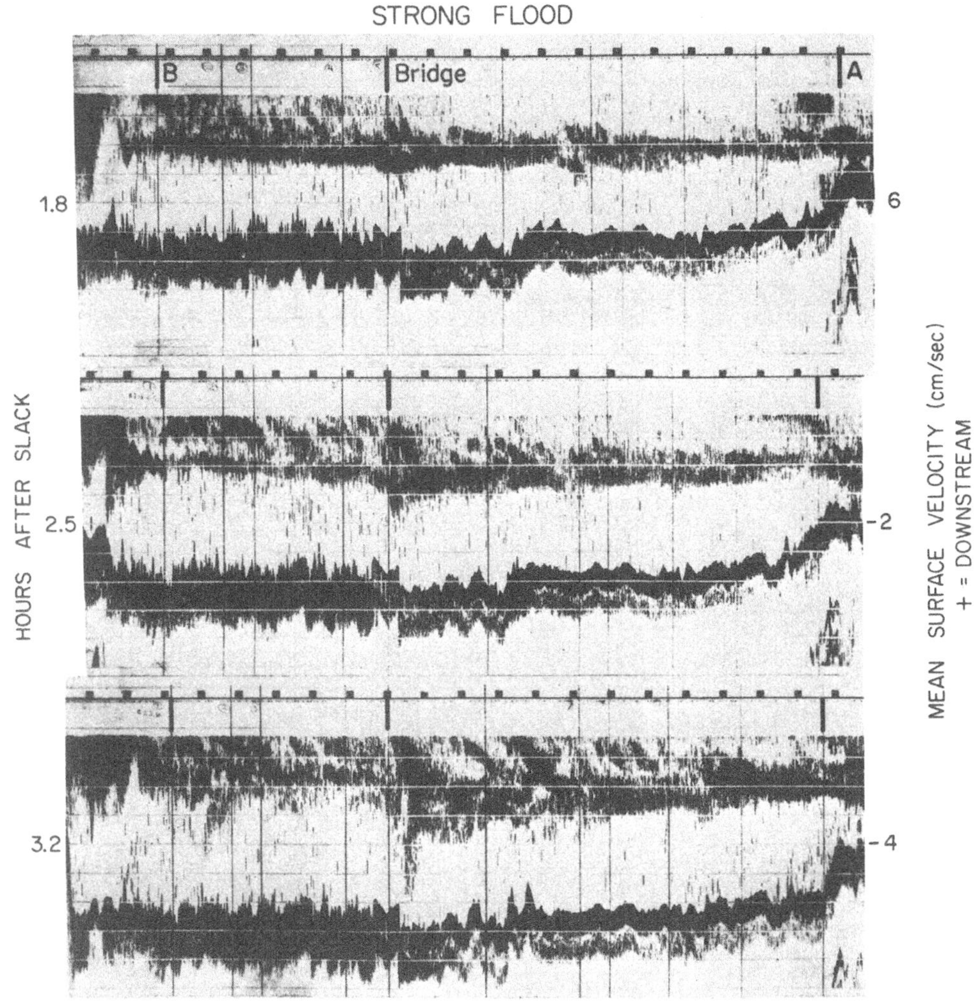

Figure 13. Echo-sounder records in the Duwamish during the first half of a strong flood. Note that there is considerably less internal wave activity than on the ebb and that the large lee wave again occurs at the bridge.

The internal hydraulics associated with the disturbed stratified flow produces mixing at the sill, on both the flood and ebb tides

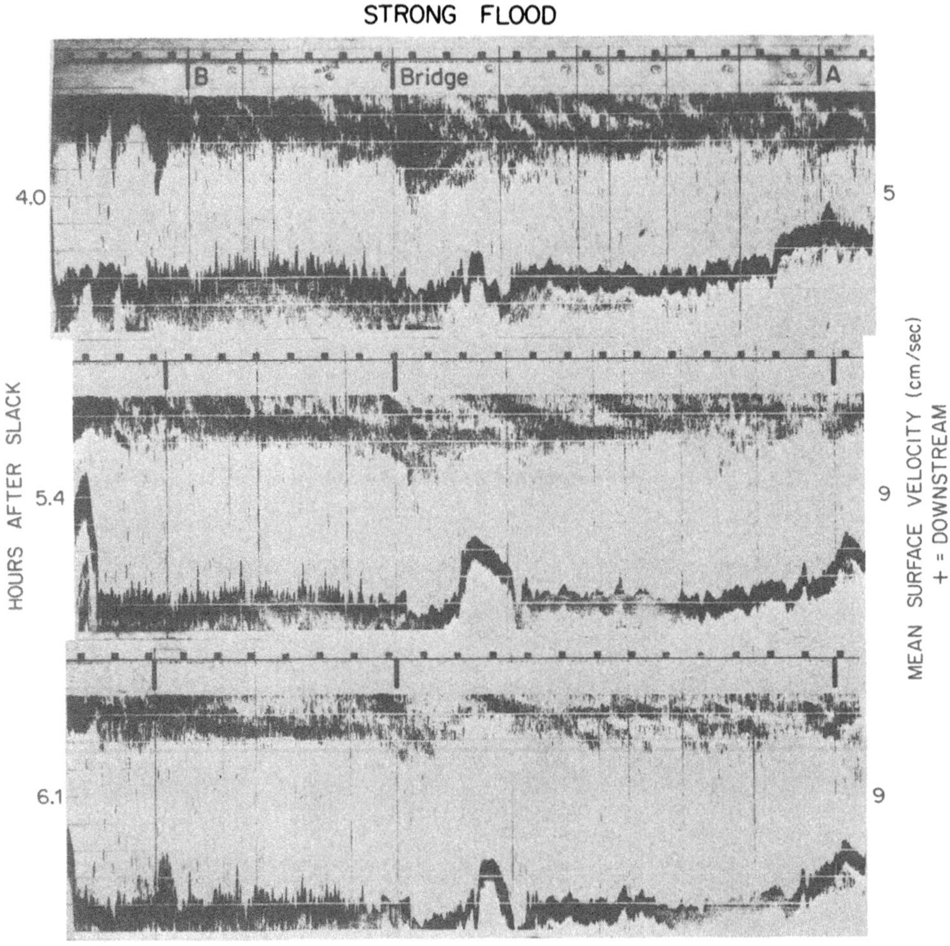

Figure 14. Echo-sounder records made in the Duwamish during the last half of a strong flood.

through various shear instabilities, and for several kilometers up-inlet of the sill, as the propagating nonlinear internal waves lose their energy. Echo soundings have provided substantive records of the effects of both these processes. Figure 16, taken on a flood tide shows large shear instabilities over the sill at Hoeya Head. Note that the instabilities occur both in the pycnocline and in the bottom layer. On the ebb, the flow over the sill becomes supercritical, returning to subcritical in an

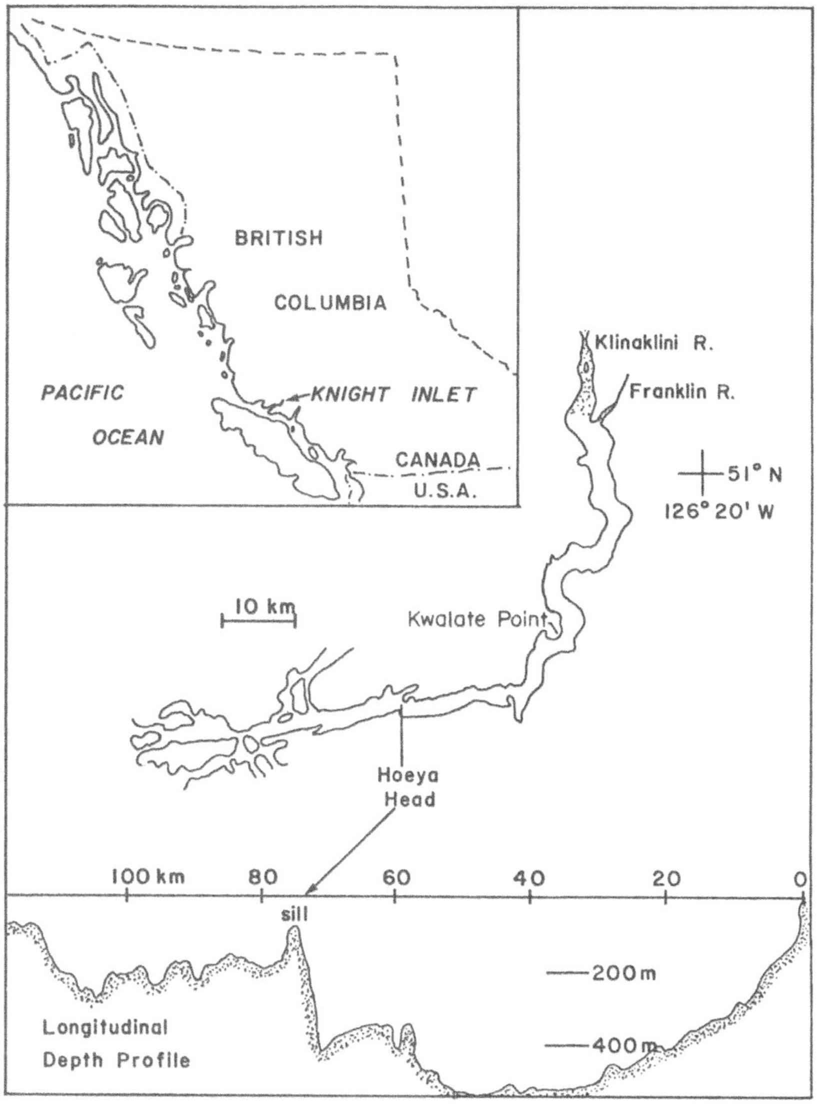

Figure 15. Map of Knight Inlet. Internal waves are generated by flow over the sill at Hoeya Head and propagate upstream to Kwalate Point. The Klinaklini and Franklin Rivers provide most of the fresh-water input to the fjord (after Farmer and Smith, in press).

internal hydraulic jump down-inlet of the sill (Figure 17). Note that as the flow goes supercritical over the crest of the sill,

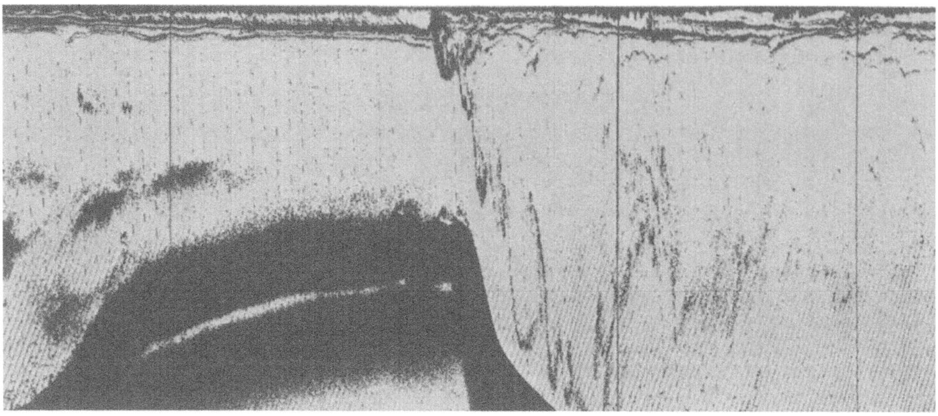

Figure 16. Echo-sounder record over the sill in Knight Inlet during flood tide. Note the shear instabilities both on the pycnocline and in the interior region. Up-inlet is to the left. The sill depth is approximately 65 meters and the plateau is approximately 1 km long.

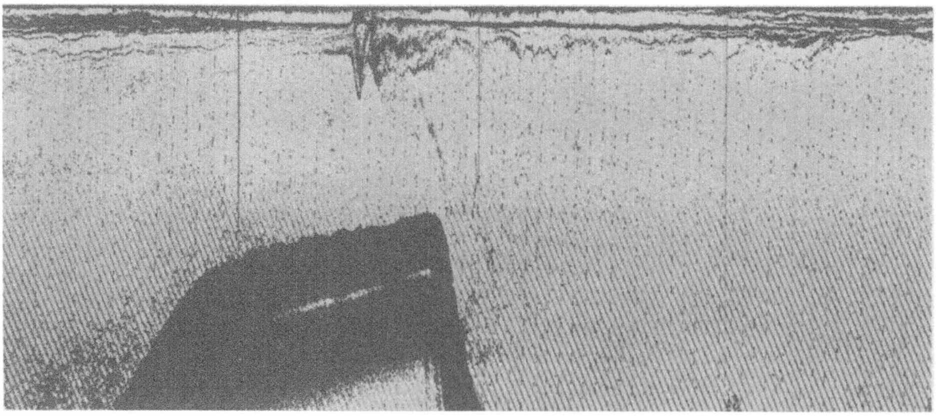

Figure 17. Echo-sounder record over the Knight Inlet sill during an ebb tide. Note the region of supercritical flow just down-inlet of the sill and the hydraulic jump further downstream.

the upper reflecting layer of the pycnocline rises, while the lower pycnocline deepens. This behaviour is characteristic of second-mode internal waves. Just downstream of the crest, the deep reflector drops steeply as the supercritical flow follows the slope of the sill. Large shear instabilities are apparent in this region. Farther down-inlet the jump is seen as the rather irregular rise of the deep reflecting layer.

Then the flow reverses, the pycnocline disturbance described above propagates up-inlet as a second-mode internal bore (Figure 18). As this bore propagates it evolves into a train of solitary waves which has been followed with the echo sounder up-inlet to Kwalate Point. In spite of extensive surveys, no evidence has been found that the waves propagate beyond Kwalate Point. This Point also serves as a boundary between two types of oceanographic systems. Up-inlet the surface layer is nearly fresh, the gradient in surface salinity is very low, and the pycnocline is nearly horizontal. Down-inlet of Kwalate Point the surface salinity increases rapidly, and isopycnals slope down-inlet. The down-inlet slope of the pycnocline further increases over the sill.

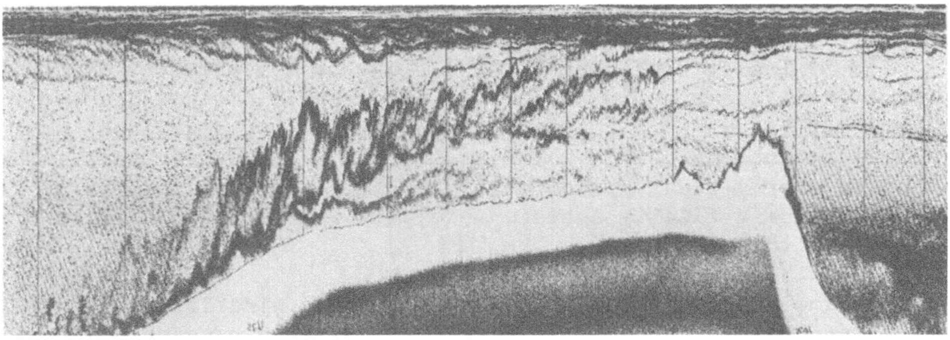

Figure 18. Echo-sounder record over the Knight Inlet sill at the end of the ebb showing that the hydraulic jump has evolved into an upstream propagating second-mode internal bore. This bore further evolves into a train of solitary waves which have been followed as far as Kwalate Point.

The two examples of flow in highly stratified estuaries presented here clearly demonstrate the importance of time dependent processes at the pycnocline and the local nature of these processes. Current understanding is not adequate to model quantitatively the mixing produced; however, an awareness of these effects is important in understanding the limitations of existing models. Hopefully, further work will allow the pycnocline processes to be incorporated in future models. For many environmental quality studies, a box model similar to that of Pearson and Winter may be adequate but a more detailed representation of mixing processes is vital when modification of an estuary's characteristics is contemplated or when short period fluctuation of pollutant concentrations is of interest.

Conclusions

1) Turbulence in well mixed estuaries may be predicted using a range of boundary layer models. To be useful in a variety of embayments, the effects of topographic variability must be included. The roughness length, velocity profile, and especially the stress field are affected by topography, but this incluence can be accounted for if a spatial average over the significant length scales of the topography is computed. At present, no turbulence model includes intermittency and there is conflicting evidence as to its presence, let alone its importance, in rough wall geophysical flows. The success of even simple eddy viscosity models in neutral or weakly stratified flows suggests that the details of intermittency may not be significant to the gross prediction of turbulent mixing.

2) Mixing in highly stratified flows is dominated by distinct events; their location is subject to topographic control and their effects may be non-local. Energy for mixing is supplied, at least in part, by internal waves which are forced by flow over topography. Averaging over a tidal period will miss these internal hydraulic effects; hence a tidally average model may reproduce poorly the density and velocity field.

3) The exact physics of mixing in highly stratified estuaries is not yet fully understood. The visual evidence presented in the echo sounder records, and the calculations of instantaneous gradient Richardson number profiles indicate that spatially identifiable, repeated events dominate the processes of turbulent mixing. Only the detailed measurement of velocity and density fields from a moving vessel provide the spatial density of samples necessary to illuminate the scales and temporal characteristics of such internal hydraulic events.

This work was supported by NSF grants numbers OCE-78-20452 and OCE-77-09945.

REFERENCES

Boeing Symposium on Turbulence, Seattle, Wa., 22-27 June, 1969. J. Fluid Mech., 41, 1970.

Bowden, K. F., Observations of turbulence in a tidal current, J. Fluid Mech. 17, 271-293, 1963.

Bowden, K. F., Mixing processes in estuaries, Estuarine Transport Processes, ed. B. Kjerfve, U. of S. Carolina Press, 11-36, 1977.

Bradshaw, P., An Introduction to Turbulence and its Measurement, Pergamon Press, 1972.

Bradshaw, P., Turbulence, Springer Verlag, 1976.

Brown, G. L. and A. Roshko, On density effects and large structure in turbulent mixing layers, J. Fluid Mech., 64, 775-816, 1974.

Chen, C. P. and R. F. Blackwelder, Large scale motion in a turbulent boundary layer: a study using temperature contamination, J. Fluid Mech., 78, 525-560, 1978.

Demotakis, P. E. and G. L. Brown, The mixing layer at high Reynolds number: large structure dynamics and entrainment, J. Fluid Mech., 78, 525-560, 1976.

Edgerton, H. E., Sonic detection of a fresh water-salt water interface, Science, 154, 1555, 1966.

Farmer, D. and J. D. Smith, Nonlinear internal waves in a fjord, Hydrodynamics of Estuaries and Fjords, ed. J. C. J. Nihoul, Elsevier Oceanographic Series 23, 465-494, 1978.

Farmer, D., and J. D. Smith, Tidal interaction of stratified flow with a sill in Knight Inlet, Submitted to Deep-Sea Research.

Fukushima, H., M. Kashiwamura and I. Yakuwa, Studies on salt wedge by ultrasonic method, Proceedings of Tenth Conference on Coastal Engineering, ASCE, 1435-1447, 1966.

Gardner, G. B. and J. D. Smith, Turbulent mixing in a salt wedge estuary, Hydrodynamics of Estuaries and Fjords, ed. J. C. J. Nihoul, Elsevier Oceanography Series 23, 79-106, 1978.

Gibson, M. M. and B. E. Launder, On the calculation of horizontal non-equilibrium turbulent shear flow under gravitational influence, J. Heat Trans., 81, 980-993, 1976.

Gordon, C. M., Intermittent momentum transport in a geophysical boundary layer, Nature, 248, 392-394, 1974.

Gordon, C. M. and C. F. Dohne, Some observations of turbulent flow in a tidal estuary, J. Geophys. Res., 78, 1971-1978, 1973.

Gordon, C. M. and J. Witting, Turbulent structure in a benthic boundary layer, Bottom Turbulence, ed. J. C. J. Nihoul, Elsevier Oceanography Series 19, 59-81, 1977.

Grant, W. D. and O. S. Madsen, Combined wave and current interaction with a rough bottom, J. Geophys. Res., 84, 1797-1808, 1979.

Grass, A. J., Structural features of turbulent flow smooth and rough boundaries, J. Fluid Mech., 50, 233-255, 1971.

Heathershaw, A. D., Bursting phenomena in the sea, Nature 248, 394-395, 1974.

Hopfinger, E. J., Buoyancy effects on the large scale structure of free turbulent flows, Structure and Mechanism of Turbulence, ed. H. Fiedler, Springer Verlag, 65-85, 1978.

Kato, H. and O. M. Phillips, On the penetration of a turbulent layer into a stratified fluid, J. Fluid Mech. 37, 643-655, 1969.

Koop, C. G., Instability and turbulence in a stratified shear flow, Ph.D. Thesis, U. of Southern California, San Diego, 1976.

Koop, C. G. and F. K. Browand, Instability and turbulence in a stratified fluid with shear, J. Fluid Mech., 93, 135-159, 1979.

Launder, B. E., On the effects of a gravitational field on the turbulent transport of heat and momentum, J. Fluid Mech., 67, 569-588, 1975.

Launder, B. E., Heat and mass transport, Topics in Applied Physics, 12, ch. 6, Springer Verlag, 1976.

Liebovich, S., The structure of vortex breakdown, Ann. Review of Fluid Mech., 10, 221-246, 1978.

Liepmann, H., The use and fall of ideas in turbulence, American Scientist, 67, 221-228, 1979.

Lighthill, J. J., Turbulence, Osborne Reynolds and Engineering Science Today, eds. D. M. McDowell and J. D. Jackson, Manchester University Press, 83-146, 1970.

Long, C. E. and J. D. Smith, A simple model of the time dependent, stably stratified planetary boundary layer, Pollutant Transport and Sediment Dispersal in the Washington-Oregon Coastal Zone, University of Washington Tech. Report, Ref: A 79-17, ch VI, 1979.

Long, R. R., Circulation and density distribution in a deep, strongly stratified, two-layer estuary, J. Fluid Mech., 71, 529-540, 1975a.

Long, R. R., the influence of shear on mixing across density interfaces, J. Fluid Mech., 70, 305-320, 1975b.

Long, R. R., some aspects of turbulence in geophysical systems, Adv. in App. Mech., 17, 2-90, 1977.

Lumley, J. L. and H. A. Panofsky, The Structure of Atmospheric Turbulence, John Wiley and Sons, New York, 1964.

Maxworthy, T., and F. K. Browand, Experiments in rotating and stratified flows: oceanographic applications, Ann. Rev. Fluid Mech., 7, 273-305, 1975.

McLean, S. R. and J. D. Smith, Turbulence measurements in the boundary layer over a sand wave field, J. Geophy. Res., in press, 1979.

McLean, S. R., Mechanics of the turbulent boundary layer over sand waves in the Columbia River, Doctoral Dissertation, University of Washington, Seattle, 1976.

Nikuradse, J., Laws of flow in rough pipes, N.A.C.A. Tech. Memo. 1292, 62, p., 1933.

Oster, D., I. Wyganski, B. Dziumba, and H. Fiedler, On the effect of initial conditions on the two dimensional mixing layer, Structure and Mechanisms of Turbulence, ed. H. Fiedler, Springer Verlag, 48-64, 1978.

Partch, E. N. and J. D. Smith, Time dependent mixing in a salt wedge estuary, Estuarine and Coastal Marine Science, 6, 3-19, 1978.

Pearson, C. C. and D. F. Winter, Two layer analysis of steady circulation in stratified fjords, Hydrodynamics of Estuaries and Fjords, ed. J. D. J. Nihoul, Elsevier Oceanography Series 23, 495-514, 1978.

Rattray, M. Jr., Some aspects of the dynamics of circulation in fjords, Estuaries, ed. G. H. Lauff AAAS Publ. No. 83-52-62, 1967.

Rattray, M. Jr., and E. Mitsuda, Theoretical analysis of conditions in a salt wedge, Estuarine and Coastal Marine Science, 2, 375-394, 1974.

Rotta, J. C., statistische Theorie nichthomogener Turbulenz, Z. Phys., 129, 547-572, 1951.

Sabot, J., I. Saleh and G. Comte-Bellot, Effect of roughness on the intermittent maintenance of Reynolds shear stress in pipe flow, Phys. Fluids, 20, S 150-S 155, 1977.

Sherman, E. C., I. Imberger, and G. M. Corcos, Turbulence and mixing in stably stratified waters, Ann. Rev. Fluid Mech., 10, 267-288, 1978.

Smith, J. D., Modeling of sediment transport on continental shelves, The Sea, 6, ed. E. D. Goldberg, John Wiley, New York, 1977.

Smith, J. D., Measurement of turbulence in ocean boundary layers, Proceedings of a Working Conference on Current Measurement, eds. W. Woodward, C. N. K. Mooers, and K. Jensen, Tech. Report Del-SG-3-78m College of Marine Studies, University of Delaware, 95-128, 1978.

Smith, J. D. and S. R. McLean, Spatially averaged flow over a wavy boundary, J. Geophy. Res., 82, 1735-1746, 1977a.

Smith, J. D. and S. R. McLean, Boundary layer adjustments to bottom topography and suspended sediment, Bottom Turbulence, ed. J. D. J. Nihoul, Elsevier Oceanography Series 19, 123-151, 1977b.

Tennekes, H., and J. L. Lumley, A First Course in Turbulence, MIT Press, 1972.

Tochko, J. S., A study of the velocity structure in a marine boundary layer - instrumentation and observations, Ph.D. Thesis, WHOI/MIT, 1978.

Townsend, A. A., Measurements in the turbulent wake of a cylinder, Proceedings Royal Society London, Ser. A., 190, 551-561, 1948.

Townsend, A. A. Turbulent flow in a stably stratified atmosphere, J. Fluid Mech., 3, 361-372, 1957.

Turner, J. S., Buoyancy Effects in Fluids, Cambridge Univ. Press, 1973.

Willmarth, W. W., Structure of turbulence in boundary layers, Adv. in Applied Mech., 15, 159-254, 1975.

OBSERVATION AND MODELING OF THE CIRCULATION IN THE CHESAPEAKE BAY

Dong-Ping Wang

Chesapeake Bay Institute, The Johns Hopkins University,

Baltimore, Maryland, 21218

ABSTRACT

In recent years, our understanding of circulation in the Chesapeake Bay has been greatly improved through long-term current meter measurement, and sophisticated numerical modeling. The circulation mainly consists of tide, river, density and wind-driven flow. The tidal current has larger amplitude, and is mainly responsible for the mixing and sediment re-suspension. On the other hand, because of its longer duration, the nontidal current determines the transport of salt, sediment and pollutant.

Vertical variations of the salinity and velocity distribution play the most important role in mixing and transport. Therefore, vertical dimension has to be included in the circulation model. By adapting the semi-implicit and mode-split method, efficient two-dimensional (in a vertical plane) and three-dimensional models have been developed, which reduce the computer time by orders of magnitude. Effective numerical modeling, coupled with sound observational basis, is crucial to the understanding of mixing and transport in a partially mixed estuary.

INTRODUCTION

In a partially mixed estuary, circulation is mainly driven by the tide, river run-off, salinity gradient and wind. Knowledge of the circulation and its variability, and particularly the ability to predict changes in it, is of central importance to ever-recurring problems of power plant siting and sewage disposal, to the maintenance and development of ship channels and other sedimentation processes, and to the effects of control of fresh water supply on the salinity distribution. At present, most of the water quality, pollutant transport and sediment studies are based on either the observed flow field or simplistic circulation models. Neither approach will be able to predict changes in circulation, and therefore, their applicability to impact assessment is doubtful.

In recent years, our understanding of the estuarine circulation has progressed rapidly, through extensive field observation (particularly, the long-term current measurement) and sophisticated numerical modeling. In the following, we briefly summarize some important results from our studies in the Chesapeake Bay (Figure 1). More detailed descriptions can be found in Blumberg (1977a, 1977b), Elliott (1976, 1978), Elliott, Wang and Pritchard (1978), Wang (1979a, 1979b), Wang and Elliott (1978), and Wang and Kravitz (1979).

OBSERVATION

Figure 2 shows a typical 40 day velocity record near Smith Pt. (37°53'N, 76°11'W) at 15 m below the surface, obtained from the moored current meter. The original record can be decomposed into the tidal and nontidal (tidally averaged) part, through a lowpass filter. The tidal part consisted of semi-diurnal and diurnal constituents; the semi-diurnal M_2 tide is dominant with an amplitude of 30 cm/s. The nontidal part was predominantly up-bay (landward), with a mean speed of 11 cm/s. The mean landward current in the lower layer is consistent with the pattern of two-layer, density-driven circulation (Pritchard, 1956). In addition, the nontidal part had velocity fluctuations at 2-7 day time scales. The fluctuating, nontidal current whose amplitude is comparable to the mean flow, is mainly induced by the wind.

The tidal part usually dominates the total velocity record; the tidal velocity ranges from about 50 cm/s at the Bay mouth, to 20 cm/s in the Mid-Bay. Because of its large amplitude, the tidal current is mainly responsible for the vertical and horizontal mixing. Tidal current is also important for the sediment re-suspension (Schubel, 1968). On the other hand, the tidal excursion length is only 3-8 km, which is small compared to the major geometry features in the Bay. Thus, mass transport due to the rectified (non-linear) tidal circulation is probably negligible in the Bay-wide scale.

The nontidal current, despite its smaller amplitude, is primarily responsible for the mass transport. For example, during the event of 1-3 January, a strong current of 30 cm/s lasted for about 2 days (Figure 2). The total displacement would be about 50 km is the entire Bay responded similarly to the wind forcing. In other words, the mass transport, and hence, movement of the

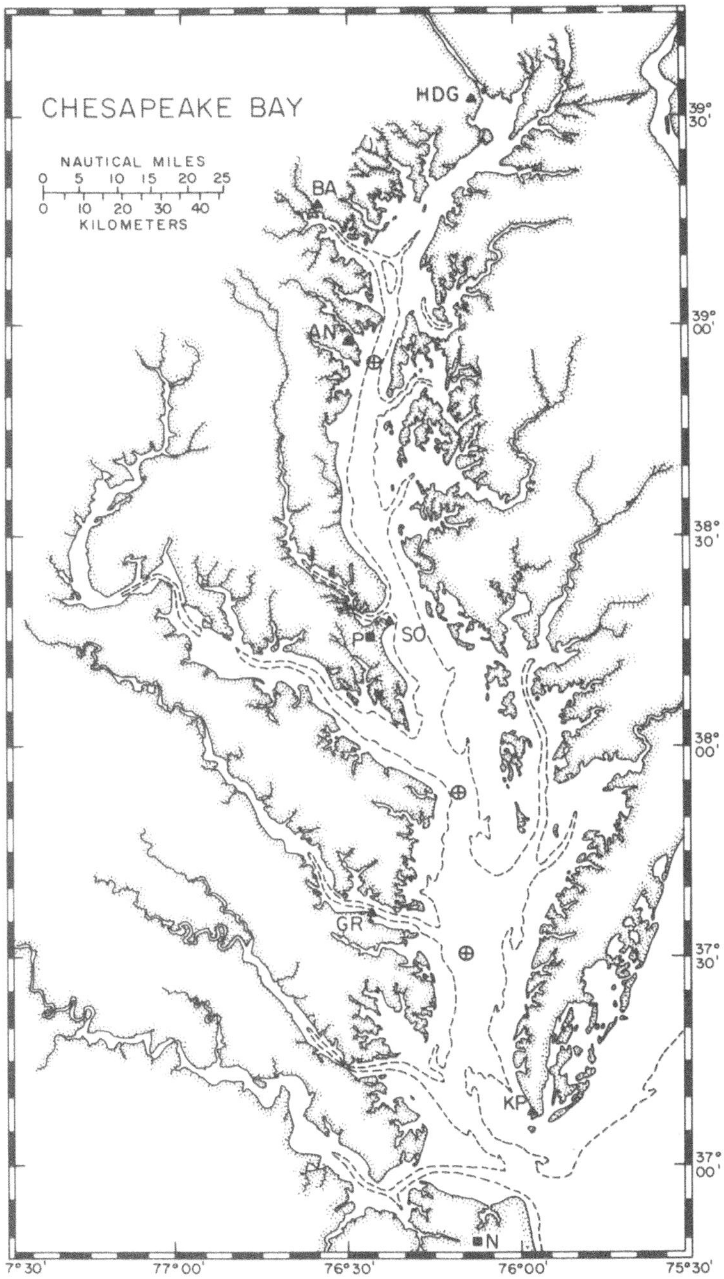

Figure 1. Map of the Chesapeake Bay showing the 9 m depth contour (dashed) and current meter (⊗), sea level (▲) and wind (■) stations.

pollutant and sediment, is mainly due to the nontidal circulation. The response of nontidal circulation to river run-off and wind forcing, is examined separately in the following:

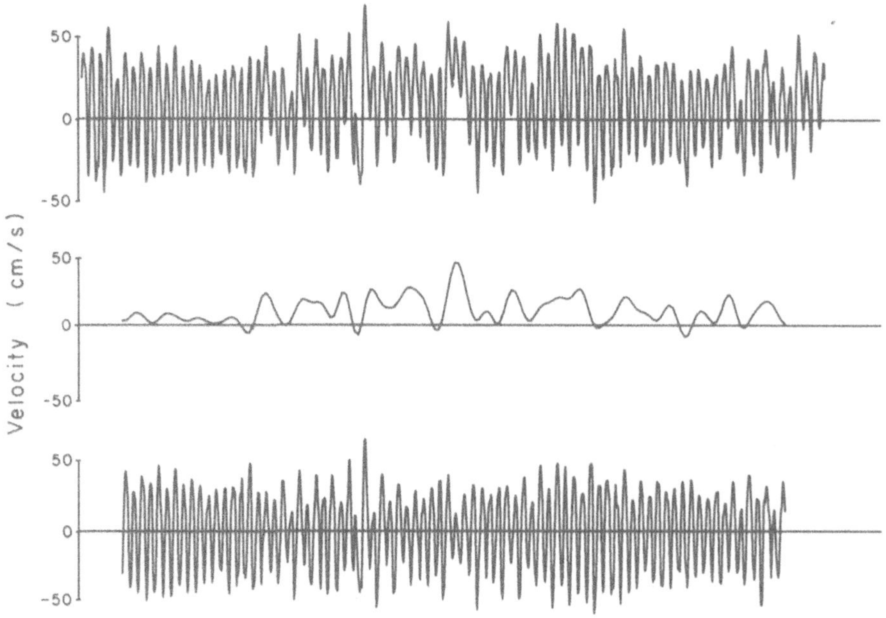

Figure 2. The original (top panel), nontidal (middle panel) and tidal (lower panel) velocity record, near Smith Pt. (the positive direction is landward).

(A) River run-off

The direct river influence is mainly confined to the upper 40 km of the Bay. Figure 3 shows nontidal velocity records near Howell Pt. (39°22'N, 76°10'W) at 7(2.1) and 15(4.6) ft(m) below the surface (Elliott, Wang and Pritchard, 1978). The mean flow in August and September was 1.6 cm/s seaward near the surface and -2.7 cm/s landward near the bottom, which was typical of the two-layer circulation. However, a surge of river flow occurred during the "wet" tropical storm Eloise in late September, with a peak runoff of 16,000 m^3s^{-1} (560,000 cfs). The corresponding nontidal flows were over 60 cm/s at a mooring position. The mean flow

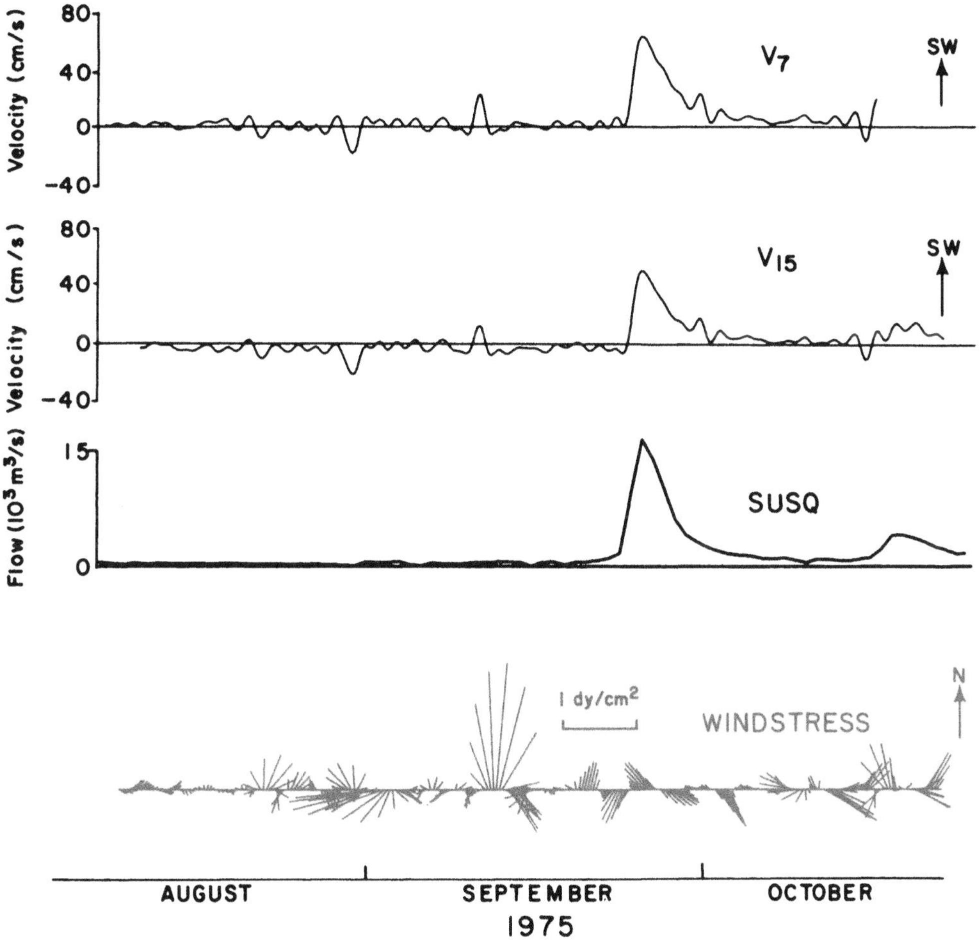

Figure 3. The lowpass time series of current, windstress and Susquehanna discharge.

remained seaward with a speed of 10 cm/s during the remainder of the study period.

The change of mean flow from the two-layer estuarine circulation to river circulation, was due to the shift of the fresh water front. During the first part of the observation, prior to Tropical Storm Eloise, the mooring was located at the fresh water front, and consequently, the salinity effect was important. After the peak discharge, the fresh water front was advected about 20 km downstream, reducing salinities at the mooring to near zero. Salinity in the head area remained insignificant for the remainder of the study, and the estuarine circulation did not re-establish it.

The indirect effect of river run-off, however, is far-reaching. A persistent (> 1 month) high discharge will significantly decrease the salinities throughout the entire Bay. For example, the 5 year (1968-1972) mean position of 16°/$_{\infty}$ surface isohaline was located at about 110 km upstream from the mouth (Figure not shown). During the high flow season, the surface 16°/$_{\infty}$ isohaline reached to within 20 km from the mouth; in contrast, it was at about 220 km from the mouth, during the "dry" spring season of 1969.

(B) Wind effect

The wind influence was visible in Figure 3, though it was shadowed by the high river discharge event. The direct river effect becomes negligible away from the head, as the cross-sectional area broadens considerably. Figure 4 shows the velocity records at upper and lower Bay, and the wind record, for a one-month period (Wang, 1979b). (The mooring position is marked in Figure 1). In the lower Bay, the velocity fluctuations were homogeneous. They were driven by the longitudinal (north-south) wind at the 2-3 day period, i.e., the northward wind drove an inflow, and an outflow was driven by the southward wind. However, at longer time scales, current and north-south wind were 180° out-of-phase. For example, during 24-25 November, a strong inflow was associated with a southward wind. The discrepancy from the local response was due to the coastal Ekman effect (Wang, 1979a). In other words, the earth rotation tends to deflect the wind-driven coastal current to the right, and hence, an onshore transport (inflow) was induced during a southward wind event.

In the upper Bay, the surface current was determined by the north-south wind, with no evidence of coastal influence (Figure 4). For example, during the strong southward wind (24-25 November), the surface current was seaward in the upper Bay, despite large inflows in the lower Bay. The bottom current, on the other hand, seems to indicate some nonlocal effects. The difference in response between the upper and lower Bay, appears to be the result of increasing bottom stress contribution in the lower Bay (Wang, 1979b).

A strong wind event will lead to large salinity change throughout the entire Bay. For example, two salinity sections were taken on 9 December and 11-12 December (Figures 5a and 5b), before and after the strong outflow event of 10-11 December (Figure 4). In the lower Bay, velocities were seaward at all depths, and hence, the high salinity (> 24°/$_{\infty}$) water was advected away. In the upper Bay, the low salinity water was also advected downstream by the seaward flow. On the other hand, the bottom current was landward,

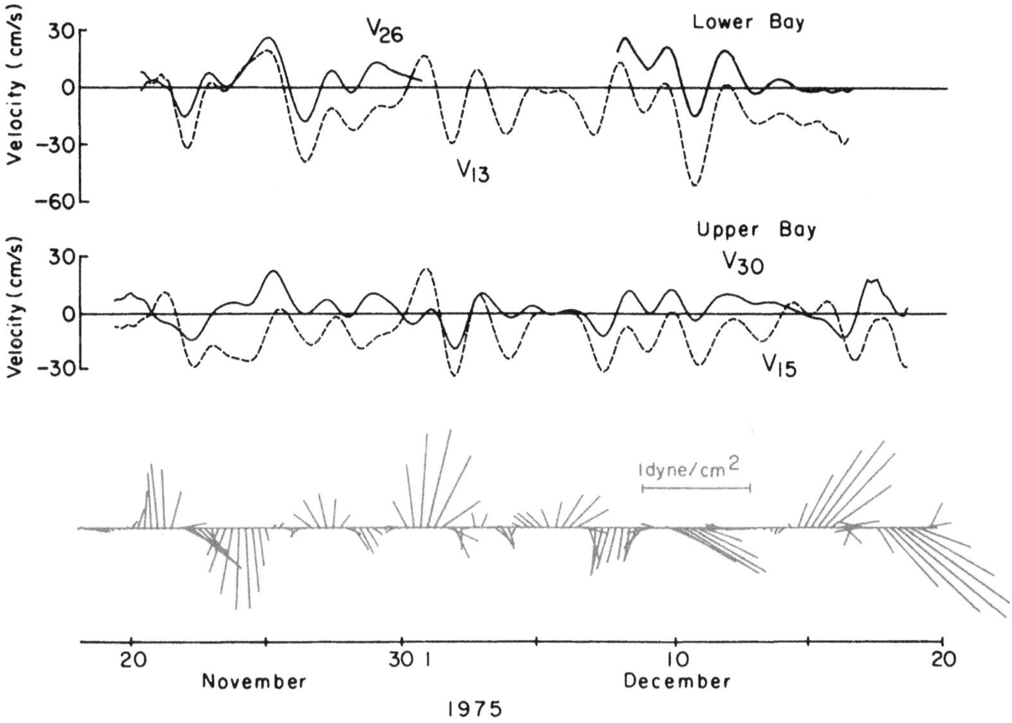

Figure 4. The nontidal velocity in the lower and upper Bay, and the windstress (the positive direction is landward).

and hence, saltier water was advected upstream in the lower layer.

MODELING

Models of estuarine circulation have been mostly based on the vertically integrated equations. While application of the homogeneous system to tide and storm surge, has been rather successful, its applicability to the water quality, salt intrusion and sediment transport is doubtful. Observations indicate large vertical shears in the density and wind-driven circulation. Vertical mixing also plays a crucial role in the diffusion of salt, nutrients and dissolved oxygen. Thus inclusion of vertical dimension appears to be essential in order to properly model the transport and mixing. The vertical variation of tidal and wind-driven circulation is examined in the following, with the two-dimensional (in a vertical plane) and three-dimensional model:

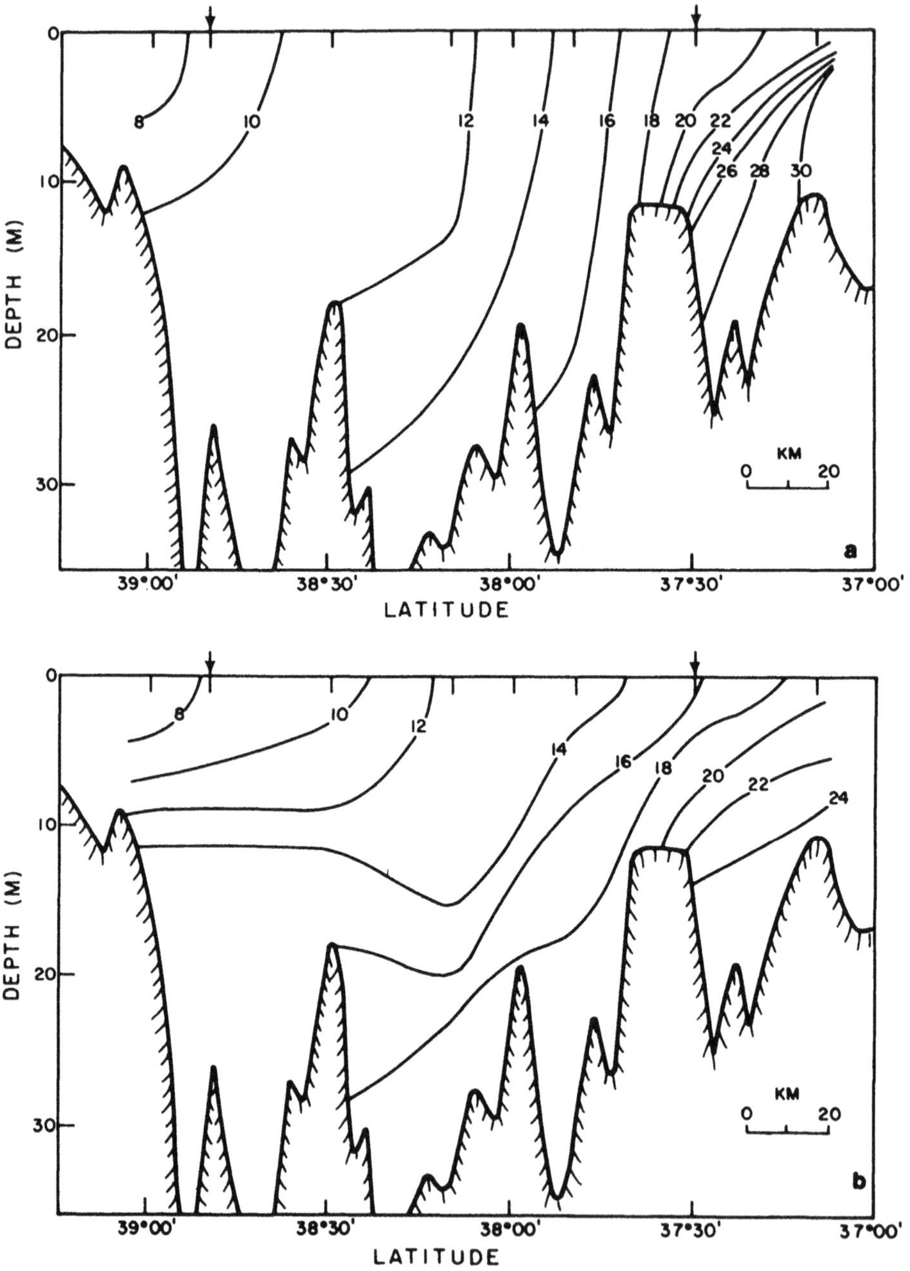

Figure 5. Salinity section (°/∞) along the Bay axis: (a) 9 December 1975, (b) 11-12 December 1975.

(A) Two-dimensional model

A two-dimensional, semi-implicit model which includes effects of the tide, wind, salinity gradient and river run-off, has been developed (Wang and Kravitz, 1979). The model also allows non-uniform geometry and variable eddy coefficients. The semi-implicit method is not restricted by the Courant-Friedrichs-Lewy condition, and therefore, it admits a much larger time step (than the explicit method) in the case of a fine horizontal resolution. For example, with a horizontal spacing of 1 km, the two-dimensional semi-implicit model having 10 vertical levels, is comparable in computer time to the one-dimensional explicit model.

Figures 6a and 6b show the amplitude and phase of the semi-diurnal tidal current in the Potomac River. Tide is an up-river progressive wave, due to frictional dissipation. It also has significant vertical, upward phase propagation. The tidal amplitude is large ($\sim$ 50 cm/s) in the mid-River, and it is small ($\sim$ 10 cm/s) near bottom. Due to the vertical change in amplitude and phase, there are considerable variations in velocity shear and density gradient within a tidal cycle.

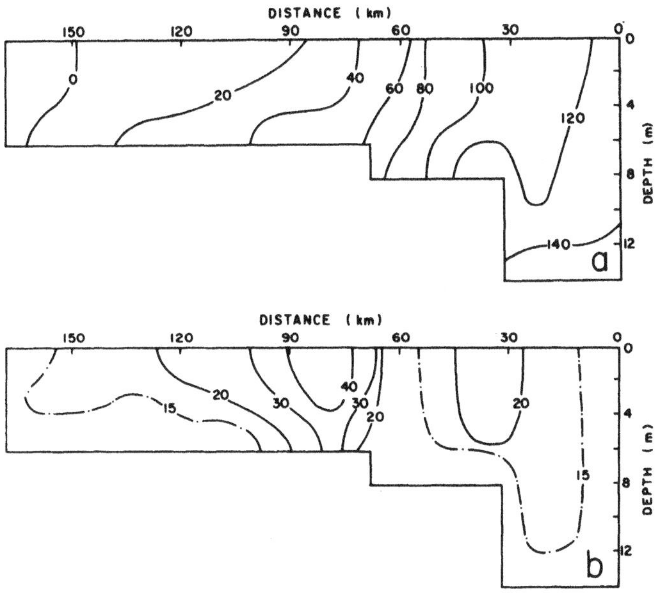

Figure 6. The semi-diurnal tidal velocity from the two-dimensional, Potomac estuary model: (a) amplitude (cm/s); (b) phase (degree).

Figures 7a and 7b show the salinity and nontidal (tidally averaged) velocity, through a wind event, at 40 km from the mouth of the Potomac River. An oscillating (sine wave) windstress with amplitude of 0.25 dyne/cm^2 and a period of 5 days, starts at Day 0. The local response is non-homogeneous: the surface current moves with the wind, while an opposite flow returns in the lower layer (Figure 7a). The wind-driven currents are comparable to the mean, density-induced circulation, that reversals of the two-layer circulation are frequent; these results agree with the long-term (one-year) observation (Elliott, 1978).

Due to the differential advection of nontidal currents, there are substantial changes in salinity distribution (Figure 7b). The stratification (vertical salinity gradient) is large at the end of the down-river wind, and is small at the end of the up-river wind. In fact, the water column becomes completely homogeneous between Day 4 and Day 5, and presumably, intense vertical mixing would occur during this period. In addition, within a tidal cycle, the stratification is large at the end of ebbing (when the total salinity is small), and is small at the end of flooding. The inter-tidal stratification change is mainly due to the near bottom velocity shear (Figure 6).

(B) Three-dimensional model

The three-dimensional, explicit model is too expensive for long-term simulation (Leendertse et al., 1973). An efficient scheme combining the semi-implicit and mode-split method is being developed. The semi-implicit scheme removes the constraint of cross-channel seiches, and the mode-split scheme separates the external mode from internal mode computation (Simons, 1974). Consequently, two orders of magnitude saving in computer time can be achieved. The three-dimensional model includes effects of the tide, wind, salinity gradient, river run-off and earth rotation.

Figure 8 shows the amplitude of semi-diurnal tidal current in the Potomac River, in the surface layer and at a cross-section about 50 km from the mouth. Large tidal current occurs in the mid-River with amplitude of 60 cm/s. At a cross-section, the current is stronger at surface, and also, the high velocity core seems to follow the deep channel.

A down-river wind generates seaward flow at the surface, and landward flow in the lower layer (Figure not shown). The outflow is stronger in the shallow part, where the direct frictional effect is more important. In addition, a cross-river circulation is induced by the earth rotation. Water moves toward the south in the surface layer, and it returns to the north in the lower layer. The cross-river circulation is stronger in the deep part.

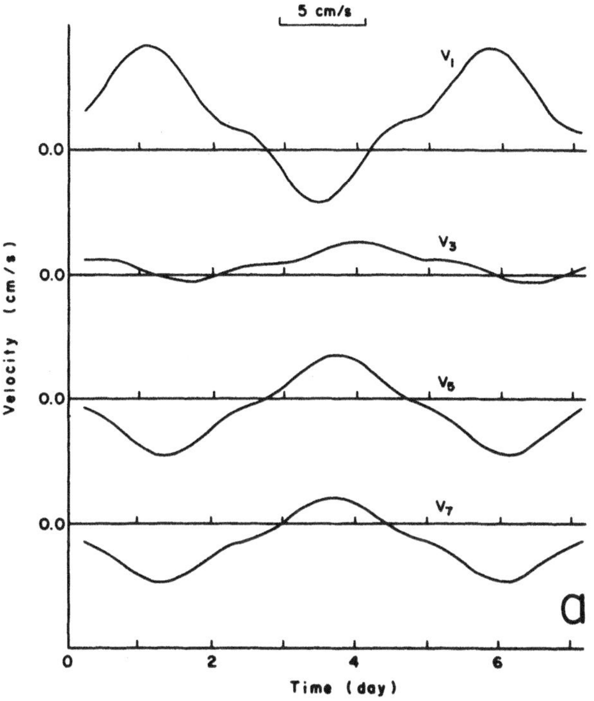

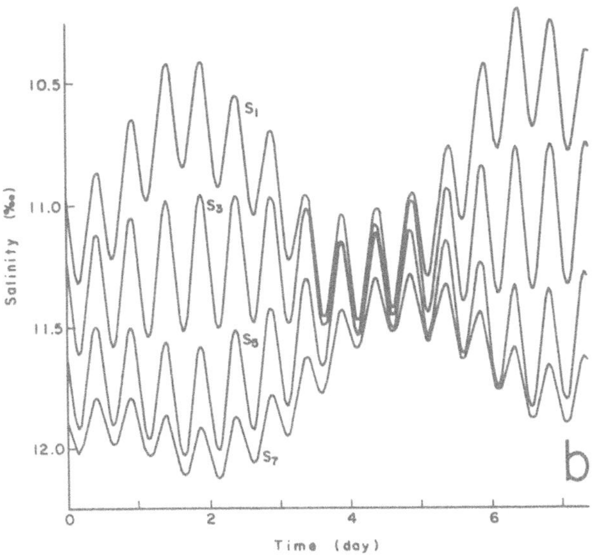

Figure 7. The wind event from the two-dimensional, Potomac estuary model: (a) tidally averaged velocity profile; (b) salinity profile (the subscript indicates depth below surface in meters).

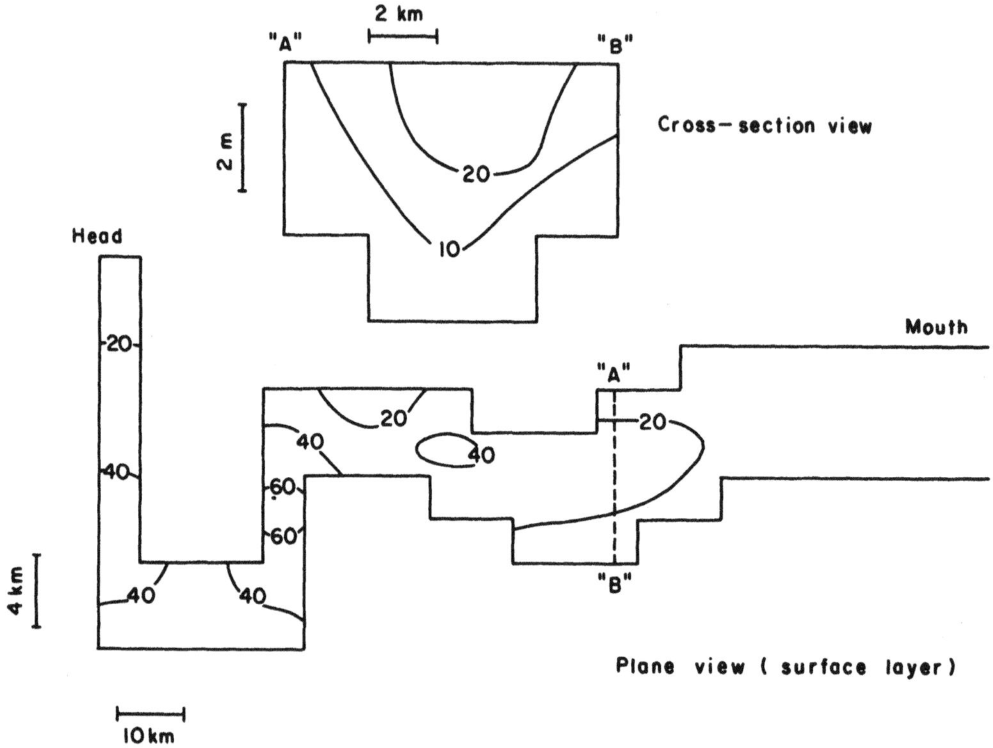

Figure 8. The tidal velocity distribution in the surface layer, and over a cross-section, from the three-dimensional model (velocity in cm/s).

DISCUSSION

Observations in the Chesapeake Bay clearly indicate that mass transport is mainly due to the wind and density induced circulation. The direct river influence is confined to the head area, and the tidal effect is limited to the very shallow water. Similar results were also obtained in Narragansett Bay (Weisberg, 1976) and Corpus Christi Bay (Smith, 1977). Thus, estimates of estuary flushing based on the tidal prism and river discharge can be quite misleading.

The traditional model approach based on the vertically integrated system, also appears to be overly naive. Vertical variations in the salinity and current, in fact, are mainly responsible for the transport and mixing in a partially mixed estuary. Inclusion of vertical dimension in the model generally requires a considerable increase of computer time. The difficulty is resolved by adapting the semi-implicit and mode-split method. Consequently, with a more efficient scheme, the long-term, two-dimensional (in a vertical plane) and three-dimensional model simulation seems now economically feasible. It is also noted that a branched, two-dimensional model (Elliott, 1976) can be applied efficiently to a complex tributary system in lieu of a three-dimensional model.

Our studies strongly suggest that, effective numerical modeling coupled with sound observational basis, is essential to fully understand the estuarine circulation.

REFERENCES

Blumberg, A. F. 1977a. "Numerical model of estuarine circulation," J. Hydrau. Div. ASCE, 103:295-310.

Blumberg, A. F. 1977b. "Numerical tidal model of Chesapeake Bay," J. Hydrau. Div. ASCE, 103:1-10.

Elliott, A. J. 1976. "A numerical model of the internal circulation in a branching estuary," Special Report 54, Chesapeake Bay Institute, 85 pp.

Elliott, A. J. 1978. "Observations of the meteorologically induced circulation in the Potomac estuary," Est. and Coastal Mar. Sci., 6:285-300.

Elliott, A. J., D-P Wang, and D. W. Pritchard. 1978. "The circulation near the head of Chesapeake Bay," J. Mar. Res. 36:643-655.

Leendertse, J. J., R. C. Alexander and S. K. Liu 1973. "A three-dimensional model for estuaries and coastal sea," The RAND Corp., R-1417-OWRR, 52 pp.

Pritchard, D. W. 1956. "The dynamic structure of a coastal plain estuary," J. Mar. Res. 15:33-42.

Schubel, J. R. 1968. "Turbidity maximum of the northern Chesapeake Bay," Science, 161:1013-1015.

Simons, T. J. 1974. Verification of numerical models of Lake Ontario. I. circulation in spring and early summer. J. Phys. Oceangr., 4:507-523.

Smith, N. P. 1977. "Meteorological and tidal exchanges between Corpus Christi Bay, Texas, and the northwestern Gulf of Mexico," Est. and Coastal Mar. Sci., 5:511-520.

Wang, D-P. 1979a. "Sub-tidal sea level variations in the Chesapeake Bay and relations to atmospheric forcing," J. Phys. Oceanogr. 9:413-421.

Wang, D-P. 1979b. "Wind-driven circulation in the Chesapeake Bay, winter 1975," J. Phys. Oceanogr. 9:563-572.

Wang, D-P and A. J. Elliott. 1978. "Nontidal variability in the Chesapeake Bay and Potomac river: evidence for nonlocal forcing," J. Phys. Oceanogr. 8:225-232.

Wang, D-P and D. W. Kravitz. 1979. "A semi-implicit two-dimensional model of estuarine circulation," To appear in J. Phys. Oceanogr.

THE TRANSPORT OF FRESHWATER OFF A MULTI-INLET COASTLINE

J. O. Blanton

Skidaway Institute of Oceanography, Savannah, Georgia

ABSTRACT

The inner continental shelf waters between South Carolina and northern Florida are weakly stratified by the many sources of freshwater ejected from the land. The mixing of freshwater in this zone is qualitatively similar to that in a partially mixed estuary. However, many inlets along the coast result in a complex orientation to the principal axes of the tidal currents offshore and the resulting mixing processes are non-homogeneous in the alongshore direction.

These complexities must be faced in any realistic model of similar oceanic regions. Data are presented to show the nature of the non-homogeneous transport process and to demonstrate that models which neglect alongshore gradients of momentum and properties are unrealistic.

INTRODUCTION

The Atlantic coast along the southeastern United States is frequently indented with inlets that connect low-lying coastal marshes to the ocean. Off Georgia and South Carolina, inlets occur each 10-15 km (Figure 1). Many rivers, some of them quite large, pass through the marshes, draining large areas. The Altamaha, Savannah, Pee Dee and Cooper/Santee Rivers provide over 80% of the total freshwater discharge (Atkinson, Blanton and Haines, 1978) between Cape Romain, South Carolina, and Jacksonville, Florida. The remainder of the flow in this region comes from many small rivers and creeks. As a consequence, freshwater inputs to the

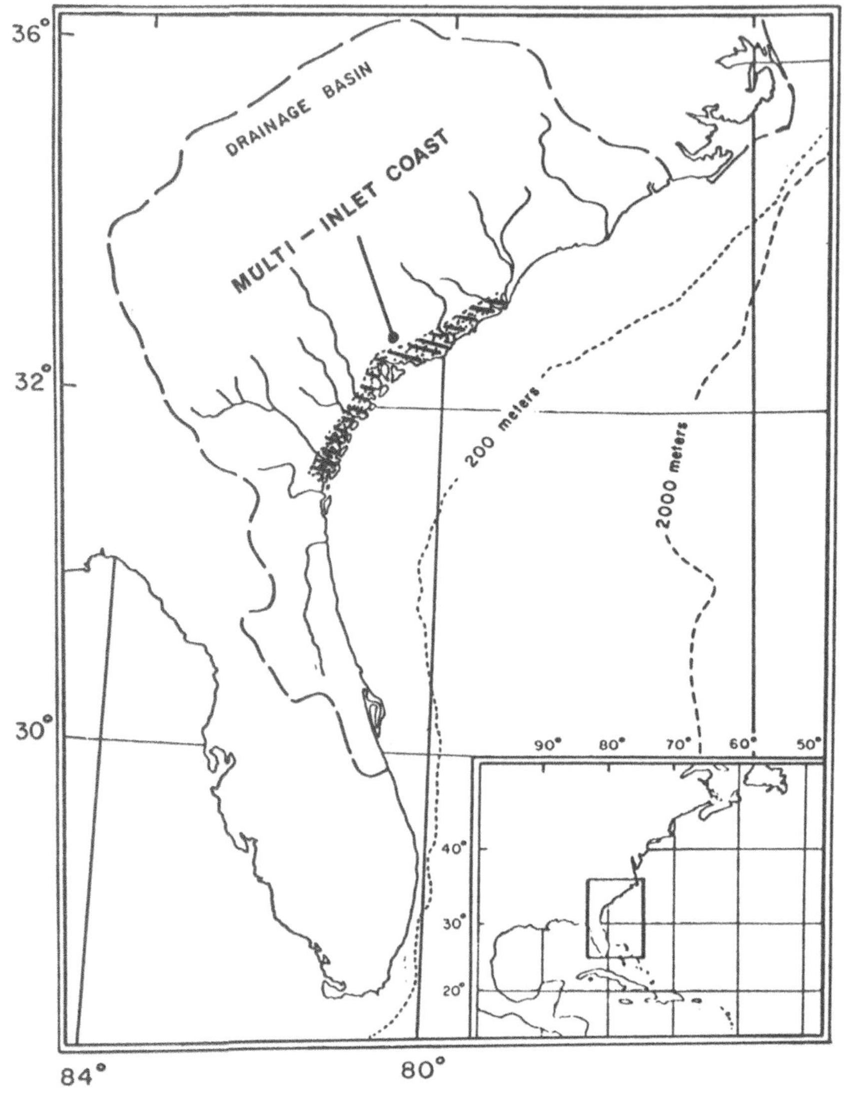

Figure 1. Location map showing the drainage basin and its major outlets to the inner continental shelf. The stippled area marks the coastline off South Carolina and Georgia that has inlets each 10-15 km.

coastal shelf water occur each 10-15 km with large variations in source strength. The inputs carry clay minerals, other suspended and dissolved material, and nutrients that show up as elevated concentrations in the coastal waters immediately offshore (Manheim, Meade and Bond, 1970; Bigham, 1973).

Not all material comes from rivers draining the coastal plain. Vast coastal marshes behind the barrier islands of Georgia and South Carolina are flooded twice daily by the incoming tide. Sediments are picked up; trace elements and nutrients are mobilized in dissolved and particulate form. As the tide ebbs, this material is collected and channeled by many small tidal creeks to the main rivers and sounds, after which the material is ejected to the continental shelf. The transport and diffusion processes that distribute this material in the shelf water near the coast also distribute the freshwater that enters the area. Thus transport of material from the coastal wetlands to the inner continental shelf contributes substantially to the total load of suspended and dissolved material.

Satellite imagery illustrates that large loads of suspended sediment ejected to the coastal zone form a heterogeneous band of turbid water (Figure 2). The complex structure of the band indicates multiple river plumes and fronts. The dynamics associated with this structure are unevenly distributed. Hydrodynamical modeling of such a region would not be realistic if a solid boundary were assumed.

The freshwater discharge is manifested by a band of low salinity water close to shore. We define the nearshore zone as the seaward extent of this band and have estimated that the freshwater represented by the low salinity band is renewed each 2-3 months (Atkinson, Blanton, and Haines, 1978). The freshwater discharge to shelf water is biomodal in character with a major peak in late summer with a range throughout the year of 1-5 km^3/month (Atkinson, Blanton, and Haines, 1978). The semi-diurnal tide strongly modulates the flux of shelf water to the continental shelf. Maximum tidal ranges are 2-3 meters at Savannah, the largest in the southeastern United States.

Advection by the tides off the coast is a major distributing process for the freshwater and its dissolved and suspended materials. While alongshore currents generated by wind seem to have a dominant role in advecting freshwater along the coast, transfer of freshwater across the nearshore zone appears to be controlled by the tides acting in conjunction with the vertical density gradients induced by the freshwater runoff (Blanton and Atkinson, 1978).

The purpose of this paper is to show that tidal currents nearshore have large alongshore variations that transport the freshwater in a complex manner. This, together with the differing amounts of freshwater ejected over a 10-15 km distance, insures incomplete mixing of freshwater.

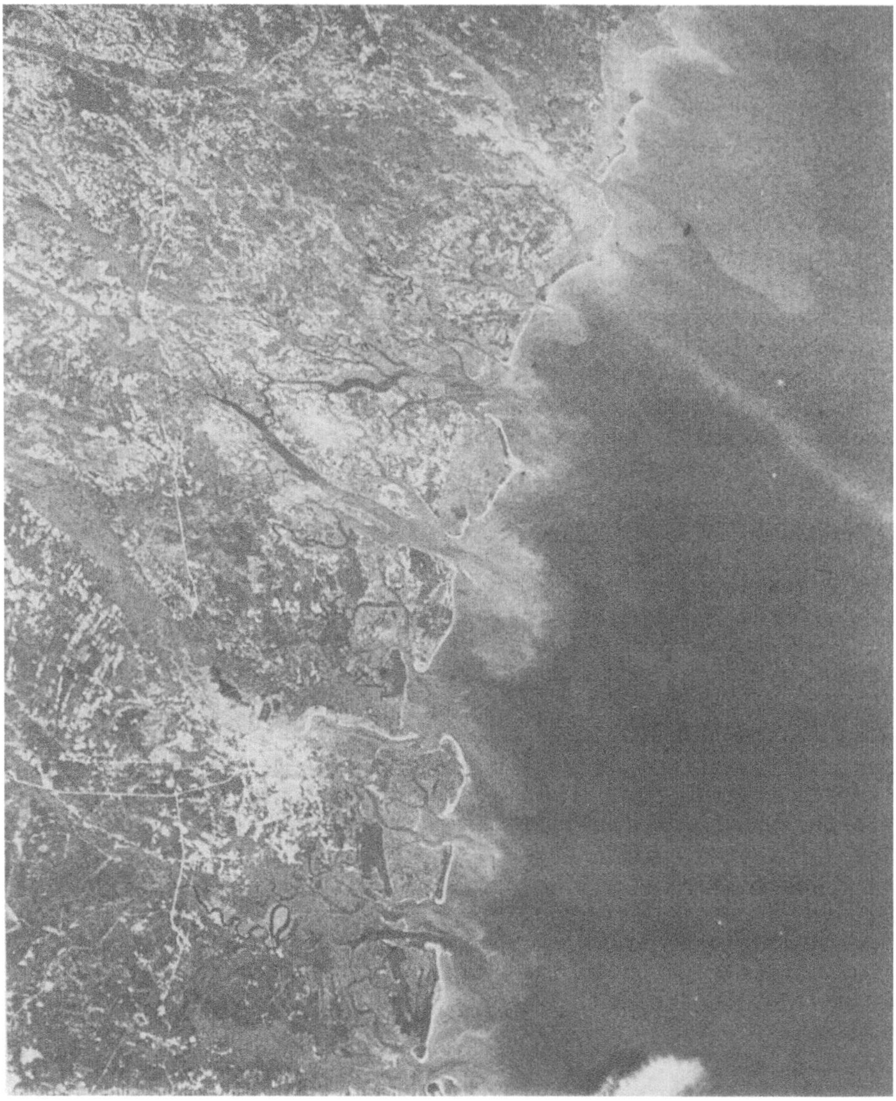

Figure 2. Satellite photograph illustrating the heterogeneous mixing of the turbid water off the coast.

HYDROGRAPHY

The continental shelf off Georgia is shallow, and the nearshore zone as we define it here seldom extends offshore to depths of 20 m out on the shelf. The bottom consists of an irregular array of channels (Figure 4), usually only 1-2 m deeper than the surrounding ridges. These channels cut across the general alongshore trend of the bottom contours. Their influence on circulation is not clear but they may partially funnel tidal currents toward the inlet mouths.

The band of shelf water diluted by freshwater runoff forms a complex frontal zone as less saline water overrides the ambient and usually vertically homogeneous shelf water (Figure 3). The width of the low salinity band is measured to the offshore position where the shelf water is vertically isohaline. This distance extended seaward about 15-20 km in November, 1976 (Figure 3a). In May, 1978, the distance was over 20 km (Figure 3b). The lower salinity in May is probably due to the larger freshwater influx to the continental shelf during spring. The dynamics maintaining this structure are qualitatively similar to those within a partially mixed estuary, but the alongshore heterogeneities within the nearshore zone suggest that the lateral mixing processes (often neglected in estuarine dynamical models) may predominate (Blanton and Atkinson, 1978). We conducted an experiment in May 1978 to assess the importance of alongshore variations in tidal currents and to assess their importance in mixing and transporting less saline water in the nearshore zone.

EXPERIMENTAL CONDITIONS

Two ships were anchored off Wassaw Sound in water 12-13 m deep for four consecutive tidal cycles (Figure 4). The R/V BLUE FIN (location "A") was anchored at a distance of 13 km offshore. The R/V KIT JONES (location "B") was anchored about 9 km to the southwest. The ships measured simultaneously vertical profiles of salinity, temperature and currents each 2 hours. The measured values were accurate to ± 1 cm/s, ± 10° magnetic respectively. The Savannah River mouth is the next inlet to the north. Ossabaw Sound to the south carries the freshwater runoff of the Ogeechee River. The salinity structure at the beginning of this experiment (Figure 3b) showed a halocline sloping upward in the offshore direction. The temperature structure reinforced the halocline to form a pycnocline with a strength of 1 sigma-t unit/m.

Winds

The wind as measured from one of the anchored ships is shown for the duration of the experiment (Figure 5a). One day before the experiment, wind speeds of 8-12m/s exerted a stress toward the N. These speeds continued another 12 hours for the first tidal cycle of the experiment and the wind direction shifted to

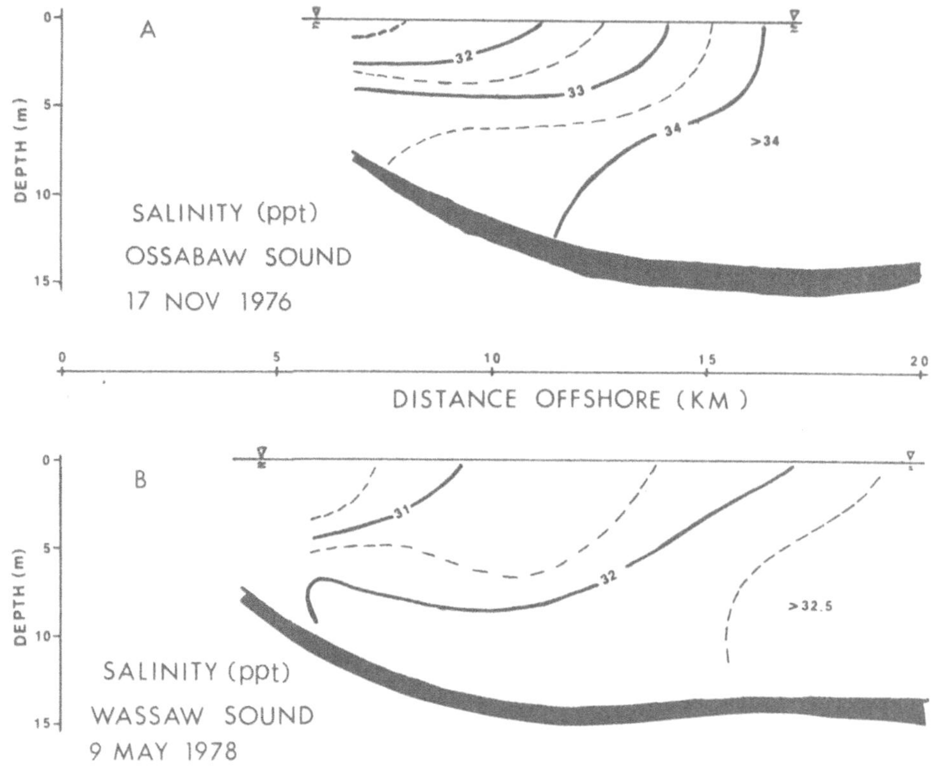

Figure 3. Salinity distribution along sections perpendicular to shore off Georgia. (a) Ossabaw Sound, 18 November 1976 - a time of low autumn runoff; (b) Wassaw Sound, 9 May 1978 - a time of high spring runoff.

produce stress toward the NE. The progressive vector diagram indicated that wind stress shifted to the ESE during the second tidal cycle as a weak cold front passed through. As the third tidal cycle began, the wind stress diminished and changed to the SE, and by the fourth cycle, the wind stress was weak and toward the SW. In summary, the wind stress was steadily toward the NE over the first tidal cycle. After the cold front passage, stress changed to SE and S and weakened during the second tidal cycle. During the third and fourth cycles, wind stress was weak toward the SW and W.

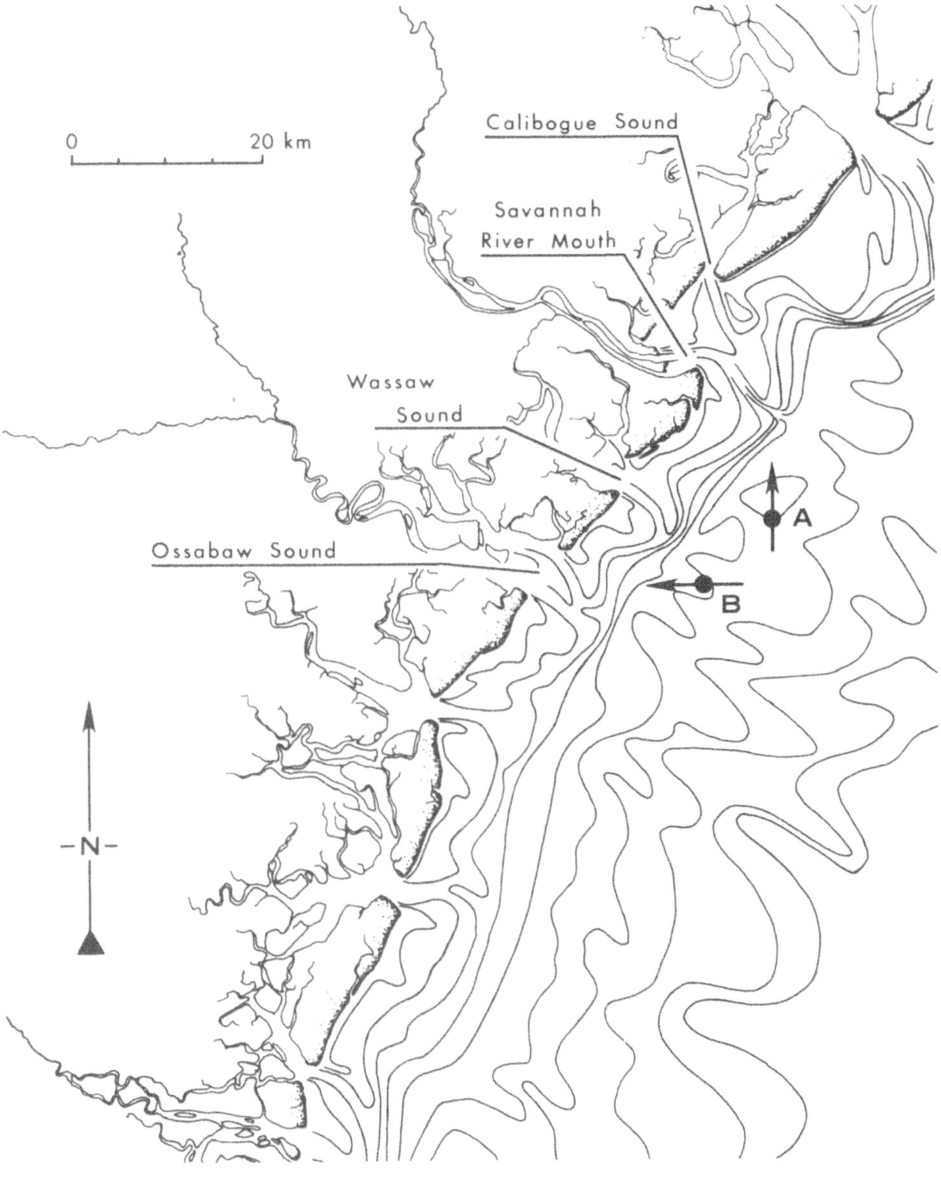

Figure 4. Map showing experiment locations A and B. The major axes of the tidal currents are indicated. See text for details.

Currents

The halocline in the salinity structure (Figure 3b) coincided with a pycnocline whose central salinity remained at near 32 ppt for the entire experiment. We reduced the profiles of currents and salinity by vertically averaging all data from the surface down to and including the depth where the salinity was 32 ppt. The averages of these data over time are labeled "upper;" the averages of the data where salinity was greater than 32 ppt are labelled "lower." Thus, the data discussed below refer to vertical averages above the pycnocline and below the pycnocline.

The upper and lower currents (Figure 5) show the semi-diurnal tidal oscillations superimposed on coastal currents, probably wind generated. The orientation of tidal oscillations was determined on the last two tidal cycles after the wind weakened. The major axis of the tidal currents (Figure 5) was approximately north-south at "A" with rms amplitudes of about 32 cm/s. During the four tidal cycles, there was net velocity of 9 cm/s to the E. Below the pycnocline, the net velocity was 4 cm/s with a cyclonic shift of about 10° from upper currents. The major axis of the tidal oscillations for lower currents was oriented north-south with rms amplutides of ± 22 cm/s.

At location B the effect of the wind stress was more evident, particularly during the first tidal cycle (Figure 5). The net drift above the pycnocline was 13 cm/s at a direction of 73° from north. During the first tidal cycle, there was little direction change since the major axis of the tidal current was east-west. Only after the wind diminished is there clear evidence of tidal oscillations. The rms east-west amplitudes were 29 cm/s. The tidal speeds were not significantly different from those at "A" but the orientation differed by 90°. Below the pycnocline, the major axis of the tidal currents was oriented 110°-290° with rms amplitudes of 17 cm/s. The net drift was slower with a speed of 8 cm/sec with a cyclonic shift of 37° from upper currents.

These results are summarized in Table 1. The significant point is the different orientation of the principal axis of the tidal components. At location "A" the orientation was 360°-180° both above and below the pycnocline. Thus, the flood and ebb currents flowed toward and away from the Savannah River - Calibogue Sound inlets (Figure 4). At "B" the orientation was 90°-270° above the pycnocline, flowing toward and away from Ossabaw Sound. Below the pycnocline, the tidal currents point 20° clockwise from those above the pycnocline. Obviously, transport of freshwater by tidal currents must be different at the two locations.

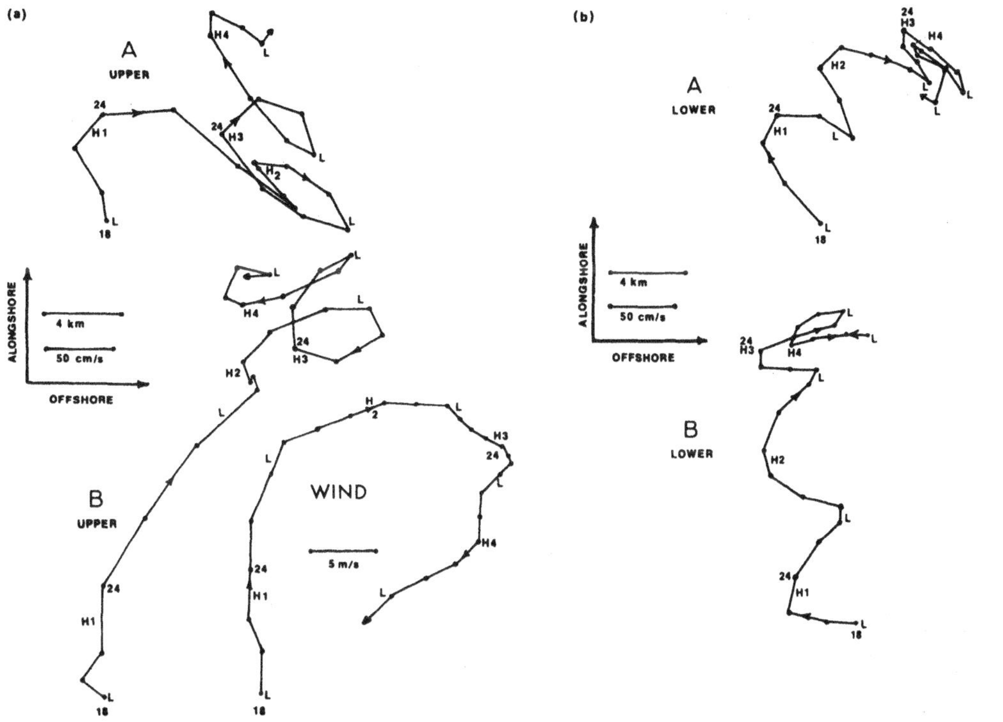

Figure 5. Progressive vector diagrams of currents measured each 2 hrs above and below the pycnocline at location A and B (Figure 4). All diagrams begin at 18 hr (EDT) on 9 May and end at 18 hr on 11 May 1978. (a) winds and "above" currents; (b) "below" currents. "L" and "H" represent times of low and high water respectively. The numbers by "H" designate the four successive tidal cycles. Times at midnight (24h) are also designated.

The salinity data show that different amounts of freshwater passed each location. The ship at "A" began its collection of data with a gradient of salinity from the surface down to about 7 m after which the salinity changed slowly down to the bottom (Figure 3b). The vertical section extended offshore at a direction of 135° and was measured two hours before the ships were anchored. Only if tidal currents have principal axis along that direction will the sloping halocline as viewed by Figure 3b pass back and forth below the anchored ship. Our current profiles demonstrate that the tidal currents at "A" would advect the salinity structure at an oblique angle to the section in Figure 3b; thus in the presence of the expected alongshore variations, we do not expect

Table 1. Summary of current measurements averaged for 4 tidal cycles above and below the pycnocline at locations "A" and "B." Direction and speed of tidal currents apply to the component oriented along the major axis as determined by the vectors for the last two tidal cycles (Figure 5).

Location	Speed	Direction	Speed	Direction
Above net current	9 cm/s	86°	13 cm/s	72°
Tidal current (major axis)	32	360-180°	29	90°-270°
Tidal current (minor axis)	22	90°-270°	17	180°-360°
Below net current	4	77°	8	54°
Tidal current (major axis)	22	360°-180°	17	110°-290°
Tidal current (minor axis)	10	90°-270°	12	200°-020°

to see a simple picture of the halocline going up and down with the tide. The same is true for location "B". We now consider simultaneous plots of the salinity over four tidal cycles (Figure 6). The waters above the halocline contained a complex sequence of gradients representing different amounts of freshwater moving past each location. Below the halocline, the salinity distribution changed more slowly. Note that there was no obvious pattern to the salinity changes that could be correlated with the tidal currents. Patches of low-salinity water passed "A" and "B" in an irregular procession.

These data were vertically averaged in the same manner as the current profiles (Figure 7). The depth of the 32 ppt isohaline is taken as an index of halocline depth. The greater variability of the water above the halocline is apparent. The depth of the halocline is also different at "A" and "B". Except for some correspondence between two perturbations that occurred on the first falling tide of 11 May, there is no obvious correlation. The amplitudes of the halocline depth fluctuations were almost 2 times greater at "A" (Table 2).

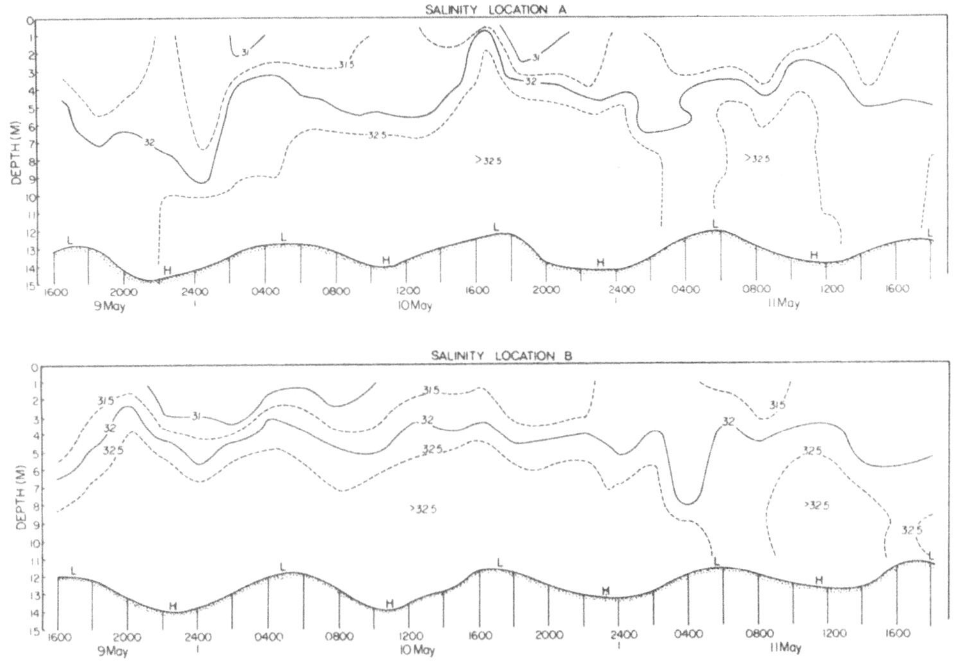

Figure 6. Time plots of the salinity profiles at locations "A" and "B" (Figure 4). "L" and "H" represent times of low and high water respectively.

The actual freshwater content of the upper layer is of primary interest here and a simple comparison can be made by integrating the vertical salinity profiles for the freshwater over and above that below the pycnocline. The maximum bottom salinity was about 32.8 ppt. With S_{max} = 32.8 ppt, we calculate the freshwater content of the water column by

$$Q_{fw} = \int_h^0 \frac{S_{max} - S(z)}{S_{max}} \, dz$$

Q_{fw} is given in $M^3 H_2O/m^2$ and includes the freshwater fraction above S_{max} = 32.8. Salinities below the pycnocline will contribute to this quantity but the quantity is small compared to the relative amounts of low salinity water above the pycnocline. The contributions below the pycnocline were estimated as follows:

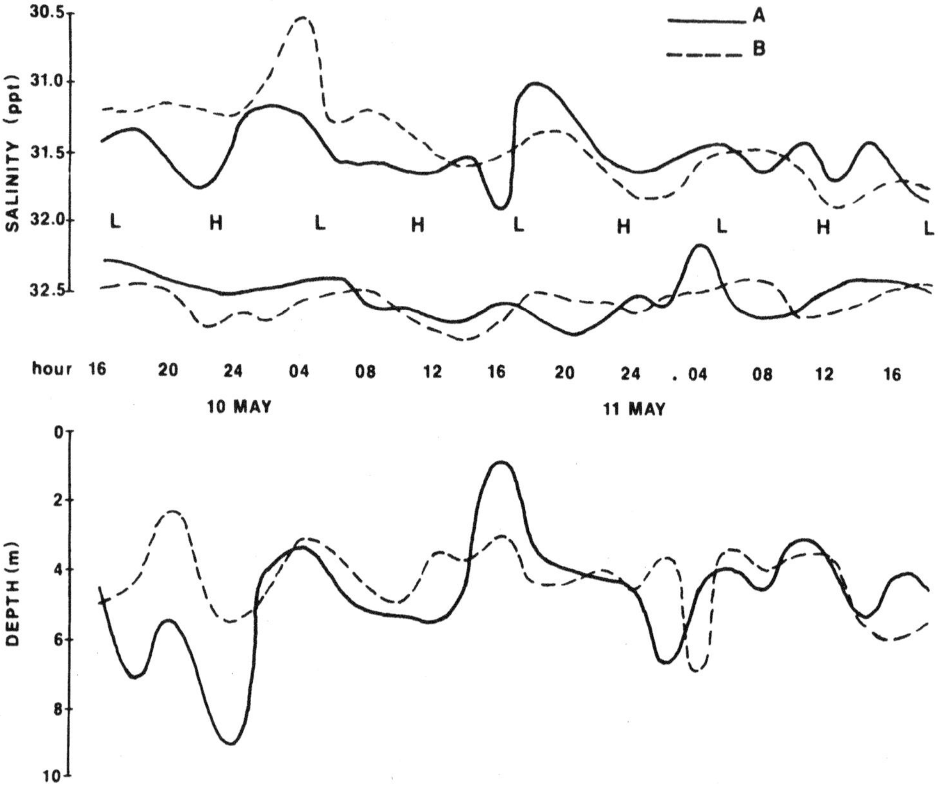

Figure 7. Average salinity above and below the pycnocline at locations "A" and "B" (Figure 4). The upper two curves represent the salinity above the pycnocline; the lower two, the salinity below the pycnocline. The lower panel represents the depth of the pycnocline as shown by the depth of the 32 ppt isohaline.

Table 2. Summary of salinity in ppt averaged for 4 tidal cycles above and below the pycnocline at locations "A" and "B". The pycnocline depth is based on the depth of the 32 ppt isohaline.

Location	A	B
Salinity above pycnocline	31.5 ± 0.2	31.4 ± 0.3
Salinity below pycnocline	32.5 ± 0.1	32.6 ± 0.1
Pycnocline depth (m)	4.7 ± 1.6	4.3 ± 1.0

Average ΔS below pycnocline = 32.5 - 32.0 = 0.5

Average Δh below pycnocline = 13 m - 5 m = 8 m

$$Q_{fw} = \frac{(0.5)}{32.5} \times 8 = 0.1 \; M^3H_2O/m^2$$

This must be kept in mind when interpreting the freshwater content above the pycnocline. The total Q_{fw} was usually 2-4 times the amount below the pycnocline. Q_{fw} for "A" and "B" are shown in Figure 8.

Compare Figure 8 with the Figure 7. The curves are practically mirror images because troughs in the pycnocline contain the greatest volumes of freshwater; the opposite is true when the pycnocline is shallow. The greatest fluctuations of freshwater occur during the first tidal cycle. "A" contained 25% more freshwater during this cycle. During the second tidal cycle, the freshwater content decreased to its lowest values at both locations. For the third and fourth cycles, there were slight increases.

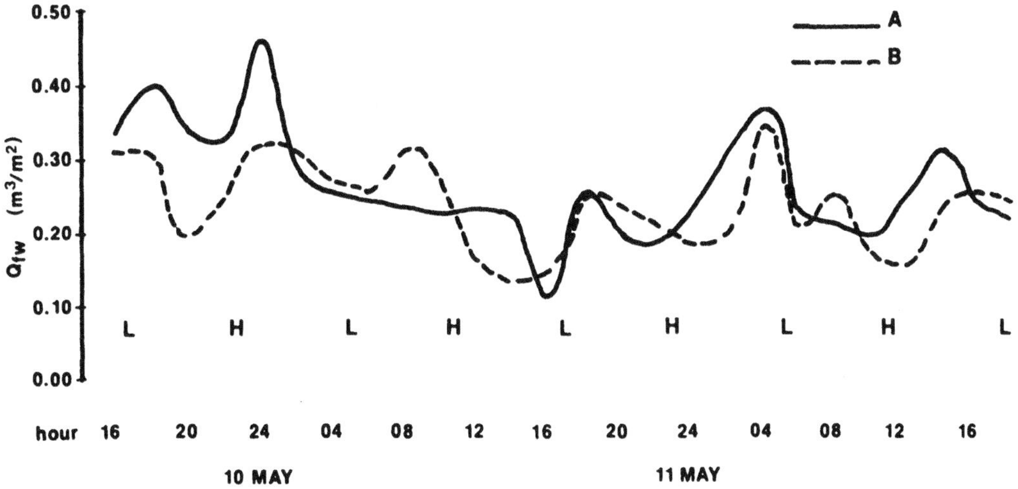

Figure 8. Freshwater (Q_{fw}) content relative to 32.8 ppt at locations A and B (Figure 4). See text for details.

DISCUSSION

If one compares Figure 5a with Figure 8, the reasons for the differing salinity regime are apparent. The ebbing currents at "B" were obviously augmented by the northeastward wind stress during the first cycle. The currents at location "A" were bringing water in from the north while those at "B" were bringing water in from the west. It is tempting to conclude that ebbing currents at "A" were bringing in water from the Savannah River mouth while those at "B" were transporting water from Ossabaw Sound. The experiment cautions against such a conclusion which assumes that Eulerian measurements yield Lagrangian information. The different orientation of the tidal currents within only 9 km demonstrates that alongshore gradients in velocity at tidal frequencies are too large to neglect. The wind stress augments those tidal currents where major axes are oriented along the wind. Net currents at the two locations demonstrate this clearly..

The average salinity for the experiment was not significantly different between "A" and "B". The similar averages are misleading. The data show significant differences in salinity as well as currents at a given time indicating that the flow of salt through each location differed.

The data confirm our expectations from turbidity in satellite imagery that the different sources of freshwater are incompletely mixed over several tidal cycles. One hypothesis we wished to test was that ebbing currents caused water masses of different freshwater content to pass through "A" and "B". The flooding currents advected water that had undergone tidal mixing, thus tending to blend together water with different freshwater content. In other words, if S_A and S_B represent the salinity above the pycnocline as functions of time at the two locations, $S_A - S_B$ for ebbing currents should be greater than $S_A - S_B$ for flood currents. Inspection of Figure 7 shows no such difference. We conclude that more than one tidal cycle are required to destroy the integrity of freshwater discharged to the nearshore zone.

The Q_{fw} values (Figure 8) can be interpreted as "pools" of diluted continental shelf water advected by each ship. The pools have varying thicknesses and probably originated from thinner plumes of riverine water that became detached and vertically mixed with ambient shelf water. Thus, our measurements above the pycnocline are not the offshore extremities of continuous plumes stretching shoreward to the Savannah or the Ogeechee River but detached pools of riverine water that have been mixed with shelf water through one or more tidal cycles.

The complex boundary conditions off multi-inlet coasts present obstacles to the use of hydrodynamical models. The coastline is not solid but perforated thus representing a non-uniform source and sink for momentum. The strengths of each source are likely to be proportional to the tidal prism of each plus the amount of freshwater output through each. The residual circulation induced by the non-uniform momentum sources is an important question. One result of the non-uniform sources is to alter the orientation of the principal axes of the tidal currents (Figure 4). The complex bottom topography near each inlet is known to consist of flood and ebb channels for each tidal cycle (Oertel and Howard, 1972). The channels farther offshore may serve the same role, but the problem has not been addressed.

The vertically integrated equations of motion contain a "tidal stress" term (Nihoul and Ronday, 1975) arising from the non-linearity of the advective terms. The "stress" term is important in regions of intensive tides. The irregular bottom off Georgia and South Carolina together with the asymmetrics in the tidal flow over short distances are likely to induce residual circulation patterns important for the distribution of materials from inlet to inlet once they are ejected out to the coastal water.

ACKNOWLEDGMENTS

This work is a result of research sponsored by the U. S. Department of Energy (DOE/EV/00889-2) and by the Georgia Sea Grant Program, Office of Sea Grant, NOAA (R/EE-3). I am grateful for the help of J. Singer, W. Chandler, P. O'Malley and B. Blanton for obtaining and processing the CTD and current meter data used in this study. I thank D. McIntosh for drafting the figures and L. Land for typing the manuscript.

REFERENCES

Atkinson, L. P., Blanton, J. O. and Haines, E. 1978 Shelf flushing rates based on the distribution of salinity and freshwater in the Georgia Bight. Estuarine and Coastal Marine Science 7, 465-472.

Bigham, G. N. 1973 Zone of influence--inner continental shelf of Georgia. Journal of Sedimentary Petrology 43, 207-214.

Blanton, J. O. and Atkinson, L. P. 1978 Physical transfer processes between Georgia tidal inlets and nearshore waters. In Estuarine Interactions (Wiley, M., ed), 514-532.

Manheim, F. T., Meade, R. H. and Bond, G. C. 1970 Suspended matter in surface waters of the Atlantic continental margin from Cape Cod to the Florida Keys. Science 167, 371-376.

Nihoul, J. D. J. and Ronday, F. C. 1975 The influence of "tidal stress" on the residual circulation. Tellus 27, 484-489.

Oertel, G. F. and Howard, J. D. 1972 Water circulations and sedimentations at estuary entrances on the Georgia coast. In Shelf Sediment Transport (Swift, Duane, and Pilkey, eds), 411-427.

BOX MODELS REVISITED

Charles B. Officer

Earth Sciences Department, Dartmouth College

Hanover, New Hampshire, 03755

ABSTRACT

A methodology in terms of box models has been reexamined for the investigation of conservative and nonconservative quantities in estuaries. Both one and two dimensional models and both tidal exchange and circulation effects are included. Various types of loss relations, sources and sinks, and vertical exchanges are considered. The box model results are tested against analytic solutions of the same problems where available and against two more refined, hydrodynamic numerical model results for a nonconservative loss problem and for a suspended sediment distribution.

The required inputs are salinity, estuary geometry and river flow, which are often known quantities. There are no undetermined or undefined hydrodynamic coefficients. In each case the relations are given by a set of linear algebraic equations. They can be solved by computer matrix algebra procedures or because of their particular form by successive approximations with a hand calculator.

The methods presented do not pretend to add to our physical oceanographic knowledge of estuarine circulation, mixing and the like. It is, however, hoped that they may be of use to those examining biological, chemical, engineering and geological distributions, transformations and other effects, which depend, in part, on estuarine hydrodynamics for their explanation.

INTRODUCTION

The intention in this paper is to delineate certain numerical procedures that can be used for the investigation and analysis of observed, hydrodynamically miscible quantities in estuaries. The procedures involve certain simplifications in the estuarine hydrodynamics, principally in terms of the consideration of finite element sections or box models. Such simplifications are necessary in order to be able also to include the other biological, chemical or geological transformations that may be important and still retain analytical forms that are subject to straight forward numerical calculations.

Estuarine hydrodynamics in itself is extraordinarily complex. When other transformations of a biological, chemical or geological nature must also be considered, the problem is compounded. Often the more refined hydrodynamic solutions do not permit easy assimilation of these other effects. Numerical modeling procedures are quite useful, but there are limitations here as well not only in terms of the time and cost involved but also in terms of the limitations in the defining numerical equations and in the introduction of coefficients which are sometimes poorly known. One of the advantages in the procedures given here is that within their limitations they do permit a direct assessment of the variations in an observed quantity which may be associated with changes in estuarine geometry, river flow, salt intrusion, transformations, effluent input and the like.

In a general way we may consider four different but related approaches to the investigation of estuarine hydrodynamic phenomena. First, there is the methodology of physical oceanography, or geophysical fluid dynamics, in which analytical solutions are given, e.g., Rattray and Hansen (1962) and Hansen and Rattray (1965) for stratified estuaries, Rattray (1967) and Winter (1973) for fjords, and Rattray and Mitsuda (1974) for salt wedges. Second, there is numerical modeling. This can be subdivided into those investigations of a scientific nature, e.g., Hamilton (1975), Bowden and Hamilton (1975), Festa and Hansen (1976), and Blumberg (1977), and those of an engineering or management nature, e.g., Harleman (1977) and O'Connor et al (1977). Third, there is the use of salt, or salinity, as a tracer in which the concentrate continuity equation is solved in terms of the salt continuity equation. In essence this is the approach taken by Ketchum (1950, 1954, 1955) and Stommel (1953) and more recently by Rattray and Officer (1978). Fourth, there is the box model, or finite element, approach in which the exchange and advection coefficients are given directly in terms of the salinity, e.g., Pritchard (1969), Elliott (1976), and Schubel and Carter (1976). The methodologies of three and four are closely related.

Box models have been successfully applied to the study of radiocarbon distributions in the world oceans, the exchanges of carbon between atmosphere and ocean, and abyssal circulation. Particular reference may be made to the excellent articles by Craig (1957, 1963), Broecker (1961, 1966), and Bolin and Stommel (1961). Their procedures and results enhanced our interest in the application of box models to what we considered a similar type of problem, natural and man made distributions in estuaries.

Parallel to the use of box models for scientific investigations in estuaries there have been engineering applications of the same principles. Here, the purpose has been to obtain reliable, and verifiable, results on various water quality parameters for management decisions with regard to pollution abatement or other procedures. A discussion of such engineering box modeling with regard to stream pollution has been given by Bella and Dobbins (1968). Of more pertinence here is the application of the box model approach by the Water Pollution Research Laboratory (U.K.) to estuaries, as summarized by Downing (1971), Gilligan (1972), and Barrett and Mollowney (1972) and also discussed by McDowell and O'Connor (1977). The case history for the Thames estuary is an outstanding example of a logical continuance of effort from first recognizing an estuarine pollution problem, to a scientific and engineering study of it and evaluation of possible corrective measures, and finally to action toward its abatement. Much of this success has been directly related to the box modeling approach taken by the Water Pollution Research Laboratory. This success is, in our opinion, one of the strongest arguments for the use of box models in estuarine investigations. In their investigations the various exchange coefficients were determined through computer procedures to give a best fit to the observed, conservative, salinity distribution. In our case we determine the exchange coefficients box by box from the observed salinity distribution; the two procedures are essentially equivalent, but our permits the coefficient determinations to be made with a hand calculator without recourse to a computer.

Much of what follows in this paper is an elaboration or extension of ideas originally presented by Pritchard (1969) and Pritchard et al (1971). Over the years the physical insight of Pritchard into an understanding of various estuarine phenomena has been among the most significant contributions that have been made thereto.

Box model constraints

There are a number of features that are pertinent to the box model approach that, in essence, define both its limitations and advantages. For this discussion we follow in a general way that given by Pritchard et al (1971).

We use averaged values for all observable quantities within each box, and we assume that both advective and nonadvective exchanges occur only at the boundaries of the boxes. This approximation is the same as the finite element approach used in numerical modeling; it reduces the continuity equations from a differential to a finite difference form. The boundaries between boxes do not, in general, represent natural boundaries such as the case in radiochemistry investigation in the ocean where the boundaries are the air-sea interface and the thermocline. An exception to this is the extention from one to two dimensional models in which the boundary between the upper and lower boxes represents that of the halocline in stratified estuaries. Further, the exchanges are both advective and nonadvective; the boxes represent flow through, or flux, conditions rather than static conditions.

We use nonadvective exchange coefficients rather than dispersion or diffusion coefficients. Instead of expressing, for example, the nonadvective salt transfer by a term of the form $K_x A\ (ds/dx)$ where K_x is the longitudinal dispersion coefficient, A the cross sectional area, s the salinity, and x the longitudinal distance, we use the terms $(K_x A/\Delta x)\ s_j = E_{ji}\ s_j$, and $-\ (K_x A/\Delta x)\ s_i = E_{ij}\ s_i$, separately. It is to be noted that implicit in this reduction is the relation that the nonadvective exchange coefficient E_{ij}, from box i to box j is equal to the exchange coefficient E_{ji} from box j to box i; this is different than the radiochemistry approach for the oceans in which these exchange coefficients are not assumed to be equal. It is, however, in keeping with the usual assumptions taken in the definition of nonadvective, hydrodynamic exchanges for estuarine circulation. The advantage in the separation of the dispersive exchange into its two components is that it allows us to consider separately the nonadvective exchanges into and out of a given box so that we can arrive at values for residence times of quantities in that box; this is not possible when the combined dispersive term is used.

We use the observed values of salinity in the estuary and the salt continuity equation to determine all the hydrodynamic advective and nonadvective exchange coefficients. In this we are using salt as a natural tracer. There are no undefined or undetermined exchange coefficients.

As illustration of these features let us consider the simple case of boxes defined by the natural geometry of the estuary, the water surface, and arbitrary vertical sections normal to the net river flow, R. Instead of the usual one dimensional salt continuity equation, as given, for example, in Officer (1976),

$$Rs = K_x A \frac{ds}{dx} \tag{1}$$

expressing the balance of the net advective, river flow of salt downestuary to the combined tidal exchange and net circulation dispersion of salt upestuary, we use

$$Rs_i + E_{ij}\, s_i = E_{ji}\, s_j \tag{2}$$

where these various advective and nonadvective exchange quantities are as given in Fig. 1. The nonadvective exchange coefficient is then given by

$$E_{ij} = E_{ji} = \frac{s_i}{s_j - s_i} R \tag{3}$$

in terms of the salinity distribution and the river flow similar to the usual expression for the longitudinal dispersion coefficient

$$K_x = \frac{Rs}{A\ (ds/dx)} \tag{4}$$

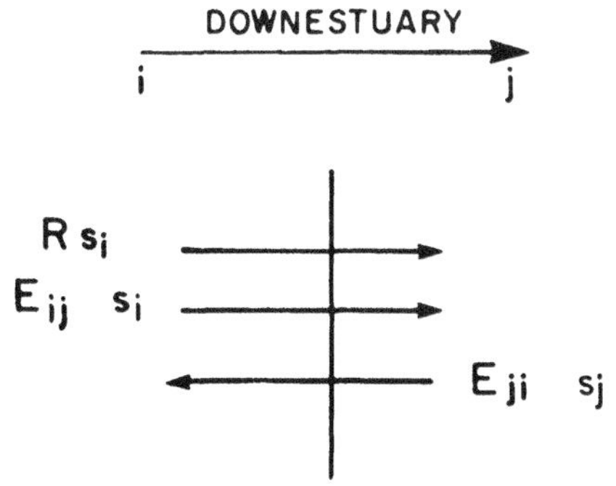

Fig. 1 - Exchanges across a vertical boundary.

$$C' = C\, e^{-\frac{Ft}{C}} \qquad (7)$$

Substituting for dC' and C' from expressions (6) and (7) into equation (5) and integrating, we obtain

$$\tau = \frac{C}{F} = \frac{cV}{F} \qquad (8)$$

where c is the concentration of the index substance in the reservoir and V the volume. We see, then, that the hydrodynamic residence time, or average life of a conservative constituent in a particular volume of an estuary, is given by the quotient of the steady state contents and the flux, which is precisely what we might have expected.

From expressions (7) and (8) we further see that the time required for the original particles, C_o, to be reduced to 1/e of their original value is τ. τ is, thus, formally equivalent to the decay, or radioactive, mean life in first order kinetic reactions. Although we could have defined τ directly as the quotient of the steady state contents and the flux, which is the usual hydrodynamic methodology, we see that this procedure gives us an additional insight into the relationship between hydrodynamic transformations and first order decay transformations, which we shall consider later.

An additional comment should be made with regard to residence time. The residence time used here is for a flux of particles into a box without regard to the origin of these particles. On occasion we shall want to consider a residence time with reference to a flux of identifiable particles from a distant river, ocean or effluent source. For any given box these two residence times will, in general, not be the same.

Box models, one dimensional, conservative

The nomenclature and box enumeration system that will be used for one dimensional considerations are given in Fig. 2. The boundaries of the boxes are the estuary bottom configuration, the water surface, and vertical sections normal to the net advective flow. All observed quantities are taken as tidal and volume averaged values within each box. No restrictions are made as to

Residence time

Residence time is a convenient, if sometimes misused, quantity in the description of estuarine phenomena. We shall restrict ourselves here to the hydrodynamic residence time, which is the same as the residence time for a conservative, miscible quantity. The hydrodynamic residence time is, in a sense, the time constant to which other biological, chemical or geological transformations must be referred. If the residence time is small with respect to the time of any of these transformations, little reaction will take place within the estuarine volume under consideration; if the residence time is large, the reaction will be essentially completed.

We shall follow the rigorous definition of residence time given by Craig (1957). In this the residence time is defined as the average life of a particle in a given volume of the estuary.

Consider a reservoir, or box, with a steady state content, C, of particles of a hydrodynamically miscible, conservative quantity and a continuous flux, F, of particles into and out of the reservoir. At a particular time, t = 0, we will have $C_o = C$ original particles, and at a later time, t, we will have C' of these original C particles still present. Then, we define the average life of a particle in the reservoir in the usual way as

$$\tau = \frac{\Sigma_i t_i C_i}{\Sigma_i C_i} = \frac{1}{C_o} \int_{t=0,\ C'=C_o}^{t=\infty,\ C'=0} t\,dC' \qquad (5)$$

where C_i is the number of particles of the original C_o which remain in the reservoir for each time t_i and dC' is the number of original particles removed in the interval between t and t + dt, i.e., the number of particles with a reservoir life time equal to t.

Now, the number of particles of the original set C_o which are removed in any interval dt will be given simply by the relative concentration of such particles in the reservoir multiplied by the total flux from the reservoir, assuming complete mixing,

$$-dC' = \frac{C'}{C} F dt \qquad (6)$$

which yields on integration

the boxes being of equal volume or longitudinal distance increments. For convenience the river flow, R, is taken as constant, i.e., no tributary rivers along the section investigated; but this condition may be relaxed for any particular numerical calculations. In the following illustrations the section of the estuary to be investigated from x = 0 to x = ℓ does not necessarily correspond to the entire length from its head, river source or sink, to its mouth, ocean source or sink.

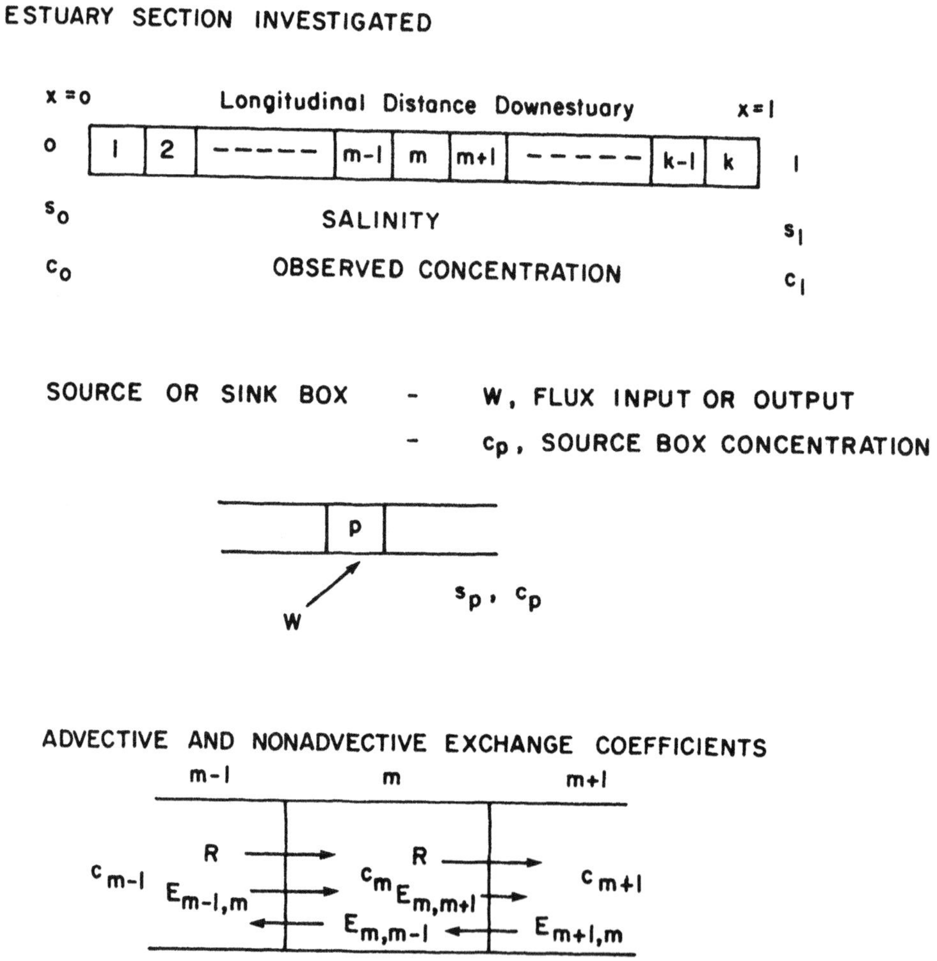

Fig. 2 - One dimensional box model nomenclature.

From Figs. 1 and 2 the defining equation for salt exchanges across the vertical boundary between box m and m + 1 is

$$Rs_m + E_{m,m+1}s_m = E_{m+1,m}s_{m+1} \tag{9}$$

from which we get for the exchange coefficients

$$E_{m,m+1} = E_{m+1,m} = \frac{s_m}{s_{m+1} - s_m} R \tag{10}$$

and similarly for the exchange coefficients between boxes m - 1 and m.

From the defining relation (8) we have for the residence time, τ_m, of box m, using either the flux into or out of box m, that

$$\tau_m = \frac{s_m V_m}{Rs_{m-1} + E_{m-1,m}s_{m-1} + E_{m+1,m}s_{m+1}}$$

$$= \frac{(s_m - s_{m-1})(s_{m+1} - s_m)}{s_m (s_{m+1} - s_{m-1})} \frac{V_m}{R} \tag{11}$$

after substituting for the exchange coefficients from expression (10) and where V_m is the volume of box m. We see, then, that in general the residence time is proportional to the relative salinity difference, Δs/s, between adjacent boxes. For boxes of equal volume the residence time will be larger in the salinity gradient zone of an estuary than upestuary or downestuary from this zone, which is what we might have anticipated.

The concentration, c_m, of a conservative quantity in box m can be given in terms of the corresponding concentrations in the adjacent boxes from the flux continuity relation

$$Rc_m + E_{m,m-1}c_m + E_{m,m+1}c_m = Rc_{m-1} + E_{m-1,m}c_{m-1} + E_{m+1,m}c_{m+1} \tag{12}$$

which reduces to

$$(s_{m+1} - s_{m-1})\, c_m = (s_{m+1} - s_m)\, c_{m-1} + (s_m - s_{m-1})\, c_{m+1}$$

or (13)

$$c_m = \frac{(s_{m+1} - s_m)\, c_{m-1} + (s_m - s_{m-1})\, c_{m+1}}{(s_{m+1} - s_{m-1})}$$

This expression can also be obtained from the definition (8) for residence time, giving

$$c_m = \frac{(Rc_{m-1} + E_{m-1,m}c_{m-1} + E_{m+1,m}c_{m+1})\, \tau_m}{V_m} \tag{14}$$

which reduces to (13).

Let us next consider a case in which there is a source, c_ℓ, of constant concentration at the ocean end and in which the concentration, c_o, at the river end has been reduced to zero. We will, then, have sequentially from expressions (13), starting at the river end,

$$c_1 = \frac{s_1 - s_o}{s_2 - s_o}\, c_2$$

$$c_2 = \frac{(s_3 - s_2)\, c_1 + (s_2 - s_1)\, c_3}{s_3 - s_1}$$

$$= \frac{s_2 - s_o}{s_3 - s_o} c_3$$

$$c_3 = \frac{s_3 - s_o}{s_4 - s_o} c_4 \qquad (15)$$

$$c_{k-1} = \frac{s_{k-1} - s_o}{s_k - s_o} c_k$$

$$c_k = \frac{s_k - s_o}{s_l - s_o} c_l$$

which reduce to

$$c_m = \frac{s_m - s_o}{s_l - s_o} c_l \qquad (16)$$

We could have alternatively considered a source, c_o, of constant concentration at the river end and zero concentration at the ocean end, giving after reduction,

$$c_m = \frac{s_l - s_m}{s_l - s_o} c_o \qquad (17)$$

As all the equations are linear, the case for sources at both ends is given by the sum of expressions (16) and (17), or

$$c_m = c_o - (c_o - c_l) \frac{s_m - s_o}{s_l - s_o} \qquad (18)$$

We see that these relations are the same as those obtained from simple dilution considerations, as expected.

With reference to Fig. 2 consider next a source box p with a flux input, W. The flux continuity relation will then be

$$Rc_p + E_{p,p-1}c_p + E_{p,p+1}c_p = W + Rc_{p-1} + E_{p-1,p}c_{p-1} + E_{p+1,p}c_{p+1} \tag{19}$$

which reduces to

$$c_p = \frac{(s_{p+1} - s_p)\, c_{p-1} + (s_p - s_{p-1})\, c_{p+1}}{(s_{p+1} - s_{p-1})} + \frac{(s_p - s_{p-1})\,(s_{p+1} - s_p)}{s_p\,(s_{p+1} - s_{p-1})}\,\frac{W}{R} \tag{20}$$

If we now assume that the concentrations at the river and ocean ends are zero, we have for c_{p-1} and c_{p+1} in terms of c_p from expressions (15) and the corresponding expressions leading to relation (17) that

$$c_{p-1} = \frac{s_{p-1} - s_o}{s_p - s_o}\, c_p \quad , \quad c_{p+1} = \frac{s_l - s_{p+1}}{s_l - s_p}\, c_p \tag{21}$$

Substituting these values into expression (20) gives

$$c_p = \frac{W}{R}\, \frac{(s_p - s_o)\,(s_l - s_p)}{s_p\,(s_l - s_o)} \tag{22}$$

Finally, for the concentration upestuary from the source box we will have from expression (16), where we now in this expression consider c_l to be c_p, that

$$c_m = \frac{W}{R} \frac{s_l - s_p}{s_p} \frac{s_m - s_o}{s_l - s_o} \quad (23)$$

and correspondingly in a downestuary direction using expression (17) that

$$c_m = \frac{W}{R} \frac{s_p - s_o}{s_p} \frac{s_l - s_m}{s_l - s_o} \quad (24)$$

which are the same as the relations obtained from Ketchum (1955) from the salt and concentrate continuity relations.

Alternatively we could take the case in which the concentrations, c_o and c_l, at the boundaries of the estuary segment being investigated are not zero. As all the box model defining relations are linear we will have from expressions (16), (17), and (23) in an upestuary direction that

$$c_m = \frac{s_l - s_m}{s_l - s_o} c_o + \frac{s_m - s_o}{s_l - s_o} c_l + \frac{W}{R} \frac{s_l - s_p}{s_p} \frac{s_m - s_o}{s_l - s_o} \quad (25)$$

and from expressions (16), (17) and (24) in a downestuary direction that

$$c_m = \frac{s_l - s_m}{s_l - s_o} c_o + \frac{s_m - s_o}{s_l - s_o} c_l + \frac{W}{R} \frac{s_p - s_o}{s_p} \frac{s_l - s_m}{s_l - s_o} \quad (26)$$

The expressions (25) and (26) illustrate a difficulty common to all approaches--analytical, numerical modeling, box models--for the solution of estuarine distributions. This difficulty is the determination of the boundary, or distant source, conditions. We never deal with a closed system; we always have a finite length of estuary with both an upestuary and downestuary boundary and with contributions across these boundaries from external reservoirs. It is usually desirable to continue the calculations beyond the immediate segment of interest in the estuary to ameliorate the effects of distant contributions, but often this is not possible either because of geometrical constraints or because of the magnitude of these distant contributions.

There are two categories of distant boundary effects. One, there are distant source contributions from external reservoirs to the estuarine segment of interest. For example, in a study of eutrophication we may know the nutrient flux input, W, from a sewage treatment plant but may have only a vague notion of the nutrient input from agricultural runoff, reflected in part in the value c_o. Another example is the study of suspended sediment distributions in estuaries for which the principal unknown is sometimes the ocean source contribution. Two, there are the effects of the internal sources, W, on the distant, river and ocean, reservoirs. The boundary conditions are not fixed but depend, in part, on the magnitude of the internal sources. There is no simple method to determine what contributions to the boundary values, c_o and c_1, come from internal and distant sources except through the use of a tracer for the known, internal sources which is alien to the distant sources.

The importance of boundary and distant source conditions is, in our opinion, critical. A given determination for an estuarine distribution will often stand or fall on how well these conditions are known.

Box models, one dimensional, nonconservative

Let us consider the case in which there is a loss phenomenon within the estuary which is time dependent only. This could correspond to situations such as the silica uptake in phytoplankton production in which the silica loss rate, B, is in units of mg/l. day. We shall also take a river and an ocean reservoir maintained at constant concentrations, c_o and c_1, and shall be interested in determining the concentration loss, Δc_m, over a conservative distribution (18).

With reference to Fig. 3 consider a box p with a flux loss, $W = B_p V_p$. Then, for a box m upestuary of box p there will be from expression (23) a corresponding concentration loss given by

$$\Delta c_m = \frac{B_p V_p}{R} \frac{s_l - s_p}{s_p} \frac{s_m - s_o}{s_l - s_o} \tag{27}$$

and for a box m downestuary of box p there will be from expression (24) a loss given by

$$\Delta c_m = \frac{B_p V_p}{R} \frac{s_p - s_o}{s_p} \frac{s_l - s_m}{s_l - s_o} \tag{28}$$

To obtain the total concentration loss for box m we add all the box p contributions (28) upestuary of box m and all the box p contributions (27) downestuary from box m, giving

$$\Delta c_m = \frac{s_l - s_m}{s_l - s_o} \left[\frac{B_1 V_1 (s_1 - s_o)}{s_1} + \frac{B_2 V_2 (s_2 - s_o)}{s_2} + - - - + \frac{1}{2} \frac{B_m V_m (s_m - s_o)}{s_m} \right] \frac{1}{R} \tag{29}$$

$$+ \frac{s_m - s_o}{s_l - s_o} \left[\frac{1}{2} \frac{B_m V_m (s_l - s_m)}{s_m} + - - - \frac{B_k V_k (s_l - s_k)}{s_k} \right] \frac{1}{R}$$

We note that this summation is equivalent in finite difference form to the integral expression obtained by Rattray and Officer (1979) from the analytic solution to the same problem.

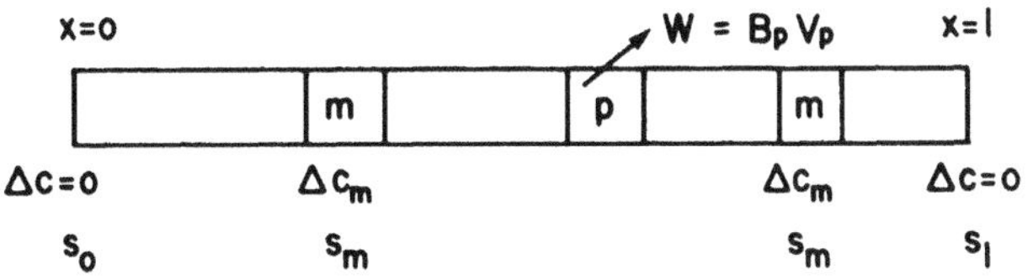

Fig. 3 - Box model time dependent loss relations.

For the methodology of the preceeding paragraph we have used a defining equation of the form (19) with a flux input, $W = -B_pV_p$. The sum of the coefficients on the left hand side of this expression can be written in terms of the residence time, τ_p, from expressions (14) and (12) as

$$R + E_{p,p-1} + E_{p,p+1} = \frac{c_pV_p}{\tau_p} \tag{30}$$

We may, then, rewrite (19) as

$$c_p = \frac{(Rc_{p-1} + E_{p-1,p}c_{p-1} + E_{p+1,p}c_{p+1} - B_pV_p)\ \tau_p}{V_p} \tag{31}$$

Alternatively we could have considered for each box p a decrease in concentration associated with the loss rate, B. This will be directly related to the residence time for box p, and the concentration decrease will be $B_p\tau_p$. The defining equation then becomes the expression (14) plus this decrease, or

$$c_p = \frac{(Rc_{p-1} + E_{p-1,p}c_{p-1} + E_{p+1,p}c_{p+1})\ \tau_p}{V_p} - B_p\tau_p \tag{32}$$

which is the same as (31).

Let us next consider a first order, exponential decay type loss in which the loss rate is related to the concentration of the substance being investigated. For an exponential decay constant, λ, and a steady state concentration, c_m, the flux loss from box m will be λc_mV_m. Our defining equation will then be

$$Rc_m + E_{m,m-1}c_m + E_{m,m+1}c_m + \lambda c_mV_m = Rc_{m-1} + E_{m-1,m}c_{m-1} + E_{m+1,m}c_{m+1} \tag{33}$$

Substituting from the relations (10) we obtain the equations

$$\left[1 + \frac{\lambda V_m}{R} \frac{(s_m - s_{m-1})(s_{m+1} - s_m)}{s_m (s_{m+1} - s_{m-1})}\right] c_m =$$

$$\frac{s_{m+1} - s_m}{s_{m+1} - s_{m-1}} c_{m-1} + \frac{s_m - s_{m-1}}{s_{m+1} - s_{m-1}} c_{m+1} \tag{34}$$

We have k equations in the k unknowns, c_1, ---- c_k, with the boundary values c_o and c_l. We note that the coefficient multiplying λ in the second term of the brackets is the residence time, τ_m, of expression (11), as we might have expected.

The k equations (34) are all linear with numerical coefficients. They can be solved by computer matrix algebra procedures. Because of their particular form they can also be solved by successive approximations using the jury, or marching, method, see for example Jeffreys and Jeffreys (1956, pg. 306), with a hand calculator. In this method a first approximation is taken from the conservative distribution and a rough estimate of the magnitude of the non-conservative losses, or by any other suitable procedure. The boundary value c_o and the first approximation for c_2 are used to determine a second approximation for c_1. This second approximation for c_1 and the first approximation for c_3 are used to determine a second approximation for c_2, and so on out to the boundary value c_l. The entire procedure is then repeated again as many times as is necessary to reduce the successive differences between approximations to a suitable level.

We can extend this type of analysis to sources with flux, W_m, in any or all of the boxes along the estuary segment being investigated. The defining equation then becomes

$$Rc_m + E_{m,m-1}c_m + E_{m,m+1}c_m + \lambda c_m V_m = W_m$$

$$+ Rc_{m-1} + E_{m-1,m}c_{m-1} + E_{m+1,m}c_{m+1} \tag{35}$$

which reduces to

$$\left[1 + \frac{\lambda V_m}{R} \frac{(s_m - s_{m-1})(s_{m+1} - s_m)}{s_m (s_{m+1} - s_{m-1})}\right] c_m = \frac{s_{m+1} - s_m}{s_{m+1} - s_{m-1}} c_{m-1}$$

$$+ \frac{s_m - s_{m-1}}{s_{m+1} - s_{m-1}} c_{m+1} + \frac{(s_m - s_{m-1})(s_{m+1} - s_m)}{s_m (s_{m+1} - s_{m-1})} \frac{W_m}{R} \qquad (36)$$

The added term in equations (36) over those in equations (35) is a numerical value, and these equations may be solved by the same methods as discussed in the previous paragraph.

At times our interest is not with distribution of a non-conservative quantity in an estuary, or segment of an estuary, but only with the flux balance. We shall consider an estuary with sources of constant concentration, c_o and c_l, in the reservoirs at each end. Then with reference to Fig. 2 the flux loss, L_m, within any box m will be

$$L_m = NF_{m-1,m} - NF_{m,m+1}$$

$$= Rc_{m-1} + E_{m-1,m}c_{m-1} - E_{m,m-1}c_m - (Rc_m$$

$$+ E_{m,m+1}c_m - E_{m+1,m}c_{m+1}) \qquad (37)$$

$$= \frac{s_m c_{m-1} - s_{m-1} c_m}{s_m - s_{m-1}} R - \frac{s_{m+1} c_m - s_m c_{m+1}}{s_{m+1} - s_m} R$$

after substitution from relation (10). The corresponding expression for the flux loss, L_e, for the estuary as a whole will be

$$L_e = NF_{o,1} - NK_{k,l}$$
$$= \frac{s_1 c_o - s_o c_1}{s_1 - s_o} R - \frac{s_l c_k - s_k c_l}{s_l - s_k} R \tag{38}$$

which is equivalent in finite difference form to the analytic expression given by Officer (1978) for the same problem.

At the river end with $s_o = 0$, there will be a constant net advective flux Rc_o. At the ocean end there will be a net flux out of or into the estuary depending on the relative magnitude of the two terms in the second portion of expression (38).

The loss, L_m, determined from equation (37), for any box m can be used to arrive at loss coefficient values. For an assumed time dependent loss condition we can calculate a B_m value from $L_m = B_m V_m$. For an assumed first order decay condition we can calculate a λ_m value from $L_m = \lambda_m c_m V_m$.

The type of flux balance analysis of the preceeding paragraphs can be easily extended for sources within the estuary, treating them as additional flux inputs.

Box models, one dimensional, coupled transformations

We can extend these nonconservative box model relations to coupled transformations in which the feedforward and feedback terms are all first order reactions. The simplest such example, which we shall take here, is the biological oxygen demand (BOD) and dissolved oxygen (DO) coupled relations. These can easily be extended to nitrogen, phosphorous, or other nutrient cycle relations.

The biological oxygen demand represents a straight forward, nonconservative transformation. The result is given by equations (36) in which c_m is the BOD concentration in mg/l, λ the BOD decay coefficient, and W_m the BOD flux inputs from sewage treatment plants, outfalls, nonpoint sources, and the like.

The dissolved oxygen deficit from an assumed saturated condition, d_m, also in mg/l, is directly dependent on the BOD condition and reaeration. The term, $\lambda c_m V_m$, is a flux input to the dissolved oxygen deficit, and the reaeration term is a flux output to the dissolved oxygen deficit. The reaeration term is taken in the form $\alpha_m d_m V_m$, indicating an oxygen flux to the estuary proportional to the undersaturated condition of the estuarine waters; the reaeration coefficient, α_m, is depth dependent, among other things. The defining equation for d_m is, then,

$$Rd_m + E_{m,m-1}d_m + E_{m,m+1}d_m + \alpha_m d_m V_m = \lambda c_m V_m$$
$$+ Rd_{m-1} + E_{m-1,m}d_{m-1} + E_{m+1,m}d_{m+1} \tag{39}$$

which reduces to

$$\left[1 + \frac{\alpha_m V_m}{R} \frac{(s_m - s_{m-1})(s_{m+1} - s_m)}{s_m (s_{m+1} - s_{m-1})}\right] d_m$$
$$= \frac{s_{m+1} - s_m}{s_{m+1} - s_{m-1}} d_{m-1} \tag{40}$$
$$+ \frac{s_m - s_{m-1}}{s_{m+1} - s_{m-1}} d_{m+1} + \frac{(s_m - s_{m-1})(s_{m+1} - s_m)}{s_m (s_{m+1} - s_{m-1})} \frac{\lambda c_m V_m}{R}$$

The solution procedure is to first solve equations (36) for c_m. Then, using the resulting numerical values for c_m, solve in the same manner equations (40) for d_m.

Box models, two dimensional, conservative

It is of interest to extend our box model considerations to two dimensions. The division that we wish to make is in the vertical direction, sectioning box m into an upper and a lower box. This division allows some consideration to be given to the density gradient, gravitational effects in estuaries. In the upper box the net circulation flux is downestuary and in the lower box upestuary, as discussed for example in Officer (1977). This division corresponds, in a general way, with the halocline in stratified estuaries; and we shall use the averaged values of salinity, s_m and s'_m, above and below the halocline for each box.

There are a number of reasons for considering two dimensional models. One, there are natural vertical variations of the salinity in stratified estuaries. We may expect these natural variations to be expressed in vertical variations of an observed substance, and such vertical variations may not be adequately represented by the one dimensional averaged conditions discussed in the previous sections. Two, the one dimensional models include both the tidal exchange and net circulation effects in the longitudinal exchange coefficients of the form, $E_{m,m+1}$. For two dimensional models we can make some separation of these two effects, one of which, tidal exchange, is essentially depth independent and the other of which, net circulation, is depth dependent. Hansen and Rattray (1966) in their discussion of estuary classification introduce the useful index, ν, defined as the ratio of the tidal exchange and possible lateral circulation fluxes to the total longitudinal dispersion flux, to distinguish the relative importance of these two contributions. For estuaries in which ν is nearly unity the tidal exchange plus possible lateral effects will be dominant. We shall use this index in the discussions that follow. Three, there are certain effects that are peculiar to the vertical dimension which are completely ignored in one dimensional considerations. One such important effect is the vertical flux, of particles, which leads to consideration of nutrient traps and suspended particulate matter distributions.

To start let us consider the case in which the net circulation effects are dominant, i.e., $\nu \sim 0$. Then, following Pritchard (1969) the exchanges across the vertical boundaries between boxes may be represented by an advective quantity, Q, representing the combined net circulation and river flows. The nomenclature and exchange coefficients to be used are as shown in Fig. 4. We now must also include the advective and nonadvective exchanges across the halocline.

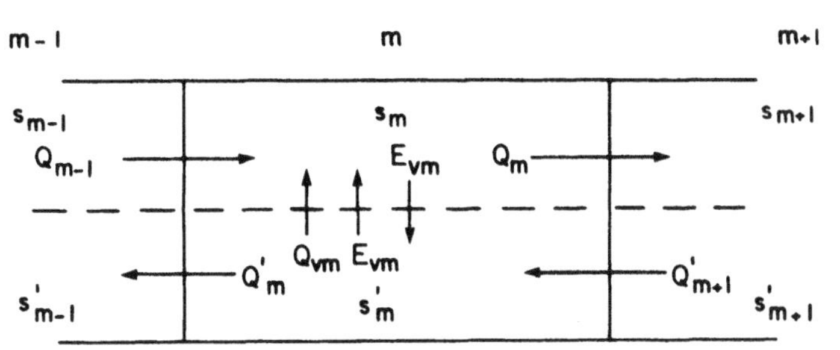

Fig. 4 - Nomenclature and exchange coefficients for two dimensional box models with negligible longitudinal tidal exchange.

With reference to Fig. 4 there are six defining equations for salt flux and flow continuity across the two vertical boundaries and across the halocline for the upper box m in the six unknown coefficients, Q_{m-1}, Q_m, Q'_m, Q'_{m+1}, Q_{vm}, E_{vm}. It is to be noted that there are two additional equations across the halocline for the lower box, which are redundant. We, then, have

$$Q_{m-1}s_{m-1} = Q'_m s'_m \qquad (41)$$

$$Q_m s_m = Q'_{m+1} s'_{m+1} \qquad (42)$$

$$Q_{m-1} - Q'_m = R \qquad (43)$$

$$Q_m - Q'_{m+1} = R \qquad (44)$$

$$Q_{m-1} + Q_{vm} = Q_m \qquad (45)$$

$$Q_{m-1}s_{m-1} + Q_{vm}s'_m + E_{vm}s'_m = Q_m s_m + E_{vm}s_m \qquad (46)$$

Solving equations (41) and (43) for Q_{m-1} and Q'_m and equations (42) and (44) for Q_m and Q'_{m+1}, we obtain

$$Q_{m-1} = \frac{s'_m}{s'_m - s_{m-1}} R \qquad (47)$$

$$Q'_m = \frac{s_{m-1}}{s'_m - s_{m-1}} R \qquad (48)$$

$$Q_m = \frac{s'_{m+1}}{s'_{m+1} - s_m} R \tag{49}$$

$$Q'_{m+1} = \frac{s_m}{s'_{m+1} - s_m} R \tag{50}$$

From equation (45) we, then, have

$$Q_{vm} = \frac{s_m s'_m - s_{m-1} s'_{m+1}}{(s'_{m+1} - s_m)(s'_m - s_{m-1})} R \tag{51}$$

and in turn from equation (46)

$$E_{vm} = \frac{s_m (s'_{m+1} - s'_m)}{(s'_m - s_m)(s'_{m+1} - s_m)} R \tag{52}$$

From the defining relation (8) we have for the residence time, τ_m, of upper box m, using either the salt flux into or out of the box, that

$$\tau_m = \frac{s_m V_m}{Q_m s_m + E_{vm} s_m}$$

$$= \frac{s'_m - s_m}{s'_m} \frac{V_m}{R} \tag{53}$$

and correspondingly for the lower box m that

$$\tau'_m = \frac{s'_m - s_m}{s_m} \frac{V'_m}{R} \tag{54}$$

It should be noted that since we have ignored longitudinal tidal exchange effects in the above considerations these residence times reduce to zero rather than to the expression (11) for unstratified conditions, $s_m = s'_m$.

Continuing as before we can obtain expressions for the concentrations, c_m and c'_m, of a conservative quantity from either the flux continuity relation or from the residence time relation. We have the defining relation for the upper box m that

$$c_m = \frac{(Q_{m-1}c_{m-1} + Q_{vm}c'_m + E_{vm}c'_m)\tau_m}{V_m} \tag{55}$$

which reduces to

$$c_m = \frac{s'_m - s_m}{s'_m - s_{m-1}} c_{m-1} + \frac{s_m - s_{m-1}}{s'_m - s_{m-1}} c'_m \tag{56}$$

Correspondingly we have for the lower box m

$$c'_m = \frac{(Q'_{m+1} c'_{m+1} + E_{vm}c_m)\tau'_m}{V'_m} \tag{57}$$

which reduces to

$$c'_m = \frac{s'_m - s_m}{s'_{m+1} - s_m} c'_{m+1} + \frac{s'_{m+1} - s'_m}{s'_{m+1} - s_m} c_m \tag{58}$$

We can combine the expressions (56) and (58) into a somewhat more useful form as

$$c_m = \frac{s'_{m+1} - s_m}{s'_{m+1} - s_{m-1}} c_{m-1} + \frac{s_m - s_{m-1}}{s'_{m+1} - s_{m-1}} c'_{m+1} \tag{59}$$

and

$$c'_m = \frac{s'_{m+1} - s'_m}{s'_{m+1} - s_{m-1}} c_{m-1} + \frac{s'_m - s_{m-1}}{s'_{m+1} - s_{m-1}} c'_{m+1} \tag{60}$$

We see that these solutions are similar in form to the expression (13) obtained for the one dimensional case.

We can now follow the same procedure, as before, and consider an ocean source, c'_l, of constant concentration and a concentration, c_o, at the river end of zero, obtaining

$$c_m = \frac{s_m - s_o}{s'_l - s_o} c'_l \ , \quad c'_m = \frac{s'_m - s_o}{s'_l - s_o} c'_l \tag{61}$$

Correspondingly we have for a river source of constant concentration, c_o, and a concentration, c'_l, at the ocean end of zero that

$$c_m = \frac{s'_l - s_m}{s'_l - s_o} c_o \ , \quad c'_m = \frac{s'_l - s'_m}{s'_l - s_o} c_o \tag{62}$$

For sources of constant concentration at both ends the expressions (61) and (62) are simply additive.

Let us next consider a source in upper box p with a flux input, W. Following the same procedures as before, we obtain for the concentration, c_p, in the source box

$$c_p = \frac{W}{R} \frac{(s'_p - s_o)\ (s'_l - s_p)}{s'_p\ (s'_l - s_o)} \tag{63}$$

corresponding to expression (22) and also for the lower box p

$$c'_p = \frac{W}{R}\,\frac{(s'_p - s_o)\,(s'_l - s'_p)}{s'_p\,(s'_l - s_o)} \tag{64}$$

In an upestuary direction from box p we, then, have

$$c_m = \frac{W}{R}\,\frac{s'_l - s'_p}{s'_p}\,\frac{s_m - s_o}{s'_l - s_o} \tag{65}$$

and

$$c'_m = \frac{W}{R}\,\frac{s'_l - s'_p}{s'_p}\,\frac{s'_m - s_o}{s'_l - s_o} \tag{66}$$

In a downestuary direction from box p we also have

$$c_m = \frac{W}{R}\,\frac{s'_p - s_o}{s'_p}\,\frac{s'_l - s_m}{s'_l - s_o} \tag{67}$$

and

$$c'_m = \frac{W}{R}\,\frac{s'_p - s_o}{s'_p}\,\frac{s'_l - s'_m}{s'_l - s_o} \tag{68}$$

For a source, W', in lower box p we obtain

$$c_p = \frac{W'}{R} \frac{(s_p - s_o)(s'_1 - s_p)}{s_p (s'_1 - s_o)} \tag{69}$$

$$c'_p = \frac{W'}{R} \frac{(s'_p - s_o)(s'_1 - s_p)}{s_p (s'_1 - s_o)} \tag{70}$$

$$c_m = \frac{W'}{R} \frac{s'_1 - s_p}{s_p} \frac{s_m - s_o}{s'_1 - s_o} \tag{71}$$

$$c'_m = \frac{W'}{R} \frac{s'_1 - s_p}{s_p} \frac{s'_m - s_o}{s'_1 - s_o} \tag{72}$$

$$c_m = \frac{W'}{R} \frac{s_p - s_o}{s_p} \frac{s'_1 - s_m}{s'_1 - s_o} \tag{73}$$

$$c'_m = \frac{W'}{R} \frac{s_p - s_o}{s_p} \frac{s'_1 - s'_m}{s'_1 - s_o} \tag{74}$$

corresponding to the expressions (59) through (68).

Let us now consider two dimensional box models with both net circulation and tidal exchange across the vertical boundaries between boxes. From Fig. 5 we will have for salt flux and flow balances

$$Q_{m-1}s_{m-1} + E_{m-1,m}s_{m-1} + E'_{m-1,m}s'_{m-1}$$
$$= Q'_m s'_m + E'_{m,m-1}s'_m + E_{m,m-1}s_m \tag{75}$$

$$Q_{m-1} - Q'_m = R \tag{76}$$

where the E coefficients now represent only the nonadvective tidal exchange. From the definition of net circulation we have that

$$Q_{m-1} = Q_{m-1,m} + \frac{R}{2} \tag{77}$$

$$Q'_m = Q'_{m,m-1} - \frac{R}{2} \tag{78}$$

where $Q_{m-1,m}$ and $Q'_{m,m-1}$ are, respectively, the net circulation flows and are equal and where we have assumed that the river flow is essentially equally divided between the two circulation flows. We can rewrite (75) in terms of these quantities as

$$\frac{R}{2}\, s_{m-1} + \frac{R}{2}\, s'_m = Q_{m-1,m}(s'_m - s_{m-1}) + E_{m-1,m}(s_m - s_{m-1}) + E'_{m-1,m}(s'_m - s'_{m-1}) \tag{79}$$

The terms on the left hand side represent a measure of the total longitudinal salt flux for the upper and lower layers respectively, the first term on the right hand side a measure of the net circulation salt flux for both layers, and the second and third terms a measure of the net tidal exchange salt flux for the upper and lower layers, respectively. From the definition of the index quantity, ν, and as appropriate to this two dimensional box model approach we shall define the tidal exchange coefficients in terms of ν as

$$E_{m-1,m} = E_{m,m-1} = \frac{1}{2}\, \frac{\nu R s_{m-1}}{s_m - s_{m-1}} \tag{80}$$

and

$$E'_{m-1,m} = E'_{m,m-1} = \frac{1}{2} \frac{\nu R s'_m}{s'_m - s'_{m-1}} \tag{81}$$

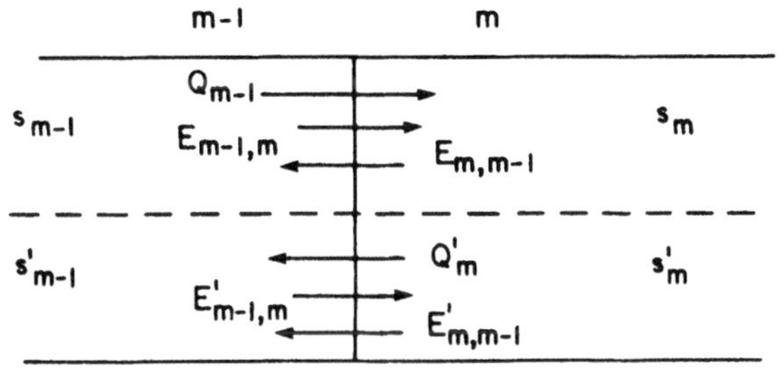

Fig. 5 - Net circulation and tidal exchange coefficients across a vertical boundary for two dimensional box models.

Using the relations (90) and (81) in equations (75) and (76) we, then, have

$$Q_{m-1} = \frac{s'_m - \nu \dfrac{s_{m-1} + s'_m}{2}}{s'_m - s_{m-1}} R \tag{82}$$

and

$$Q'_m = \frac{s_{m-1} - \nu \dfrac{s_{m-1} + s'_m}{2}}{s'_m - s_{m-1}} R \tag{83}$$

For $\nu = 1$, tidal exchange only and no net circulation effects, equations (82) and (83) reduce to R/2 and -R/2, as expected; for $\nu = 0$, they reduce to (47) and (48). From equation (45), which still applies in this case, we obtain

$$Q_{vm} = \frac{s_m s'_m - s_{m-1} s'_{m+1}}{(s'_{m+1} - s_m)(s'_m - s_{m-1})} (1 - \nu) R \quad (84)$$

The salt balance relation corresponding to equation (46) now becomes

$$Q_{m-1} s_{m-1} + E_{m-1,m} s_{m-1} + Q_{vm} s'_m + E_{vm} s'_m + E_{m+1,m} s_{m+1} \quad (85)$$

$$= Q_m s_m + E_{m,m-1} s_m + E_{vm} s_m + E_{m,m+1} s_m$$

which reduces to

$$E_{vm} = \frac{s_m (s'_{m+1} - s'_m)}{(s'_m - s_m)(s'_{m+1} - s_m)} (1 - \nu) R \quad (86)$$

The reduction in the relations (84) and (86) for $\nu = 1$ and $\nu = 0$ are as expected.

The defining equation for the residence time, τ_m, of upper box m now is

$$\tau_m = \frac{s_m V_m}{Q_m s_m + E_{m,m-1} s_m + E_{vm} s_m + E_{m,m+1} s_m} \quad (87)$$

After substitution for the quantities in the denominator and reduction, we obtain

$$\tau_m = \frac{(s'_m - s_m)(s_m - s_{m-1})(s_{m+1} - s_m)(V_m/R)}{s'_m (s_m - s_{m-1})(s_{m+1} - s_m)(1-\nu) + s_m (s'_m - s_m)(s_{m+1} - s_{m-1})(\nu/2)} \quad (88)$$

and a corresponding expression for τ'_m. Again it should be noted that (88) reduces to expression (53) for $\nu = 0$ and to an expression corresponding to (11) for $\nu = 1$.

The analysis given above for conditions of $\nu \neq o$ has been included here mainly as an exercise for completeness. The assumptions with regard to the division of the river flow between the two circulation flows and the definition of relations (80) and (81) are certainly subject to question.

Box models, two dimensional, nonconservative

We can continue the two dimensional box model approach for nonconservative quantities in much the same manner as we did for the one dimensional box models. This will not be repeated here. Rather we shall be interested in examining a nonconservative relation which is peculiar to the two dimensional approach, i.e., the settling of particles from the upper to the lower box. This corresponds to the dynamics for suspended particulate matter distributions in estuaries. We now add to Fig. 4 a downward advective exchange coefficient, G_m, across the halocline in addition to the vertical exchange coefficients Q_{vm} and E_{vm}. G_m is given by $G_m = A_m w_m$ where A_m is the area of the halocline within box m and w_m is the settling velocity.

The defining equations for the upper and lower boxes m will then be, assuming also that ν is nearly zero,

$$Q_{m-1} c_{m-1} + Q_{vm} c'_m + E_{vm} c'_m = Q_m c_m + E_{vm} c_m + G_m c_m \qquad (89)$$

and

$$Q'_{m+1} c'_{m+1} + E_{vm} c_m + G_m c_m = Q'_m c'_m + Q_{vm} c'_m + E_{vm} c'_m \qquad (90)$$

Following the same procedures as for equations (55) through (60) these reduce to

$$-\frac{s'_m - s_m}{s'_m - s_{m-1}} c_{m-1} + \left[1 + \frac{G_m}{R}\frac{s'_m - s_m}{s'_m}\right] c_m - \frac{s_m - s_{m-1}}{s'_m - s_{m-1}} c'_m = 0 \tag{91}$$

and

$$-\left[\frac{s'_{m+1} - s'_m}{s'_{m+1} - s_m} + \frac{G_m}{R}\frac{s'_m - s_m}{s_m}\right] c_m + c'_m - \frac{s'_m - s_m}{s'_{m+1} - s_m} c'_{m+1} = 0 \tag{92}$$

We, then, have 2k equations in the 2k unknowns, c_m and c'_m, with boundary values c_o and c'_l.

Let us return briefly, as we did at the end of the section on one dimensional, nonconservative relations, to a consideration of the flux balance and flux loss, $L_m + L'_m$, within any box m. From Fig. 4 we have simply for $\nu = o$

$$L_m + L'_m = Q_{m-1}c_{m-1} - Q_m c_m - (Q_m c_m - Q'_{m+1}c'_{m+1})$$

$$= \frac{s'_m c_{m-1} - s_{m-1}c'_m}{s'_m - s_{m-1}} R - \frac{s'_{m+1}c_m - s_m c'_{m+1}}{s'_{m+1} - s_m} R \tag{93}$$

corresponding to expression (37). For $\nu \neq o$ we can obtain from Fig. 5 and the relations (80) through (86)

$$L_m + L'_m = \left[\frac{s'_m c_{m-1} - s_{m-1}c'_m}{s'_m - s_{m-1}} - \frac{s'_{m+1}c_m - s_m c'_{m+1}}{s'_{m+1} - s_m}\right](1-\nu) R$$
$$+\left[\frac{s_m c_{m-1} - s_{m-1}c_m}{s_m - s_{m-1}} + \frac{s'_m c'_{m-1} - s'_{m-1}c'_m}{s'_m - s'_{m-1}} - \frac{s_{m+1}c_m - s_m c_{m+1}}{s_{m+1} - s_m} - \frac{s'_{m+1}c'_m - s'_m c'_{m+1}}{s'_{m+1} - s'_m}\right]\frac{\nu}{2} R \tag{94}$$

Box models, mixed one and two dimensional

We wish to examine here the relations for an estuary whose salinity distribution can be described as follows. In the middle and lower reaches of the estuary the conditions are stratified with a halocline and with increasing salinity in the longitudinal direction toward the ocean. In the upper reaches of the estuary the conditions are well mixed with low salinity values, generally less than 1°/$_{\infty}$.

Following Elliott (1976) we shall approximate the upper reaches with a one dimensional box model and the middle and lower reaches with a two dimensional box model with, for simplicity, $\nu = 0$. For the transition box m we shall have exchange coefficients of a one dimensional nature at the boundary between boxes m - 1 and m and of a two dimensional nature between boxes m and m + 1 and with the requisite vertical exchange coefficients for continuity within box m, as shown in Fig. 6. We, then, have the five defining equations

$$Rs_{m-1} + E_{m-1,m}s_{m-1} = E_{m,m-1}\frac{s_m + s'_m}{2} \tag{95}$$

$$Q_m s_m = Q'_{m+1}s'_{m+1} \tag{96}$$

$$Q_m - Q'_{m+1} = R \tag{97}$$

$$\frac{R}{2} + Q_{vm} = Q_m \tag{98}$$

$$\frac{R}{2}s_{m-1} + \frac{E_{m-1,m}}{2}s_{m-1} + Q_{vm}s'_m + E_{vm}s'_m = Q_m s_m + E_{vm}s_m + \frac{E_{m,m-1}}{2}s_m \tag{99}$$

in the five unknowns, $E_{m-1,m}$, Q_m, Q'_{m+1}, Q_{vm}, E_{vm}. Solving we obtain

$$E_{m-1,m} = E_{m,m-1} = \frac{s_{m-1}}{\frac{s_m + s'_m}{2} - s_{m-1}} R \tag{100}$$

$$Q_m = \frac{s'_{m+1}}{s'_{m+1} - s_m} R \tag{101}$$

$$Q'_{m+1} = \frac{s_m}{s'_{m+1} - s_m} R \tag{102}$$

$$Q_{vm} = \frac{s'_{m+1} + s_m}{s'_{m+1} - s_m} \frac{R}{2} \tag{103}$$

$$E_{vm} = \left[\frac{2s_m s'_{m+1} - s'_m s'_{m+1} - s_m s'_m}{(s'_{m+1} - s_m)(s'_m - s_m)} - \frac{s_{m-1}}{s_m + s'_m - 2s_{m-1}} \right] \frac{R}{2} \tag{104}$$

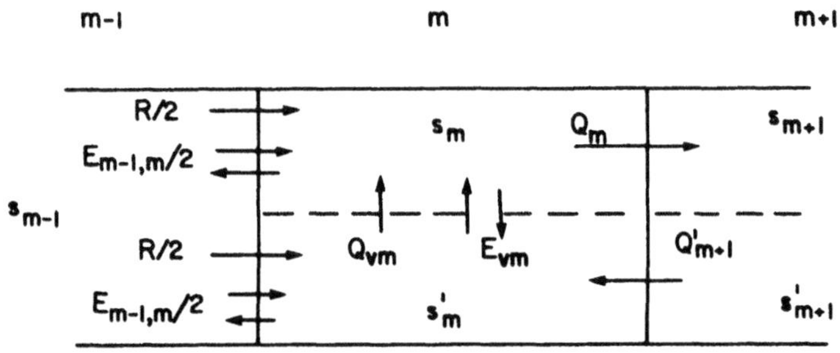

Fig. 6 - Nomenclature and exchange coefficients for mixed one and two dimensional box models.

For the flux balance equations for a conservative or non-conservative quantity, c_m, we use the one dimensional forms up through box m - 1 and the two dimensional forms for box m + 1 onward. For the flux balance equation for box m we have equations of the form (99) in terms of c_m for the upper and lower layers. Often at the upper end of the estuary we may make the approximation that $s_{m-1} = 0$. The exchange from box m - 1 to box m is, then, by river advection only with $E_{m-1,m} = 0$ and with a consequent reduction in the form of E_{vm}.

Numerical model comparisons

Wherein possible it is instructive to make comparisons of box model calculations for the distribution of nonconservative quantities in an estuary with more sophisticated numerical model calculations for the same problem. We make two such comparisons here.

Peterson et al (1978) modeled the dissolved silica distribution in San Francisco bay. The silica modeling was based on the Festa and Hansen (1976) hydrodynamic model for two dimensional, gravitational circulation; and the silica loss, or uptake related to phytoplankton production, was taken at a time dependent rate, B, in units of μg at/liter. day. In the model calculations the silica concentrations in the river and ocean reservoirs at each end were taken to be constant, c_o = 300 μg at/l and c_l = 20 μg at/l; and the vertical and horizontal diffusion coefficients and estuary cross section were also constant. Calculations were made for various uptake rates, B, taken as constant for the estuary for a given set of calculations, and for various river flow rates, R = 100, 200, and 400 m^3/sec. The estuary length, or location of the river silica reservoir, was also varied for these three determinations, being, respectively, ℓ = 160, 120, and 64 km. The longitudinal salinity distributions at middepth, determined from the model for these various river flow conditions, were also included in the Peterson et al (1978) paper as Fig. 11D.

Rattray and Officer (1979) made a comparison using the analytic solution in reduced form for constant coefficients. As has been pointed out, this solution is the same as the box model expression (29). The longitudinal salinity distributions necessary for these simplified calculations, were taken from Fig. 11D of Peterson et al (1978).

The comparison for various river flow and uptake rate conditions is repeated here as Figs. 7, 8, and 9. The agreement is quite good. The curves of Figs. 7B and 8B essentially duplicate those of Figs. 7A and 8A. Fig. 9B slightly underestimates the corresponding curve of Fig. 9A.

Festa and Hansen (1978) modeled the suspended sediment turbidity maximum. As in the previous example the turbidity maximum modeling was based on the Festa and Hansen (1976) hydrodynamic model

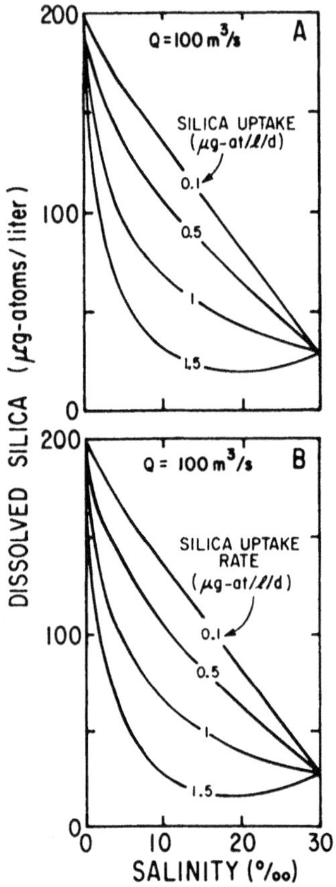

Fig. 7 - Simulated silica-salinity correlation for river flow rate, R = 100 m^3/sec, and various silica uptake rates. A) Numerical model results from Peterson et al (1978). B) Analytic results from Rattray and Officer (1978).

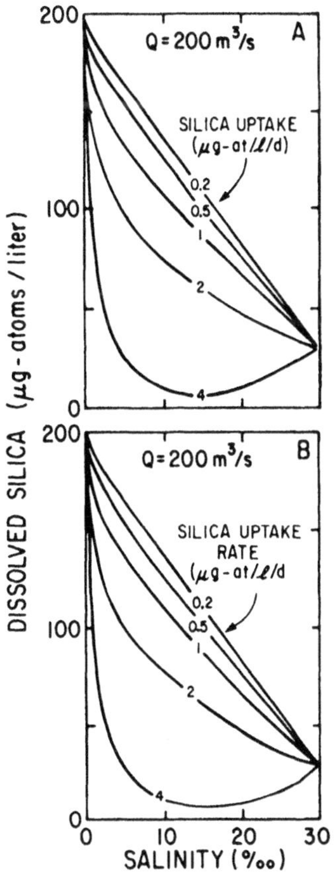

Fig. 8 - Simulated silica-salinity correlation for river flow rate, R = 200 m^3/sec, and various silica uptake rates. A) Numerical model results from Peterson et al (1978). B) Analytic results from Rattray and Officer (1978).

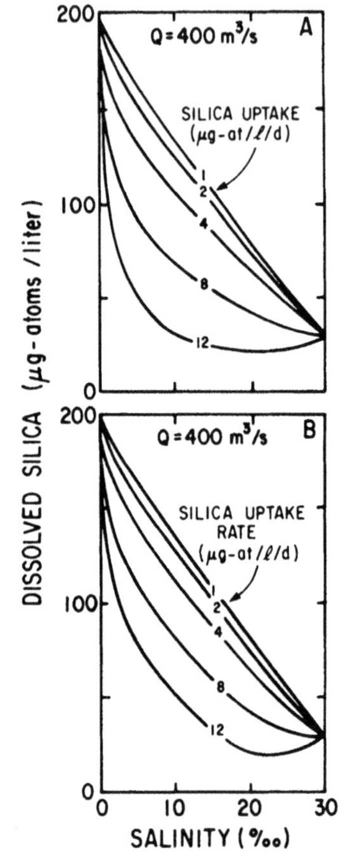

Fig. 9 - Simulated silica-salinity correlation for river flow rate, R = 400 m^3/sec, and various silica uptake rates. A) Numerical model results from Peterson et al (1978). B) Analytic results from Rattray and Officer (1978).

for two dimensional, gravitational circulation. The model includes longitudinal and vertical diffusion coefficients in addition to the longitudinal and vertical circulation velocities. The model also assumes no net deposition or erosion of bottom sediments. At the river end an exponential profile is maintained, expressing the balance between the upward vertical diffusion and the downward settling of suspended sediment. At the ocean end the gravitational dynamics are considered to be well developed and dominant over the longitudinal diffusion dynamics. The only ocean boundary condition that is specified is the ratio of the ocean-to-river concentration values at the bottom.

The resultant estuarine dynamics as determined by the hydrodynamic model are shown in Fig. 10. The salinity values are scaled to an ocean bottom salinity value of 30°/$_{\infty}$. The longitudinal velocities are in units of cm/sec, measured positive in an upestuary direction; the river flow advective velocity used in the model was 2 cm/sec. The vertical velocities are in units of 10^{-3} cm/sec and are measured positive in an upward direction.

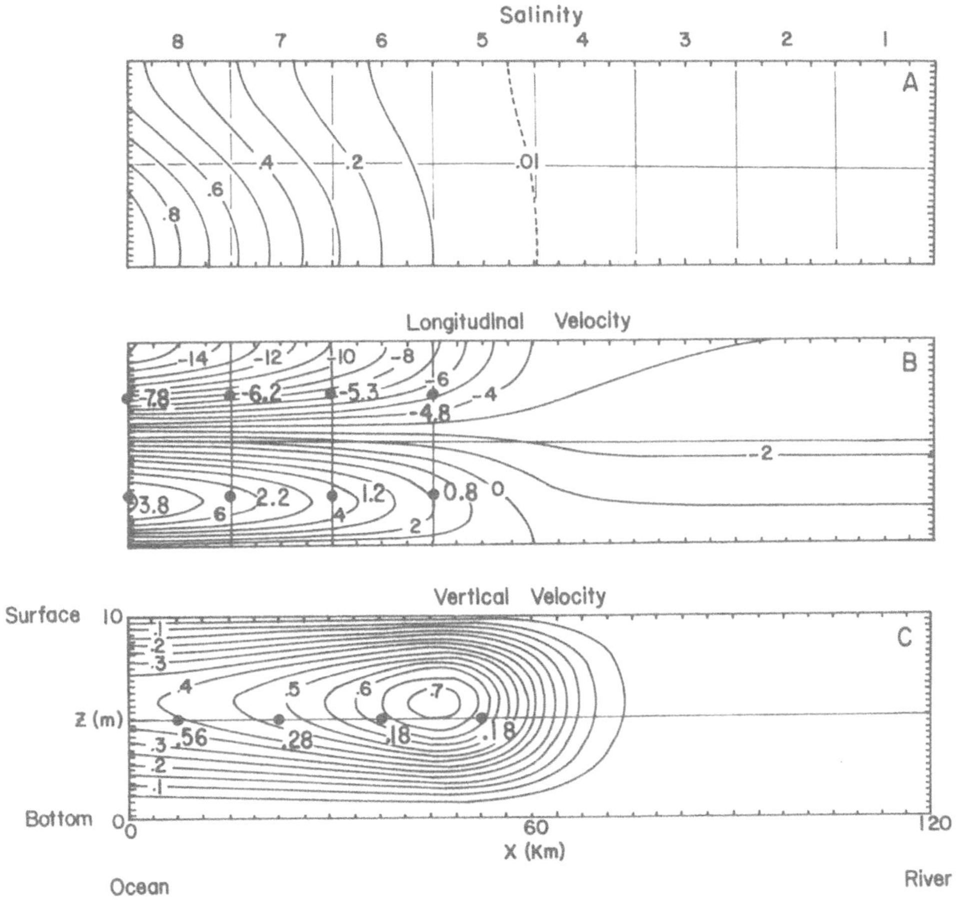

Fig. 10 - Resultant estuarine dynamics from numerical model. A) Salinity scaled by 30°/$_{\infty}$. B) Longitudinal velocity in units of cm/sec. C) Vertical velocity in units of 10^{-3} cm/sec. Repeated from Festa and Hansen (1978).

For our box model approach to the same problem we have divided the estuary into eight equal vertical segments and have taken the division between the upper and lower boxes at middepth corresponding to the longitudinal velocity distribution. From Fig. 10 we see that for the first three boxes the salinity is essentially zero, the longitudinal velocity that of the advective river flow only, and the vertical velocity zero.

Accordingly we move our upestuary boundary to that between boxes three and four. The flux balance for box four will now be of a mixed one and two dimensional nature, as shown in Fig. 11. The defining flux balance equations for the upper and lower layers of box four are, respectively

$$\frac{R}{2}\, c_{m-1} + Q_{vm} c'_m + E_{vm} c'_m = Q_m c_m + E_{vm} c_m + G_m c_m \tag{105}$$

and

$$Q'_{m+1} c'_{m+1} + E_{vm} c_m + G_m c_m + \frac{R}{2}\, c'_{m-1} = Q_{vm} c'_m + E_{vm} c'_m \tag{106}$$

Substituting for the exchange coefficients and solving, as before, for c_m and c'_m we obtain

$$-\frac{s'_m - s_m}{s'_m}\, c_{m-1} + \left[1 + \frac{2G_m}{R}\,\frac{s'_m - s_m}{s'_m}\right] c_m - \frac{s_m}{s'_m}\, c'_m = 0 \tag{107}$$

and

$$- \frac{s'_m - s_m}{s_m} c'_{m-1} - \left[\frac{2s_m s'_{m+1} - s'_m s'_{m+1} - s_m s'_m}{s_m (s'_{m+1} - s_m)} + \frac{2G_m}{R} \frac{s'_m - s_m}{s_m} \right] c_m$$

$$+ c'_m - \frac{2(s'_m - s_m)}{s'_{m+1} - s_m} c'_{m+1} = 0 \qquad (108)$$

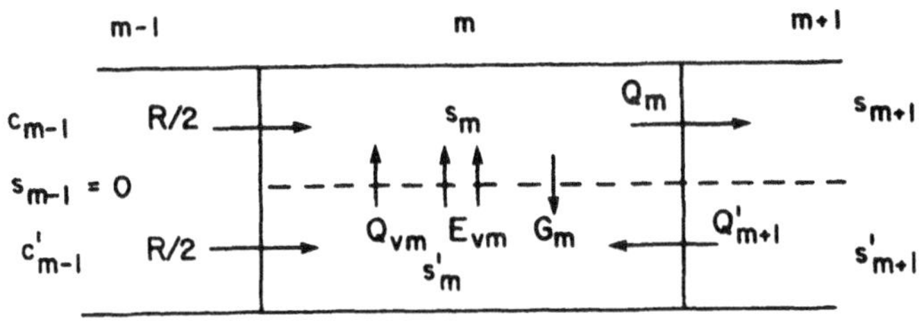

Fig. 11 - Upestuary boundary conditions for suspended sediment distribution.

For our box model calculations we, then, use equations (107) and (108) for the upper and lower layers of box four and equations (91) and (92) for the upper and lower layers of boxes five through eight. The salinity values used in the calculations were taken from Fig. 10A as the midbox value in each case; the salinity value along the vertical boundary between boxes three and four was taken to be zero. The quantity G_m/R is given by $G_m/R = w_m \Delta x/uh$ where w_m is the settling velocity, Δx the longitudinal extent of each box, u the river flow velocity and h the water depth. In the numerical model w_m was taken variously as 1, 2 and 3 x 10^{-3} cm/sec, and h was 10 m; for the box models Δx was 15 km. At the river end the boundary value exponential profile used in the numerical model was averaged for the upper and lower layers, respectively, as boundary values for the box model. At the ocean end the box model boundary value for the lower layer was taken, somewhat arbitrarily, as 0.8 of the bottom value used in the numerical model.

The ten linear, box model equations in the ten unknowns were solved by a computer procedure for the set of simultaneous linear, algebraic equations. They were also checked using the jury method described by Jeffreys and Jeffreys (1956, pg. 306) with a hand calculator. The calculated values settled to small variations after six to seven successive approximations taking a total calculation time of around fifteen minutes for each set of equations. The results, as expected, were the same as those obtained with the computer procedure with only slight and insignificant variations in values. If available, the computer solution is, of course, to be preferred on the basis both of accuracy and time. The time for the computer solution is essentially only that taken in entering the equational coefficients into the computer program for the solution of such simultaneous linear equations. Using the computer solution procedure, the number of boxes was doubled; with the new equational coefficients the results were not significantly different from those obtained for the box configuration shown in Fig. 10A.

A comparison of the numerical and box model results is shown in Figs. 12 and 13. It should be kept in mind that in addition to the various approximations associated with the box model approach the box model calculations used here include only circulation effects, $\nu = 0$, whereas the numerical model determinations include horizontal and vertical diffusion effects in addition to the circulation effects.

Fig. 12, taken from Festa and Hansen (1978), is a plot of suspended sediment concentration values obtained from the numerical model for an assumed settling velocity of 1×10^{-3} cm/sec and with ocean-to-river bottom concentration ratios varying from 0.1 to 1 to 10. The box model concentration values, determined for the same conditions as the numerical model, have been superimposed on this figure and have been plotted at the midpoint of each box. The agreement is reasonable. The box model values predict the same approximate location for the turbidity maximum; and the magnitude of the box model values, taken point by point, is much the same as the numerical model values.

Fig. 13, also taken from Festa and Hansen (1978), is a plot of suspended sediment concentration values for an ocean-to-river bottom concentration ratio of 0.1 and with settling velocities varying from 1×10^{-3} to 2×10^{-3} to 3×10^{-3} cm/sec. The box model concentration values have been superimposed as in the previous figure. The agreement, here, becomes progressively poorer for the higher settling velocities in the immediate vicinity of the turbidity maximum. The dimensions of the turbidity maximum in Fig. 13C are much the same as that of the boxes, and we might expect from these dimensional considerations that the vertical division into two boxes would be a limitation under such circumstances. Further, there is the statement in Festa and Hansen (1978) that results are not presented for their numerical model for settling velocities beyond 3×10^{-3} cm/sec because the choice of mixing coefficients and the vertical resolution used give erroneous values when the settling velocity exceeds this value, which limitation may also progressively effect the comparison of the two approaches in Figs. 13B and 13C.

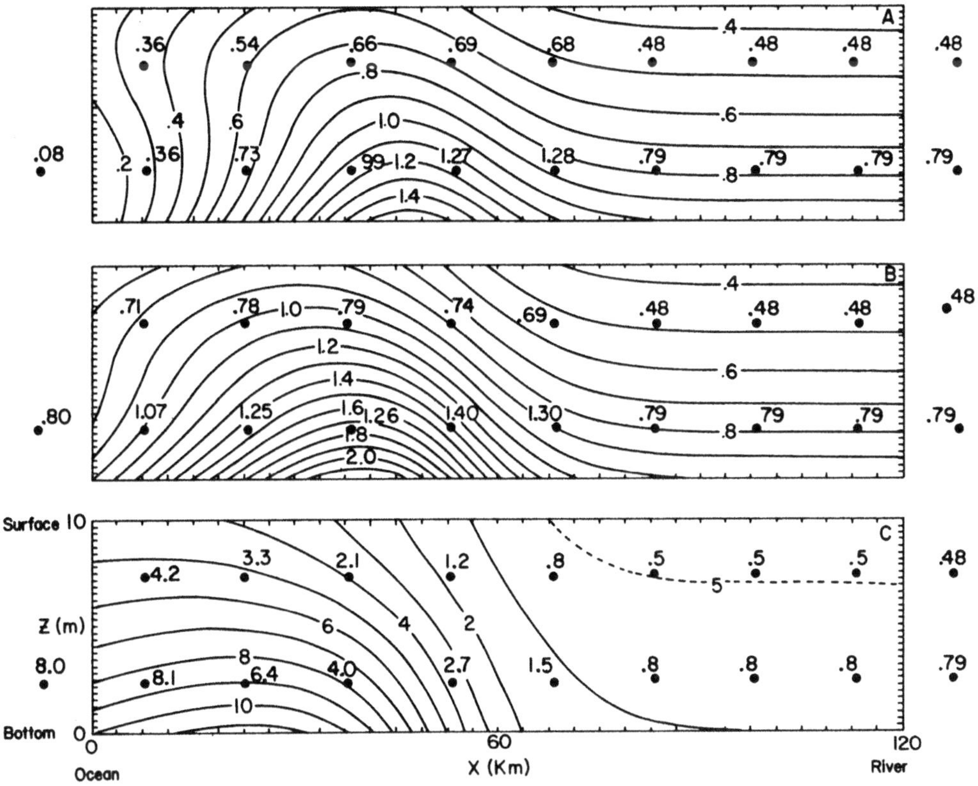

Fig. 12 - Suspended sediment distribution for a particle settling velocity of 1 x 10^{-3} cm/sec and ocean-to-river source concentration ratios of A) 0.1, B) 1, and C) 10 from Festa and Hansen (1978) numerical model with superimposed box model values.

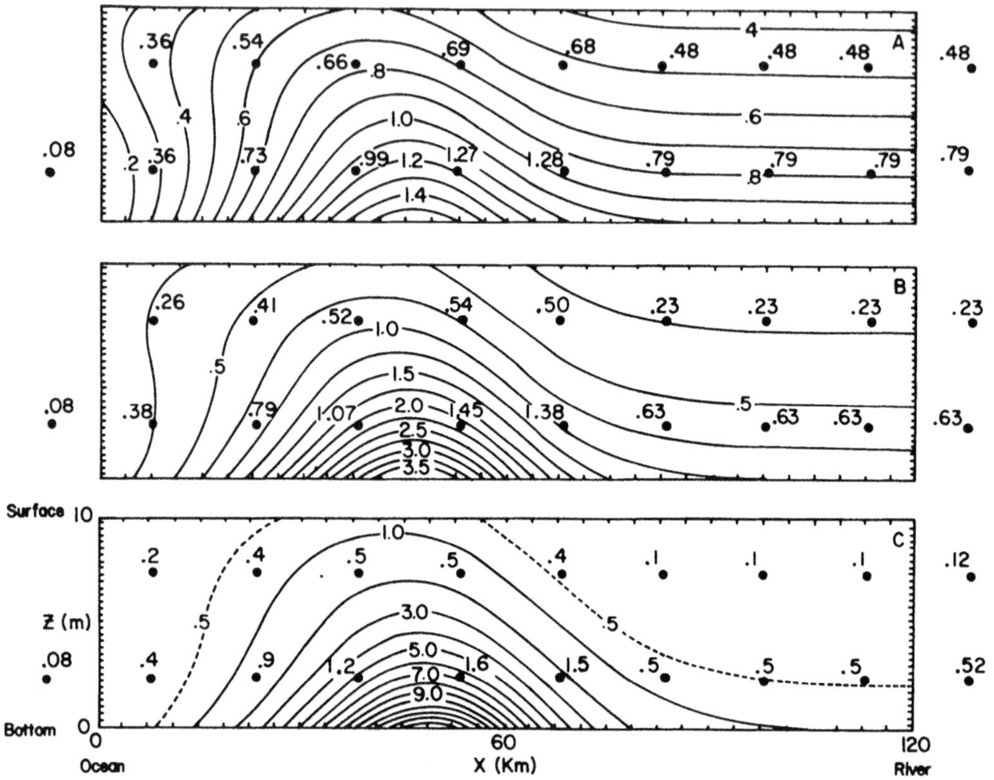

Fig. 13 - Suspended sediment distribution for an ocean-to-river source concentration ratio of 0.1 and particle settling velocities of A) 1 x 10^{-3}, B) 2 x 10^{-3}, and C) 3 x 10^{-3} cm/sec from Festa and Hansen (1978) numerical model with superimposed box model values.

We may make a further comparison between the numerical and box model results along the following lines. In the numerical model the assumption was made that the estuary acted as a conservative system with regard to the suspended sediment, i.e., that there was neither deposition of suspended sediment nor erosion of the bottom sediments acting as a source of suspended sediment supply. This assumption was preserved in the box model equations used here. The net flux, NF, of suspended sediment into the estuary from both the river and ocean reservoirs should, then, be zero. From the relation similar to equation (93) for this case we have

$$NF = \left[\frac{c_o + c_o'}{2} - \frac{s_9'c_8 - s_8c_9'}{s_9' - s_8} \right] R \tag{109}$$

where the first term represents the net flux in at the river end and the second term the net flux out at the ocean end. From the box model values we obtain that the difference between these two fluxes is 2, 1, 4, 1 and 2% of the net flux in from the river end for the conditions shown in Figs. 12A, 12B, 12C, 13C and 13D, respectively, as reasonable an agreement as could be expected.

We may also make a comparison for the hydrodynamic variables of longitudinal and vertical velocity between the numerical and box models. From the relations (49) and (50) we have that the average longitudinal velocities, u_m and u'_{m+1}, for the upper and lower boxes at the vertical boundary between boxes m and m+1 are given by

$$u_m = 2u \frac{Q_m}{R} = 2u \frac{s'_{m+1}}{s'_{m+1} - s_m} \tag{110}$$

and

$$u'_{m+1} = 2u \frac{Q'_{m+1}}{R} = 2u \frac{s_m}{s'_{m+1} - s_m} \tag{111}$$

where, as before, u = 2 cm/sec is the river flow advective velocity. From the box model salinity values we obtain the average longitudinal velocities shown in Fig. 10B for the upper and lower boxes. The agreement is quite reasonable with the corresponding averaged longitudinal velocities from the numerical model.

From the relation (51) we have for the vertical velocity, v_m, between boxes m and m' that

$$v_m = \frac{uh}{\Delta x} \frac{Q_{vm}}{R} = \frac{uh}{\Delta x} \frac{s_m s'_m - s_{m-1} s'_{m+1}}{(s'_{m+1} - s_m)(s'_m - s_{m-1})} \tag{112}$$

Again using the box model salinity values we obtain the vertical velocities shown in Fig. 10C. The box model results show the same order of magnitude for the vertical velocities in units of 10^{-3} cm/sec as for the numerical model. The distribution, however, is quite different; the box model results show an increase in vertical velocity toward the ocean end of the estuary whereas the numerical model results show an increase in the opposite direction with a maximum at the boundary between boxes 5 and 6.

This difference between the results for the vertical velocities is considered to be related to the gross averaging procedures used in the box model approach and to the fact that longitudinal diffusion has been neglected and thereby states an implicit limitation to the box model approach. This discrepancy also appears in the manner in which the nonadvective vertical exchanges are handled in the two methods. In the numerical model the vertical diffusion coefficient, K_v, was assumed to be constant. In the box model the corresponding nonadvective vertical coefficients, E_{vm}, are determined by the salinity values and are variable. Using the box model salinity values the corresponding values for E_{vm}/R from the relation (52) are .80, 1.24, 1.05 and .79 for boxes 5, 6, 7 and 8, respectively. These values do show much the same type of longitudinal variation as the numerical model salinity values. Thus, the box model incorrectly models the division between vertical advective and nonadvective exchanges and correspondingly the effects related to particle settling.

Discussion

Box models are a gross simplification of the hydrodynamic effects that occur in estuaries. In addition to their inherent limitations the presentation given here does not include lateral and temporal variations; both can be important. As more data and analyses become available, the temporal variations related to the semidiurnal tidal cycle, neap to spring tidal cycle, variable wind stresses, seasonal river flow changes and other time variable forcing functions appear as determining factors. The correct analytic and numerical modeling of such phenomena will undoubtedly provide some of the major advances in understanding estuarine phenomena.

The advantage of the box model is that within its limitations it can be applied to actual conditions in well mixed estuaries and in partially mixed estuaries in which the horizontal net circulation exchanges are dominant over the horizontal nonadvective exchanges. In particular it can be applied to the study of nonconservative distributions and transformations and can give reasonable, first order approximations to the rate, exchange or loss coefficients.

A comparison has been given here of box model results against two more refined, hydrodynamic numerical model results, one for a nonconservative loss problem and the other for a suspended sediment distribution. An example of its application to actual field data has been given by Taft et al (1978) in their study of ammonium and nitrate fluxes in Chesapeake bay.

References

Barrett, M. J. and B. M. Mollowney, 1972, Pollution problems in relation to the Thames barrier, Philosophical Transactions of the Royal Society, London, A, 272, 213-221.

Bella, D. A. and W. E. Dobbins, 1968, Difference modeling of stream pollution, American Society of Civil Engineers, Journal of the Sanitary Engineering Division, 94, SA 5, 995-1016.

Blumberg, A. F., 1977, A two dimensional numerical model for the simulation of partially mixed estuaries, 323-331. In Estuarine Processes, Volume II, Adacemic Press, New York.

Bolin, B. and H. Stommel, 1961, On the abyssal circulation of the World ocean, IV, Origin and rate of circulation of deep ocean water as determined with the aid of tracers, Deep Sea Research, 8, 95-110.

Bowden, K. F. and P. Hamilton, 1975, Some experiments with a numerical model of circulation and mixing in a tidal estuary, Estuarine and Coastal Marine Science, 3, 281-301.

Broecker, W. S., 1966, Radioisotopes and the rate of mixing across the main thermoclines of the ocean, Journal of Geophysical Research, 71, 5827-5836.

Broecker, W. S., R. D. Gerard, M. Ewing, and B. Heezen, 1961, Geochemistry and physics of ocean circulation, 301-322. In Oceanography, American Association for the Advancement of Science, Washington.

Craig, H., 1957, The natural distribution of radiocarbon and the exchange time of carbon dioxide between atmosphere and sea, Tellus, 9, 1-17.

Craig, H., 1963, The natural distribution of radiocarbon: Mixing rates in the sea and residence times of carbon and water, 103-114. In Earth Science and Meteorites, North Holland Publishing, Amsterdam.

Downing, A. L., 1971, Forecasting the effects of polluting discharges on natural waters, II, Estuaries and coastal waters, International Journal of Environmental Studies, 2, 221-226.

Elliott, A. J., 1976, A mixed dimension kinematic estuarine model, Chesapeake Science, 17, 135-140.

Festa, J. F. and D. V. Hansen, 1976, A two dimensional numerical model of estuarine circulation: The effects of altering depth and river discharge, Estuarine and Coastal Marine Science 4, 309-323.

Festa, J. F. and D. V. Hansen, 1978, Turbidity maxima in partially mixed estuaries: A two dimensional numerical model, Estuarine and Coastal Marine Science, 7, 347-359.

Gilligan, R. M., 1972, Forecasting the effects of polluting discharges on estuaries, Chemistry and Industry, 865-874, 909-916, and 950-958.

Hamilton, P., 1975, A numerical model of the vertical circulation of tidal estuaries and its application to the Rotterdam waterway, Geophysical Journal of the Royal Astronomical Society, 40, 1-21.

Hansen, D. V. and M. Rattray, 1965, Gravitational circulation in straits and estuaries, Journal of Marine Research, 23, 104-122.

Hansen, D. V. and M. Rattray, 1966, New dimensions in estuary classification, Limnology and Oceanography, 11, 319-326.

Harleman, D. R. F., 1977, Real time models for salinity and water quality analysis in estuaries, 84-93. In Estuaries, Geophysics and the Environment, National Academy of Sciences, Washington.

Jeffreys, H. and B. S. Jeffreys, 1956, Methods of Mathematical Physics, Cambridge University Press, Cambridge.

Ketchum, B. H., 1950, Hydrographic factors involved in the dispersion of pollutants introduced into tidal waters, Journal of the Boston Society of Civil Engineers, 37, 296-314.

Ketchum, B. H., 1954, Relation between circulation and planktonic populations in estuaries, Ecology, 35, 191-200.

Ketchum, B. H., 1955, Distribution of coliform bacteria and other pollutants in tidal estuaries, Sewage and Industrial Wastes, 27, 1288-1296.

Ketchum, B. H., J. C. Ayers, and R. F. Vaccaro, 1952, Processes contributing to the decrease of coliform bacteria in a tidal estuary, Ecology, 33, 247-258.

McDowell, D. M. and B. A. O'Connor, 1977, Hydraulic Behaviour of Estuaries, John Wiley and Sons, London.

O'Connor, D. J., R. V. Thomann, and D. M. Di Toro, 1977, Water quality analyses of estuarine systems, 71-83. In Estuaries, Geophysics and the Environment, National Academy of Sciences, Washington.

Officer, C. B., 1976, Physical Oceanography of Estuaries and Associated Coastal Waters, John Wiley and Sons, New York.

Officer, C. B., 1977, Longitudinal circulation and mixing relations in estuaries, 13-21. In Estuaries, Geophysics and the Environment, National Academy of Sciences, Washington.

Officer, C. B., 1978, Discussion of the behaviour of nonconservative dissolved constituents in estuaries, Estuarine and Coastal Marine Science, in press.

Peterson, D. H., J. F. Festa, and T. J. Conomos, 1978, Numerical simulation of dissolved silica in San Francisco bay, Estuarine and Coastal Marine Science, 7, 99-116.

Pritchard, D. W., 1969, Dispersion and flushing of pollutants in estuaries, American Society of Civil Engineers, Journal of the Hydraulics Division, 95, HY1, 115-124.

Pritchard, D. W., R. O. Reid, A. Okubo, and H. H. Carter, 1971, Physical processes of water movement and mixing, 90-136. In Radioactivity in the Marine Environment, National Academy of Sciences, Washington.

Rattray, M., 1967, Some aspects of the dynamics of circulation in fjords, 52-62. In Estuaries, American Association for the Advancement of Science, Washington.

Rattray, M. and D. V. Hansen, 1962, A similarity solution for circulation in an estuary, Journal of Marine Research, 20, 121-133.

Rattray, M. and E. Mitsuda, 1974, Theoretical analysis of conditions in a salt wedge, Estuarine and Coastal Marine Science, 2, 375-394.

Rattray, M. and C. B. Officer, 1979, Distribution of a nonconservative constituent in an estuary with application to the numerical simulation of dissolved silica in the San Francisco bay, Estuarine and Coastal Marine Science, 8, 489-494.

Schubel, J. R. and H. H. Carter, 1976, Suspended sediment budget for Chesapeake bay, 48-62. In Estuarine Processes, Volume II, Academic Press, New York.

Stommel, H., 1953, Computation of pollution in a vertically mixed estuary, Sewage and Industrial Wastes, 25, 1065-1071.

Taft, J. L., A. J. Elliott and W. R. Taylor, 1978, Box model analysis of Chesapeake bay ammonium and nitrate fluxes, 115-130. In Estuarine Interactions, Academic Press, New York.

Winter, D. F., 1973, A similarity solution for steady state gravitational circulation in fjords, Estuarine and Coastal Marine Science, 1, 387-400.

NUMERICAL HYDRODYNAMICS OF ESTUARIES

John Eric Edinger and Edward M. Buchak

J. E. Edinger Associates, Inc.

37 West Avenue, Wayne, Pennsylvania, 19087

ABSTRACT

Classically, estuaries have been classified dimensionally on the basis of the dominant salinity gradients. Following Pritchard (1958) the general classifications based on spatial averaging of the constituent transport relationship are: (1) three-dimensional; (2) laterally homogeneous with longitudinal and vertical spatial gradients dominant; (3) vertically homogeneous with longitudinal and lateral spatial gradients dominant, and (4) sectionally homogeneous with longitudinal gradients dominant. Development of the hydrodynamic (momentum transport) relationships follow similar spatial averaging and classification.

In general, the momentum balances determine the flow field by which the constituent is transported. The momentum and constituent transport are interrelated in estuaries through the horizontal density gradient as determined from the constituent distribution. Only the fourth case, sectional homogeneity, is solvable for a few limiting situations without use of the hydrodynamic relationships, and are situations for which the advective flow field can be inferred from fresh water inflow.

The development of numerical hydrodynamics for estuaries begins with a presentation of the equations of motion and constituent transport in three dimensions. The basic equations are: (1) the u-velocity or longitudinal momentum balance; (2) the v-velocity or lateral momentum balance; (3) the pressure distribution, p, as determined from the vertical momentum balance as the hydrostatic approximation; (4) the w or vertical velocity as determined from local continuity; (5) the salinity, S, constituent transport;

(6) the equation of state relating density, ρ, to the constituent concentration, and (7) the free water surface elevation, η, as determined from vertically integrated continuity. The general numerical problem is, therefore, to spatially integrate numerically over time seven equations for the seven unknowns of u, v, w, P, S, ρ, η given appropriate geometry and time-varying boundary data. The seven equations are interrelated with the constituent distribution, S, determining density, ρ; with density and the free water surface elevation, η, determining pressure, P, and with the pressure distribution entering the momentum balance.

The two-dimensional and one-dimensional cases are derived from the three-dimensional relationships by spatial averaging. The laterally homogeneous estuary dynamics include a majority of the interrelationships of density, pressure and surface elevation incorporated in the three-dimensional equations. Explicit and implicit solution procedures can be illustrated for the laterally homogeneous relationships as they depend upon the inclusion of vertically integrated velocities in the surface elevation computations. Laterally averaged hydrodynamic solution procedures that utilize simplifying assumptions for the longitudinal density gradient are also examined. The sectionally homogeneous hydrodynamics is shown to be a reduced case of the laterally homogeneous relationships.

The two-dimensional vertically homogeneous dynamics is presented as a reduced form of the vertically integrated three-dimensional case. The vertically homogeneous case has spatially explicit and implicit solution procedures, the properties of which can be illustrated from the basic equations. It will be shown that a surface elevation relationship exists for this case that has a variational statement leading to spatial finite element description.

INTRODUCTION

The development and use of numerical models in estuaries requires an understanding of the basic hydrodynamic equations and the steps involved in their manipulation to numerical form. It is useful at this stage of development to systematically review the equations of motion and transport for the four basic cases of (1) three-dimensional hydrodynamics; (2) two-dimensional laterally averaged hydrodynamics; (3) one-dimensional sectionally homogeneous hydrodynamics; and, (4) two-dimensional vertically mixed hydrodynamics, in order to examine their common numerical features and to provide a basis for examining different computational codes or models developed for these four basic hydrodynamic cases. It is not the purpose of this presentation to review and compare different codifications or models, but to present the features of the numerical formulations that should be common to the numerous codes and models. Thus, it is the purpose of this presentation to give the basis from which the model developer, model user and model selector or reviewer must begin in order to understand the relationship between the basic nonlinear partial differential equations of fluid motion and their numerical computational analogs for use in estuaries.

The general interest is in numerical estuary models that represent the detailed geometry and boundaries of semi-closed water-bodies, and that utilize time varying boundary data over extended periods of time. Relative to the time varying boundary data of tide heights; salinity and constituent concentrations; fresh water inflows; constituent sources, sinks, and interactions; wind derived surface shear and surface and bottom exchange process, the numerical equations become a complex interpolator of velocities and constituent distributions within the boundaries of the water-body.

In general a numerical model is first configured to represent a particular waterbody and is initially exercised over a period of known conditions within the waterbody for verification. Configuration includes not only detailing the geometry of the waterbody but also formulating the boundary equations that link the time varying boundary data to the numerical equations. Formulation of the boundary equations is by far the most difficult task since it requires insight to the waterbody dynamics beyond the numerical formulation of the basic equations and is presently an area of much needed research and investigation now that numerical formulation of the basic equations has generally been mastered.

Once a numerical model has been configured and verified for a particular application, there are two interests in its use. First, the most direct use is to determine how the model results are changed by changes in boundary data and then to infer how this computed change might be related to conditions within the real waterbody. Two examples are the study of the response of constituent concentrations within the waterbody to changes in constituent source strengths; and, the changes in flow fields due to geometry modification by dredging, filling or diking with inferences about sediment transport.

The second and more indirect use of the numerical model is to investigate how well it represents the hydrodynamic features of the waterbody beyond those parameters normally used for verification. This is the stage at which one learns what is wrong with the model results and determines what is necessary to improve them. This stage of model use is described by Pritchard (1978). Careful analysis of the boundary and verification data is likely to teach more about real waterbody dynamics than the model results themselves. Analysis of velocity records by Pritchard (1978), for example, has shown the importance of wind stress in controlling barotropic bottom currents in deep estuaries. Thus, data base analysis and identification of its important features provides the next step for model improvement.

The hydrodynamic relationships for each of the four descriptions of estuaries are examined and expanded by an algebraic approach to differencing equations without commitment to a particular differencing scheme. It is found that numerical integration of the equations over time implies, either directly or indirectly, solving the free surface long-wave equation. The latter can, therefore, be used to determine the character of the numerical solutions.

Three-Dimensional Hydrodynamics and General Principles

The three-dimensional equations of waterbody dynamics can be stated after appropriate ensemble averaging as the horizontal longitudinal (x) and lateral (y) momentum balances:

$$\frac{\partial u}{\partial t} + \frac{\partial uu}{\partial x} + \frac{\partial uv}{\partial y} + \frac{\partial uw}{\partial z} - fv = -\frac{1}{\rho}\frac{\partial P}{\partial x} + \frac{\partial}{\partial x}(A_x \frac{\partial u}{\partial x})$$

$$+ \frac{\partial}{\partial y}(A_y \frac{\partial u}{\partial y}) - \frac{\partial \tau_x}{\partial z} \qquad (1)$$

$$\frac{\partial v}{\partial t} + \frac{\partial vu}{\partial x} + \frac{\partial vv}{\partial y} + \frac{\partial vw}{\partial z} + fu = -\frac{1}{\rho}\frac{\partial P}{\partial y} + \frac{\partial}{\partial x}(A_x \frac{\partial v}{\partial x})$$

$$+ \frac{\partial}{\partial y}(A_y \frac{\partial v}{\partial y}) - \frac{\partial \tau_y}{\partial z} \qquad (2)$$

where u, v, and w are the x, y and z velocity components; ρ is the density and P is the pressure; A_x and A_y are the horizontal eddy dispersion coefficients; and, τ_x and τ_y are the vertical shear stresses. The terms on the left hand side of the momentum balances are the local temporal and spatial fluid accelerations and the Coriolis acceleration where f is the latitude dependent Coriolis parameter of the earth's rotation. The terms on the right hand side are the horizontal pressure gradient, the horizontal shear stresses or horizontal dispersion of momentum and the vertical shear stress.

In equation 1 and 2 the cross-shear terms are expressed as eddy coefficients and velocity gradients. The relationship between ensemble averaged velocity fluctuation products and eddy coefficients are given in Pritchard (1971).

The vertical equation of motion can be reduced to the hydrostatic approximation as:

$$\frac{\partial P}{\partial z} = \rho g \tag{3}$$

where the assumption is made that local temporal and spatial vertical accelerations are small relative to g, the gravitational acceleration. The vertical z axis is taken positive downward. Continuity is stated as:

$$\frac{\partial u}{\partial x} + \frac{\partial v}{\partial y} + \frac{\partial w}{\partial z} = o \tag{4}$$

which applies below the free water surface. Vertical integration of continuity and application of Liebniz's rule for differentiation of an integral with appropriate kinematic boundary conditions gives the free water surface equation of:

$$\frac{\partial \eta}{\partial t} = \frac{\partial}{\partial x} \int_{\eta}^{h} udz + \frac{\partial}{\partial y} \int_{\eta}^{h} vdz \tag{5}$$

where η is the free water surface coordinate and h is the bottom coordinate ($h > \eta$) written for z positive downward.

Next is the constituent transport written for salinity in the estuary as:

$$\frac{\partial S}{\partial t} + \frac{\partial uS}{\partial x} + \frac{\partial vS}{\partial y} + \frac{\partial wS}{\partial z} - \frac{\partial}{\partial x}\left(D_x \frac{\partial S}{\partial x}\right)$$

$$- \frac{\partial}{\partial y}\left(D_y \frac{\partial S}{\partial y}\right) - \frac{\partial}{\partial z}\left(D_z \frac{\partial S}{\partial z}\right) = 0 \qquad (6)$$

where S is the salinity at x, y, z and t and D_x, D_y and D_z are the dispersion coefficients. Other constituents such as temperature, dissolved oxygen or sediment concentrations have a similar form but with appropriate spatial and temporal source and sink terms. Last is the equation of state relating density and constituent concentrations, for example, as a function of salinity:

$$\rho = R(S) \qquad (7)$$

and which can be expanded to include other constituents such as temperature and sediment concentrations. The details of ensemble and spatial averaging required to arrive at the above relationships are given in Pritchard (1971) for the hydrodynamic relationships and in Pritchard (1958) for constituent transport relationships.

The numerical hydrodynamic estuary free water surface problem is thus stated in terms of seven equations to be solved for the unknowns of u, v, w, P, η, S, and ρ. It is presumed that the dispersion coefficients and vertical shear can be expressed in terms of the velocity and density field, (Pritchard 1958, 1971). The vertically integrated equation of continuity (5), defines the equations as uniquely applicable to free water surface problems. The pressure can be related to the free water surface by vertically integrating (3) to any depth z as:

$$P = g \int_{\eta}^{z} \rho dz \qquad (8)$$

which is required in the momentum equations as the horizontal pressure gradient. Thus for x-momentum in Equation 1:

$$\frac{1}{\rho}\frac{\partial P}{\partial x} = -g\frac{\partial \eta}{\partial x} + \frac{g}{\rho}\int_{\eta}^{z}\frac{\partial \rho}{\partial x}\,dz \tag{9}$$

and for y-momentum in Equation 2:

$$\frac{1}{\rho}\frac{\partial P}{\partial y} = -g\frac{\partial \eta}{\partial y} + \frac{g}{\rho}\int_{\eta}^{z}\frac{\partial \rho}{\partial y}\,dz \tag{10}$$

after appropriate application of Liebnitz's rule. The horizontal pressure gradient is thus divided into the two components of the surface slope ($\partial\eta/\partial x$) and the vertical integral of the horizontal density gradient. These two components of the horizontal pressure gradient are referred to as the barotropic and baroclinic pressure gradient respectively.

The numerical integration of the above seven equations for seven unknowns can be set up in a unique fashion for the time varying boundary condition problem. Finite temporal integration allows one to choose to take certain of the individual terms forward in time while evaluating other terms from results available at earlier time steps. This choice, which presently appears to be without rigorous mathematical guidance as to its hydrodynamic implications, is the most fundamental yet most unestablished procedure in numerical hydrodynamics of estuaries. Typically, the local horizontal accelerations are taken forward in time after substitution for the horizontal pressure gradients to give:

$$u' = u + g\frac{\partial \eta}{\partial x}\Delta t + F_x\,\Delta t \tag{11}$$

and:

$$v' = v + g\frac{\partial \eta}{\partial y}\Delta t + F_y\,\Delta t \tag{12}$$

where u' and v' are the new velocity components at $t + \Delta t$ and F_x and F_y are the remaining terms in the momentum equations to be evaluated from results at earlier time steps. The barotropic pressure gradient term is retained for illustration of its role in the time integration procedure.

The computation begins at t = o with initial values for all of the variables throughout the waterbody and boundary values for $t = \Delta t$. The u' and v' from (11) and (12) are algebraically substituted into (5) to compute an η' at $t = \Delta t$. The u' and v' are also used in (4) rewritten as:

$$w(z-\Delta z) = w(z) + \left(\frac{\partial u'}{\partial x} + \frac{\partial v'}{\partial y}\right) \Delta z \tag{13}$$

to compute w from the bottom upward where w(h) = o. Given the flow field, the constituent transport is computed from (6) and the density field is computed from (7). Old values of the variables are updated to new values and the cycle is repeated.

The key step in the computation is numerical substitution of u' and v' in the free surface relation (5). This step can be performed algebraically to give:

$$\eta' = \eta + g\Delta t^2 \frac{\partial}{\partial x}\left(H \frac{\partial \eta}{\partial x}\right) + g\Delta t^2 \frac{\partial}{\partial y}\left(H \frac{\partial \eta}{\partial y}\right)$$

$$+ \Delta t \int_\eta^h \left(\frac{\partial u}{\partial x} + \frac{\partial u}{\partial x}\right) dz + \Delta t^2 \int_\eta^h (F_x + F_y)\, dz \tag{14}$$

where $H = (h-\eta)$ is the waterbody depth and where the surface slopes $\partial\eta/\partial x$ and $\partial\eta/\partial y$ are constant over z. Equation (14) is a numerical form of the frictionally damped long wave equation with the horizontal density gradient as a driving force. Expanding the spatial derivatives into a finite difference spatial (i, j) horizontal grid gives approximately:

$$\eta'(i,j) = \eta(i,j) + g\frac{\Delta t^2}{\Delta x^2} H(\eta(i+1,j) - 2\eta(i,j) + \eta(i-1,j))$$

$$+ g\frac{\Delta t^2}{\Delta y^2} H(\eta(i,j+1) - 2\eta(i,j) + \eta(i-1,j)) + F \tag{15}$$

where F is the remainder of terms. Collecting the $\eta(i,j)$ terms, the integration implied by the computational substitution is of the form:

$$\eta'(i,j) = \eta(i,j)\ (1-2g\ H\ \Delta t^2/\Delta s^2) + \ .\ .\ . \tag{16}$$

where for illustrative purposes $\Delta s = \Delta x = \Delta y$. Thus, the computation reduces to a forward in time integration of a simple first order difference equation for which:

$$\frac{\Delta s}{\Delta t} > \sqrt{2gH} \tag{17}$$

is required for numerical stability. Obviously the maximum water-body depth governs. The condition goes by several names including the explicit solution gravity wave stability criterion and the Courant stability criterion. This condition limits the integration time step, Δt, once the grid size, Δs, is chosen. The stability criterion states physically that computations must be made faster at a point than the time of travel of the gravity wave between two points.

The above description of numerical procedures is for a spatially explicit computational scheme where the variables are determined one at a time at each spatial position from previous time step values. If, however, variables are computed simultaneously at a number of positions, a spatially implicit computational scheme is derived. The spatially implicit scheme uses (15) directly in the computational procedure with the spatial gradients of η taken forward in time to give:

$$-g\left(\frac{\Delta t}{\Delta s}\right)^2 H\,(\eta'(i+1,j) + \eta'(i-1,j)) + \left(1+4g\left(\frac{\Delta t}{\Delta s}\right)^2 H\right)\eta'(i,j)$$

$$-g\left(\frac{\Delta t}{\Delta s}\right)^2 H\,(\eta'(i,j+1) + \eta'(i,j-1)) = \eta(i,j) + F \tag{18}$$

which is a family of simultaneous equations for all $\eta(i,j)$ at $t + \Delta t$ that can be solved by matrix algebra. Their solution is unconditionally stable at least to the first order since the main-diagonal term is always positive while the off diagonal terms are always negative. Thus given $\eta'(i,j)$ from a spatially implicit computation at $t + \Delta t$, they can be substituted for the $\partial\eta/\partial x$ and $\partial\eta/\partial y$ in (11) and (12) to compute u' and v' at the same time level as η'.

A spatially implicit computational scheme for the water surface elevation eliminates the gravity wave speed criterion but does not eliminate all of the numerical stability criteria implied in evaluating F_x and F_y in (11) and (12) from previous time step results. Schematically extracting u(i,j,k) multipliers in (11) from (1) for the advection and dispersion of momentum gives:

$$u'(i,j,k) = u(i,j,k)\left(1 - \frac{u\Delta t}{\Delta x} - A_x \frac{\Delta t}{\Delta x^2} - \ldots\right) + \ldots \tag{19}$$

which is a simple first order difference equation requiring at least that:

$$\frac{\Delta x}{\Delta t} > u \tag{20}$$

and

$$\frac{\Delta x^2}{\Delta t} > A_x/2 \tag{21}$$

The first condition, known as the Torrence condition states that grid length, Δx, must be longer than the velocity travel time across it and appears to be a necessary condition for any numerical scheme. The second condition requires that no more momentum be dissipated in an Δx space than is stored in that space. It can be eliminated by taking dispersion terms forward in time similar to the spatially implicit manner for evaluating $\eta(i,j)$ in (18) but may be a numerically inconvenient method when evaluating A_x from previously computed velocity results. The conditions of (20) and (21) have their counterparts in evaluating various numerical forms of the constituent transport equation (6), where salt rather than momentum is transported.

Satisfying the numerical stability criteria for the set of numerical equations does not necessarily assure accurate solutions. For estuarine tidal situations, spatial differencing and averaging of various terms in the basic equations can result in computational oscillations spatially and temporally that are not properties of the real waterbody dynamics. Investigation of the numerical equations is quite tedious as illustrated by Leendertse (1971) even for the simple one dimensional advective dispersive transport equation. It is almost impossible to perform algebraically for the full set of seven equations common to the waterbody dynamics problem. Rather, the approach is to learn from previous numerical formulations, make inferences from the simpler one-dimensional analyses, and to exercise the numerical model for a problem with simple geometry for which an analytic solution is known. The latter is a necessary step in model validation and should also be utilized in comparative studies between models.

Another set of criteria for testing the numerical formulations are the conservation principles. The computations should certainly conserve overall waterbody volume and overall constituent balances. A more rigorous test is of the balance between kinetic energy, potential energy and dissipation. It is first necessary to form the energy balance from the momentum balance and then to express the energy balances in terms of the numerical variables that are used at different time levels. Leendertse and Liu (1977) have developed the testing of numerical energy relationships to the point of examining the transfer of energy from one scale to the next for comparison with results from the statistical theory of turbulence.

Leendertse (1973) has presented a complete development of the three-dimensional numerical form of (1) to (7) with emphasis on the spatial differencing and averaging of the various terms in the equations of motion. An explicit integration scheme is chosen based partly on computer storage requirements for the arrays of an implicit scheme and experience with implicit numerical forms of vertically homogeneous hydrodynamics. Leendertse indicates that although there is no gravity wave stability criterion for the implicit method, the propagation of long waves may be affected if the integration time step exceeds 3 to 5 times the gravity wave stability limit. The implicit method discussed by Leendertse (1973), was for a complex simultaneous solution for velocity and surface elevation. The advantage of beginning each iteration from the surface wave relationship, (18), has apparently not been fully explored for the three dimensional equations.

Demonstration applications of three-dimensional hydrodynamics to estuaries has been performed by Leendertse and Liu (1975) for Chesapeake Bay, San Francisco Bay and the Strait of Juan de Fuca. Extensive horizontal and vertical spatial detail was represented in each of the demonstration applications, but simulations were run only over a few tidal cycles at time steps on the order of 10 seconds. The simulations showed the problems of specifying proper surface slope conditions at the open boundary, specifying vertical momentum and mass exchange, and the need to consider more detail in specifying wind forces and atmospheric pressure differences.

Laterally Averaged Hydrodynamics

Many coastal plain drowned river mouth estuaries have dominant variations in their longitudinal and vertical structure. The three dimensional hydrodynamic relationships can be laterally averaged to arrive at the free surface longitudinal and vertical hydrodynamics. The laterally averaged equations of fluid motion and transport are the horizontal momentum balance:

$$\frac{\partial UB}{\partial t} + \frac{\partial UUB}{\partial x} + \frac{\partial WUB}{\partial z} = -\frac{B}{\rho}\frac{\partial P}{\partial x} + \frac{\partial}{\partial x}\left(BA_x \frac{\partial U}{\partial x}\right) + \frac{B\partial \tau_y}{\partial z} \tag{22}$$

where B is the estuary width as a function of x and z; and, U and W are the laterally averaged horizontal and vertical velocity components. The vertical equation of motion reduces to the hydrostatic approximation as:

$$\frac{\partial P}{\partial z} = \rho g \tag{23}$$

The equation of continuity becomes:

$$\frac{\partial UB}{\partial x} + \frac{\partial WB}{\partial z} = qB \tag{24}$$

where q is the side or tributary inflow per unit ($\Delta x \Delta z B$) volume. Vertically integrated continuity gives the free water surface relationship of:

$$\frac{\partial B_\eta \eta}{\partial t} = \frac{\partial}{\partial x} \int_\eta^h UBdz - \int_\eta^h qBdz \tag{25}$$

where B_η is the time and spatially varying surface width and η is the free water surface elevation. The constituent transport for salt becomes:

$$\frac{\partial BS}{\partial t} + \frac{\partial UBS}{\partial x} + \frac{\partial WBS}{\partial z} - \frac{\partial}{\partial x} (BD_x \frac{\partial S}{\partial x}) - \frac{\partial}{\partial z} (BD_z \frac{\partial S}{\partial z}) = S_q B \tag{26}$$

where S_q is the tributary source flux per unit ($\Delta x \Delta z B$) volume, and:

$$\rho = R(S) \tag{27}$$

is the equation of state. As for the three-dimensional case, each additional constituent has a balance similar to (26) with specific source and sink terms. The equation of state is similarly modified for constituents such as temperature and salinity that have a significant effect on density.

Six equations (22) to (27) are to be solved for the six unknowns of U, P, W, η, S and ρ. Lateral averaging of the three-dimensional relationships eliminates the lateral momentum balance, the lateral velocity component, and the Coriolis acceleration. The computational problem is reduced to six equations in six unknowns and most important to two coordinate directions. The reduction to two coordinate directions is the main feature that reduces numerical computational time and storage over the three-dimensional case.

The horizontal pressure gradient is evaluated from (23) to give:

$$B \frac{\partial P}{\partial x} = -\rho g B \frac{\partial \eta}{\partial x} + gB \int_{\eta}^{z} (\partial \rho / \partial x) dz \tag{28}$$

at any depth z. As for the three-dimensional case, the horizontal pressure gradient is divided into the two components of the surface slope and the vertical integral of the horizontal density gradient. The horizontal density gradient is the major driving force for the density circulation exhibited in coastal plain estuaries.

The basic characteristics of the longitudinal and vertical free water surface hydrodynamics can be examined through evaluation of the water surface relationship (25). The vertical integral of the horizontal flow required in (25) can be determined from the algebraic forward time difference of the local acceleration of horizontal momentum in (22). Formulation of the forward time difference of UB is the first step in evaluating the numerical equations. It gives:

$$U'B' = UB + g\, B\Delta t \partial \eta / \partial x - \frac{gB\Delta t}{\rho} \int_{\eta}^{z} (\partial \rho / \partial x) dz + F_x \Delta t \tag{29}$$

where (28) has been substituted for the horizontal pressure gradient and F_x is:

$$F_x = \frac{\partial}{\partial x} (BA_x \frac{\partial U}{\partial x}) - \frac{\partial UUB}{\partial x} - \frac{\partial WUB}{\partial z} \tag{30}$$

The vertical integrals of the various terms in (29) can be further evaluated for insertion into the vertical integral of the flow required in the free water surface balance, (25).

The vertical integral of the horizontal pressure gradient can be evaluated from (28) to give:

$$\frac{1}{\rho}\int_{\eta}^{h} \frac{B\partial P}{\partial x}\, dz = -\frac{\partial \eta}{\partial x}\, g \int_{\eta}^{h} B dz + \frac{g}{\rho}\int_{\eta}^{h} \{B \int_{\eta}^{z} \frac{\partial \rho}{\partial x}\, dz\}\, dz \qquad (31)$$

The first term on the right hand side results from the fact that $\partial\eta/\partial x$ is a function only of x and is constant over z. The integral of width, B, over depth is the total cross-sectional area across which the surface slope contribution to the horizontal pressure gradient acts. The second term is the force due to the horizontal density gradient:

The vertical integral of the horizontal shear stress can be expanded from the derivatives of $\partial B\tau_x/\partial z$ to give:

$$\int_{\eta}^{h} \frac{B\partial \tau_x}{\partial z}\, dz = B_h \tau_h - B_\eta \tau_\nu - \int_{\eta}^{h} \tau_x \frac{\partial B}{\partial z}\, dz \qquad (32)$$

The first term is the bottom shear evaluated at z = h and can be evaluated from bottom velocity friction relationships. The surface shear, $B_\eta \tau_\eta$ is the surface wind shear component parallel to the x axis. The third term is the wall or bottom shear due to the horizontal projection of the sloping sides of the waterbody ($\partial B/\partial z$). It can be evaluated as bottom shear over the projected width ΔB at each elevation. The internal velocity shear cancels out of the vertical integration.

Collecting the various terms of (29) into (25) gives the surface elevation equation of:

$$\frac{\partial B_\eta \eta}{\partial t} - g\Delta t \frac{\partial}{\partial x} \{\frac{\partial \eta}{\partial x} \int_\eta^h B dz\} = \frac{\partial}{\partial x} \int_\eta^h U\ B\ dz$$

$$- \frac{g\Delta t}{\rho} \frac{\partial}{\partial x} \int_\eta^h \{B \int_\eta^z \frac{\partial \rho}{\partial x} dz\} dz$$

$$+ \frac{\partial}{\partial x} \{B_h \tau_h - B_\eta \tau_\eta - \int_\eta^h \tau_x \frac{\partial B}{\partial z} dz\} \Delta t$$

$$+ \frac{\partial}{\partial x} \{\int_\eta^h F_x dz\} \Delta t + \{\int_\eta^h qB\ dz\}\Delta t \qquad (33)$$

With the η or surface elevation terms collected on the left hand side, (33) is a numerical form of the frictionally dampened long wave equation for an irregular geometry stratified waterbody. An evaluation of (33) similar to that presented for the three-dimensional case in equations (14) to (16) demonstrates the gravity wave stability criterion for the laterally homogenous case when U'B' is evaluated for a lower time level $\partial\eta/\partial x$.

For the laterally homogenous estuary, (33) can be simply evaluated implicitly from:

$$\frac{-g\Delta t^2}{\Delta x^2} \sigma(i-1)\eta'(i-1) + B_\eta + \frac{g\Delta t^2}{\Delta x^2} (\sigma(i-1)$$

$$+ \sigma(i))\eta'(i) \frac{-g\Delta t^2}{\Delta x^2} \sigma(i)\eta'(i+1) =$$

$$B_\eta \eta(i) + \Delta t G(i) + \Delta t^2 \int_\eta^h qBdz \qquad (34)$$

Where $\eta'(i)$ is the new time level value of the surface elevation at a finite x location i; $\sigma(i)$ is the total cross-sectional area between i and i+1 locations of $\eta'(i)$ and $\eta'(i+1)$; $\eta(i)$ is the water surface elevation at the previous time step; and G(i) is the sum of terms on the right hand side of (33) except for lateral inflows and outflows. Equation (34) is a spatially implicit surface relationship that eliminates the gravity wave stability criterion. It can be evaluated on each time step using the efficient Thomas algorithm for a tridiagonal matrix. Development and use of (34) in laterally averaged spatially implicit hydrodynamics has been presented in Hamilton (1975) and in Edinger and Buchak (1975, 1977).

The use of (22) to (27) in the numerical evaluation of laterally averaged estuarine circulation has been presented by Blumberg (1975). Blumberg uses essentially an explicit computational scheme limited by the gravity wave speed criterion. Blumberg presents test of the computations by comparisons to analytic results for an ideal oscillating basin and to the high and low tide water surface elevations of an experimental flume. An application is made to the Potomac River estuary. However, the main emphasis of Blumberg's numerical experiments is on evaluating vertical mixing relationships.

A somewhat reduced form of the basic laterally averaged hydrodynamics has been used to study estuarine circulation by Hamilton (1975). The numerical form assumes that the horizontal density gradient, $\partial\rho/\partial x$, is constant with depth. Thus from (28) the horizontal pressure gradient is reduced to:

$$B \frac{\partial P}{\partial x} = - \rho g B \frac{\partial \eta}{\partial x} + g B \overline{\frac{\partial \rho}{\partial x}} (z-\eta) \qquad (35)$$

Where $\overline{\partial\rho/\partial x}$ is an average density gradient that can be approximated from the longitudinal salinity profile. Hamilton (1975) indicates that this approximation saves a number of numerical integration steps and is valid as long as the longitudinal salinity gradient is much larger than the vertical salinity gradient. The additional integration steps on the horizontal density gradient for vertically varying $\partial\rho/\partial x$ are shown in (33) to result in the horizontal gradient of a double integral that is implied by the numerical substitution of U'B' in the surface elevation equation.

The numerical laterally averaged estuarine hydrodynamics of Bowden and Hamilton (1975) and Blumberg (1977a) have allowed examination of different Richardson number formulations for the vertical mixing coefficients of momentum (A_z) and salinity (D_z).

The Richardson number criterion for vertical momentum and constituent dispersion was proposed by Munk and Anderson (1948). Using dimensional analysis they arrived at the relationship of:

$$A_z = A_0 \ (1 + 10 \ Ri)^{-1/2} \tag{36}$$

for the vertical eddy momentum viscosity and:

$$D_z = D_0 \ (1 + 3.33 \ Ri)^{-3/2} \tag{37}$$

for the vertical constituent eddy viscosity, where Ri is the Richardson number defined as:

$$Ri = \frac{g\partial\rho/\partial z}{\rho(\partial U/\partial z)^2} \tag{38}$$

The Richardson number formulations have the property that when $\partial\rho/\partial z$ is zero or there is no density gradient, then A_z and D_z have the values of A_0 and D_0 in the presence of vertical velocity shear. When $\partial\rho/\partial z > 0$ or the water column is stable, then A_z and D_z are less than A_0 and D_0, implying a suppression of turbulent mixing by vertical stratification. For the case of no shear, $\partial U/\partial z = 0$, in the presence of a density gradient A_z and D_z approach minimum values with molecular viscosity and diffusivity being the lower limit. Other exponents and multipliers of Ri in (36) and (37) have been examined by Bowden and Hamilton (1975) and Hamilton and Rattray (1978).

Blumberg (1975) has examined the Richardson number relationship of:

$$D_z = D_1 z^2 \ (1-z/h)^2 \ \left|\partial U/\partial z\right| \ (1 + Ri/Ri_c)^{-1/2} \tag{39}$$

where Ri_c is a critical Richardson number to be determined experimentally but thought to be ≈ 10. This formulation is based on the use of turbulent flow mixing length concepts. Leendertse and Liu (1975) used relationships of the form:

$$D_z = D_0 \; e^{-k \; Ri} \tag{40}$$

which have the advantage that D_z can be evaluated for the full range of negative to positive Richardson numbers.

The Richardson number criterion is chosen to represent A_z and D_z over the whole spectrum of vertical mixing ranging from molecular diffusion for a very stable density gradient ($\partial\rho/\partial z > o$) to rapid vertical convective mixing for an unstable density gradient ($\partial\rho/\partial z < o$). Both Bowden and Hamilton (1975) and Blumberg (1977a) show that different Richardson number formulations significantly alter the resulting longitudinal and vertical salinity and velocity profiles. The numerical hydrodynamic formulations used by Bowden and Hamilton (1975) become unstable for Richardson number formulations that are related to depth and the local vertical velocity and density gradient but stability can be satisfied for a bulk Richardson number criterion that depends on average velocities and density differences in the water column. The numerical hydrodynamics utilized by Blumberg (1977a) allows use of a local Richardson number formulation for the vertical mixing coefficients, however, the Richardson number formulation is a function of the computational scheme and particularly the vertical resolution of the numerical model. Both studies show that additional combined numerical and experimental investigations are required to determine the interrelationships between the numerical hydrodynamic formulations and Richardson number formulations. It is also probable that the values of A_0 and D_0 are related to the scale of the computational grid relative to the waterbody (Edinger and Buchak, 1979).

One Dimensional Sectionally Homogenous Hydrodynamics

The one-dimensional sectionally homogenous relationships have been used in numerical tidal hydraulics for a long time. The basic relationships can be developed by integrating the three-dimensional equations across a cross-sectional area and defining sectional homogenous averages of each variable. The longitudinal momentum equation reduces to:

$$\frac{\partial \sigma U}{\partial t} + \frac{\partial UU\sigma}{\partial x} = -\frac{1}{\rho}\int_{\eta}^{h} \frac{B\partial P}{\partial x}\, dz + \frac{\partial}{\partial x}\left(\sigma A_x \frac{\partial U}{\partial x}\right) + (\tau_\eta - \tau_h) \tag{41}$$

where σ is the cross-sectional area as a function of x and t; U is the sectionally homogenous horizontal velocity; τ_η is the surface (wind) stress; and, τ_h is the velocity related bottom stress. The vertical integral of the horizontal pressure gradient becomes from (31):

$$\frac{1}{\rho}\int_{\eta}^{h} B\,\frac{\partial P}{\partial x}\, dz = -\, g\sigma\,\frac{\partial \eta}{\partial x} + \frac{g}{\rho}\,\frac{\partial \rho}{\partial x}\int_{\eta}^{h} B(z-\eta)\, dz \tag{42}$$

where ρ is the sectionally homogenous density. The vertical integral of the horizontal pressure gradient is often reduced to the more convenient width averaged form of:

$$\frac{1}{\rho}\int_{\eta}^{h} B\,\frac{\partial P}{\partial x}\, dz = -\, g\sigma\,\frac{\partial \eta}{\partial x} + \frac{g}{\rho}\,\frac{\partial \rho}{\partial x}\, B\,\frac{(h-\eta)^2}{2} \tag{43}$$

which shows the dependence of the baroclinic component on the square of the total depth.

Continuity becomes:

$$\frac{\partial B_\eta \eta}{\partial t} = \frac{\partial U\sigma}{\partial x} - q_\sigma \tag{44}$$

where q_σ is the lateral inflow per $\sigma\Delta x$ volume and B_η is the surface width. The constituent transport is:

$$\frac{\partial \sigma S}{\partial t} + \frac{\partial U \sigma S}{\partial x} - \frac{\partial}{\partial x}\left(\frac{\sigma D_x \partial S}{\partial x}\right) = S_q \sigma \tag{45}$$

and the equation of state is:

$$\rho = R(S) \tag{46}$$

It is presumed in the above relationships that B_η and σ are known functions of the surface coordinate, η. The numerical problem therefore reduces to solving for U, η, S and ρ from four equations after substituting the horizontal pressure gradient from (42) into the horizontal momentum equation (42).

The surface wave formulation can be developed as for the two previous cases or it can be inferred directly from (33) to be:

$$\frac{\partial B_\eta \eta}{\partial t} - g\Delta t \frac{\partial(\sigma \partial \eta)}{\partial x\, \partial x} = - \frac{\partial \sigma U}{\partial x} - \frac{g\Delta t}{\rho} \frac{\partial(\sigma \partial \rho)}{\partial x\, \partial x} + \Delta t \frac{\partial}{\partial x} \sigma(\tau_\eta - \tau_h)$$

$$+ \Delta t \frac{\partial F_x}{\partial x} + q_\sigma \Delta t \tag{47}$$

where:

$$F_x = - \frac{\partial UU\sigma}{\partial x} + \frac{\partial}{\partial x}\left(\sigma A_x \frac{\partial U}{\partial x}\right) \tag{48}$$

are the remainder of terms. In (47), $\partial UU\sigma/\partial x$ is implied to be at a previous time step while the time level of $\partial\eta/\partial x$ remains unspecified. If the spatially implicit integration is chosen for η, then the one-dimensional sectionally homogenous wave equation becomes identical to (34). The gravity wave stability criterion can be shown to exist for explicit spatial integration by previous methods.

There is an extensive literature on the numerical forms of the one dimensional hydrodynamic equations. Numerous computational schemes have been tested by Fread (1973, 1975) for stability, accuracy, and introduction of computational waves. Much of the one dimensional literature has served as a basis for inferences on the computational properties of two and three dimensional numerical schemes.

Various forms of the sectionally homogenous relationships have been presented by Harleman (1971) with discussions on the evaluations of bottom stress terms and longitudinal dispersion coefficients. Harleman (1971) also presents algebraic solutions to the one-dimensional constituent transport, equation (45), for various assumptions about the horizontal velocity. Pritchard (1958), however, has shown that the horizontal velocity component, U, cannot be evaluated independently of cross-sectional area, σ, particularly when forming tidal averages of the transport equation.

A recent application of the full set of the combined sectionally homogenous momentum and transport relationships to the Delaware River Estuary has been presented by Harleman and Thatcher (1978). The one-dimensional relationships were extended through Delaware Bay to an ocean boundary. Since the main purpose of the analysis was to compute accurate time varying salinity up the estuary, the ocean boundary salinity condition was examined in detail. It was found that the estuary discharge into the coastal waters and coastal transport had to be accounted for in establishing the ocean boundary condition.

Vertically Homogenous Hydrodynamics

Two-dimensional vertical homogenous hydrodynamics have found wide application in shallow bays and estuaries over the past years and have even been used to compute the tidal hydraulics of an estuary as deep as Chesapeake Bay (Blumberg, 1977b). The two-dimensional vertically homogenous relationships can be derived from a vertical integration of the three dimensional equations. The vertically integrated longitudinal and lateral momentum balances are:

$$\frac{\partial UH}{\partial t} + \frac{\partial UUH}{\partial x} + \frac{\partial UVH}{\partial y} - fVH = -\frac{1}{\rho}\int_{\eta}^{h} \frac{\partial P}{\partial x}\,dz + \frac{\partial (HA_x \partial U)}{\partial x\ \ \partial x}$$

$$+ \frac{\partial(HA_y \partial U)}{\partial y \quad \partial y} + \tau_\eta - \tau_h \tag{49}$$

for x momentum where H is the total water column depth (H=h-η). The y-momentum balance is:

$$\frac{\partial VH}{\partial t} + \frac{\partial VUH}{\partial t} + \frac{\partial VVH}{\partial y} + fUH = -\frac{1}{\rho}\int_\eta^h \frac{\partial P}{\partial y} dz$$

$$+ \frac{\partial(HA_x \partial V)}{\partial x \quad \partial x} + \frac{\partial(HA_y \partial V)}{\partial y \quad \partial y} + \tau_\eta - \tau_h \tag{50}$$

The vertically integrated horizontal pressure gradients are respectively:

$$\frac{1}{\rho}\int_\eta^h \frac{\partial P}{\partial x} dz = -gH\frac{\partial \eta}{\partial x} + \frac{g}{\rho}\frac{\partial \rho}{\partial x}\frac{H^2}{2} \tag{51}$$

and:

$$\frac{1}{\rho}\int_\eta^h \frac{\partial P}{\partial y} dz = -gH\frac{\partial \eta}{\partial y} + \frac{g}{\rho}\frac{\partial \rho}{\partial y}\frac{H^2}{2} \tag{52}$$

and continuity is:

$$\frac{\partial H}{\partial t} + \frac{\partial UH}{\partial x} + \frac{\partial VH}{\partial y} = qH \tag{53}$$

Constituent transport is:

$$\frac{\partial HS}{\partial t} + \frac{\partial UHS}{\partial x} + \frac{\partial VHS}{\partial y} - \frac{\partial (HD_x \partial S)}{\partial x \quad \partial x} - \frac{\partial (HD_y \partial S)}{\partial y \quad \partial y} = S_q H \tag{54}$$

and the equation of state is as given previously. After substitution of (51) and (52) into x and y momentum there are the five unknowns of U, V, H, S and ρ to be solved from five equations.

The general computational procedure is to compute UH and VH from local acceleration in the momentum balances then to substitute into continuity to determine a new depth, H. The remainder of terms are evaluated at some previous time step. The constituent distribution, S, is determined once the flow is known. As shown previously, the substitution of UH and VH into continuity is the numerical equivalent of integrating the surface wave equation forward in time. The surface wave equation of the vertically homogenous hydrodynamics is identical to that developed for the three-dimensional case which is (15) for explicit integration and (18) for implicit integration.

One of the first explicit numerical forms of the vertically homogenous hydrodynamics was by Reid and Bodine (1963) who neglected the spatial accelerations, horizontal dispersion of momentum and horizontal density gradient forces to essentially produce the long wave approximations for the computation of storm surges. The more complete equations were assembled by Leendertse (1970, 1971) in explicit form and with complex spatial averaging. These two works have been the basis of many two dimensional codes.

There have been enough applications of two-dimensional vertically homogenous hydrodynamics that many numerical problems beyond equation formulation are being found which apply to this and other cases. One problem is the proper dynamic boundary conditions to use at openings and inlets in semi-enclosed waterways. Usually tide height boundary conditions are specified, often from tidal records. However, there can be a sea surface slope between inlets due to wind stresses that usually cannot be detected from the tidal records. Wind stress on the interior waterbody without proper accounting for elevation differences of the inlets can set up an artificial circulation.

A second concern, examined by Lean and Weare (1979), is proper representation of boundary stresses in the horizontal momentum equations and lateral transfer of momentum. It is suggested that vorticity needs to be introduced into the flow by a no slip boundary condition and that lateral eddy dispersion coefficients (A_x and A_y) be computed as a function of distance from the boundaries to properly represent secondary flows.

One of the more important applications of the two-dimensional vertically homogenous relationships is to estuarine sediment transport in shallow bays. In shallow waterbodies bottom sediment scavenging is produced by secondary bottom currents due to wind-waves which have no net horizontal transport, and the horizontal transport is produced by the larger scale flow represented by the vertically homogenous hydrodynamics (Grant and Madsen, 1979). The wind effect on the large scale flow as represented by a surface wind shear and results in a surface slope set up in the direction of wind. Since wind waves and their resultant orbital bottom currents are not represented in the large scale hydrodynamics, a separate wave forecast is required to identify the amount of bottom scour at different regions throughout the waterbody. The wind wave generated bottom scour, or lack of it, becomes the source or sink of bottom sediment relative to the vertically homogenous transport. Thus, a complete solution to time varying sediment transport requires coupling a wind wave forecasting scheme to the vertically homogenous transport computations.

Finite Element Formulations

The four estuarine numerical hydrodynamic cases have been examined as spatial finite difference formulations derived directly from the differential equations for spatially uniform but different Δx, Δy and Δz. Spatially uniform Δx and Δy presents some limitations on gridding the geometry to real waterbody topography particularly for the vertically homogenous case. Uniformly varying Δx and Δy has been used in some formulations but requires extensive spatial averaging of the computed variables that can lead to numerical oscillations and errors not representative of a real waterbody. A further method is to use coordinate transformations which map the surface area of the waterbody onto a rectangular uniform grid. The finite element formulation provides a numerical method for accurate spatial representation of the equations over a spatially varying grid.

In finite difference schemes, the spatial representation of the equations is derived directly from the governing differential equations for numerical integration. In finite element formulations an equivalent variational integral to the basic relationships is first sought and the variational integral is evaluated over space using finite elements. The theory is that minimization of the variational integral produces the solution to the basic differential equations.

The finite element theory has its roots in variational calculus which developed as a basis for expressing many physical problems (Franklin, 1944). In the original form, the physical problem, such as minimizing the balance between kinetic and potential energy in a stretched membrane, is first expressed in integral form and the governing differential equations are deduced using variational principles that minimize the integral. The finite element problem is just the reverse; that is, given the basic differential equations then infer the equivalent variational integral for numerical evaluation. Many different methods have been used for casting the basic hydrodynamic relationships into integral finite element form (Gray, et al. 1977). However, the basic variational problem for free water surface hydrodynamics has not been clearly defined.

The simplest hydrodynamics for which to examine the variational statement for free water surface problems is the vertically homogenous long wave approximation to two-dimensional horizontal flow. Making the approximation that the water surface perturbation, η, is much smaller than the total depth ($\eta < H$) then for η measured vertically upwards the momentum balance reduces to:

$$\frac{\partial U}{\partial t} = - g\frac{\partial \eta}{\partial x} + F_x \tag{55}$$

$$\frac{\partial V}{\partial t} = - g\frac{\partial \eta}{\partial y} + F_y \tag{56}$$

and continuity becomes:

$$\frac{\partial \eta}{\partial t} + \frac{H\partial U}{\partial x} + \frac{H\partial V}{\partial y} = q \tag{57}$$

It is presumed that shear stress and spatial momentum terms are incorporated in F_x and F_y.

The long wave surface relationship is found directly from equations (50) to (52) by differentiation of x-momentum with respect to $\partial/\partial x$, and y-momentum with respect to $\partial/\partial y$ and substitution into the time derivative, $\partial/\partial t$ of continuity. This gives:

$$\frac{\partial^2 \eta}{\partial t^2} - gH\left(\frac{\partial^2 \eta}{\partial x^2} + \frac{\partial^2 \eta}{\partial y^2}\right) = \frac{\partial q}{\partial t} - \frac{\partial F_x}{\partial x} - \frac{\partial F_y}{\partial y} \tag{58}$$

which is the long wave surface equation. Comparison of (58) to (14, 33 and 47), where the latter are derived by numerical substitution of time forward differenced momentum into vertically integrated continuity, shows that the latter are a first time integral approximation to the complete long wave surface equation. Performing the previous manipulations of:

$$U' = U - g\Delta t \frac{\partial \eta}{\partial x} + F_x \Delta t \tag{59}$$

and

$$V' = V - g\Delta t \frac{\partial \eta}{\partial y} + F_y \Delta t \tag{60}$$

which are identical to (11) and (12) for the three-dimensional case except for the sign convention on η, gives on substitution into (57):

$$\frac{\partial \eta}{\partial t} - gH\Delta t \left(\frac{\partial^2 \eta}{\partial x^2} + \frac{\partial^2 \eta}{\partial y^2}\right) = - H \left(\frac{\partial U}{\partial x} + \frac{\partial V}{\partial y}\right)$$

$$+ q - \left(\frac{\partial F_x}{\partial x} + \frac{\partial F_y}{\partial y}\right) \Delta t \tag{61}$$

Thus, (61) is an approximation to (58) similar to the previous surface relationships in (14, 33 and 47). The basis of the finite element solution to two dimensional vertically homogenous hydrodynamics can be found by identifying the variational integral problem statement that leads to (58) or (61).

The equations of motion actually define an extremal for the problem of minimizing Hamilton's integral for the difference between the kinetic energy and potential energy of a particle displacement (Franklin, 1944). For a surface displacement, η, Hamilton's integral for waves is:

$$I = \int_t \left\{ \int_x \int_y \frac{1}{2} \left[\left(\frac{\partial \eta}{\partial t}\right)^2 - gH \left(\left(\frac{\partial \eta}{\partial x}\right)^2 + \left(\frac{\partial \eta}{\partial y}\right)^2 \right) \right] dxdy \right\} dt \tag{62}$$

where $\partial\eta/\partial t$ is the vertical velocity of the particle and hence the first term is the kinetic energy and where $\partial\eta/\partial x$ or $\partial\eta/\partial y$ is the spatial displacement or the potential energy of the particle. The differential equation resulting from minimizing the integral I, is the left hand side of (58).

For the free water surface problem, it is not convenient to minimize I over all time of integration. The problem can be made stationary within a time step by taking η forward in time one additional time step from (51) to give:

$$\eta' - gH\Delta t^2 \left(\frac{\partial^2 \eta'}{\partial x^2} + \frac{\partial^2 \eta'}{\partial y^2}\right) = \eta - H\left(\frac{\partial U}{\partial x} + \frac{\partial V}{\partial y}\right) \Delta t + G \tag{63}$$

where G represents the remainder of terms. Equation (63) results from minimizing:

$$I = \int_x \int_y \frac{1}{2} \left[gH\Delta t^2 \left(\left(\frac{\partial \eta'}{\partial x}\right)^2 + \left(\frac{\partial \eta'}{\partial y}\right)^2 \right) - \eta'^2 \right] dxdy$$

$$+ \int_x \int_y \frac{1}{2} \left[\eta'\eta - \eta' H\left(\frac{\partial U}{\partial x} + \frac{\partial V}{\partial y}\right) \Delta t + \eta' G \right] dxdy \tag{64}$$

which can be cast into finite element form for η' once the horizontal velocity components, U and V, are identified relative to the nodal points at which η' are determined.

The location of the U and V components for an element relative to the nodal locations of η' can be identified from (64) where Green's Theorem gives:

$$\int_x \int_y \left(\frac{\partial U}{\partial x} + \frac{\partial V}{\partial y}\right) dxdy = \int_{B_{xy}} (Udy - Vdy) \tag{65}$$

to approximation that η' and H are constant within the element, and where B_{xy} signifies a line integral around the boundary of the element. The line integral on the right hand side of (60) is the net flow into or out of the element, across the sides of the element, in a Δt time step. Thus, U and V must be evaluated along the sides of the element between node points from the ∂η/∂x and ∂η/∂y or x and y components of the surface slope between node points. Evaluation of U and V along each side of an element allows writing the constituent transport for the flow and dispersion normal to the sides of an element which has the advantage that conservation of constituent transport can be accounted for within the element volume rather than at a node point.

Conclusions

The basic differential equations for time varying free surface estuarine hydrodynamics in three, two and one dimensions have been presented and their use in numerical computations have been examined. It is found that the usual method of numerical integration implies directly or indirectly the numerical integration of the free surface long wave equation. The long wave equation can be used as the basis for identifying spatially explicit and implicit numerical solution techniques and for identifying stability characteristics of the solutions. Although not demonstrated in the paper, the free surface wave equation of the numerical integration is also useful for identifying convergence properties of the numerical solutions (Edinger and Buchak, 1975). The long wave equation derived from algebraic substitutions similar to those implied by numerical substitutions is also shown to be the basis for identifying a variational statement of free surface hydrodynamics that leads to finite element solution techniques.

Acknowledgments

The authors gratefully acknowledge the careful preparation of the manuscript by Linda Shinn. Some ideas included in this report were initiated by Contract No. DACW-78-C-0057 U. S. Army Corps of Engineers Waterway Experiment Station.

REFERENCES

Bowden, K. F. and P. Hamilton (1975), "Some Experiments with a Numerical Model of Circulation and Mixing in a Tidal Estuary", Estuarine and Coastal Marine Science 3, 281-301, (1975).

Blumberg, A. F. (1975), "A Numerical Investigation into the Dynamics of Estuarine Circulation", The Johns Hopkins University, Chesapeake Bay Institute, Technical Report 91, Baltimore, MD, October, 1975.

Blumberg, A. F. (1977b), "Numerical Tidal Model of Chesapeake Bay", Jr. Hyd. Div. ASCE, Vol. 103, No. HY1, January, 1977.

Blumberg, A. F. (1977a), "Numerical Model of Estuarine Circulation", Jr. Hyd. Div. ASCE, Vol. 103, No. HY3, March, 1977.

Edinger, J. E. and E. M. Buchak (1975), "A Hydrodynamic, Two-Dimensional Reservoir Model: The Computational Basis", Contract No. DACW 27-74-C-0200, U. S. Army Engineer Division, Ohio River, Cincinnati, OH, September, 1975.

Edinger, J. E. and E. M. Buchak (1977), "A Hydrodynamic Two-Dimensional Reservoir Model: Development and Test Application", Contract No. DACW 27-76-C-0089, U. S. Army Engineer Division, Ohio River, Cincinnati, OH, August, 1977.

Edinger, J. E. and E. M. Buchak (1979), "Preliminary LARM Simulations of the WES GRH Flume", Report to the Reservoir Water Quality Branch, Hydraulic Structures Division, Hydraulics Laboratory, Corps of Engineers Waterways Experiment Station, Vicksburg, MS, April, 1979.

Franklin, P. (1944), Methods of Advanced Calculus, McGraw-Hill, NY, 1944.

Fread, D. L. (1973), "Technique for Dynamic Routing in Rivers with Tributaries", Jr. Water Resources Research, 9 (4) (1973).

Fread, D. L. (1975), Discussion of "Comparison of Four Numerical Methods for Flood Routing", Jr. Hyd. Div. ASCE, Vol. 101, No. HY3, March, 1975.

Grant, W. D. and O. S. Madsen (1979) "Combined Wave and Current Interaction With a Rough Bottom", Jr. Geo. Res. AGU, Vol. 84 No. C4, April 20, 1979.

Gray, W. G., G. F. Pinder, and C. A. Brebbia (1977) Finite Elements in Water Resources, Peutech Press, London (1977).

Hamilton, P. (1975) "A Numerical Model of the Vertical Circulation of Tidal Estuaries and its Application to the Rotterdam Waterway", Geophs. J. R. Astr. Soc., 40, 1-21, (1975).

Hamilton, P. and M. Rattray (1978). "Theoretical Aspects of Estuarine Circulation", in Estuarine Transport Processes, Edited by Bjorn Kjerfve, University of South Carolina Press, Colombia, SC, 1978.

Harleman, D. R. F. (1971), "One-Dimensional Models", in Estuarine Modeling: An Assessment by Tracor, Inc. for the Water Quality Office Environmental Protection Agency, Project 16070DZV. U. S. Government Printing Office, Stock No. 5501-0129, Washington, D. C., February, 1971.

Harleman, D. R. F. and M. L. Thatcher (1978), "Development of a Deterministic Time Varying Salinity Intrusion Model for the Delaware Estuary (MIT-TSIM), "Report to the Delaware River Basin Commission, Trenton, NJ, May, 1978.

Lean, G. H. and T. J. Weare (1979), "Modeling Two Dimensional Circulating Flow", HYD. Div. ASCE Vol. 105, No. HYD 1, January, 1979.

Leendertse, J. J. (1970), "A Water-Quality Simulation Model for Well-Mixed Estuaries and Coastal Seas: Vol. I, Principles of Computation", RAND Report RM-6230-RC, February, 1970.

Leendertse, J. J. (1971), "Solution Techniques: Finite Differences: in Estuarine Modeling: An Assessment by Tracor, Inc. for the Water Quality Office Environmental Protection Agency, Project 16070DZV. U. S. Government Printing Office, Stock No. 5501-0129, Washington, D. C., February, 1971.

Leendertse, J. J. and E. C. Gritton (1971), "A Water Quality Simulation Model for Well Mixed Estuaries and Coastal Seas: Vol. II, Computational Procedures", RAND Report R-708-NYC, July, 1971.

Leendertse, J. J. (1973), "A Three-Dimensional Model for Estuaries and Coastal Seas: Vol. 1, Principles of Computation", RAND Report R-1414-OWRR, Santa Monica, CA, December, 1973.

Leendertse, J. J. and S-K Liu (1975), "A Three-Dimensional Model For Estuaries and Coastal Seas: Vol. II, Aspects of Computation", RAND Report R-1764-OWRT, Santa Monica, CA, June, 1975.

Leendertse, J. J. and S-K Liu (1977), "A Three-Dimensional Model For Estuaries and Coastal Seas: Vol. IV, Turbulent Energy Computation", RAND Report 2-2187-OWRT, Santa Monica, CA, May, 1977.

Munk, W. H. and E. R. Anderson (1948), "Note on the Thermocline", Jr. Mar. Res. 7: 276-295.

Pritchard, D. W. (1958), "The Equations of Mass Continuity and Salt Continuity in Estuaries", Jr. Mar. Res. 17: 412-423.

Pritchard, D. W. (1971), "Hydrodynamic Models", in Estuarine Modeling: An Assessment by Tracor, Inc. for the Water Quality Office Environmental Protection Agency, Project 16070DZV. U. S. Government Printing Office, Stock No. 5501-0129, Washington, D. C., February, 1971.

Pritchard, D. W. (1978), "What Have Recent Observations Obtained for Adjustment and Verification of Numerical Models Revealed About the Dynamics and Kinematics of Estuaries?" in Estuarine Transport Processes, Edited by Bjorn Kjerfve, University of South Carolina Press, Colombia, SC, 1978.

Reid, R. O. and B. R. Bodine (1963), "Numerical Model for Storm Surges in Galveston Bay", Proc. ASCE WW1, February, 1963.

EVOLUTION OF A NUMERICAL MODEL FOR SIMULATING LONG-PERIOD WAVE BEHAVIOR IN OCEAN-ESTUARINE SYSTEMS

H. Lee Butler

U. S. Army Corps of Engineers, Waterways Experiment Station, P. O. Box 631, Vicksburg, Mississippi, 39180

ABSTRACT

Numerical modeling of water-wave behavior has progressed rapidly in the last several years and is now generally recognized as a useful tool capable of providing solutions to many coastal engineering problems. This paper discusses the evolution of a numerical hydrodynamic model including its applications to a variety of problems in which long-wave theory is valid. To achieve a solution to the governing equations, finite difference techniques are employed on a stretched rectilinear grid system. The most recent version of the model permits a selection of solution schemes. Choices include both implicit and explicit formulations written in terms of velocity or transport dependent variables. The model predicts vertically integrated flow patterns as well as the distribution of water surface elevations. Code features include the treatment of regions which are inundated during a part of the computational cycle, subgrid barrier effects, variable grid, and a variety of permissible boundary conditions and external forcing functions. Reproduction of secondary flow effects is an important aspect for a hydrodynamic model. Discussion of methods which are appropriate for treating the nonlinear terms in the governing equations (terms which cause secondary flow effects) is given. Direction of future code developments also is discussed.

Applicability of the numerical model is demonstrated through a presentation of various ocean-estuarine system problems for which the model was applied. These include simulations of tidal circulation as well as coastal flooding from hurricane surges and tsunami waves.

INTRODUCTION

The Corps of Engineers has had to address the problem of providing reliable estimates of estuarine circulation and coastal flooding from tides and hurricane surges in order to make sound engineering decisions regarding the design, operation, and maintenance of various coastal projects. The need for a generalized numerical model for treating these hydrodynamic problems has been apparent for some time. This paper reports on a two-dimensional finite difference model developed and improved over recent years and applied in a variety of Corps of Engineers studies.

The model described herein, known as the Waterways Experiment Station (WES) Implicit Flooding Model (WIFM), was first devised for application in simulating tidal hydrodynamics of Great Egg Harbor and Corson Inlets, New Jersey (Butler, 1978). Popular approaches for solving the governing equations of fluid flow have included explicit and implicit finite difference techniques as well as finite element schemes. A series of papers by Hinwood and Wallis (1975, 1976) and follow-up discussions by Abbott (1976) and Abraham and Karelse (1976) have classified or reviewed many of these models.

Program WIFM originally employed an implicit solution scheme similar to that developed by Leendertse (1970). The model's reliability and economics were tested by comparing applications of WIFM and an explicit formulated model (Masch, et al., 1973) to Masonboro Inlet, North Carolina (Butler and Raney, 1976). A variable grid system was developed for simulating tidal hydrodynamics of the Coos Bay-South Slough, Oregon, complex (Butler, 1978). Storm-surge applications have included hindcasts of surge and coastal flooding from Hurricane Eloise at Panama City, Florida (Butler and Wanstrath, 1976) and Hurricane Carla at Galveston, Texas (Butler, 1978). Simulation of tsunami inundation at Crescent City, California (Houston and Butler, 1979), was carried out to demonstrate that the model could be used for delineating inundation limits of tsunamis in the Hawaiian Islands based upon predicted elevations near the shoreline (Houston, et al., 1977).

Basic features of the model include inundation simulation of low-lying terrain, treatment of subgrid barrier effects and a variable grid option. Included in the model are actual bathymetry and topography, time and spatially variable bottom roughness, inertial forces due to advective and coriolis accelerations, rainfall, and spatial and time-dependent wind fields. Horizontal diffusion terms in the momentum equations are optionally present and can be used, if desired, for aiding stability of the numerical solution. Current model development permits a selection of solution schemes. Choices include both implicit and explicit formulations written in terms of velocity or transport dependent variables. The model also employs a one-dimensional channel routine which can be coupled dynamically to any cell in the grid system. The channels can extend outside the two-dimensional domain or can remain within the grid system. No lateral flow into or from the channel is permitted.

Future model developments will include subgrid-scale channels which may laterally interact with the two-dimensional model, embedded or patched grid systems for detailed resolution of important smaller scale features within the model limits, and addition of the salt conservation equation for determining horizontal salinity gradients.

HYDRODYNAMIC EQUATIONS

The equations of fluid flow used in WIFM are derived from the classical Navier-Stokes equations in a Cartesian coordinate system (Figure 1). By assuming vertical accelerations are small and the

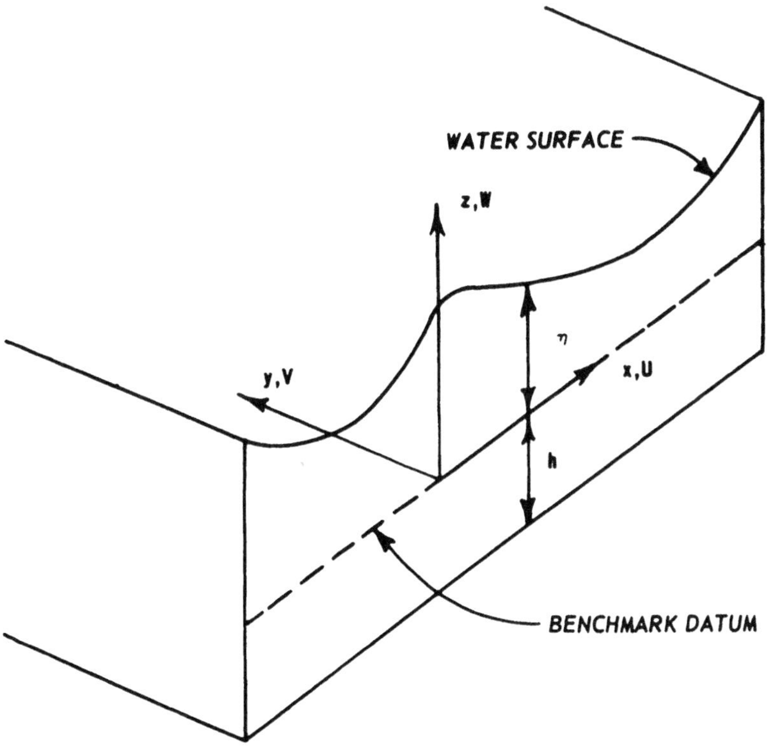

Figure 1. Cartesian coordinate system.

fluid is homogeneous and integrating the flow from sea bottom to water surface, the usual two-dimensional form of the equations of momentum and continuity are obtained. A major advantage of WIFM is the capability of applying a smoothly varying grid to the given study region permitting simulation of a complex landscape by locally increasing grid resolution and/or aligning coordinates along physical boundaries. For each direction, a piecewise reversible transformation is independently used to map prototype or real space into computational space. The transformation takes the form

$$x = a + b\alpha_1^c \tag{1}$$

where a, b, and c are arbitrary constants. By applying a smoothly varying grid transformation, whose functional as well as first derivatives are continuous, many stability problems commonly associated with variable grid schemes are eliminated provided that all derivatives are centered in α-space. By applying transformations patterned on Equation 1 to both the x and y coordinates, the equations of motion in α-space can be written as

Momentum:

$$u_t + \frac{1}{\mu_1} uu_{\alpha_1} + \frac{1}{\mu_2} vu_{\alpha_2} - fv$$

$$+ \frac{g}{\mu_1} (\eta - \eta_a)_{\alpha_1} + \frac{gu}{C^2 d} (u^2 + v^2)^{1/2}$$

$$-\varepsilon\left(\left(\frac{1}{\mu_1}\right)^2 u_{\alpha_1\alpha_1} + \frac{1}{\mu_1} \left(\frac{1}{\mu_1}\right)_{\alpha_1} u_{\alpha_1} + \left(\frac{1}{\mu_2}\right)^2 u_{\alpha_2\alpha_2}\right.$$

$$\left. + \frac{1}{\mu_2} \left(\frac{1}{\mu_2}\right)_{\alpha_2} u_{\alpha_2}\right) = F_{\alpha_1} \tag{2}$$

$$v_t + \frac{1}{\mu_1} uv_{\alpha_1} + \frac{1}{\mu_2} vv_{\alpha_2} + fu + \frac{g}{\mu_2} (\eta - \eta_a)_{\alpha_2}$$

$$+ \frac{gv}{C^2 d} (u^2 + v^2)^{1/2} - \varepsilon \left(\left(\frac{1}{\mu_1}\right)^2 v_{\alpha_1\alpha_1} + \frac{1}{\mu_1} \left(\frac{1}{\mu_1}\right)_{\alpha_1} v_{\alpha_1}\right.$$

$$\left. + \left(\frac{1}{\mu_2}\right)^2 v_{\alpha_2\alpha_2} + \frac{1}{\mu_2} \left(\frac{1}{\mu_2}\right)_{\alpha_2} v_{\alpha_2} \right) = F_{\alpha_2} \qquad (3)$$

Continuity:

$$\eta_t + \frac{1}{\mu_1} (du)_{\alpha_1} + \frac{1}{\mu_2} (dv)_{\alpha_2} = R \qquad (4)$$

where

$$\mu_1 = \frac{\partial x}{\partial \alpha_1} \text{ and } \mu_2 = \frac{\partial y}{\partial \alpha_2}$$

and η is the water-surface elevation; η_a is the hydrostatic elevation corresponding to the atmospheric pressure anomaly; u and v are the vertically integrated velocities at time t in the α_1 and α_2 directions, respectively; $d = \eta - h$ is the total water depth; h is the still-water elevation; f is the Coriolis parameter; C is the Chezy frictional coefficient; g is the acceleration of gravity; ε is a generalized eddy viscosity coefficient; R represents the rate at which additional water is introduced into or taken from the system (for example, through rainfall and evaporation); and F_{α_1} and F_{α_2} are terms representing external forcing functions such as wind stress in the α_1 and α_2 directions. Quantities μ_1 and μ_2 define the stretching of the regular-spaced computational grid in α-space to approximate a study region in real space. Directions α_1 and α_2 correspond to x and y, respectively.

FINITE DIFFERENCE FORMULATIONS

Since obtaining a solution to the governing nonlinear equations on a region with highly complex geometry and topography is intractable using a pure analytical approach, a numerical technique is employed. A variable rectilinear grid is first developed for the model area. The appropriate variables are defined on the grid in a space-staggered fashion as depicted in Figure 2 for a typical cell.

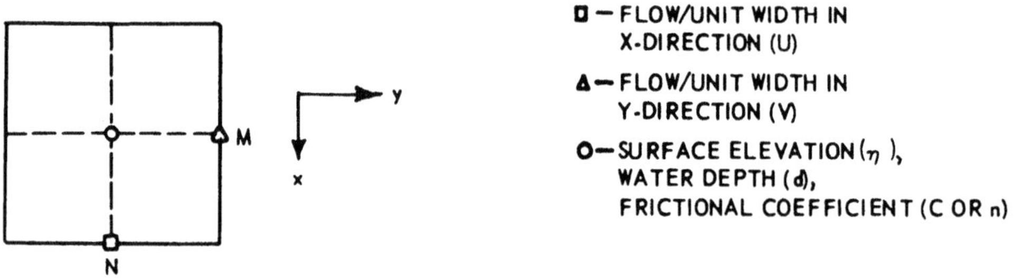

Figure 2. Computational cell definition.

The differential equations (Eqs. 2-4) are now approximated by difference equations. Various solution schemes, including implicit and explicit formulations, could be employed. WIFM permits a selection of difference formulations, but this paper will concentrate on alternating direction techniques. To illustrate how various implicit schemes can be derived consider the simplified matrix equation

$$U_t + AU_x + BU_y + KU = 0 \qquad (5)$$

where

$$U = \begin{pmatrix} \eta \\ u \\ v \end{pmatrix} \qquad A = \begin{pmatrix} o & d & o \\ g & o & o \\ o & o & o \end{pmatrix}$$

$$B = \begin{pmatrix} o & o & d \\ o & o & o \\ g & o & o \end{pmatrix} \qquad K = \begin{pmatrix} o & o & o \\ o & k & o \\ o & o & k \end{pmatrix}$$

A standard method for developing an implicit scheme is to apply the Crank-Nicholson technique to Equation 5 to obtain

$$\frac{1}{\Delta t} (U^{n+1} - U^n) + \frac{1}{2} (\frac{A}{\Delta x} \delta_x + \frac{B}{\Delta y} \delta_y + K) (U^{n+1} + U^n) = 0 \tag{6}$$

where δ_x and δ_y are centered difference operators. Equation 6 can be simplified further to read

$$(1 + \lambda_x + \lambda_y) U^{n+1} = (1 - \lambda_x - \lambda_y) U^n \tag{7}$$

where

$$\lambda_x = \frac{1}{2} \frac{\Delta t}{\Delta x} A\delta_x \text{ and } \lambda_y = \frac{1}{2} \frac{\Delta t}{\Delta y} B\delta_y$$

Adding the quantity $\lambda_x \lambda_y (U^{n+1} - U^n)$ to permit factorization, the following relation is obtained

$$(1 + \lambda_x) (1 + \lambda_y) U^{n+1} = (1 - \lambda_x) (1 - \lambda_y) U^n \tag{8}$$

By introducing an intermediate value, U*, Equation 8 can be split into a two-step operation in a variety of ways:

Standard ADI:

$$(1 + \lambda_x)\ U^* = (1 - \lambda_y)\ U^n \tag{9}$$

$$(1 + \lambda_y)\ U^{n+1} = (1 - \lambda_x)\ U^* \tag{10}$$

Approximate Factorization Scheme:

$$(1 + \lambda_x)\ U^* = (1 - \lambda_x)\ (1 - \lambda_y)\ U^n \tag{11}$$

$$(1 + \lambda_y)\ U^{n+1} = U^* \tag{12}$$

Stabilizing Correction Scheme:

$$(1 + \lambda_x)\ U^* = (1 - \lambda_x - 2\lambda_y)\ U^n \tag{13}$$

$$(1 + \lambda_y)\ U^{n+1} = U^* + \lambda_y\ U^n \tag{14}$$

The first step in each procedure is carried out by sweeping the grid in the x direction, and second step is computed by sweeping in the y direction. Completing both sweeps constitutes a full time step, advancing the solution from the nth time level to the (n + 1) time level. A more complete description of these methods can be found in a paper by Weare (1979).

Five solution schemes are presently available in program WIFM. These include the following:

a. ADI multi-operational scheme formulated in terms of transport per unit width.

b. ADI multi-operational scheme as formulated by Leendertse (1970).

c. ADI stabilizing correction scheme formulated in terms of velocities using a full three-time-level approach.

d. Explicit scheme formulated in terms of velocities using a full three-time-level approach.

e. ADI stabilizing correction scheme formulated in terms of transport per unit width using a full three-time-level approach.

As an illustration of one of these methods, the difference equations for the stabilizing correction scheme is given by

x-sweep:

$$\frac{1}{2\Delta t}(\eta^* - \eta^{k-1}) + \frac{1}{2\mu_1\Delta\alpha_1}\{\delta_{\alpha_1}(u^*\bar{d}^k + u^{k-1}\bar{d}^k)\}$$

$$+ \frac{1}{\mu_2\Delta\alpha_2}\delta_{\alpha_2}(v^{k-1}\bar{d}^k) = R \text{ at } (n,m) \tag{15}$$

$$\frac{1}{2\Delta t}(u^* - u^{k-1}) + \frac{1}{2\mu_1\Delta\alpha_1}u^k\delta_{2\alpha_1}(u^k) + \frac{1}{2\mu_2\Delta\alpha_2}\bar{\bar{v}}^k\delta_{2\alpha_2}(u^k)$$

$$-f\bar{\bar{v}}^k + \frac{g}{2\mu_1\Delta\alpha_1}\{\delta_{\alpha_1}(\eta^* + \eta^{k-1} - 2\eta_a^{\,k})\}$$

$$+ \frac{g}{(\bar{C}^2\bar{d})^k}u^*\{(u^{k-1})^2 + (\bar{\bar{v}}^{k-1})^2\}^{1/2} - \varepsilon\{\frac{1}{(\mu_1\Delta\alpha_1)^2}\delta_{\alpha_1\alpha_1}(u^k)$$

$$+ \frac{1}{(\mu_2 \Delta\alpha_2)^2} \delta_{\alpha_2\alpha_2} (u^k) + \frac{1}{2\mu_1 \Delta\alpha_1{}^2} \delta_{\alpha_1} (u^k)$$

$$+ \frac{1}{2\mu_2 \Delta\alpha_2{}^2} \delta_{\alpha_2} (\frac{1}{\mu_2}) \delta_{2\alpha_2}(u^k) \} = F^k_{\alpha_1} (n,m + \frac{1}{2}) \qquad (16)$$

y-sweep:

$$u^{k+1} = u* \quad \text{at } (n,m + \frac{1}{2} \qquad (17)$$

$$\frac{1}{2\Delta t} (\eta^{k+1} - \eta*) + \frac{1}{2\mu_2 \Delta\alpha_2} \{\delta_{\alpha_2} (v^{k+1}\bar{d}^k - v^{k-1}\bar{d}^k)\} = 0$$

$$\text{at } (n,m) \qquad (18)$$

$$\frac{1}{2\Delta t} (v^{k+1} - v^{k-1}) + \frac{1}{2\mu_1 \Delta\alpha_1} \bar{\bar{u}}^k \delta_{2\alpha_1} (v^k) + \frac{1}{2\mu_2 \Delta\alpha_2} v^k \delta_{2\alpha_2} (v^k)$$

$$+ f\bar{\bar{u}}^k + \frac{g}{2\mu_2 \Delta\alpha_2} \{\delta_{\alpha_2}(\eta^{k+1} + \eta^{k-1} - 2\eta^k_a)\}$$

$$+ \frac{g}{(\bar{c}^2\bar{d})^k} v^{k+1}\{(\bar{\bar{u}}^{k-1})^2 + (v^{k-1})^2\}^{1/2} - \varepsilon\{\frac{1}{(\mu_1 \Delta\alpha_1)^2} \delta_{\alpha_1\alpha_1} (v^k)$$

$$+ \frac{1}{(\mu_2 \Delta\alpha_2)^2} \delta_{\alpha_2\alpha_2} (v^k) + \frac{1}{2\mu_1 \Delta\alpha_1{}^2} \delta_{\alpha_1} (\frac{1}{\mu_1}) \delta_{2\alpha_1} (v^k)$$

$$+ \frac{1}{2\mu_2 \Delta\alpha_2^2} \delta_{\alpha_2} \left(\frac{1}{\mu_2}\right) \delta_{2\alpha_2} (v^k) \} = F^k_{\alpha_2} \text{ at } (n + \frac{1}{2}, m) \tag{19}$$

In these expressions, a single bar represents a two-point spatial average and a double bar a four-point spatial average. The subscripts m and n correspond to spatial locations and superscript k to time levels.

Implicit methods are characterized by a property of unconditional stability in the linear sense. Explicit schemes usually carry a stringent criterion limiting the time step. For the explicit scheme used in WIFM the stability relation

$$\Delta t \leq \{\sqrt{gd} \left(\frac{1}{\Delta x} + \frac{1}{\Delta y}\right)\}^{-1} \tag{20}$$

must be satisfied at all computational points. The stabilizing correction scheme as well as the other ADI schemes are limited by a weak condition, namely,

$$\Delta t \leq \frac{\min(\Delta x, \Delta y)}{(u^2 + v^2)^{1/2}} \tag{21}$$

if a WIFM implicit scheme is applied while neglecting the advective terms, stability is assured.

It has been shown that the stability of implicit difference schemes can be improved by time-centering of the nonlinear terms (Weare, 1976). For this reason the difference scheme outlined above employs three full time levels permitting such time centering of the appropriate terms. Optional use of the horizontal diffusion terms is permitted to aid in stabilizing the solution method. Secondary flow effects have been studied by Vreugdenhil (1973) and Kuipers and Vreugdenhil (1973). Although inclusion of these diffusion terms does not constitute proper closure of the model (Flokstra, 1977), it does aid in stabilizing the procedure. Care must be taken in choosing a reasonable value for the eddy viscosity coefficient.

BOUNDARY CONDITIONS

A variety of boundary conditions are employed throughout the computational grid. These include prescribing water levels, velocities, or flow rates, fixed or movable land-water boundaries, and subgrid barrier conditions.

a. Open Boundaries: Included in this category are seaward boundaries terminating the computational grid or channels exiting the two-dimensional grid at any point in the system. Water levels, velocities, or flow rates are prescribed as functions of location and time and are given as tabular input to the code.

b. Water-land Boundaries: These conditions relate the normal component of flow at the boundary to the state of the water level at the boundary. Hence, water-land boundaries are prescribed along cell faces. Fixed land boundaries are treated by specifying $u = 0$ or $v = 0$ at the appropriate cell face. Low-lying terrain may alternately dry and flood within a tidal cycle or surge history. Inundation is simulated by making the location of the land-water boundary a function of local water depth. By checking water levels in adjacent cells, a determination is made as to the possibility of inundation. Initial movement of water onto dry cells is controlled by using a broad crested weir formula (Reid and Bodine, 1968). Once the water level on the dry cell exceeds some small prescribed value, the boundary face is treated as open and computations for η, u, and v are made for that cell. The drying of cells is the inverse process. When the nonlinear terms are included in the solution scheme, optional choice of free-slip or no-slip conditions is available. In most circumstances, no-slip conditions are preferred in order to correctly introduce vorticity where appropriate. Mass conservation is maintained within these procedures.

c. Sub-Grid Barriers: Such barriers are defined along cell faces and are of three types: exposed, submerged, and overtopping. Exposed barriers are handled by simply specifying no-flow conditions across the appropriately flagged cell faces. Submerged barriers are simulated by controlling flow across cell faces with the use of a time-dependent frictional coefficient. The term "overtopping barrier" is used to distinguish barriers which can be submerged during one phase of the simulation and totally exposed during another. Actual overtopping is treated by using a broad-crested weir formula to specify the proper flow rate across the barrier. Water is transferred from the high to low side according to this rate. Once the barrier is submerged (or conversely exposed), procedures as described for submerged barriers (or exposed) are followed.

APPLICATIONS

Program WIFM has been applied successfully in many applications. A summary of these simulations is presented.

a. Masonboro Inlet, North Carolina: This application was used in the development phase of WIFM (Butler and Raney, 1976). A map of the Masonboro Inlet system is shown in Figure 3. A constant

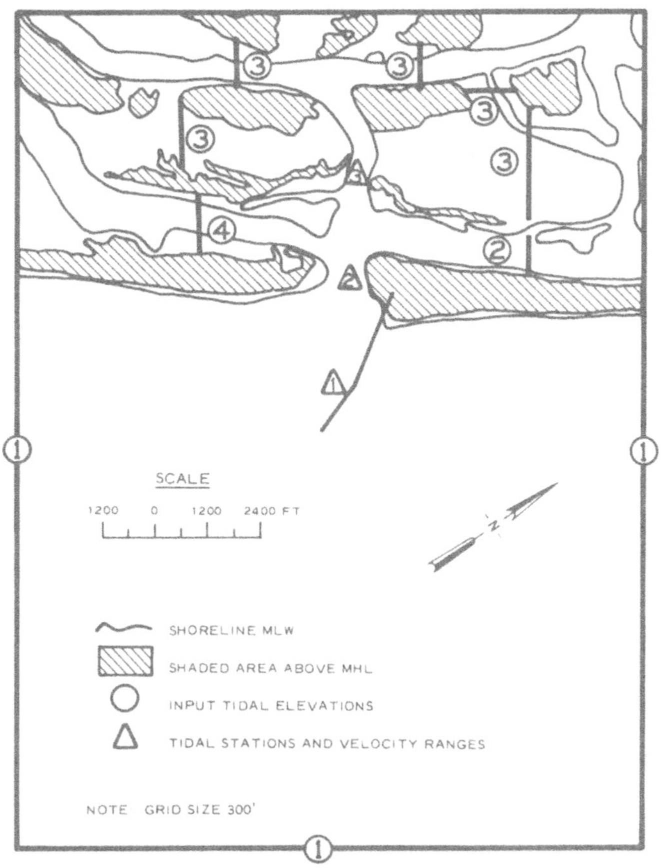

Figure 3. Computational grid limits for Masonboro Inlet.

spatial step of 300 ft was adopted resulting in a mesh of dimensions 41 by 57. Data from a physical model were used, as well as prototype data, to supply needed boundary conditions, topography, and base hydrodynamic conditions. The 3000-ft jetty system protruding from the outer barrier island is composed of a 1000-ft low weir section with the remainder of the structure being impermeable. Exposed barriers were used to represent very narrow strips of high land in the marsh area. Figures 4-5 display a comparison of tidal elevation and velocity prototype data with results from WIFM as well as from an explicit model similar to that of Reid and Bodine (1968). Both models describe the tides equally well. Less effort was undertaken to improve agreement for the explicit model velocity comparison since execution time was significantly longer ($\Delta t_e = \Delta t_i/30$).

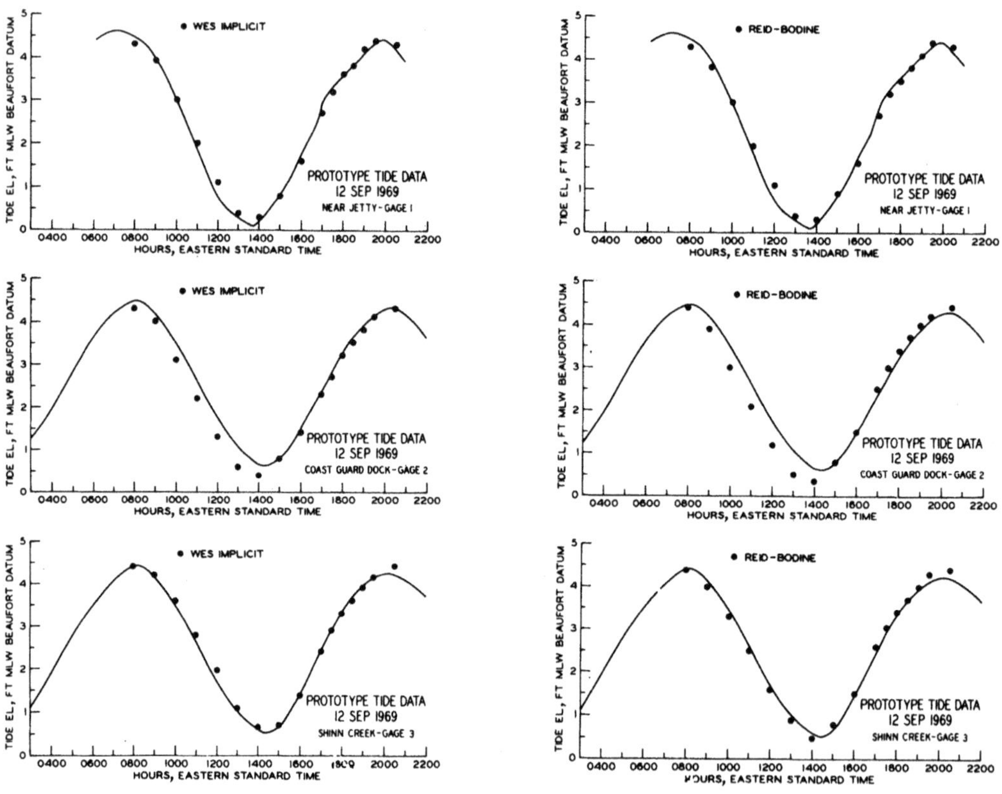

Figure 4. Comparison of surface elevation agreement with prototype for WIFM and the RB explicit model.

b. <u>Corson and Great Egg Harbor Inlets, New Jersey (GECI)</u>: Application of WIFM to these inlets was carried out to aid in evaluating proposed modifications to navigation channel entrances.

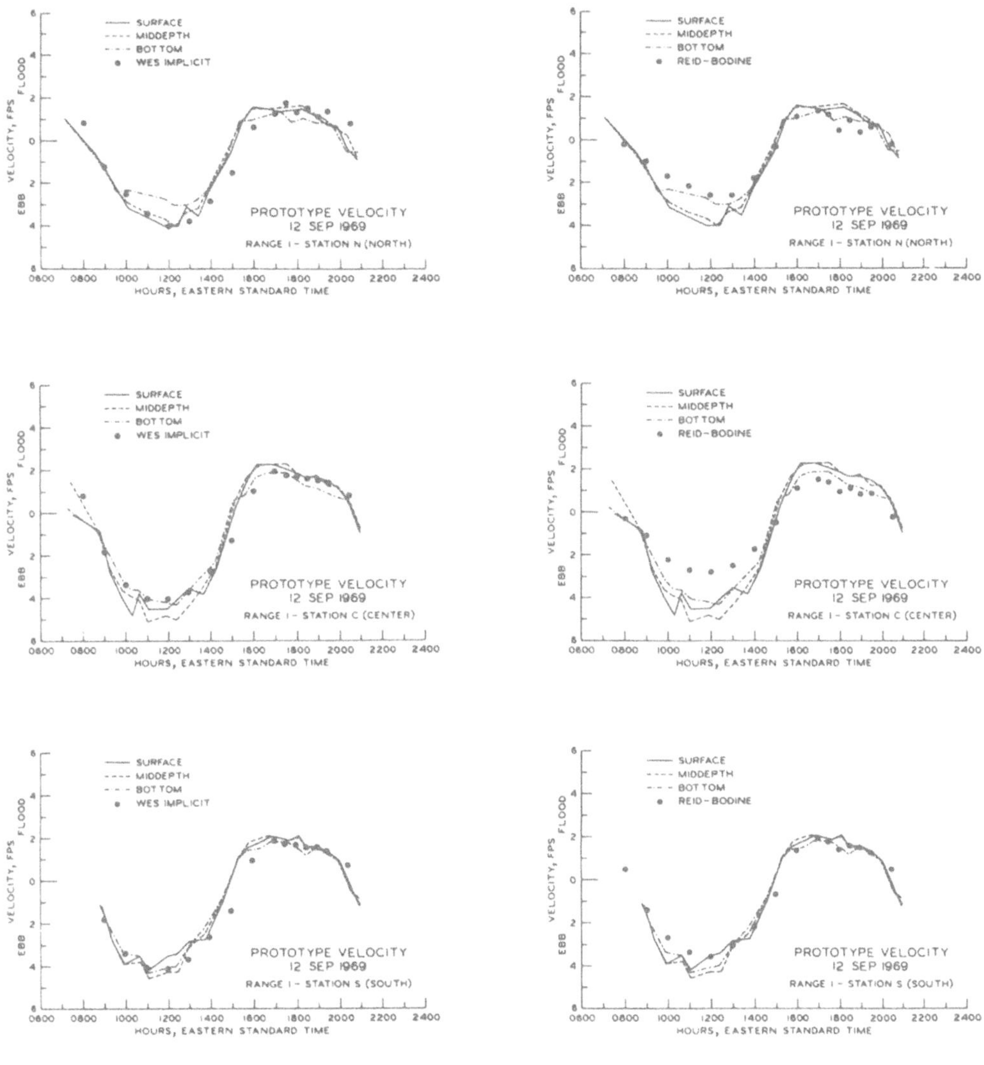

Figure 5. Comparison of current velocity agreement with prototype at range 1 for WIFM and the RB explicit model.

The numerical model was used to predict quantitatively the hydrodynamics (exclusive of sediment transport and wave action) of the tidal flow in the system and, hence, draw a comparison between existing conditions and alternate plan conditions. A prototype survey was conducted by Ahlert, et al. (1976), to provide data for model verification. As a compromise between grid resolution and computational effort, a 300-ft mesh size was selected. Grid dimensions were 110 by 91 for Corson Inlet and 152 by 106 for Great Egg Harbor Inlet. Navigation entrances to both inlets were purported to be unsafe except under ideal conditions of tide and wind. Proposed improvement plans provided for a low-weir and jetty structure on the updrift side of each inlet, including a deposition basin behind the weir, construction and maintenance of specified navigation channels, a bulkhead and jetty structure located on the downdrift side of each inlet, and optional development of a landfill area adjacent to the bulkhead for public recreational use. The models were successfully verified and subsequently applied to test alternate plans. Figures 6 and 7 show the model limits and gage locations for Corson Inlet and Great Egg Harbor Inlet, respectively. Figures 8-10 display both flow and velocity circulation patterns for existing conditions and two plans for Great Egg Harbor Inlet at one instance during the tidal cycle. In the flow plot the lightest background indicates land and the dark background water. The immediate shaded area indicates flooded terrain. A vector length equal to a cell width represents a flow of 50 ft^2 per sec or a velocity of 3 ft per sec. None of the proposed plans introduced significant changes in the tidal prism nor affected circulation within the back-bay areas except in areas local to the structures.

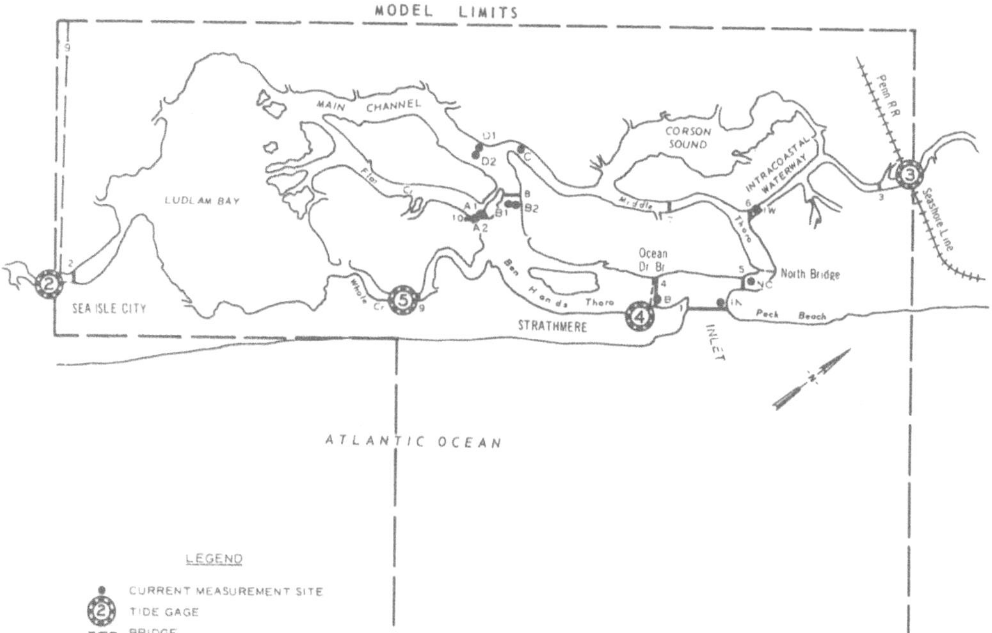

Figure 6. Extent of model area and gage locations for Corson Inlet.

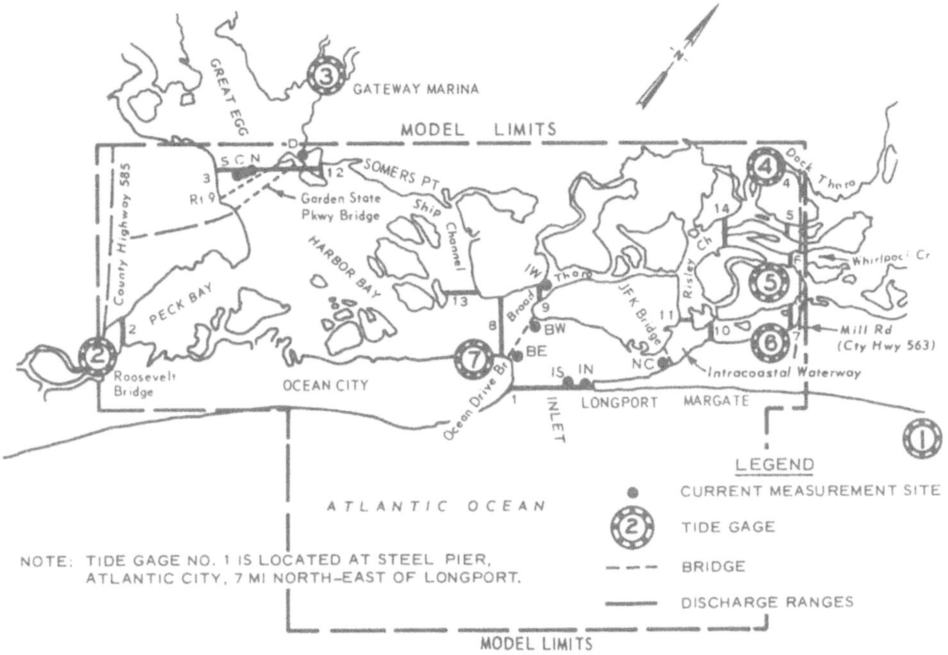

Figure 7. Extent of model area and gage locations for Great Egg Harbor Inlet.

c. Coos Bay Inlet-South Slough, Oregon: Primary objectives of this investigation are similar to those of the GECI study (Butler 1978). The inlet routes heavy shipping up the Coos Bay River to North Bend and small craft traffic (commercial fishing and pleasure craft) to Charleston Harbor just south of the inlet entrance. Continual shoaling problems exist within the entrance channel to Charleston Harbor, and cost of additional maintenance may warrant installation of structures that would alleviate the problem as well as providing a reduction in wave damage in the harbor's boat basin. South Slough is a shallow body of water south of Charleston and fed by the Charleston channel. The upper reaches of the slough constitute a National Marine Santuary administered by NOAA, the first created under the Coastal Zone Management Act. The major task was to analyze alternate plan conditions as to their impact on the tidal hydraulics of the entire system and of South Slough in particular. Prototype tidal and velocity data were gathered in a field survey conducted by WES. A variable mesh was selected with the prototype grid spacing ranging from 150 ft to 900 ft. The finer mesh was focused around the inlet entrance and Charleston Harbor where proposed structures would likely be placed.

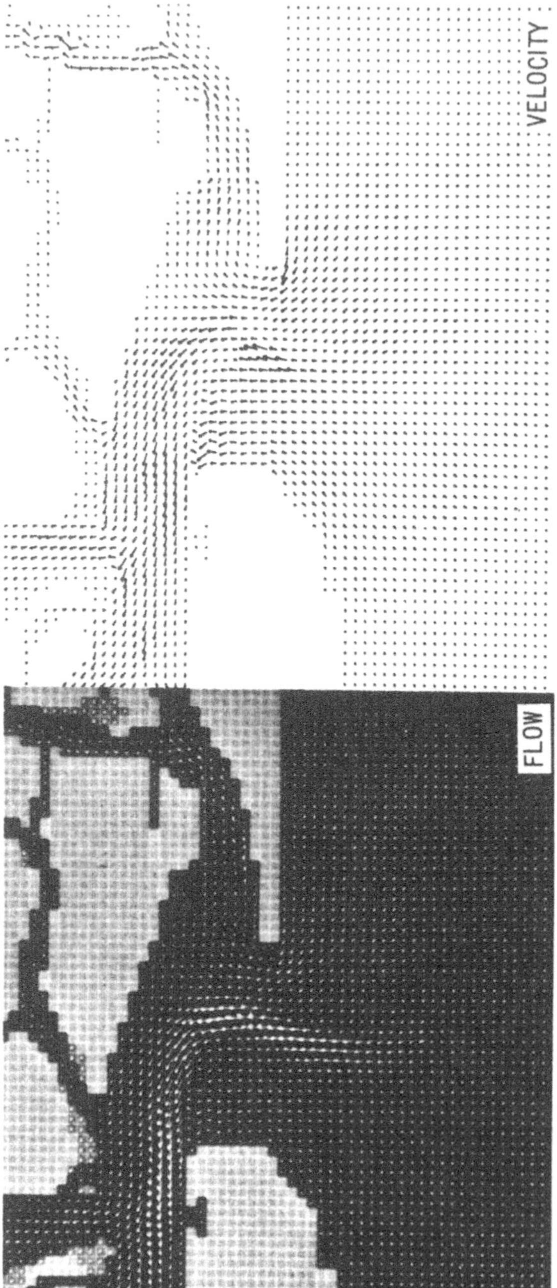

Figure 8. Comparison of unit flow and velocity patterns at 0500 EST for verification run of Great Egg Harbor Inlet.

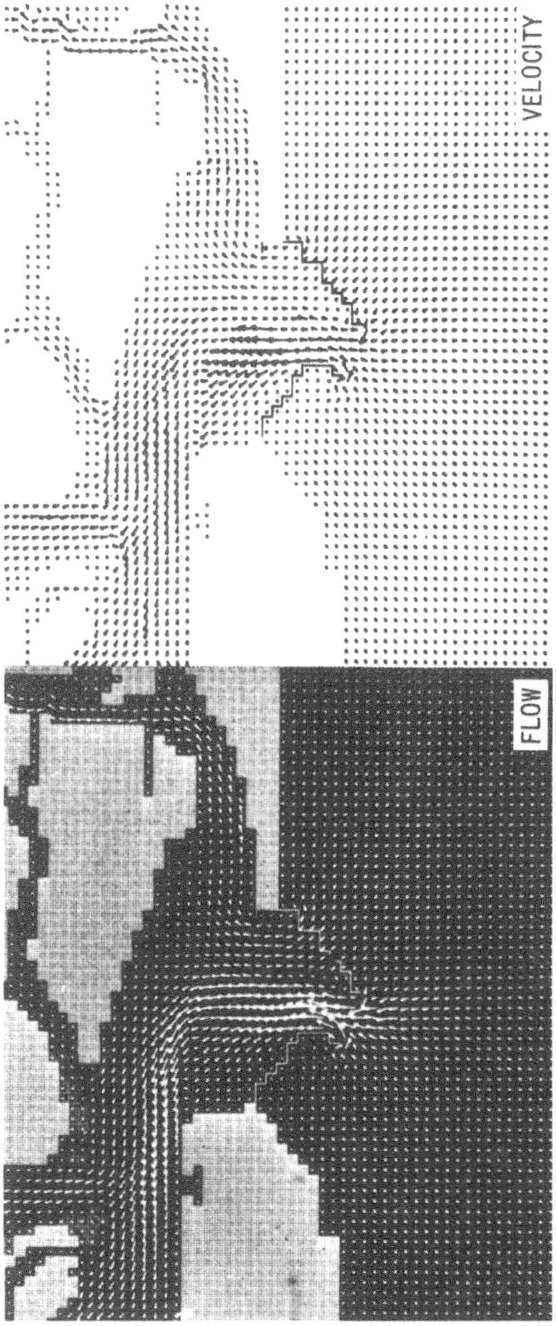

Figure 9. Comparison of unit flow and velocity patterns at 0500 EST for Plan A at Great Egg Harbor Inlet.

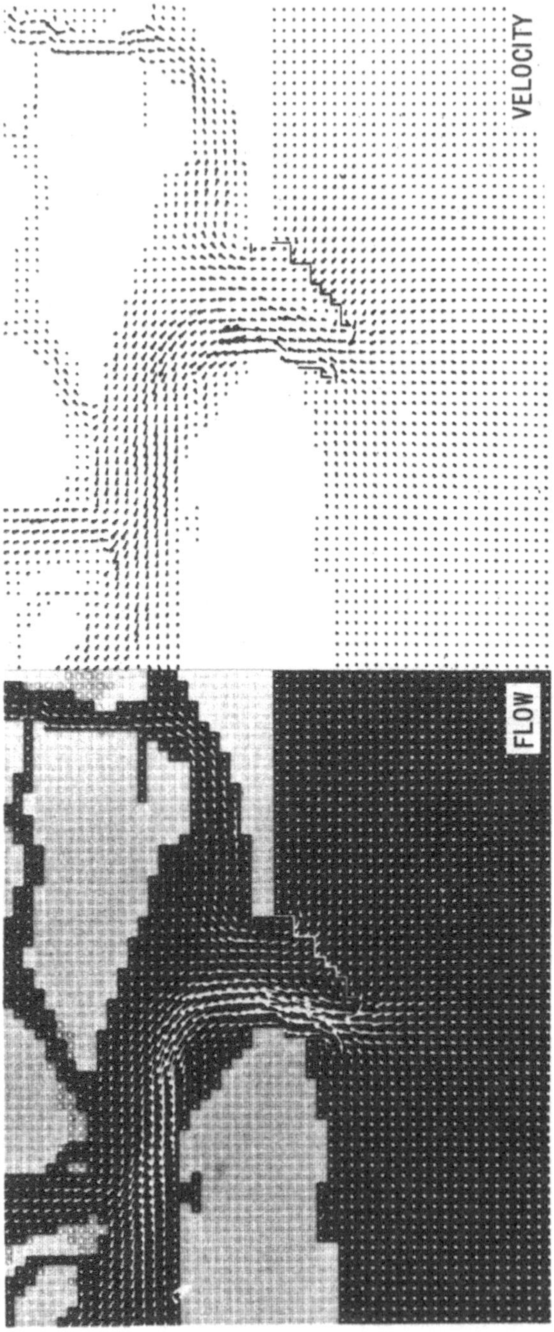

Figure 10. Comparison of unit flow and velocity patterns at 0500 EST for Plan B at Great Egg Harbor Inlet.

Figure 11 displays all plan conditions and major velocity stations used in comparing plan with base conditions. Figures 12 and 13 show typical results obtained for model verification while Figure 14 typifies the type of circulation patterns plotted at each hour during the simulation period for existing conditions and proposed plans. Figure 15 depicts the comparison between plan and existing conditions for current velocities at Station U in the entrance to South Slough. Volumetric discharge through various ranges within the system also was computed, and all results indicated that the tidal prism and circulation in South Slough would not be altered by construction of any of the plans tested.

d. Storm Surge Applications:

(1) Hurricane Eloise. In order to demonstrate WIFM's ability to hindcast coastal flooding due to hurricane surge and the feasibility of coupling it to an orthogonal curvilinear open-coast model (Wanstrath and Butler, 1976), an application to the Panama City, Florida, area was performed. Figure 16 displays the inland model area. Comparison of predicted and observed high-water elevations at various gage locations throughout the model limits is given in the following tabulation:

Location	High-Water Elevations (ft) Predicted	Observed
Panama City	4.9	5.3-5.4
Dyers Point	5.0	5.0
Lynn Haven	7.1	6.6
Ent. Grand Lagoon	5.4	4.8
St. Andrews Sound	8.4	8.0
Watson Bayou	4.4	4.9

(2) Hurricane Carla: A more detailed surge application was undertaken for hindcasting the surge and coastal flooding from Hurricane Carla in the Galveston Bay area (Butler, 1978). Prior to the Carla simulation, a tidal verification of the model for Galveston Bay was accomplished by first simulating a given tidal and design surge condition used by Brogdon (1969) in a physical model study conducted at WES. The computational grid for the bay area is displayed in Figure 17 as well as gage locations used for comparing numerical and observed data (physical model or prototype surge data). Figure 18 demonstrates the model's ability to reproduce a large radius high translation design surge having adjusted the model to predict a specified tidal condition. Again, Wanstrath's model was applied to develop time-dependent open-sea boundary values for WIFM. Both open-coast and inland models used

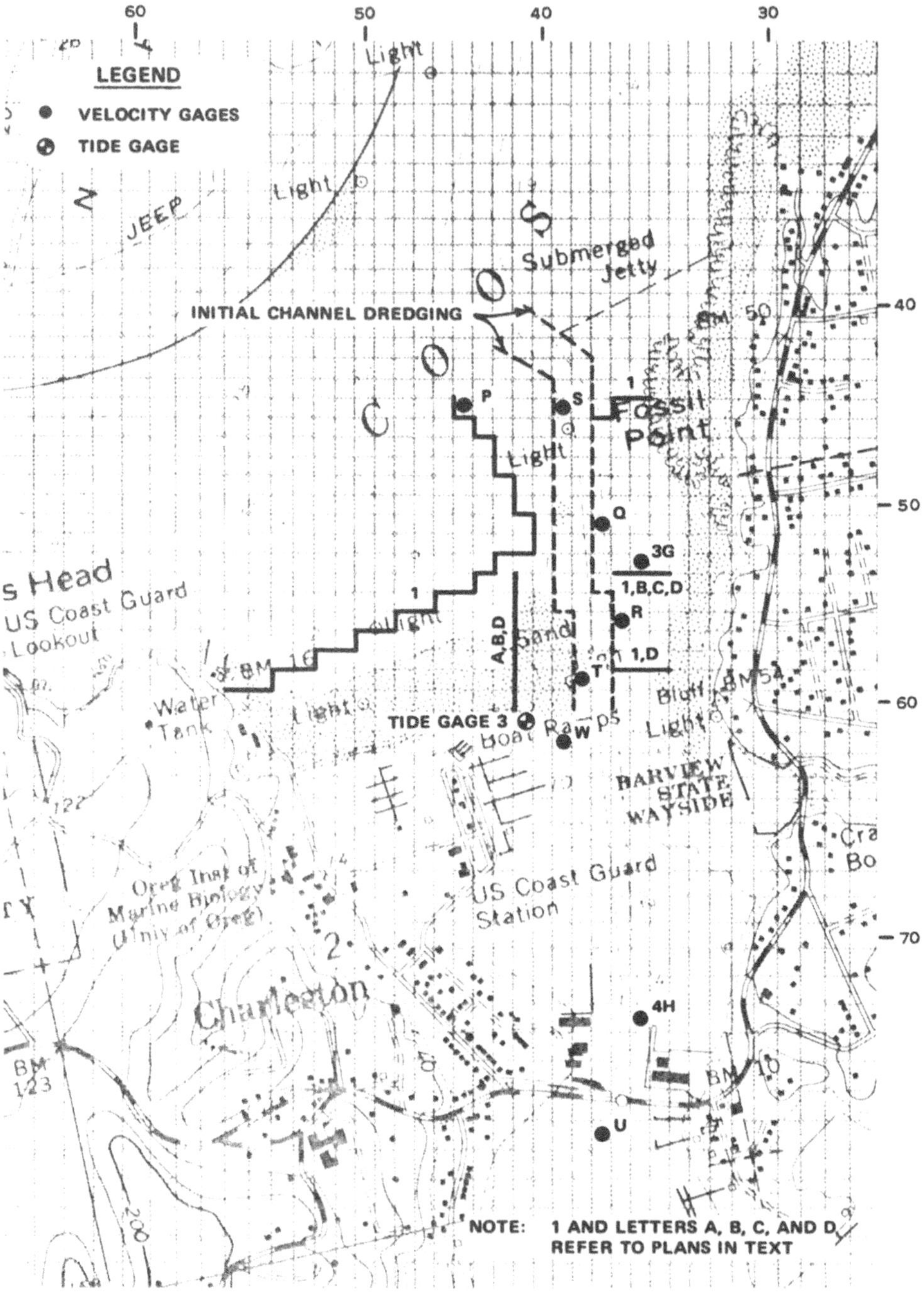

Figure 11. Proposed plans for Coos Bay Inlet with major velocity stations depicted.

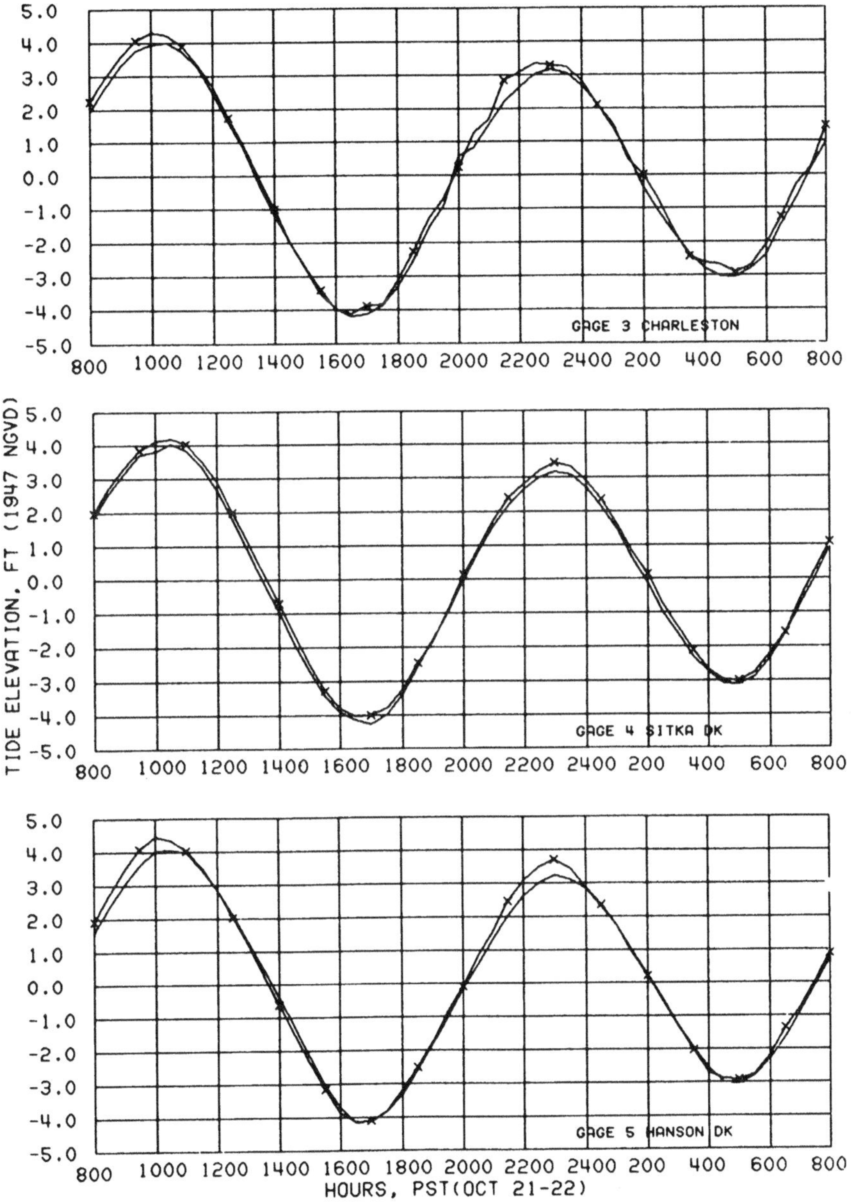

Figure 12. Verification of tidal elevations. Gages 3, 4 and 5

Test Conditions
Ocean tide range 8.2 ft
21-22 Oct. 1976

Legend
——————prototype
x————x model

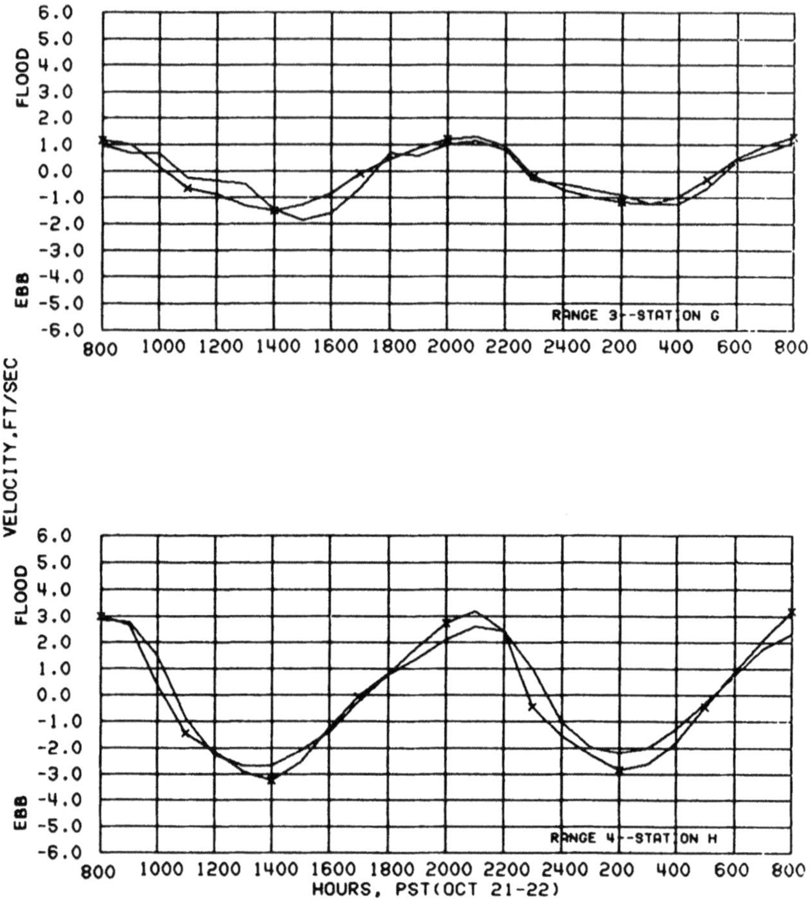

Figure 13. Verification of model velocities. Stations G and H.

Test Conditions	Legend
Ocean tide range 8.2 ft	———— prototype
21-22 Oct. 1976	x——x model

the same windfield model (Jelesnianski, 1965) adjusted to account for wind deformation due to land influence. The astronomical tide was included in the open-coast simulation (and, consequently, in the driving function for the seaward boundary conditions for WIFM) to eliminate uncertainties due to the linear superposition of tide and surge elevations. Figure 19 depicts observed and model results at the Pleasure Pier open-coast Gage 21 and at back bay Gages 24 (Pelican Bridge) and 26 (Texas City Dike-North). Figure 20 presents a comparison of observed high-water levels at various locations throughout the system with those computed by WIFM. Peak elevation data were obtained from a tide gage reading either crest of tide or still high water elevation marks. The mean absolute difference

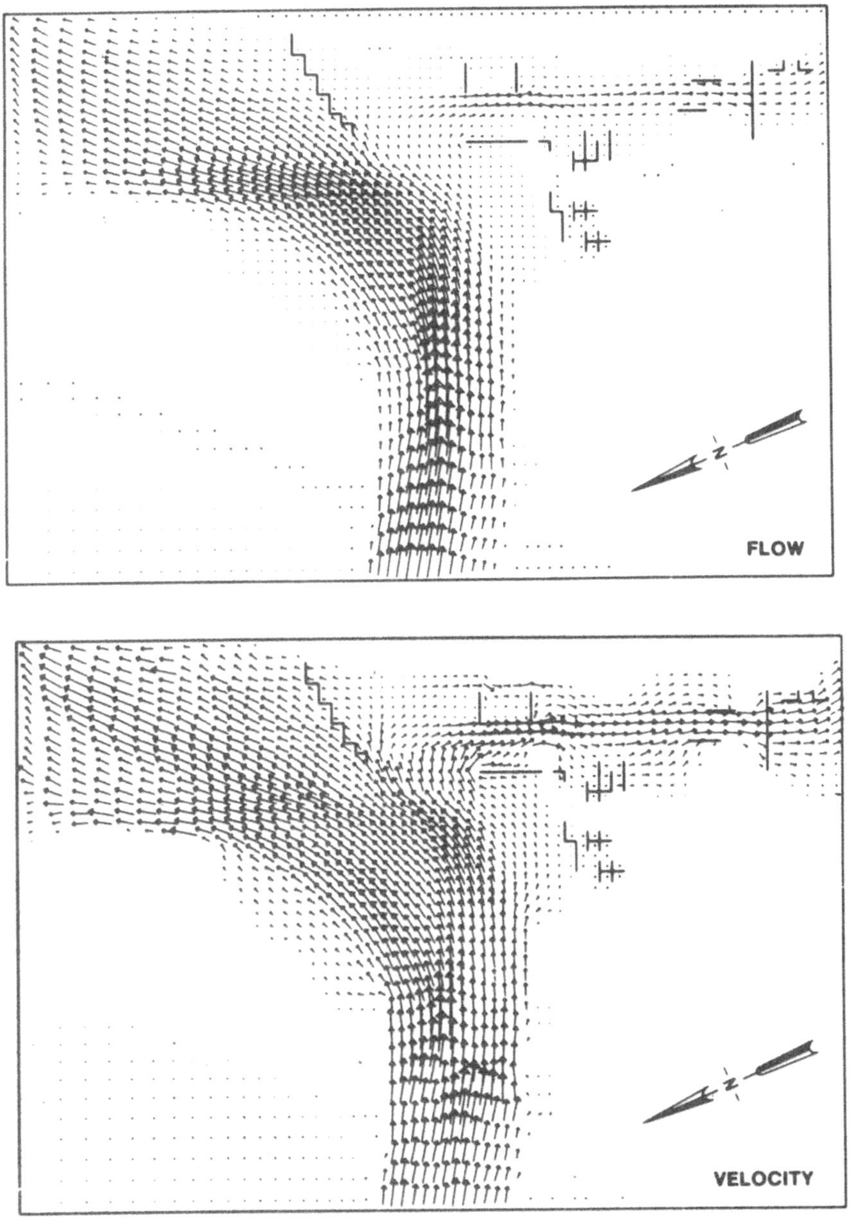

Figure 14. Current Patterns. Plan D. Hour 2100.

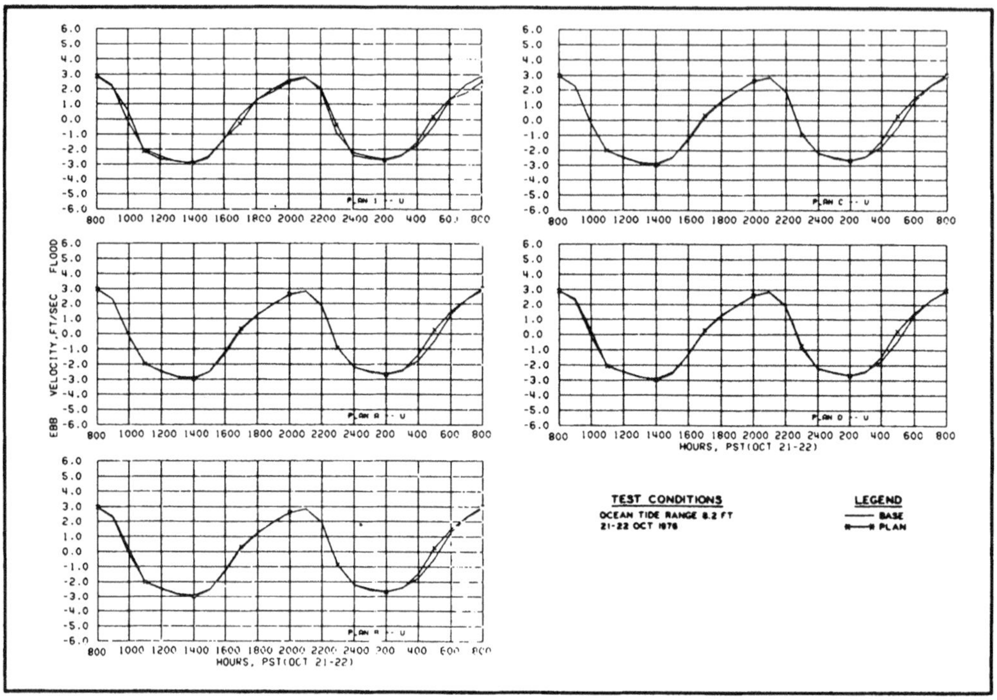

Figure 15. Velocities comparison of plan with base.

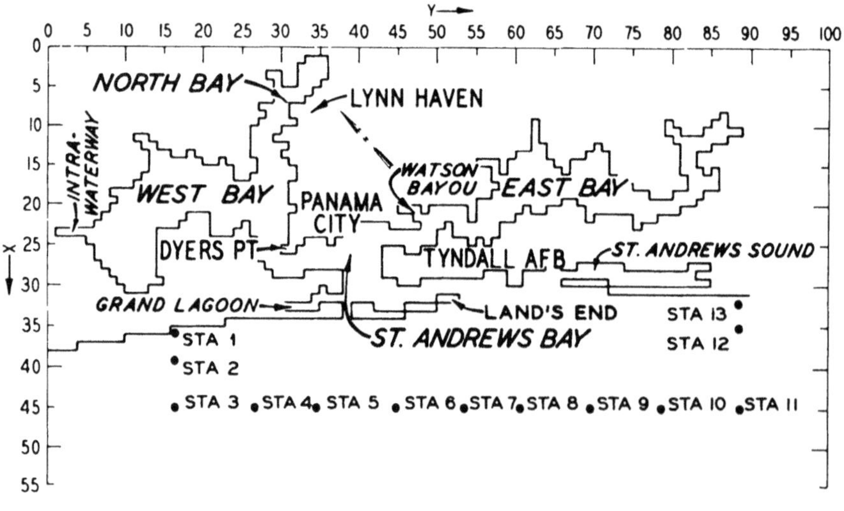

Figure 16. Model limits for Panama City, Florida.

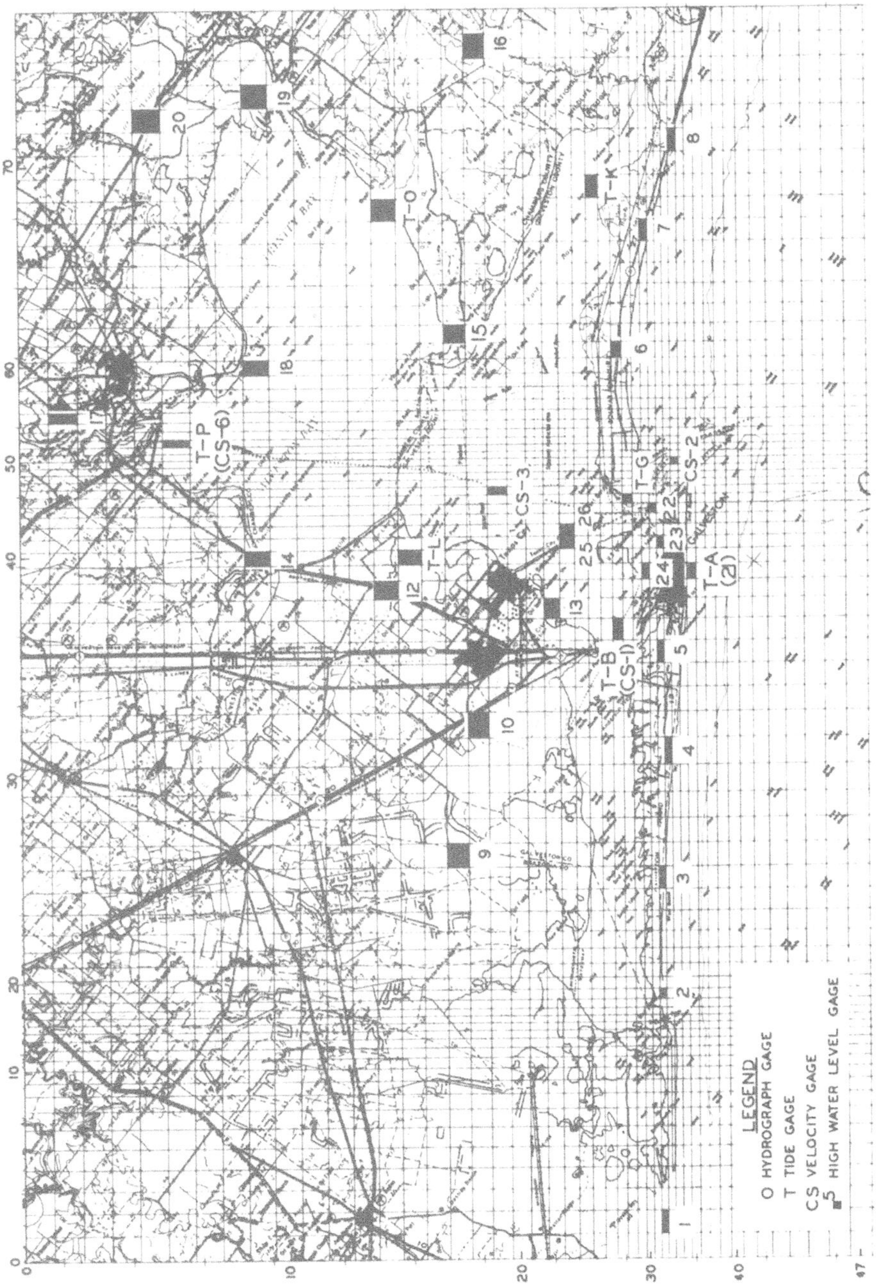

Figure 17. Computational grid and gage locations for Galveston Bay.

Legend
O Hydrograph Gage   T Tide Gage   CS Velocity Gage   ■5 High Water Level Gage

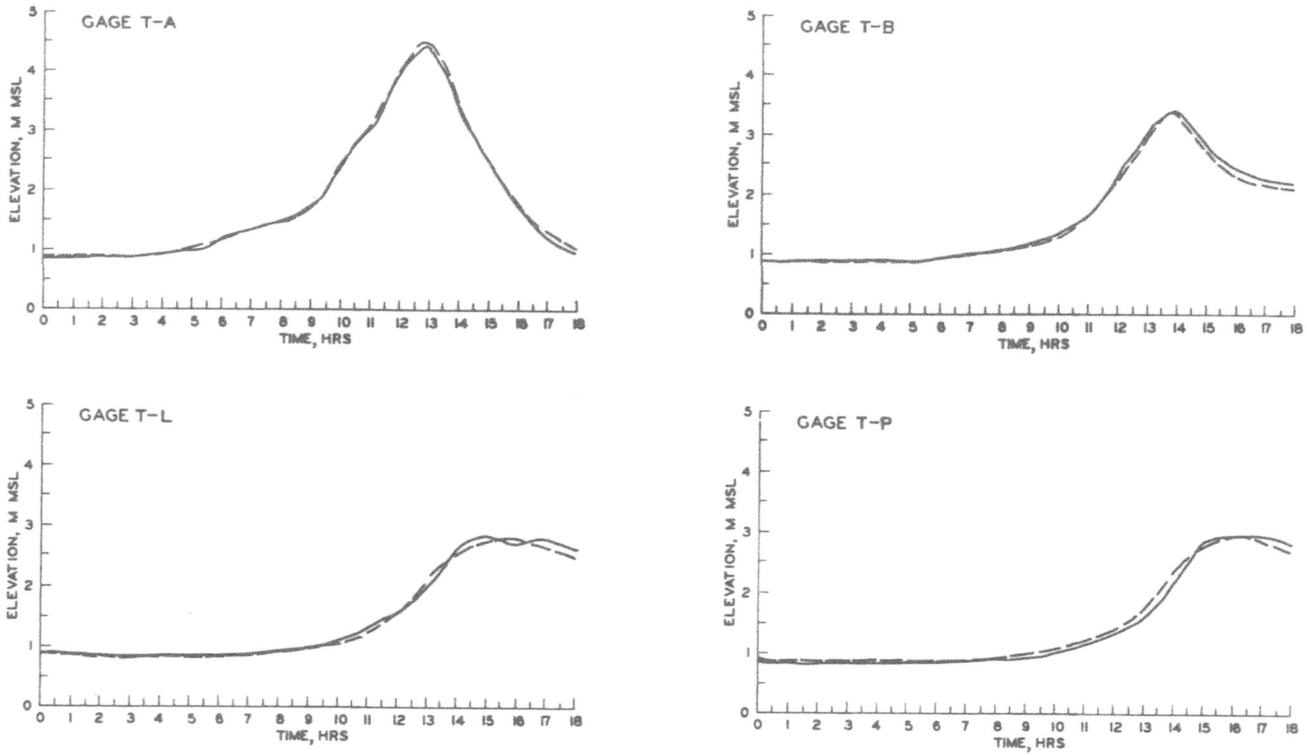

Figure 18. Sample results for large radius high translation design surge.

<u>Legend</u>
----------Physical Model ————Numerical Model (WIFM)

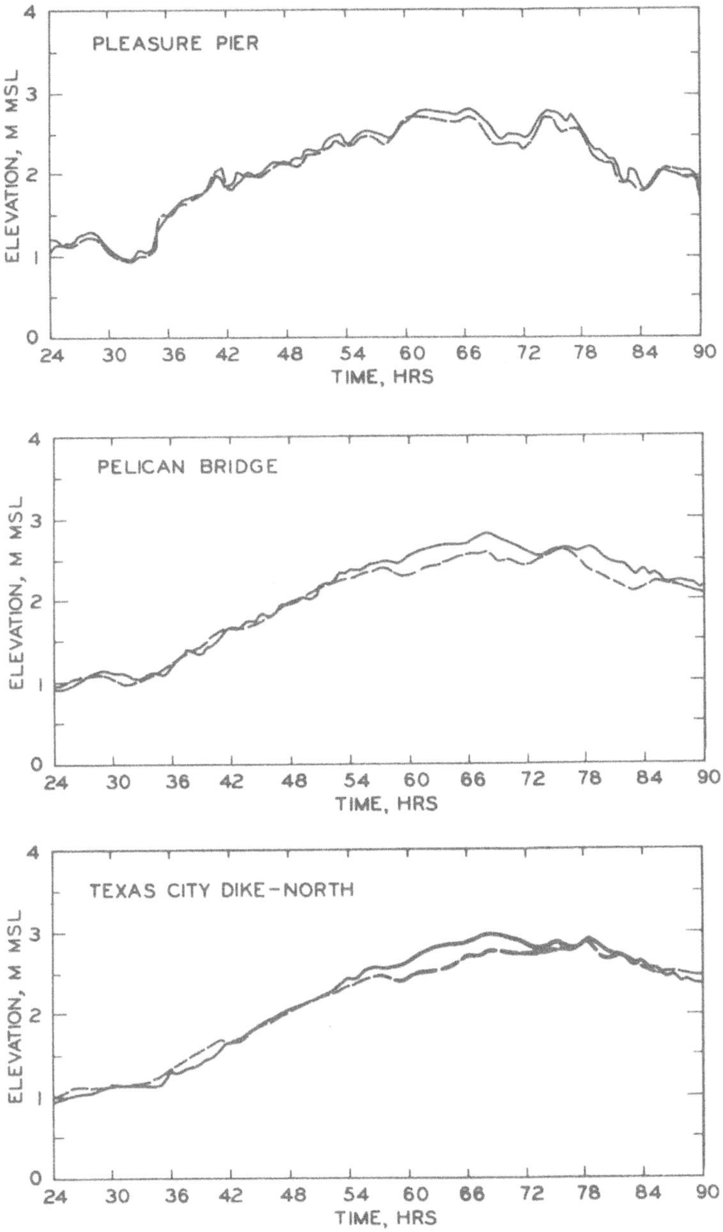

Figure 19. Comparison of computed and observed surge elevations in the Galveston Bay area for Hurricane Carla (Sept. 1961).

<u>Legend</u>

----------Observed Data

——————Numerical Model (WIFM)

COMPARISON OF HIGH WATER LEVELS AT SELECTED GAGE LOCATIONS

GAGE NO.	GAGE LOCATION	OBSERVED m	COMPUTED m	DIFFERENCE m	GAGE NO.	GAGE LOCATION	OBSERVED m	COMPUTED m	DIFFERENCE m
1	OYSTER CREEK	3.11	3.29	+0.18	15	SMITH POINT	2.99	3.17	+0.18
2	SAN LUIS PASS	3.29	3.05	-0.24	16	OYSTER BAYOU	3.20	3.35	+0.15
3	SEA ISLE BEACH	3.69	3.05	-0.64	17	SCOTT BAY	4.33	4.30	-0.03
4	BERMUDA BEACH	3.20	2.99	-0.21	18	HUMBLE DOCKS	4.18	3.84	-0.34
5	SCHOLES FIELD	2.59	2.93	+0.33	19	ANANUAC	3.78	3.87	+0.09
6	BOLIVAR BEACH	2.83	2.83	+0.00	20	WALLISVILLE	4.27	4.26	+0.00
7	CRYSTAL BEACH	2.68	2.87	+0.18	21	PLEASURE PIER	2.83	2.87	+0.03
8	ROLLOVER BEACH	2.93	2.83	-0.09	22	FORT POINT	2.74	2.90	+0.15
9	HALLS BAYOU	4.36	4.30	-0.06	23	PIER 21	2.68	2.90	+0.21
10	HIGHWAY SIX	3.84	3.87	+0.03	24	PELICAN BRIDGE	2.74	2.87	+0.12
11	SIEVERS COVE	3.23	2.83	-0.39	25	TEXAS CITY DYKE (SOUTH)	2.90	3.05	+0.15
12	DICKINSON BAYOU	3.47	3.60	+0.12					
13	CARBIDE DOCKS	3.35	3.17	-0.18	26	TEXAS CITY DYKE (NORTH)	2.96	3.05	+0.09
14	KEMAH	4.33	3.90	-0.43					

MEAN ABSOLUTE ERROR = 0.18 m.

Figure 20.

between observed and computed water levels was 0.18 m.

e. Tsunami Application: There have been few extensive surveys of tsunami high-water marks near the shoreline and detailed measurements of inundation levels in addition to recordings of tsunami wave forms. The 1964 tsunami at Crescent City, California, was quite large, and surveys of the resulting inundation were reported by Magoon (1965). Figure 21 depicts the 65 by 67 grid used in the WIFM calculations. Grid cells are concentrated in the city and harbor area and are oriented such that the incident wave approaches from the direction predicted by refraction diagrams (Roberts and Kauper, 1964). The input to the model is a wave crest that, when propagated to shore, reproduces as accurately as possible the historical maximum elevation at a tide-gage location in the harbor. The model region has many complex features: a harbor which is protected by breakwaters, some of which were overtopped and others which were not; a developed city area; mud flats and an extensive riverine flood plain. Sand dunes and elevated roads played a prominent role in limiting flooding in certain areas. Verification for such a region would indicate the model's applicability for most any region. Figure 22 shows contour lines of

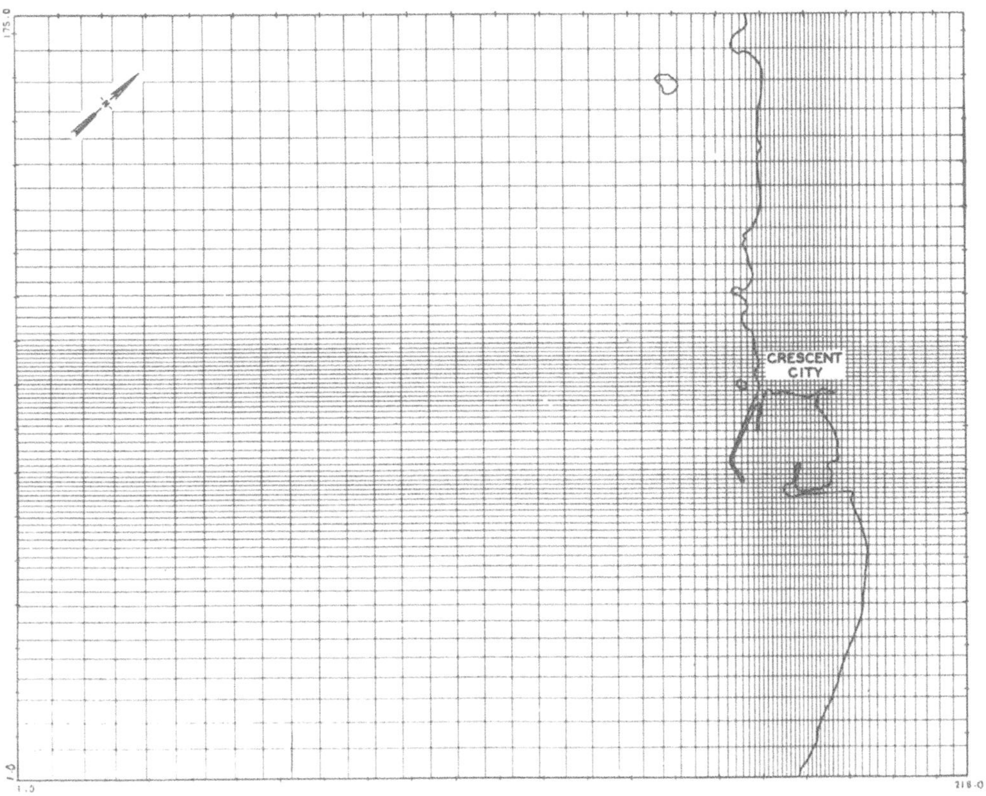

Figure 21. Numerical grid for Crescent City region.

the tsunami elevation above ground level within the developed area of Crescent City. The following tabulation presents the measured elevations at high-water marks and calculated elevations from the model:

High-Water Mark Number (Magoon, 1965)	Measured Elevation mllw (ft)	Calculated Elevation mllw (ft)	Difference (ft)
302	18.35	17.95	+0.40
305	18.74	19.39	-0.65
307	20.70	19.45	+1.25
309	19.84	19.79	+0.05
312	19.41	20.23	-0.82
316	16.29	17.09	-0.80
1	15.90	15.43	+0.47
2	17.80	19.24	-1.44
3	19.30	18.89	+0.41
4	16.50	18.94	-2.44
5	20.50	19.13	+1.37

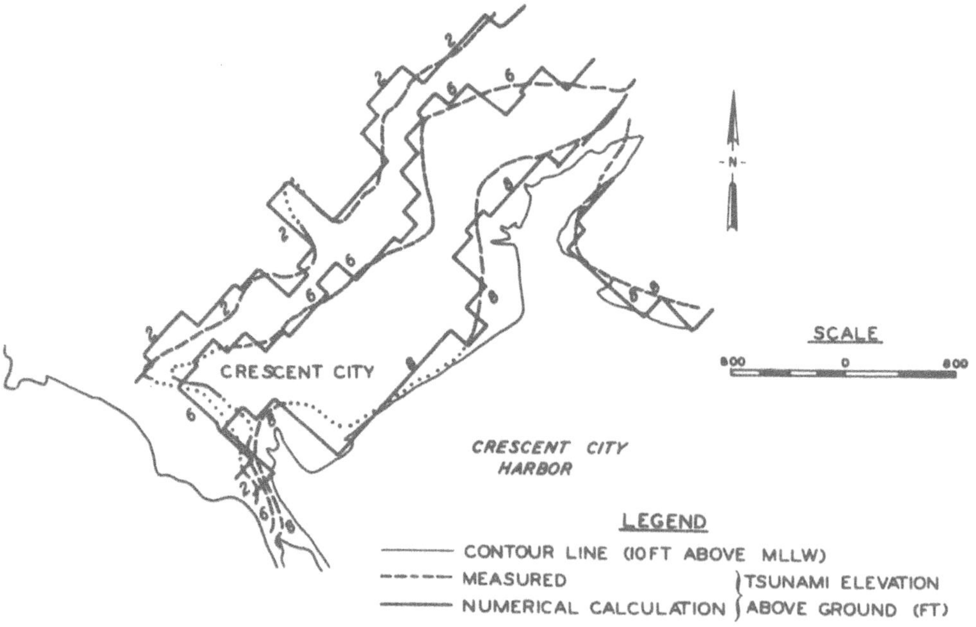

Figure 22. Inundation lines at Crescent City, California, for 1964 tsunami.

Agreement between measured and computed elevations was quite good, the root-mean-square error being 1 ft. To exemplify the use of the model, an application to the Hauula-Punaluu region of the island of Oahu was made to determine flooding produced by 50- and 100-year tsunamis. The results of these applications are reported in detail by Houston and Butler (1979).

SUMMARY AND CONCLUSIONS

This paper presents details of both the development and applications of a two-dimensional finite-difference long-period wave model. The variable grid characteristic of WIFM permits the capability to obtain finer resolution in important local areas without sacrificing economical application of the model. Other features (treatment of flooding, subgrid-scale representations, selection of solution scheme, channel modeling, and so forth) make the model quite general. The efficacy of the model has been demonstrated in a wide variety of applications. WIFM has been extensively tested and shown to be of practical value as a tool for addressing problems associated with various types of coastal projects. By including in future developments the capability of modeling dynamically coupled subgrid channels, embedded variable mesh grid systems within the global grid for detailed resolution of important smaller scale features, and horizontal salinity gradients, the model will have an even wider range of applicability.

ACKNOWLEDGMENT

The work upon which this paper is based was performed for various Corps of Engineer District Offices. The author wishes to acknowledge these Corps offices for authorizing the studies which supported the efforts presented herein.

REFERENCES

Abbott, M. B., discussion of "Review of Models of Tidal Waters," by J. B. Hinwood and I. G. Wallis, Journal of the Hydraulics Division, ASCE, Vol. 102, No. HY8, Proc. Paper 12280, Aug. 76, pp. 1145-1148.

Abraham, G. and Karelse, M., discussion of "Classification of Models of Tidal Waters" and Review of Models of Tidal Waters," by J. B. Hinwood and I. G. Wallis, Journal of the Hydraulics Division, ASCE, Vol. 102, No. HY6, Proc. Paper 12164, June 76, pp 808-811.

Ahlert, R. C., Harlukowicz, T. J., and Nordstrom, K. F., "Hydrodynamic Surveys of Corson and Great Egg Harbor Inlets," TR 76-1 WRE, Rutgers University, New Brunswick, N. J., 1976.

Brogdon, N. J., "Galveston Bay Hurricane Surge Study, Report 1, Effects of Proposed Barriers on Hurricane Surge Heights," Technical Report H-69-12, U. S. Army Engineer Waterways Experiment Station, CE, Vicksburg, MS, 1969.

Butler, H. L., "Numerical Simulation of Tidal Hydrodynamics: Great Egg Harbor and Corson Inlets, New Jersey," Technical Report H-78-11, U. S. Army Waterways Experiment Station, CE, Vicksburg, MS, June 1978.

Butler, H. Lee, "Numerical Simulation of the Coos Bay-South Slough Complex," Technical Report H-78-22, U. S. Army Engineer Waterways Experiment Station, CE, Vicksburg, MS, Dec 1978.

Butler, H. Lee, "Coastal Flood Simulation in Stretched Coordinates," 16th International Conference on Coastal Engineering, Proc. to be published, ASCE, Hamburg, Germany, 27 Aug-1 Sep 1978.

Butler, H. L. and Raney, D. C., "Finite Difference Schemes for Simulating Flow in an Inlet-Wetlands System," Proceedings of the Army Numerical Analysis and Computers Conference, The Army Mathematics Steering Committee, Durham, NC, Mar 1976.

Butler, H. Lee and Wanstrath, J. J., "Hurricane Surge and Tidal Dynamic Simulation of Ocean Estuarine Systems," ASCE 1976 Hydraulics Division Specialty Conference, Purdue University, Indiana, 4-6 Aug 1976.

Flokstra, C., "The Closure Problem for Depth-Averaged Two-dimensional Flow," Paper A 106, 17th International Association for Hydraulic Research Congress, Baden-Baden, Germany, 1977, pp. 247-256.

Hinwood, J. B., and Wallis, I. G., "Classification of Models of Tidal Waters," Journal of the Hydraulics Division, ASCE, Vol. 101, No. HY 10, Proc. Paper 11643, Oct 1975, pp. 1315-1331.

Hinwood, J. B. and Wallis, I. G., "Review of Models of Tidal Waters," Journal of the Hydraulics Division, Vol. 101, No. HY 11, Proc. Paper 11693, Nov 1975, pp. 1405-1421.

Hinwood, J. B. and Wallis, I. G., Closure to "Classification of Models of Tidal Waters," Journal of the Hydraulics Division, ASCE, Vol. 102, No. HY12, Proc. Paper 12586, Dec, 1976, pp. 1776-1777.

Houston, J. R., Carver, R. D., and Markle, D. G., "Tsunami-Wave Elevation Frequency of Occurrence for the Hawaiian Islands," Technical Report H-77-16, U. S. Army Engineer Waterways Experiment Station, CE, Vicksburg, MS., 1977.

Houston, James R. and Butler, H. Lee, "A Numerical Model for Tsunami Inundation," Technical Report HL-79-2, U. S. Army Engineer Waterways Experiment Station, CE, Vicksburg, MS, Feb 1979.

Jelesnianski, C. P., "A Numerical Calculation of Storm Tides Induced by a Tropical Storm Impinging on a Continental Shelf," MWR, 93, 1965, pp 343-358.

Kuipers, J. and Vreugdenhil, C. B., "Calculations of Two-Dimensional Horizontal Flow," Report No. S163, Part 1, Delft Hydraulics Laboratory, Delft, the Netherlands, Oct., 1973.

Lean, George H., and Weare, T. John, "Modeling Two-dimensional Circulating Flow," Journal of the Hydraulics Division, Proc. Paper 14312 ASCE, Vol. 105, No. HY 1, Feb 1979, pp. 17-26.

Leendertse, J. J., "A Water-Quality Simulation Model for Well-Mixed Estuaries and Coastal Seas., Vol. 1, Principals of Computation," RM-6230-rc, Rand Corp., Santa Monica, CA, Feb 1970.

Magoon, O. T., "Structural Damage by Tsunamis," Coastal Engineering Santa Barbara Specialty Conference of the Waterways and Harbors Division, ASCE, Oct 1965, pp. 35-68.

Masch, F. D., Brandes, R. J., and Reagan, J. D., "Simulation of Hydrodynamics in a Tidal Inlet," Water Resources Engineers, Inc., Austin, Texas, Technical Report prepared for U. S. Army Coastal Engineering Research Center, Fort Belvoir, VA, Feb 1973.

Reid, R. O., and Bodine, B. R., "Numerical Model for Storm Surges in Galveston Bay," Journal of Waterways and Harbors Division, ASCE, Vol. 94, No. WW1, Proc. Paper 5805, Feb 1968, pp. 33-57.

Roberts, J. A. and Kauper, E. K., "The Effects of Wind and Precipitation on the Modification of South Beach, Crescent City, California," ARG64 FT-168, Office of the Chief of Research and Development, Washington, D. C., 1964.

Vreugdenhil, C. B., "Secondary-Flow Computations," Delft Hydraulics Laboratory, Publication No. 114, Nov 1973.

Wanstrath, J. J., Whitaker, R. E., Reid, R. O., and Vastano, A. C., "Storm Surge Simulation in Transformed Coordinates, Vol. I - Theory and Application," Technical Report 76-3, U. S. Army Coastal Engineering Research Center, CE, Fort Belvoir, VA, Nov 1976.

Weare, T. John, "Instability in Tidal Flow Computational Schemes," Journal of the Hydraulics Division, ASCE, Vol. 102, No. HY5, Proc. paper 12100, May 1976, pp. 569-580.

Weare, T. John, "Personal Communication on Implicit Finite Difference Schemes," Paper to be published, 1979.

HYDROGRAPHY AND CIRCULATION PROCESSES OF GULF ESTUARIES

George H. Ward, Jr.

Espey, Huston & Associates, Inc.

3010 South Lamar, Austin, Texas 78704

ABSTRACT

Gulf estuaries (excluding the Mississippi) are lagoonal embayments, which, although possessing qualitative features common to most estuarine circulations, frequently exhibit these in extreme ranges or altered importance. These hydrographic features must be considered in developing or applying transport models (and hence water quality models) for these systems. In particular the following factors are generally the most important to bay hydrography: meteorological forcing, tides, freshwater inflow, and density currents. The bays are sensitive to meteorological forcing, especially relative to the feeble tidal effects. Among the important meteorological influences are windwaves, large-scale wind-driven gyres and flushing due to frontal passages. Freshwater inflows are highly transient and are important in establishing salinity gradients. Insofar as general water-quality considerations are concerned, the density current affects the large-scale circulation and transport within the bay, and is extremely important when the bay is transected by deepdraft ship channels (as are most of the Gulf estuaries). Mathematical water quality (including salinity) models usually parameterize the density-current transport by an inflated dispersion coefficient, however this approach is poorly founded theoretically, and for bays can lead to large errors in the water quality predictions. Examples are presented to display the characteristics and significance of each of these factors, and available modeling techniques (both physical and mathematical) are appraised with respect to each.

INTRODUCTION

The northern coastline of the Gulf of Mexico extends in a 2300-km roughly semicircular arc from the Florida Keys on the east to the mouth of the Rio Grande on the west. Across this coastline into the Gulf of Mexico flows about 51% of the riverine discharge of the coterminous United States, and approximately 73% of this is carried by the Mississippi River. Along the Gulf shoreline lie numerous indentations, embayments, and inlets, which collectively constitute perhaps the most valuable estuarine resource of the United States.

Most of these estuaries exhibit a surprising degree of similarity in their general hydrographic characteristics. The purpose of this paper is to review these hydrographic characteristics and the attendant circulation processes. The approach is generic, delineating those factors of prime importance to the hydrography of these waterbodies and illustrating their role and range by specific examples. The intent of the paper is a summary, so that only selected literature references are provided. (One of the greatest impediments to progress in the study of these systems -- especially -- is that much of the literature is gray, contained in reports of limited distribution or in filing cabinets of varying obscurity. Although much of this is valuable, it is uncirculated and uncatalogued, therefore basically inaccessible. An exhaustive literature review, identifying the content and availability of these sources, would be of inestimable value.)

The focus of this paper is upon the Gulf embayments. The estuary of the Mississippi River is excluded: as a continental drainage estuary, it is a unique system requiring separate treatment. The few rivers which discharge directly to the Gulf, such as the Brazos and Suwanee, are not considered, as they have estuarine reaches with properties fundamentally of the classic river-mouth estuary. Finally the sounds and bights (such as Apalachee Bay) clearly represent distinct hydrographic systems and are therefore excluded; it could be debated whether these are indeed estuaries.

MORPHOLOGY, HYDROLOGY, AND CLIMATOLOGY

The embayments of the Gulf of Mexico are broad, shallow lagoons with characteristic horizontal dimensions on the order of several tens of kilometres and depths on the order of 1 to 5 metres. Connection with the Gulf of Mexico is generally limited by the presence of barrier islands and peninsulas which fringe the Gulf shoreline. Exchange with the Gulf takes place either through relatively narrow inlets, which in their natural state exhibit significant flood and ebb bar development, or through transient storm washovers. Many of the bays are transected by narrow navigation channels with depths on the order of 10-12 m and widths in the range of 50-150 m. Although a network of channels constitutes a negligible proportion of its bay volume, it has profound hydrographic effects, as will be seen.

Figure 1 displays the northern Gulf of Mexico coast with the major embayments indicated. On a gross scale, freshwater inflow to the bays varies from a maximum approximately in the center of the Gulf coastline, near Mobile Bay, to a minimum at either end in Florida and Texas. The Mobile River, actually the confluence of the Tombigbee and Alabama Rivers, exhibits the largest discharge to a bay system, an annual average of about 1800 m^3s^{-1}. In Table 1 are tabulated indicators of the freshwater runoff to several of the major bay systems around the periphery of the Gulf, viz. the long-term mean discharge (see, e.g., A. Wilson, 1967), the discharge per unit watershed area, and the (reciprocal) discharge per unit bay volume. Discharge per unit watershed area is in fact a measure of basin aridity, being influenced by the precipitation surfeit (i.e., precipitation less evaporation), and, to a lesser extent, infiltration and evapotranspiration in the watershed. It is apparent that the Gulf coast exhibits a wide range of aridity from the extremely humid areas of Louisiana, Mississippi, and Alabama to the arid areas of south Texas and south Florida. The ratio of bay volume to discharge of course measures the relative influence of freshwater inflow on a specific bay system, and has dimensions of time; one might be tempted to regard this as the freshwater replacement time for the bay (as it would be so regarded for, say, a lake) but this simple physical interpretation is invalidated by the more complex hydrodynamics of these systems. Sabine Lake receives the largest freshwater inflow per unit bay volume on the Gulf.

The Gulf coastal climatology is characterized by southerly to southeasterly prevailing winds, governed by the intensity of the circulation about the Azores-Bermuda High. The prevailing winds are therefore onshore for much of the Gulf coastline from approximately Corpus Christi Bay in Texas to Apalachee Bay in Florida, this onshore flow providing the single most important source of atmospheric moisture to the continental United States. This prevailing flow is most pronounced in the summer when the High is strongest and furthest north. In winter, the Bermuda High weakens considerably, and with the southward migration of the climatological equator the Gulf falls under the increasing influence of disturbances in the midlatitude westerlies. Figure 1 displays prevailing wind direction and mean speeds which typify the variation of prevailing wind about the Gulf periphery.* During the equinoctial seasons and even more so in winter, the coastal zone becomes subject to the weather systems of the midlatitude westerlies. Intrusions of polar air thrust deep into the region, sometimes decelerating and stalling

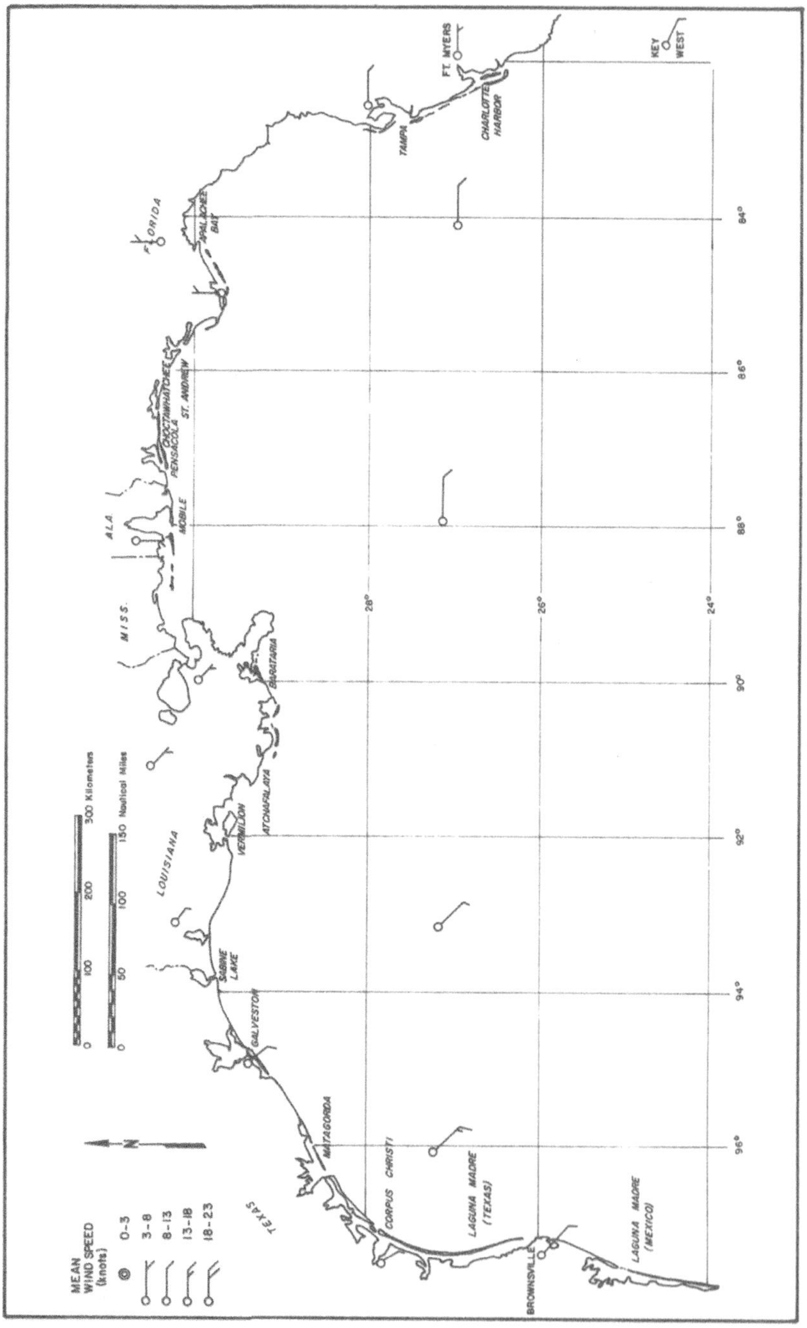

Figure 1. Gulf of Mexico, northern coastline

TABLE I

PHYSICAL, HYDROLOGIC AND TIDAL CHARACTERISTICS OF SELECTED GULF BAYS

Bay	Surface Area (km^2)	Volume $V(10^6 m^3)$	Main Rivers	Mean Discharge $Q(m^3 s^{-1})$	Drainage Area $DA(km^2)$	Q/DA (cm/yr)	V/Q (days)	Tidal Prism* $(10^6 m^3)$	$\frac{K_1+O_1}{M_2+S_2}$**
Tampa	880	2900	Hillsborough, Alafia, et al.	40	5800	22	800	-	1.3
Choctawhatchee	-	-	Choctawhatchee	210	12600	53	-	45[a]	-
Pensacola	265	840	Blackwater, Yellow and Escambia	300	22900	41	30	230[a]	9.1
Mobile	1020	3060	Tombigbee and Alabama	1820	114000	50	20	760[a]	7.5
Barataria/ Caminada	490	530	-	-	-	-	-	90[a]	10.7
Sabine Lake	240	440	Sabine and Neches	490	48900	32	10	90[b]	1.3
Galveston	1420	4300	Trinity and San Jacinto	275	56500	15	180	350	1.6
Matagorda	1070	3300	Lavaca & Navidad Colorado***	30 85	11900 110000	8 2	1300 450	250[b]	-
Corpus Christi	435	1340	Nueces	25	41100	2	620	60[a,c]	-

*Great declination

**Contiguous Gulf

***A significant - but unknown - portion of the Colorado discharge enters Matagorda Bay through Parkers Cut

[a]Average of values reported by Jarrett (1976)

[b]Cubature using tide records

[c]One-half of prism of Aransas Pass

over land to dissipate as zones of low-level convergence, and sometimes surging far over the Gulf as a stable dome of cold air. In summer the climate becomes almost tropical in character, punctuated by convective airmass thunderstorms and infrequent synoptic-scale tropical disturbances, e.g., easterly waves and hurricanes. The interaction between moist, convectively unstable Gulf air and frontal systems associated with midlatitude disturbances constitutes a primary mechanism for precipitation production. This interaction is maximal during the equinoctial seasons, especially spring. Most areas of the Gulf therefore exhibit a spring maximum in rainfall, and hence in freshwater runoff, and a few regions exhibit a bimodal annual rainfall distribution with maxima in both spring and fall. The western and northwestern segments of the Gulf also receive large amounts of precipitation from tropical disturbances in late summer and early fall, further enhancing the fall maximum. For a few areas, the eastern Gulf for instance, convective airmass showers result in a summer maximum.

*The prevailing northerlies in Florida are an artifact due to concentration of frontal winds about the N compass point but dispersion of the truly prevailing wind among S, SE, and E. A windrose is a preferred method of display for this reason, but windrose data were not available for inclusion in Fig. 1.

CIRCULATION PROCESSES

In the present context, hydrography is limited to hydrodynamic elements, i.e., currents and current structure in the embayment and the distribution of density. Density in the Gulf bays is dominated by salinity, also a valuable indicator of fresh-salt water interaction. Hydrographic characterization of these systems is approached through enumerating the principal hydrographic factors and the response of the bay systems to those factors.

Tides

Tidal variation in the Gulf of Mexico is feeble relative to that of the Atlantic and Pacific coasts. Tidal range seldom exceeds 1 metre and more typically is of the order 0.3-0.7 m. Further, the tide is dominated by the diurnal components K_1 and O_1, the result of an apparent resonance with the period of free oscillation of the Gulf (about 24 hours). The exception is the northeast coast of the Gulf approximately from above Tampa Bay to Apalachee Bay, where there is a remarkable increase in the amplitude of the M_2 component producing a predominantly semi-diurnal tide with a range of as much as 1.5 m. Some appreciation for the variability in tidal characteristics around the Gulf may be derived from the Formzahl $F = (K_1 + O_1)/(M_2 + S_2)$ typical values of which are tabulated in Table 1. More information on the harmonic analysis of Gulf tides may be found in Marmer (1954). As most tides in the Gulf are dominated by diurnal components, the variation between spring and neap is less important than the variation between tropical and equatorial tides, so that it is the declination of the moon which primarily governs tidal range.

The K_1 component appears to propagate in through the Strait of Florida thence around the Gulf cyclonically and out the Strait of Yucatan (see Grace, 1932). However, the total phase lag between Tampa Bay and the Mississippi Delta is only 0.3 radians, and is much less than this for the northwest Gulf thence to the Rio Grande. Thus, for practical purposes, the Gulf Coastline can be regarded as cotidal for the diurnal tide. The M_2 tide appears to have more the character of a standing oscillation about a nodal line across which there is a π phaseshift. Away from the nodal line (which is actually a zone of densely spaced cotidal lines), the M_2 tide appears to be virtually cotidal on either side (Grace, 1933).

As an index to their bulk tidal dynamics, Table 1 includes estimated tidal prisms for some of the Gulf bays. Within the bays, the tide propagates through the passes as a progressive wave, with both tidal stage and current in phase. As the effects of the limit of tidal influence, hence storage, begin to be felt the tide acquires a standing component which becomes increasingly prominent toward the head of the estuary. (Frequently, the correspondence between tides and tidal currents is naively described according to standing wave behavior, with high and low waters concurrent with slacks in the current. This is usually invalid, but depends upon position in the system.) The precise behavior of tidal amplitude and phase within a bay system is dependent upon its physiographic configuration. Generally, friction in the shoal water attenuates amplitude and lags phase, but if the convergence of volume is sufficiently rapid, as is the case in Mobile Bay, a zone may occur where tidal amplitude increases. Figure 2 displays the alteration in tide through Galveston Bay, which possesses a mid-bay constriction where is situated an extensive reef-shoal system, Red Fish Bar. The effect of this constriction is much like a gigantic weir, with significant attenuation in the tide across the Bar and almost complete filtering of semidiurnal components. Above Red Fish Bar, the tide becomes virtually diurnal and cotidal, though of considerably reduced amplitude.

Meteorological Influences

As a general rule, the bay systems of the Gulf of Mexico are

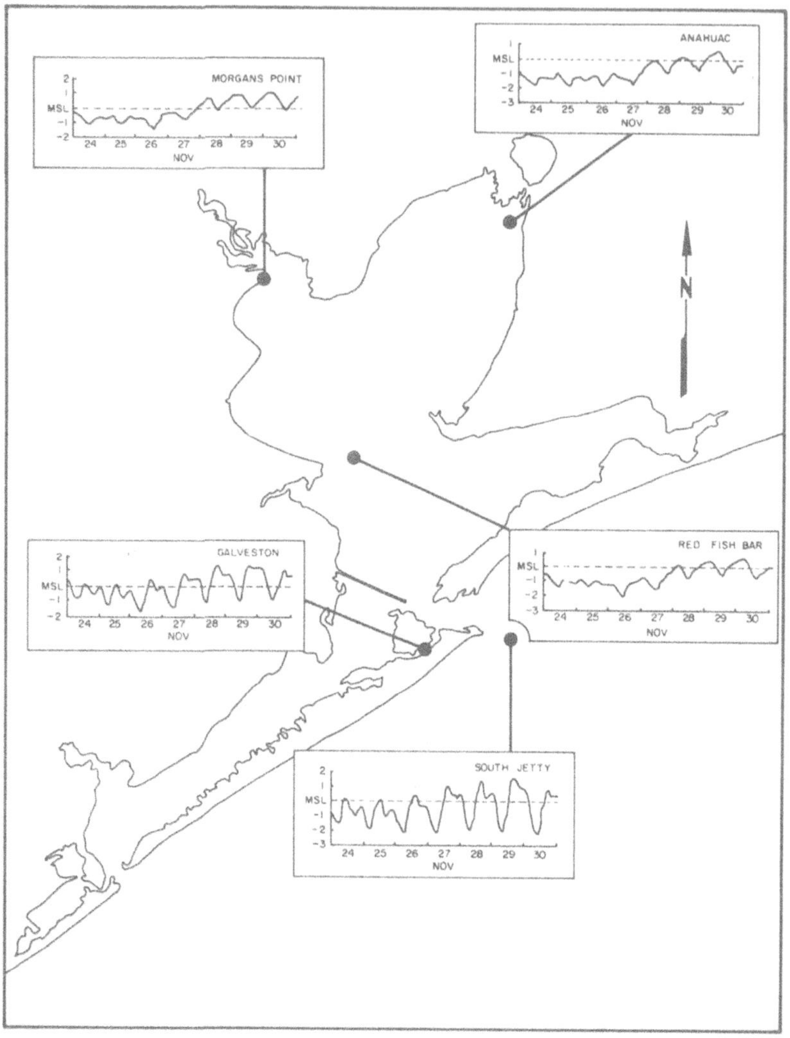

Figure 2. Tidal attenuation in Galveston Bay (1963 data, in ft)

meteorologically dominated. This is a consequence of the large surface-area-to-volume ratios of these bays, and the intensity of meteorological forcing compared to the relatively feeble astronomical tide.

Perhaps the most obvious meteorological effect is the wind-driven waves. As a consequence of their large surface areas most of the bays present long overwater fetches, permitting the development of intense surface waves under even a light-to-moderate wind. The shallow depths of the bays render these waves efficacious mixing agents, reducing or eliminating vertical gradients in constituent concentrations. It is also noteworthy that windwave action is a primary erosive agent.

Another meteorological effect is the tilting of the water surface due to transport of water by an imposed windstress. Such set-up and set-down is so common and dramatic (along the northwest Gulf Coast in particular) that the resulting "wind tides" frequently obscure the astronomical tide. The occurrence of wind set-up and set-down is usually precipitated, not so much by a strong wind *per se* as by a change in wind, notably that accompanying a frontal passage. Although the characteristics and morphology of frontal systems are highly variable, a general scenario may be sketched as follows. As a polar outbreak traverses the Great Plains (say), low-level convergence in the frontal zone enhances the normal southerly flow. The resultant onshore winds elevate water levels in the upper parts of the bays and along the coast. This forces a volume of water through the passes into the bays. With passage of the front, pressure increases inland and the wind turns to the north, depressing water levels in the upper bays and Gulf thus discharging a volume of water through the passes into the Gulf. Total excursion in water level in the bay systems can be in excess of a metre within a short period of time.

Despite the importance of the frontal phenomenon to the hydrography of the bays, little quantitative data is available. This is because, first, the encountering of a front during a scheduled current measuring program is largely serendiptitious*, and, second, the inclement weather and sea conditions render data collection difficult to the point of abandoning the program. Some information is available from the analysis of water-level records sufficient to typify the magnitude and importance of the phenomenon. For example, Figure 3 shows recorded water levels in Galveston Bay which exhibit its response to a frontal passage of 3 November 1936. Total recorded depression of the water surface at Red Fish Bar was one metre and at Anahuac nearly 1.2 m; further this set-down completely obscured the tide whose range is normally about 0.3 m at these gauges. The volume of water evacuated from Galveston Bay was computed by Gilardi (1942) from a dense network of tide gauges to be 1.14×10^9 m^3 and from detailed time series of current measurements in the bay inlet to be 1.58×10^9 m^3. This is more than triple the tidal prism of Galveston Bay.

*Field workers would probably choose a different adjective.

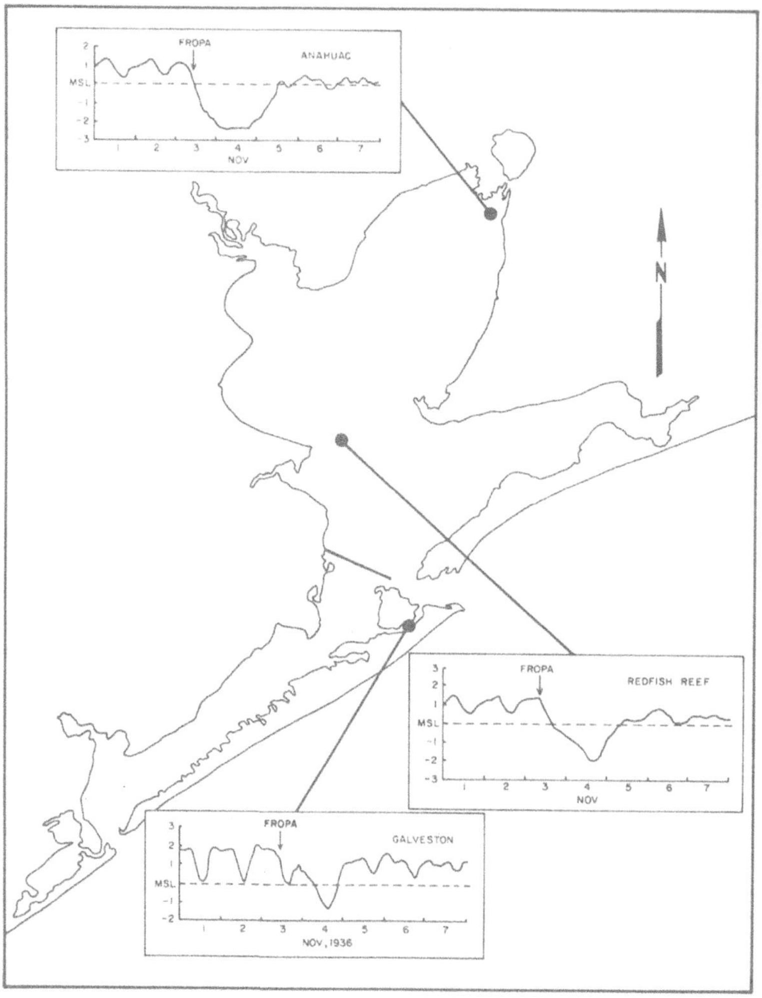

Figure 3. Frontal response in Galveston Bay (1936 data, in ft)

Ward and Chambers (1978) report a field study mounted in Sabine Pass, the 8-km inlet connecting Sabine Lake to the Gulf of Mexico, for the specific purpose of characterizing the response of this system during an energetic frontal passage. Sabine Lake is particularly well-suited for such a study since it has a prominent longitudinal dimension of about 18 km aligning roughly north-south so that the system is quite responsive to both prevailing southerly winds and frontal northerlies. From an analysis of the frontal system of 9 December 1977 and three historical frontal passages of similar morphology, the essential features of the response of Sabine Lake to the front were found to consist of three regimes: (1) a period of influx due to enhanced southerlies as the front approaches the coast, during which water levels are set-up in the northwest Gulf and in the bay; (2) immediately following the frontal passage, a plunge in water level occurring rather abruptly, in one to two hours; (3) following the minimum point of the frontal plunge, a resurgence, then another drop in water level to approximately the first minimum, followed by a second resurgence that returns the water level to approximately the prefrontal elevation. The initial plunge was concluded to be a forced response while the resurgences appeared to have more the character of a Gulf seiche superposed upon a monotonic rise in water level. Approximately one-fourth of the water level response was determined to be forced by inverse barometer, and the remainder by windstress effects. Most important perhaps was the observation that the frontal plunge and resurgences at the Gulf lead similar responses within the southern part of Sabine Lake which in turn lead those in the northern part of the bay, implying that the response of the bay is not directly induced by the windstress on the bay but rather is indirectly produced by water level variations in the Gulf. The volume of water debouching through Sabine Pass in this case was approximately 1.7×10^8 m^3, a volume nearly double the tidal prism.

The character of the inlets is important in modulating the response of a bay system to meteorological forcing. Both Galveston Bay and Sabine Lake are connected with the Gulf of Mexico by energetic tidal inlets. As an example of a different type of response, San Antonio Bay has much more restrictive communication with the Gulf of Mexico, directly only through a narrow washover inlet (Cedar Bayou) and indirectly through passes connecting the adjacent bay systems. In Figure 4, the response of water level in San Antonio Bay to frontal passages is shown by records from tide gauges in its northern and southern ends. There is a nearly simultaneous, almost antisymmetric response, consisting of water-level depression at the upper gauge and elevation at the lower gauge. In this case, then, water level is tilted across the bay by windstress, much more like the behavior of a totally enclosed waterbody.

An important meteorological effect, particularly with respect to transport of waterborne constituents, is the development of large-scale wind-generated circulation patterns. Unlike tidal currents which are essentially oscillatory and water level set-ups due to fronts which are random in time and of varying intensity, large-scale wind-driven gyres are semi-permanent elements of the circulation and can be an important determinant in the distribution of tidal-mean currents. Unfortunately, little is known about this element of bay circulation. In most of the Gulf systems, existing

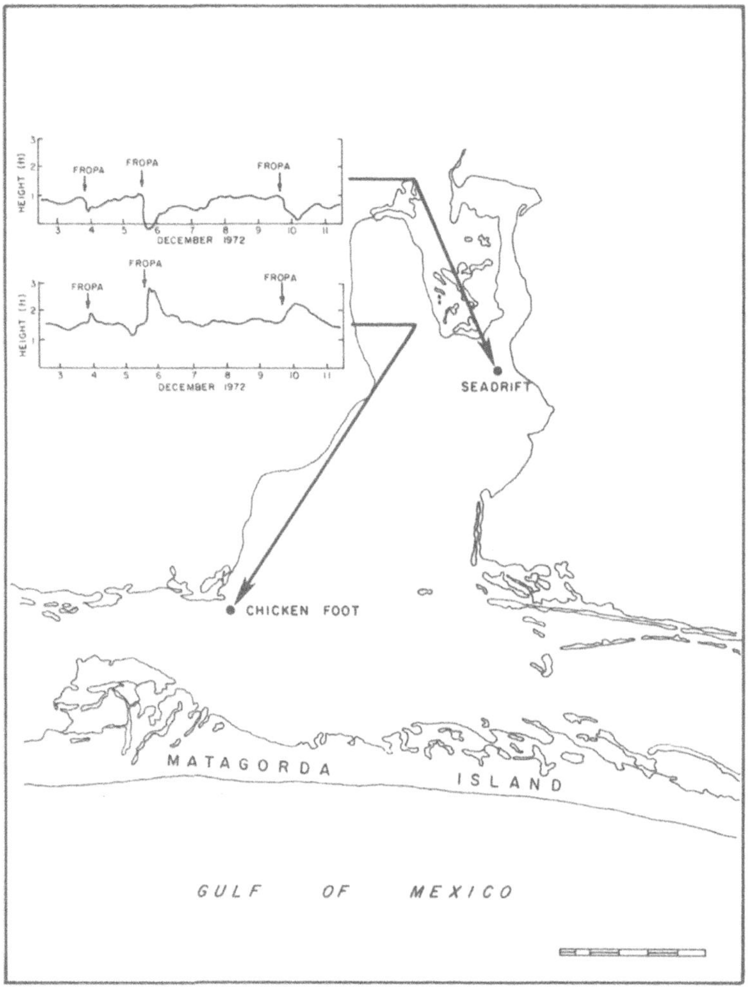

Figure 4. Frontal response in San Antonio Bay

current measurements are inadequate to establish even the sense of the large-scale circulation, much less its intensity. It is evident that the requirements of a data program with this objective should include fairly dense sampling in space sustained over a sufficient period of time to permit averaging to expose the large-scale mean circulations. Some information has been obtained from movement of indicator parameters in the bay systems and from numerical models of bay hydrodynamics. For example, von Deesten (1924) reports a program of intensive salinity measurements at a dense network of stations throughout the Sabine Lake system, from which he was able to discern patterns of salinity intrusion and extrusion to sufficient resolution to determine the sense of the main wind-driven circulations. He found that under the prevailing southeasterly winds the general circulation is clockwise in the main bay system but reverses to counterclockwise under frontal north winds. Other smaller gyres probably occur as well, but could not be resolved from the salinity data.

Freshwater Inflow and Salinity Distribution

The feeding rivers are of course responsible for the estuarine character of the Gulf bays. The most prominent feature of the inflows is their extreme variability, ranging over several orders of magnitude, from drought stage to flood. The year-to-year variability is almost as dramatic, which renders it difficult to display a "typical" year. The larger scale variations, i.e., the influxes of runoff in the spring or fall, are produced by the climatological variation of rainfall described above. Even during the dry seasons, however, local convective thundershowers generate sharp pulses of runoff.

The direct influence of the rivers on bay circulation is limited to moderate-to-high flows (relative to the volume of the bay), when the inflows can drive significant (throughflow) currents. However, important to the general circulations of all of the bay systems, even in the arid areas of the Gulf, is the role of freshwater inflow in establishing a horizontal gradient of salinity. This gradient can be maintained by even low levels of freshwater inflow, and gives rise to the development of density currents.

Quantitative interbay comparisons of salinity are difficult, if not impossible, due to the inhomogeneity of the data bases from which a "typical salinity" is to be deduced. The wide variation in the types and strategies of salinity sampling programs for the bays results in differing periods of record, horizontal and vertical spatial densities, and sampling intervals. Differences in each of these can significantly bias an average salinity, however computed. Qualitatively, the salinities of the Gulf bays vary opposite the quantity of freshwater inflow received by the bay relative to its volume (see Table 1). Increasing aridity, together with the shoal depths of the bays, makes evaporation an increasingly important part of the bay water budget. The decreasing volume of freshwater inflow and the semi-isolation of the south Texas bays, in particular, produce summer salinities ranging well above those of the contiguous Gulf of Mexico. The extreme example is of course the Laguna Madre during the summer low-flow regime, with salinity typically double that of the Gulf.

Although vertical stratification in the open areas of the Gulf bays is reduced substantially by the intensive windwave mixing, it is not altogether absent. Those bays which are deeper and also receive large quantities of freshwater inflow seem to be capable of developing stratification, for example, both Mobile and Pensacola Bays. The data of McPhearson (1970) indicate surface-bottom salinity differences (annual mean) on the order of 3-4 °/₀₀ at stations with depths about 4 m, and 0.5-1 °/₀₀ at stations in 1-m depths. Pronounced stratification is indicated in the ship channel stations. Olinger et al. (1975) report surface-bottom salinity differences on the order of 10°/₀₀ throughout Pensacola Bay during the period of January-September 1974. This much stratification may be an anomaly of the sampling period -- the prevailing wind was northerly for this sampling period -- but clearly indicates the proclivity of the system to develop stratification.

A remarkable example of the interaction of wind, tide and stratification is the Mobile Bay jubilee, in which a combination of east wind and flooding tidal current transports bottom waters of Mobile Bay to the eastern shore. Waters of depths exceeding about 4 m are sufficiently stratified that oxygen is depleted near the bed. As the oxygen-depleted water mass moves to the east, large numbers of fish and crustaceans are driven toward the shore where they are easily harvested by the locals. May (1973) describes the phenomenon in detail.

Galveston Bay, in contrast, exhibits stratification in the open bay only rarely, generally during freshets, even in 4-m depths (see, e.g., Huston, 1971). Stratification in the Huston Ship Channel is not nearly so marked as that in the Mobile Ship Channel, a typical surface-bottom salinity difference ranging 5-10 °/₀₀. A similar situation occurs in Matagorda and Corpus Christi Bays (see, e.g., Ratzlaff, 1976), though the ship channel stations in these bays appear to exhibit even less pronounced stratification than in the Galveston system, evidencing near-homogeneity a significant portion of the time. Vertical stratification is virtually absent in Barataria Bay (e.g., Gagliano et al., 1970) which receives a substantial freshwater inflow but is quite shallow. Sabine Lake rarely stratifies in the open bay, even during freshet events, with surface-bottom differences of at most 1-2 °/₀₀ being occasionally reported (e.g., Ward, 1973; Ratzlaff, 1976).

Generally, freshwater inflows to the Gulf bays are transient and the usual state of the hydrographic structure of the bays is a dynamic response to these transient inflows. This accounts for the apparently paradoxical situation that, for many bays, given any particular inflow one can find in the historical data a wide range in salinities, although one's intuition might dictate a closer association of salinity and inflow regime. This characteristic transience of the hydrographic structure renders analysis of these systems, particularly from field data, very difficult. One useful way of getting around the transient character of the system and its associated dynamic response is to focus upon an event in which the levels of freshwater inflow are sustained for a sufficient period of time to permit acquisition of a tidal-mean equilibrium structure. (Indeed, many water quality analyses are carried out on the basis of a so-called steady-state regime which implicitly assumes such an equilibrium, though these analyses rarely include consideration of the hydrologic conditions necessary to produce this equilibrium.)

Central is the concept of a sustained flow event, a period of prescribed duration during which inflows to the system remain within a certain tolerance of their mean values for the period. Details of the definition and hydrologic analyses for sustained flow events are given in Ward (1979); for present purposes, the defining parameters may be considered to be the event duration and variation tolerance. Specification of the duration is dictated by the response time of the hydrographic variables of concern (e.g., currents or salinity structure) while the tolerance criterion is dictated by the "integration capability" of the hydrographic structure, i.e., the temporal variance in inflow that can be tolerated by the equilibrium of the system. At present, there is very little theoretical basis for delineating these factors, so that hydrographic analysis of a system must proceed hand-in-hand with its hydrologic analysis. Of primary importance is the ability to explain observed behavior. Figure 5 illustrates a 30-day sustained-flow event analysis for the Lavaca River period-of-record in the Matagorda Bay system. This diagram effectively compresses twelve monthly cumulative-frequency ogives by displaying contours of exceedance frequency. The dashed curve is the relative frequency of occurrence of a sustained flow event independent of flow magnitude. Upon screening salinity data to those obtained only during sustained flow events, the correlation of Lavaca Bay salinity with inflow more than doubled. (Selection of the 30-day sustained flow period devolved from prior studies in which such a response time for the salinity structure was suggested by temporal variation.)

Density Currents

The density current is probably the least appreciated of the main elements of Gulf bay circulation, yet it is a remarkably ubiquitous factor. Fundamentally, the density current is the mean current directed from the mouth to the head of the estuary forced by the seaward gradient in salinity. The density difference entailed by this salinity gradient produces a horizontal pressure-gradient acceleration which is the basic driving mechanism. Of course the response of the water column to this acceleration is modulated by the vertical eddy fluxes of salt and momentum

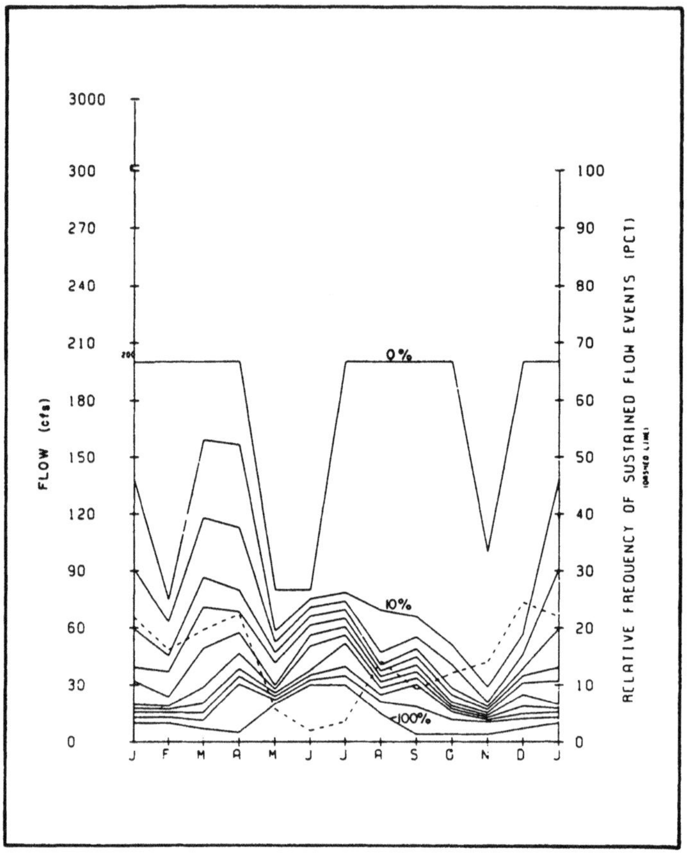

Figure 5. Sustained flow analysis (30-day event), Lavaca River, Texas

(reviewed in Ward, 1977). Generally the density current is an order of magnitude smaller in intensity than the tidal current, so can be exposed only by averaging a sequence of current profiles over a tidal cycle. Under conditions of weak tidal currents, e.g., small lunar declination or around slack water, the density current can be measured directly.

For present purposes, four principles suffice to summarize the physics of the density current:

(1) Intensity of the density current increases with the horizontal gradient of salinity.

(2) All other factors being equal, the density current intensity increases approximately as the cube of water depth.

(3) In equilibrium, the density current forces a return flow from the head of the estuary to the mouth.

(4) Vertical stratification is no index to the existence of a density current: a pronounced density current can, and frequently does, exist in a vertically homogeneous estuary.

The significance of the density current in Gulf bays derives from property (2) above, the dramatic increase in intensity with water depth. These bays typically are transected by dredged navigation channels whose greater depth relative to the surrounding water makes them effective developers of density currents. Indeed, salinity intrusion occurs preferentially up these channels, and the vertical-mean isohaline distributions characteristically evidence a tongue of higher salinity aligning with the ship channel.

It is useful to distinguish two channel configurations with respect to the resultant density current: the confined channel and the unconfined channel. The confined channel is bounded on either side by shore so that lateral currents are constrained relative to those along the longitudinal dimension. In confined channels the return flow is in the surface layers so that the net circulation is two-layered, with flow up the estuary in the lower layer and down the estuary (toward the mouth) in the upper layer. This is of course the classic estuarine density current, for which treatments may be found in the oceanographic literature back to the 19th Century (see, e.g., the excellent literature review of Hinwood, 1970). Two examples from Texas systems are shown in Figure 6, where the characteristic current reversal with depth is evident. In the example from the Sabine-Neches Canal, the density current was sufficiently strong that a current reversal with depth was exhibited at any stage of the tidal cycle.

An unconfined channel is dredged through an open bay so that the channel in effect is a deep slot within a broad, much shallower waterbody. In the extreme, intense vertical mixing and free access to open water on either side of the channel can combine to produce a unidirectional current in the channel, with the return flow in the shallower, open bay waters. An example of this is the Houston Ship Channel within the Galveston system. In Figure 7 are shown contours of flow predominance (i.e., the percentage of flow over the tidal cycle that is directed downstream) determined from current measurements over a tidal cycle at 11 stations in the Houston Ship Channel. In the open bay, the lower 25 miles, the tidal mean flow

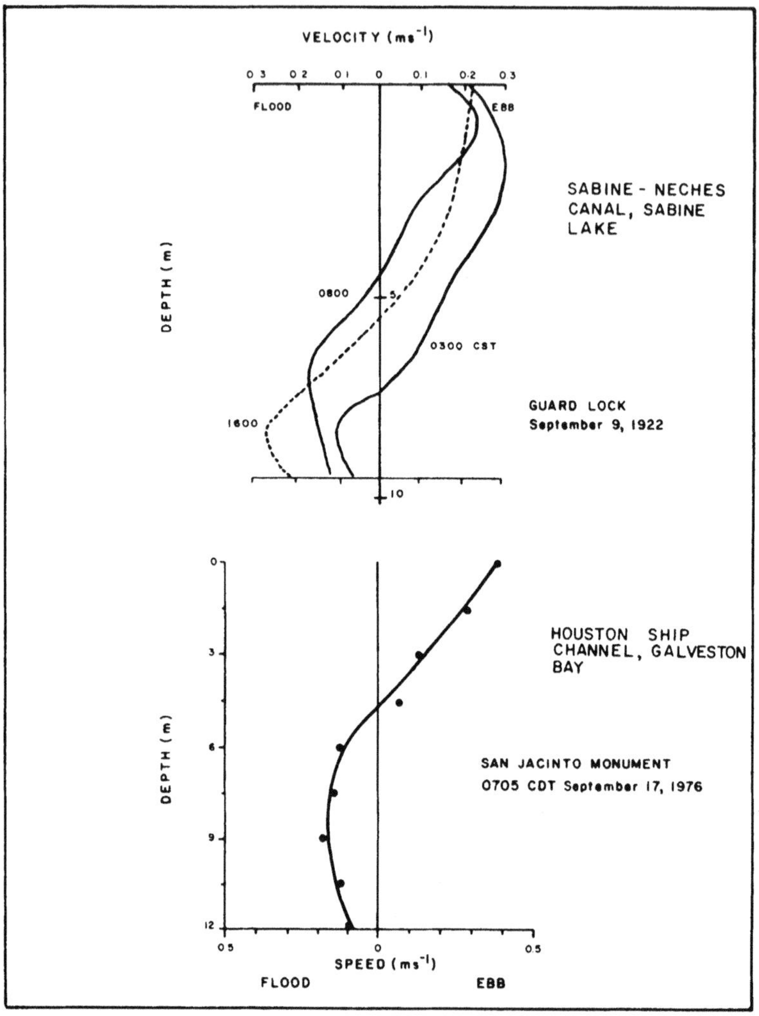

Figure 6. Confined density currents

is directed upstream throughout the depth. (Above Morgans Point, the circulation is that of a confined channel with two-layer flow.)

There seems to be little doubt that increased salinity intrusion in many of the bays has resulted from the progressively deepened networks of ship channels. Sabine Lake is an excellent example, first because of its near-fresh condition prior to the institution of navigation channels, and second because the rice industry along the Texas side of the bay utilized freshwater from the system for irrigation, so that a fair record of salinity encroachment is

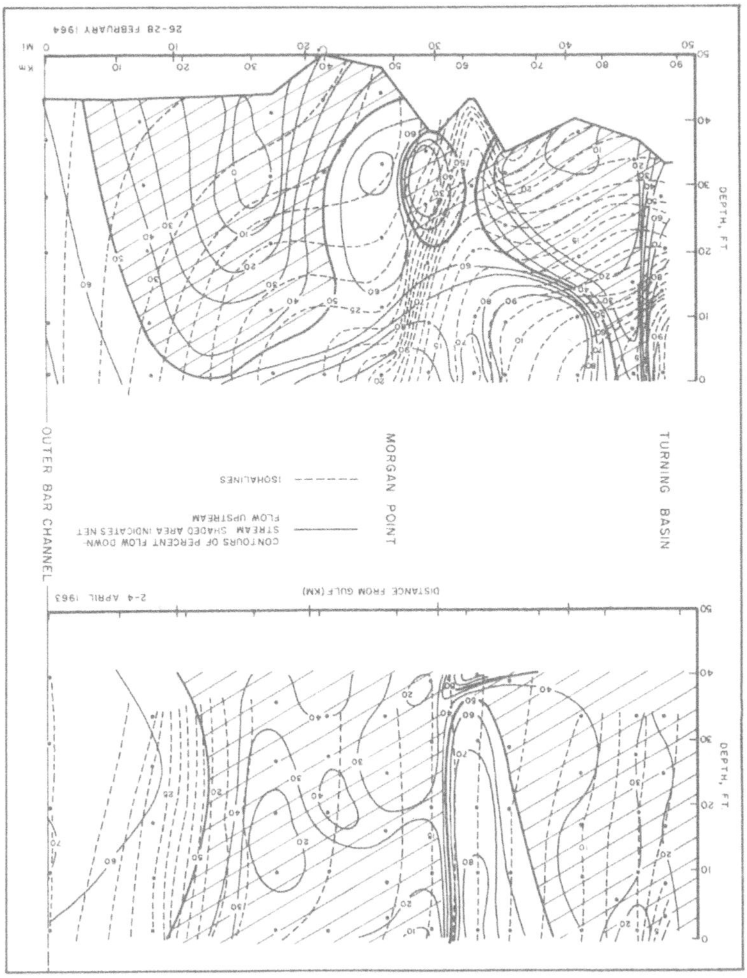

Figure 7. Confined and unconfined density currents in Houston Ship Channel, Galveston Bay

provided by the irrigation intakes. In 1883 the offshore Gulf bar was channelized and by 1900 there was a dredged channel 7.5 m deep through Sabine Pass to Port Arthur. In 1902, in association with a drought, salinity intruded up Taylors Bayou contaminating water needed for rice irrigation. By 1908 a 3 x 6-m channel extended to the mouths of the Sabine and Neches Rivers in the northern parts of the bay, and 5 years later this channel was deepened to 7.5 m and extended up the Neches to Beaumont. For the first time salinity intrusion occurred in the Neches. The lowermost irrigation intake had to be abandoned, and pumping interests further upstream quickly installed a salt barrier in the river. The decade 1924-1934 saw the navigation channel system further deepened and widened, ultimately to 10 x 100 m. Since 1925 salinity intrusion extended to Pine Island Bayou (55 km from the mouth of the Neches) almost every year, and the City of Beaumont found it necessary to move the municipal water supply intakes successively upstream. In 1950 the channel was deepened to 11 m and since then salt water has intruded 65 km up the Neches each year except 1950.

Sabine Lake also offers a rather bizarre example of salinity intrusion. The Sabine-Neches Canal was dredged around the western periphery of Sabine Lake producing an elongate island which isolates the Sabine-Neches Canal from the main body of the bay, except at its lower end and at the mouth of the Neches in its northwest corner. Because salinity intrusion, through the mechanism of the (confined) density current, proceeds much more rapidly around the periphery through the Sabine-Neches Canal than through the open body of Sabine Lake, much of the time more saline water is presented at the mouth of the Neches in the upper bay than in the open bay waters. There is, in effect, a reversed salinity gradient in the upper region of the bay, with salinity increasing toward the estuary head.

Other Factors

There are of course many other hydrographic factors not treated here. These include temperature regimes, turbulent diffusion, and sediment transport. The last is of particular significance to the Gulf systems, primarily due to the economic aspects of dredging for navigation improvement and of shoreline development. Their omission in this context is not indicative of their relative importance but rather dictated by the need for brevity.

The extreme transience of freshwater inflows to these systems was noted above and the use of the sustained-flow event in analyzing equilibrium configurations described. Of importance also is the freshet event, which is capable of sudden and dramatic alteration of bay hydrography. Much is lost if the flushing effects of these events are considered only in their contribution to long-term mean inflows. Ship traffic appears to play an important part in bay hydrography, particularly in the vertical mixing introduced within the channels by prop turbulence. Hurricanes represent infrequent cataclysmic events, but nonetheless are significant to the hydrography of the bays.

The Coriolis effect can be important in relatively slowly varying flows on large space scales, but its significance to the circulation of Gulf bays is not at all well established. The rotation of the earth has been invoked to explain the occurrence of fresher salinities on the west side of Mobile and Pensacola Bays. However, a simpler explanation may suffice, which is that the principal riverine discharges are located along the western shores. In contrast, in Galveston Bay, located on about the same latitude, fresher water occurs on the east side of the bay; and in this case the main riverine discharge is on the northeastern shore.

MODELING TECHNIQUES

As in the case for estuarine systems elsewhere, the need for a predictive capability for the Gulf embayments has necessitated a quantitative representation -- a model -- of their hydrographic processes. And, as with estuaries elsewhere, the types of models employed for the Gulf estuaries fall into three classes: statistical, where the dependency of a particular parameter is related to the presumed controlling variables through an analysis of historical data; mathematical, the formulation and solution of appropriate hydrodynamic and mass-transport equations; physical, an operating scaled replica of the real system embodying all factors presumed to be important. The utility of statistical models is ultimately delimited by the data base, and as most of the predictive problems involve fundamental alterations in the system, the historical data base strictly becomes irrelevant. Accordingly, statistical models are not considered further here.

The complexity of physiography and hydrodynamic interactions reduces the modeling alternatives for Gulf estuaries to the tidal physical model or the multi-dimensional mathematical model (implemented for numerical integration on a digital computer). General uses to which modeling has been applied in the Gulf embayments appear to fall into the three (nonexclusive) classes: storm surges, intratidal hydrodynamics and hydraulic analysis, and long-term transport of waterborne constituents (particularly water quality indicators). This last application, it should be noted, also necessitates a supporting hydrodynamic model in order to quantify the transport mechanisms.

Mathematical Models

The broad, complex physiographic configurations of the Gulf bays generally dictate that the mathematical formulation include variation along two horizontal dimensions. The simplest such model employs mathematical equations integrated in the vertical, to eliminate variation with the vertical dimension, and written in terms of vertical-mean variables. Almost universally, Gulf bay hydrodynamics have been modeled by the vertical-mean equations of momentum conservation and continuity, in which the hydrodynamic forces are the bed stress with the quadratic parameterization, the surface windstress, and the hydraulic head gradient, and the field-acceleration terms are omitted, i.e., the "storm surge" equations. (McHugh, 1976, follows the approach of Hansen, 1962, and Leendertse, 1967, and retains the nonlinear field terms.)

Reid and Bodine (1961) applied the storm-surge equations with two-nautical-mile resolution to the analysis of hurricane surge for the Galveston Bay system. Following this application, the storm-surge equations were adopted for use in more general situations, particularly in the modeling of hydrodynamic transport of constituents. A one-nautical-mile hydrodynamic model was developed for the Galveston Bay project and for most of the other bays along the Texas coast (Ward and Espey, 1971; Masch, 1971; Brandes and Masch, 1972). Similar applications of the surge equations have been made in Mobile Bay (April et al., 1976; Ng and April, 1976), Barataria Bay (McHugh, 1976) and several other systems. Increased spatial resolution may be developed by the technique of nested grids in which an interior grid system is implemented where the increased resolution is needed, and driven around its boundary by the interpolated values from the coarser grid model. Such applications have been made to wastewater outfall regions, especially power-plant cooling water returns, and to subsegments of inlets. For example, a sequence of nested-grid hydrodynamic calculations has been recently employed to evaluate tidal trajectories and potential scour in the inlet to Galveston Bay (USCE, 1979).

The vertical-mean hydrodynamic model results in a time-varying calculation of currents, which is further averaged over one or more tidal cycles for application in long-term models of the transport of waterborne constituents. This is necessitated by the complex geometry of Gulf lagoons; for longitudinal estuaries, such as those on the Atlantic seaboard, Q/A has long sufficed as a hydrodynamic model for long-term water-quality analysis. The hydrodynamic model generally assumes constant density, but as seen above, inhomogeneities in the density structure, arising from variation in salinity, give rise to the estuarine density current. Since the constant-density hydrodynamic model is intrinsically incapable of incorporating this phenomenon, its effects have been simulated in water-quality models by utilization of an inflated horizontal dispersion coefficient in the lower reaches of the estuary where the density currents are prominent. In the Gulf bays, this has been done with the enlarged dispersion coefficients aligning with the main axis of the channels. Such an artifice is basically empirical, since the value of the dispersion coefficient must be established by the analysis of data, and is strongly dependent upon the behavior of the tracer utilized to establish the dispersion coefficients.

The most common tracer employed for this purpose is salinity, and in this regard the practice follows the convention of estuarine studies elsewhere of using observed salinity distributions to "calibrate" the dispersion coefficients. For confined density currents, such as those prevalent in river-mouth estuaries, vertical shear resulting from the two-layered circulation does indeed contribute to a large upstream-and-downstream dispersion, so the practice is not without foundation (although the variation of density current with longitudinal salinity gradients should rigorously dictate a varying dispersion coefficient with freshwater inflow). Applied uncritically to the more complex open bay systems, however, this may result in large errors. The enhanced salinity intrusion along deepened channels necessitates a large dispersion coefficient to replicate the salinities. For a different tracer, such as a wasteload, injected at the upper end of the estuary, the inward-directed density current can impede movement of the waste down the channel while enhancing its transport to either side in the shallow area. This is exactly the reverse of the effect accomplished by the mathematical dispersion term. Figure 8 exhibits an example of this, comparing observed contours of nitrogen in upper Galveston Bay at the mouth of the Houston Ship Channel with modeled contours using an inflated dispersion coefficient determined from salinity data (cf. Figure 7). The latter produces convex-seaward contours, whereas the measured data contours are concave.

Clearly the means for working around this inadequacy is to incorporate density effects directly into the hydrodynamic model. The capabilities of the digital computer in principle permit this approach, though there are significant mathematical difficulties since the hydrodynamic and salinity (density) equations are strongly coupled in the nonlinear terms. The feeble tidal effects and substantial salinity gradients in the Gulf embayments in fact render it dubious whether the surge equations can be validly applied, for anything but short-term dynamic responses. Indeed, a calculation of hydraulic head gradients based upon homogeneous models and density-driven pressure gradient accelerations based upon typical salinity distributions show these terms to be of the same order in many regions of these embayments. (The model of Pitts, 1976, includes coupling of the salinity and momentum, as well as the vertical dimension. Verification is limited, however, and for salinity does not appear to be particularly good.)

Of the meteorological factors affecting bay hydrography, relatively little attention has been given to the effects of wind although the capability for this is implicit in the windstress terms. Examples of model-predicted (tidal-mean) wind-driven circulations for prevailing winds are shown in Figure 9. It is evident that the computation is capable of producing wind-generated gyres. However, the sense of these gyres in some cases is the

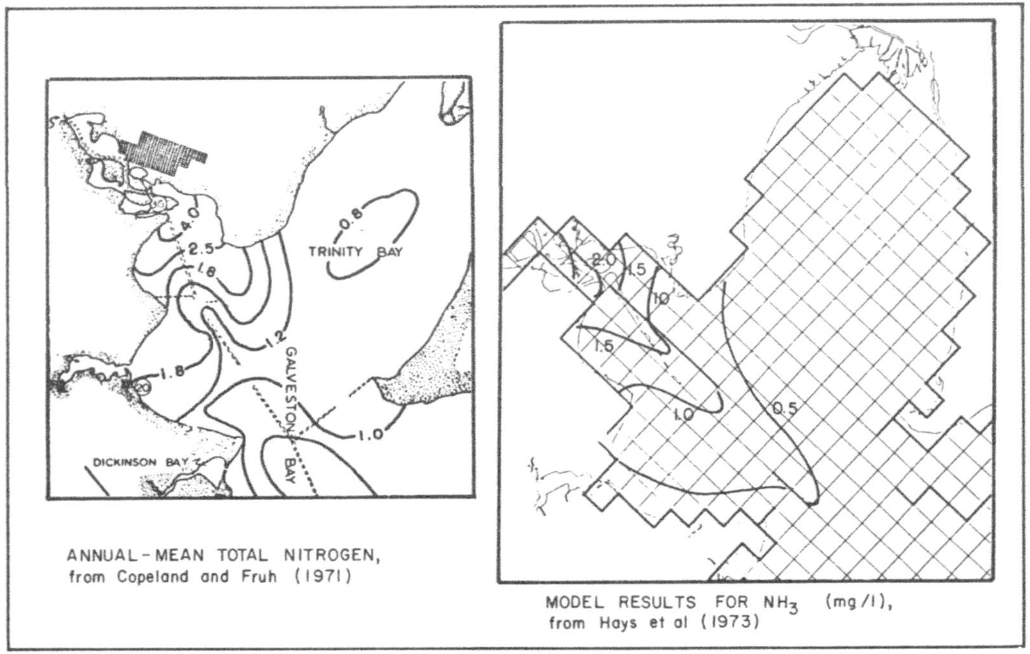

Figure 8. Comparison of shape of isopleths for transport by unconfined density current (left) and dispersive-transport model (right)

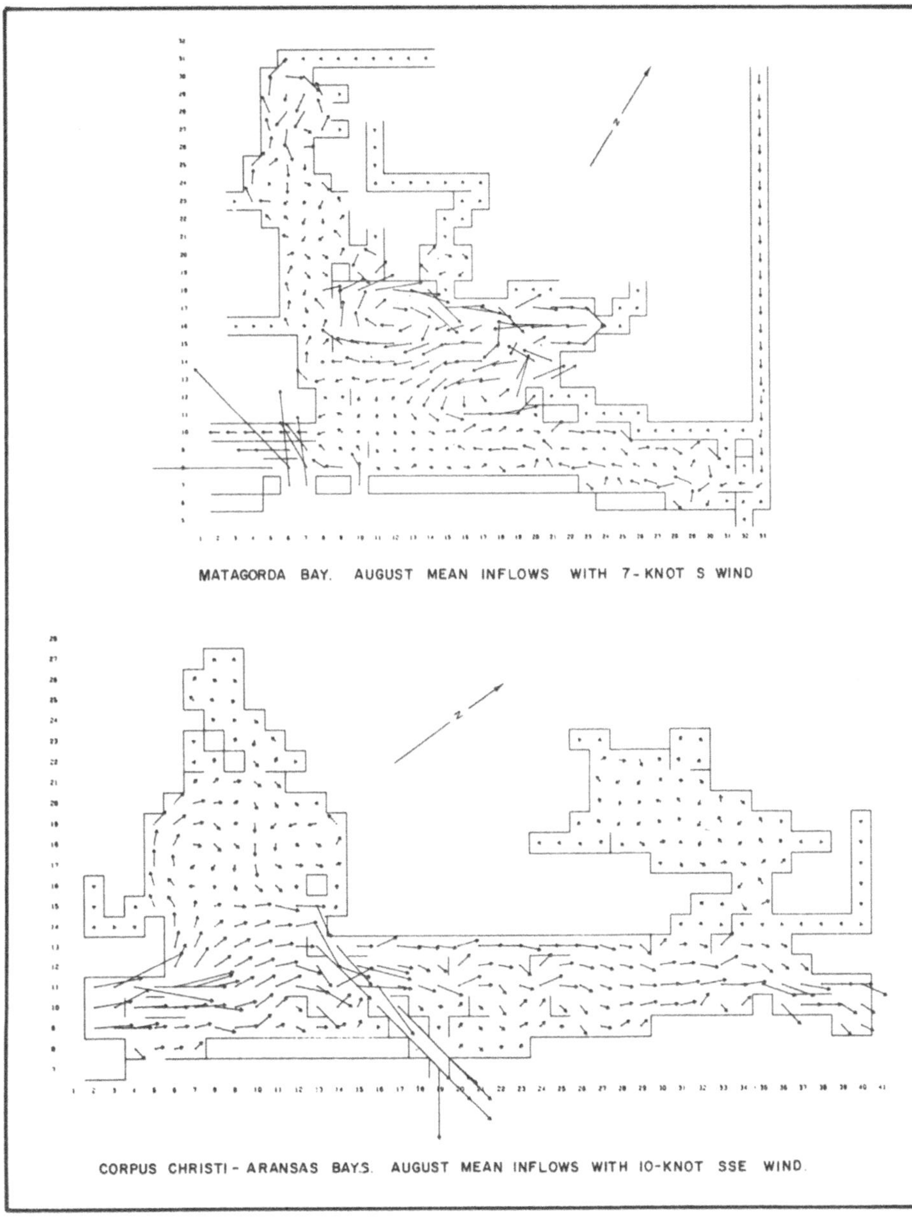

Figure 9. Model predictions of wind-driven tidal-mean circulations

reverse of that believed to be reality. Study of this factor is limited both by the physical formulation of wind phenomena in the model as well as available hydrodynamic data from the bay systems. The need for simulation of frontal events has been rather limited, consequently it is difficult to appraise the capability of these models with respect to this factor. In any event, the limited field data available would preclude extensive verification.

Physical Models

Physical models offer an obvious attraction in their capability to incorporate the complex physiography of the Gulf bays as well as the interaction of many hydrographic factors. The large surface-area-to-volume ratio, however, dictates a rather large model -- even with a 10:1 vertical distortion -- in order that the water depths in the model be sufficiently deep to minimize spurious effects such as surface tension. Thus, construction of a physical model for a Gulf bay (in particular) demands considerable resources in both staff and facilities; all of the work done in this area thusfar has been performed at the U. S. Corps of Engineers Waterways Experiment Station, Vicksburg, Mississippi.

An excellent example of these models is that constructed recently for Mobile Bay (Lawing et al., 1975). The model was built to scale ratios 1:1000 horizontal, 1:100 vertical, and based upon Froude similitude so that velocity scales 1:10, time 1:100 and salinity 1:1. Physically, the model is approximately 2800 m^2 (0.7 acres) in area, representing the area from 12-fathom water in the Gulf of Mexico to a point some 65-km upstream from Mobile, from Dauphin Island on the west to Pine Beach on the east. The model is equipped with a sump of saline water and various pumps and valves controlled by a mechanical programmer to simulate the Gulf tide. A secondary tide generator is located in Mississippi Sound where the model boundary terminates on an open-water section. River flows are simulated by constant-head tanks equipped with flowmeters. Measurement of hydrographic parameters in the model includes remote recording tide gauges, temporary tide gauges, point current measurements with miniature current meters, time-lapse photography of floating confetti, salinity measurements, and fluorescent dye tracing.

It is elementary that so long as water is the model fluid, exact similitude is not possible, presuming even perfectly scaled physiography. (A summary of the theory underlying the physical model is provided by Harleman, 1971.) Selection of the Froude number as the scaling index of necessity entails distortion of all but Froude-dominated processes. However, it can be argued whether any process is <u>totally</u> Froude-dominated (or Reynolds- or Prandtl-, etc.). For example, the propagation of the tidal wave is ultimately governed by frictional dissipation, whose similitude requires Reynolds scaling. For this reason the model must be equipped with devices to simulate the effects of non-Froude processes, most notable of which are metal strips embedded in strategic areas of the model bed, adjusted by bending and snipping, in the painstaking process of model calibration, to replicate the propagation of the real tide.

The Mobile model is typical of the fixed-bed models of Gulf estuaries. One of the significant problems encountered early in development of these models was their tendency to exhibit far too much vertical stratification within the shallow open areas of the bay. This was ascribed to neglect of windwave mixing, and was finally overcome by installation of many (fifty-three in the Mobile model) oscillating fans mounted on stands throughout the open areas of the model, which appeared to promote mixing at a sufficient rate. A similar problem was encountered in the ship channels, which also exhibited far too much stratification. It was postulated that the lack of ship traffic was responsible for this model inadequacy, and various measures were attempted to remedy the problem, including mechanical ship models moving up and down the channels. The method found most successful, and now in use, is to force air bubbles from a small pipe along the centerline of the ship channel bottom, the discharge rate and spacing of the orifices adjusted by trial-and-error.

The initial motivation for development of models of this type was basically hydraulic, e.g., in determining optimum routes of ship channels with respect to currents, shoaling, etc. The closest approach to water-quality considerations was in determining the effects upon salinity structure due to various physical structures or navigation projects (such as the proposed placement of hurricane barriers in Galveston Bay). In recent years, however, there has been an increasing emphasis on utilizing these models for the simulation of the transport of waterborne constituents. The Galveston Bay model (known as the Houston Ship Channel Model at WES to distinguish it from earlier moveable-bed and hurricane-surge models of the Bay) has been employed to simulate the disposition of power plant plumes. Dye dispersion tests have been performed in almost all of the Gulf models to quantify diffusion processes.

Clearly the extension of the application of the physical model to processes which are non-Froude (such as dye diffusion) must be regarded with suspicion. There might be some fortuitous justification for this with the Atlantic or Pacific estuaries, where the density current circulation is two-layer thus contributing to shear dispersion, and where much of the current fine-structure derives from the tide. However, this practice with the Gulf bays is tenuous; indeed, the need for devices such as oscillating fans and bubbling air may flag the limit of applicability of this type of model. The review by Ward and Espey (1971) suggests that even salinity simulation may not be performed all that well for Gulf embayments.

DISCUSSION

Table 2 summarizes the primary hydrographic processes considered in this paper and the relative capabilities of the various modeling techniques for incorporating these processes. The emphasis in model development has always been utilitarian and it is reasonable to presume that this will continue to be the case. From the water quality standpoint, long-term circulation patterns will no doubt continue to enjoy major attention, and indeed there is much remaining to be learned about these systems in this regard.

As this is intended to be a review paper it is appropriate to proffer some recommendations disguised as conclusions. First, the available data base is presently (in this writer's opinion) the greatest impediment to fruitful progress in the study of these systems. The history of interest in estuarine processes has been shorter in the Gulf Coast, by and large, than in other areas of the country, and the complexity of these systems presents a particularly acute need for intensive, carefully executed field studies. Appreciation for hydrographic factors should be invested in field strategy, particularly in a careful delineation of the hydrometeorological regime necessary to ensure satisfactory data.

Trenchancy is needed in assessing model capabilities, even to the point of ruthlessness. In too many cases the application of models overreaches their capabilities. This requires in turn a better-founded physical understanding of these estuarine systems. It is remarkable, for example, that we lack a rigorous scale analysis for estuarine motions, permitting judgments as to which factors are important at which levels of approximation and as to how the governing hydrodynamic equations might be validly simplified. We may also note that frequently numerical methods overreach the theoretical foundation of estuarine physical oceanography.

Finally an appreciation of time and space scales in the interpretation and analysis of estuarine motions is needed. For Gulf bays the time scales of response may range from a few days to many weeks depending on ambient conditions and the parameters of concern. Although hydrographers of necessity develop an intuition for response-type behavior, frequently this is discarded when analyzing the distributions of indicator constituents. As an example of the need for consideration of response time, consider the problems of verifying a salinity model. For a steady-state model, two

Table 2. Principal Hydrographic Factors and Capabilities of Present Modeling Techniques for Their Representation

	Physical	Mathematical (2-Dimensional)	Comments
Tides	Good	Good	
Freshwater inflow			
Steady	Good	Good	Throughflow in current structure
Transient	?	?	
Salinity (horizontal)	Fair	Good	Using empirical dispersion in mathematical model
Salinity (vertical	Poor	No capability	
Meteorological effects			
Windwave mixing	Poor	?	Vertical fluxes not important in 2-dimensional model, indirect effects on horizontal fluxes unknown
Wind-driven mean circulation	No capability	Poor	
Frontal	No capability	?	
Density current	Poor	Poor	Mathematical model uses parameterization in transport calculation

conditions are required for the verification data set: stabilization of freshwater inflows and elapsing of enough time to ensure that the isohaline structure has equilibrated. For Gulf bays it appears that a period of one to several months are required for the isohaline distribution to fully equilibrate. On the other hand, for a time-varying model, the initial salinity distribution, an important input, is generally determined from observations. The long response time for salinity significantly filters transient variations in freshwater inflow, which means that a short period of time integration produces only minor departures from the initial conditions, and hence apparently good verification against measured data. This is meretricious. Instead a long period of integration is required to determine whether the basic model is truly capable of predicting salinity. (Often it is postulated that a steady-state model with long-term mean inflows will calculate the long-term mean salinity, and can therefore be verified against long-term average data. At present there is absolutely no theoretical justification for this postulate. A sufficiently detailed, statistically rigorous validation of the "climatological" capability of steady-state models is needed. In view of the intrinsic nonlinearity of the transport equation and the complex hydrodynamic interactions affecting salinity in the Gulf bays, the possibility of such a validation seems remote.

In summary, the Gulf embayments do not exhibit any hydrographic phenomenon peculiar only to them. However, the combination of feeble tides, complex physiography with broad shallow areas and narrow channels, and sensitivity to meteorological factors alters the relative importance of these phenomena and the range of responses they engender vis-a-vis the more familiar Atlantic and Pacific estuaries. Evidently these differences are not a matter of detail, but rather necessitate an adjustment in the conceptual and strategic framework for the study of the hydrography of these bays, and concomitantly their water quality.

REFERENCES

April, G. C. et al. 1976. Water resources planning for rivers draining into Mobile Bay. Final Report, Contract NAS8-29100, University of Alabama.

Brandes, R. J., and F. D. Masch. 1972. Tidal hydrodynamic and salinity models for Corpus Christi and Aransas Bays, Texas. Report to Texas Water Development Board, Austin.

Copeland, B. J., and E. G. Fruh. 1969. Ecological studies of Galveston Bay. Report to Texas Water Quality Board, Austin.

Gagliano, S. M., A. J. Kwon, J. L. van Beek. 1970. Salinity regimes in Louisiana estuaries. Center for Wetland Resources, Louisiana State University, Baton Rouge.

Grace, S. 1932. The principal diurnal constituents of tidal motion in the Gulf of Mexico. Mon. Not. R. Astr. Soc. Geo. Phys. Suppl., 70.

Grace, S. 1933. The principal semidiurnal constituent of tidal motion in the Gulf of Mexico. Mon. Not. R. Astr. Soc. Geophys. Suppl., 156.

Hansen, W. 1962. Hydrodynamical methods applied to oceanographic problems. Proc. Symp. Math.-Hydro. Methods Phys. Oceans., Inst. Meeres. Univ. Hamburg.

Harleman, D. R. F. 1971. Physical hydraulic models. In: Estuarine modeling, an assessment (G. H. Ward and W. H. Espey, eds.), U. S. Government Printing Office, Washington, D. C.

Hays, A. J., R. M. Chen, and M. J. Cullender. 1973. User's manual for the nitrogen model of the Galveston Bay System. Document No. T72-AU-9564, Tracor, Inc., Austin.

Hinwood, J. B. 1970. The study of density stratified flows up to 1945 (Part I). La Houille Blanche, No. 4, 347-359.

Huston, R. J. 1971. Galveston Bay project, compilation of water quality data, July 1968-September 1971. Document No. T71-AU-9617, Tracor, Inc., Austin.

Jarrett, J. T. 1976. Tidal prism-inlet area relationships. GITI Report 3, Waterways Experiment Station, Vicksburg, Mississippi.

Lawing, R. J., R. A. Boland, and W. H. Bobb. 1975. Mobile Bay model study. Technical Report H-75-13, Waterways Experiment Station, Vicksburg, Mississippi.

Leendertse, J. J. 1967. Aspects of a computational model for long-period water-wave propagation. Memo RM-5294-PR, The RAND Corp., Santa Monica, California.

Marmer, H. A. 1954. Tides and sea level in the Gulf of Mexico. In: Gulf of Mexico, its origin, water and marine life. Fishery Bulletin 89, U. S. Fish and Wildlife Service, 55.

Masch, F. D., and Associates. 1971. Tidal hydrodynamic and salinity models for Corpus Christi and Aransas Bays, Texas. Report to Texas Water Development Board, Austin.

May, E. B. 1973. Extensive oxygen depletion in Mobile Bay, Alabama. Limn. and Ocean., 18(3), 353-366.

McHugh, G. F. 1976. Development of a two-dimensional hydrodynamic numerical model for a shallow, well mixed estuary. Sea Grant Publ. LSU-T-76-008, Center for Wetland Resources, Louisiana State University, Baton Rouge.

McPhearson, R. M. 1970. The hydrography of Mobile Bay and Mississippi Sound, Alabama. J. Mar. Sci. Alabama, 1(2):1-83.

Ng, S., and G. C. April. 1976. A user's manual for the two-dimensional hydrodynamic model. BER Report 203-112, University of Alabama.

Ollinger, L. W. et al. 1975. Environmental and recovery studies of Escambia Bay and the Pensacola Bay system, Florida. Region IV, USEPA, Atlanta.

Pitts, F. H. 1976. A three-dimensional time dependent model of Mobile Bay. Final Report, Contract NAS8-30380, Louisiana State University, Baton Rouge.

Ratzlaff, K. W. 1976. Chemical and physical characteristics of water in estuaries of Texas, October 1971-September 1973. Report 208, Texas Water Development Board, Austin.

Reid, R. O., and B. R. Bodine. 1968. Numerical model for storm surges in Galveston Bay. Proc. ASCE, 94(WWI), 33-57.

U. S. Corps of Engineers. 1979. Draft environmental impact statement, multi-purpose deepwater port and crude oil distribution system at Galveston, Texas. Permit Application 10400, USCE, Galveston District.

von Deesten, A. P. (Capt.). 1924. Sabine-Neches Waterway (salt-water guard lock), salinity survey. Document No. 12, 68th Congress, 2nd Session, Committee on Rivers and Harbours, House of Representatives, Washington, D. C.

Ward, G. H. 1973. Hydrodynamics and temperature structure of the Neches Estuary: 1, Physical Hydrography. Document No. T73-AU-9509, Tracor, Inc., Austin.

Ward, G. H. 1977. Formulation and closure of a model of tidal-mean circulation in a stratified estuary. Estuarine Processes, 2, Academic Press, New York.

Ward, G. H. 1979. The sustained flow event. In Prep. for subm. to Wat. Res. Res.

Ward, G. H., and C. L. Chambers. 1978. Meteorologically forced currents in Upper Sabine Pass, Texas. Doc. No. 7869, Espey, Huston & Associates, Inc., Austin.

Ward, G. H. and W. H. Espey. 1971. Case studies, Galveston Bay. In: Estuarine modeling, an assessment (G. H. Ward and W. H. Espey, eds.), U. S. Government Printing Office, Washington, D. C.

Wilson, A. 1967. River discharge to the sea from the shores of the coterminous United States. Hydrologic investigations atlas HA-282, U. S. Geological Survey, Washington, D. C.

PREDICTING THE EFFECTS OF STORM SURGES AND ABNORMAL RIVER FLOW ON FLOODING AND WATER MOVEMENT IN MOBILE BAY, ALABAMA

Gary C. April and Donald C. Raney

The University of Alabama, P. O. Box G, University

Alabama 35486

ABSTRACT

The threat of man-made and natural disturbances to the coastal environment is a continuing and perplexing problem. With the advent of rapid, numerical simulation models describing coastal water behavior, the ability to better understand these regions and to provide data to offset the adverse impacts caused by these disturbances, has greatly improved.

This paper discusses the recent numerical modeling activities of the Mobile Bay system under severe conditions. Results are presented in terms of changes that occur in water elevation and movement, and, in salinity distribution patterns when the bay is subjected to river flooding inflows and storm surges.

At a river flood stage of 7000 m^3/s, water behavior in the northern and central portions of the bay is totally governed by the fresh water inflow. A salinity level of 5 ppt is restricted to the lower bay at a point 15 km from the Main Pass. Usual salinity values under normal conditions are in the range of 15 to 20 ppt in this area. A critical river flow rate of 8500 m^3/s is also identified. At or above this flow, saline water intrusion in the lower bay becomes stabilized at 10 ppt on a line 6 km north of the Main Pass.

Conversely, large amounts of saline water enter the bay under storm surge conditions investigated in this study. Conditions typical of those caused by Hurricane Camille in 1969 were used in the modeling exercise. Salinity levels as high as 26 ppt were predicted for the northern bay area. This high saline water intrusion is caused by the development of the surge hydrograph at the gulf/bay interface as the storm approaches the coastline.

In both cases the model results were shown to be representative of bay behavior. Comparisons with existing field observations were made to calibrate and verify the models.

INTRODUCTION

Hurricanes and tornadoes cause extensive damage in the United States. Between the years 1955 and 1970, the average annual property damage caused by hurricanes was $500,000,000, and that caused by tornadoes was $75,000,000. In the same period the average number of deaths due to tornadoes and hurricanes were 125 and 75, respectively (National Science Board 1972). To cite some specific cases, Hurricane Camille (1969) and Hurricane Betsy (1965) each caused more than $1.4 billion in damage, while Hurricane Agnes (1972) caused $3.5 billion in damage. There were 122 deaths associated with Agnes, 255 deaths and 68 people missing with Camille (Lamoreaux and Chermock 1976). Hurricane Fifi in 1974, caused an estimated 5,000 deaths in Honduras and extensive damage to lowland plantations that seriously affected the economy of the country.

Other serious threats to coastal areas caused by severe weather are storm surges which produce water levels above the normal, astronomical tides. Currents and winds contribute to this buildup of water. Because of the relatively shallow, often converging shape of coastal embayments, buildup along shorelines occurs as a result of the inability of streams, marshes and lowlands to rapidly assimilate the increased water volume. This causes flooding of inhabited areas, property damage by high waves and winds, and upsets to the marine environment.

In addition to the destructive nature of the surges, torrential rainfall along the coastal and inland watershed area is another source of abnormal and damaging behavior within the system. Heavy rainfalls cause flooding along rivers and their tributaries and eventually contribute to high water levels in the coastal embayments. In large river systems where several days are required for runoff to reach coastal waters, the high flow rates are delayed such that the storm surge precedes the influence of flood stage discharges. Such delays usually prolong high water levels within the bay and make cleanup operations and the return of system behavior to normal levels a long process. In much shorter coastal streams, runoff accumulates rapidly and directly contributes to increased water levels produced by the surge condition.

It is evident from the above discussions that studies designed to provide supplemental information about storm surges and their impact on the coastal environment are important undertakings. These studies could be related to the development and application of new instrumentation such as weather satellites or to the method of computer simulation of coastal waters under severe weather conditions. Each could potentially yield accurate information in advance of the time when the storm or its after effects reaches the coastal region.

The central theme of this paper deals with the application of computer simulation models to describe Mobile Bay behavior under severe meteorologic (storm surge) or hydrologic (river flood stage) conditions.

Mobile Bay System

Mobile Bay is approximately 1036 km^2 in area and is located on the north eastern shoreline of the Gulf of Mexico east of the Mississippi River delta (see Figure 1). The estuary is about 50 km long and varies in width from 13 to 16 km in the northern half to about 39 km wide in the southern portion. The southeastern region of the estuary is referred to as Bon Secour Bay. The southern end is blocked from the open Gulf by land barriers; Gulf Shores to the east and Dauphin Island to the west. There are two passes located in this area; the main pass which connects with the Gulf at Mobile Point, and the pass which connects with Mississippi Sound at Cedar Point. The bay is the terminus of the Mobile River Basin which consists of more than 114,000 km^2 of drainage area; the fourth largest in the United States. Variations in river discharge rate from a normal high of 3850 m^3/s in March to a normal low of 450 m^3/s in September have been recorded. Gauging stations at Jackson (or Coffeeville), Alabama and Claiborne, Alabama provide continuous discharge records for the Tombigbee and Alabama Rivers, respectively.

Mathematical Models of the Mobile Bay System

In order to better understand the complex, interactive effects influencing water movement in the bay, several mathematical models based on the laws of conservation of mass and momentum have been formulated (Hill and April 1974). These include models describing the hydrodynamics, conservative and non-conservative species transport within the bay (Table 1).

The mathematical model describing water movement and tidal elevation within Mobile Bay is based on a two-dimensional unsteady flow equation and is referred to as a hydrodynamic model. The water mass is considered to be reasonably mixed such that integration

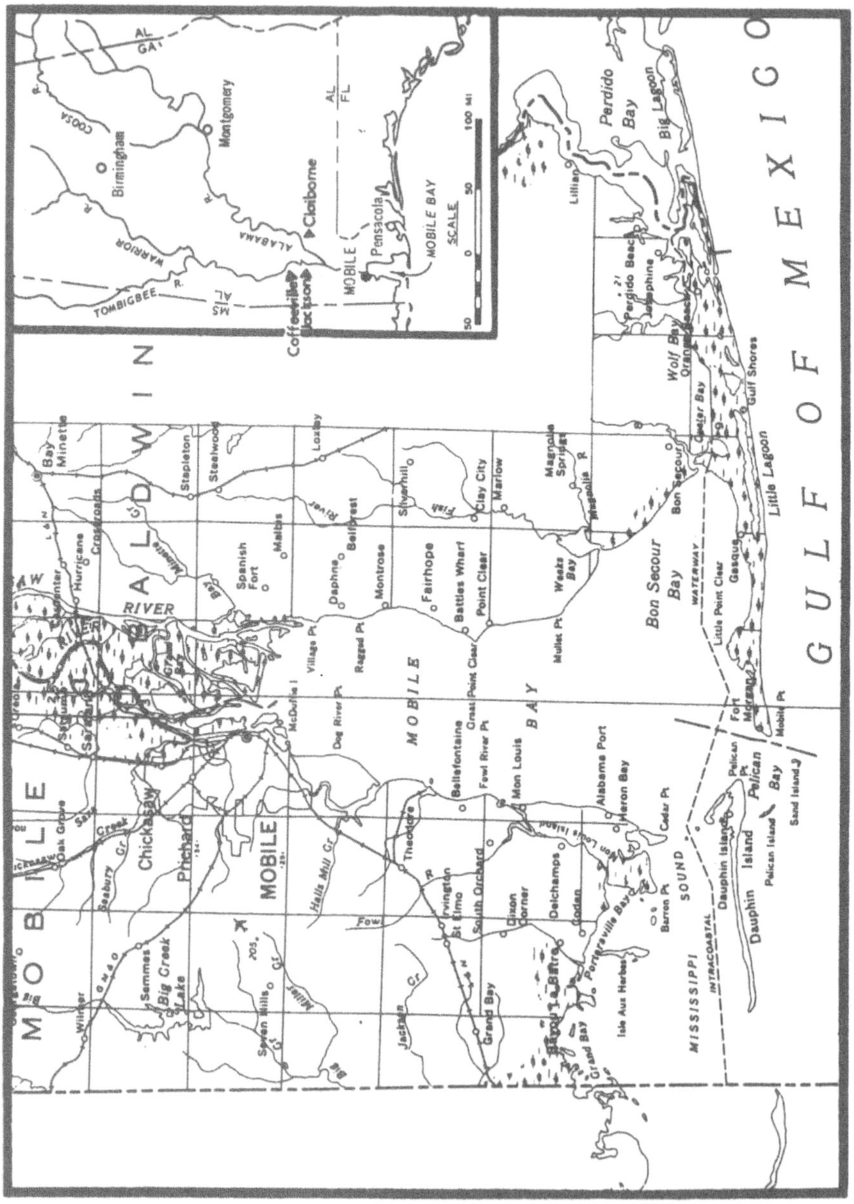

Figure 1. Mobile Bay System

TABLE I

MATHEMATICAL REPRESENTATION AND OPERATIONAL MODES OF THE PHYSICAL MODELS FOR MOBILE BAY

Name	Equation Form	Results	Modes
Continuity	$\frac{\partial Q_x}{\partial x} + \frac{\partial Q_y}{\partial y} + \frac{\partial H}{\partial t} = -(R + E)$	Tidal Height	Tidal Cycle Daily Avg. Monthly Avg. Seasonal
Momentum x-Component	$\frac{\partial Q_x}{\partial t} + gD\frac{\partial H}{\partial x} = K\eta^2\cos\Psi - fQQ_x D^{-2}$ $+ Q_x(2W\sin\phi)$ $+ D^{-1}\left(\frac{\partial(V_x^2)}{\partial x} + \frac{\partial(V_x V_y)}{\partial y}\right)$	x-Component of System Current	Tidal Cycle Daily Avg. Monthly Avg. Seasonal
y-Component	$\frac{\partial Q_y}{\partial t} + gD\frac{\partial H}{\partial y} = K\eta^2\sin\Psi - fQQ_y D^{-2}$ $+ Q_y(2W\sin\phi)$ $+ D^{-1}\left(\frac{\partial(V_y^2)}{\partial y} + \frac{\partial(V_x V_y)}{\partial x}\right)$	y-Component of System Current	Tidal Cycle Daily Avg. Monthly Avg. Seasonal
Species Continuity	$\frac{\partial C}{\partial t} + V_x\frac{\partial C}{\partial x} + V_y\frac{\partial C}{\partial y} = E\left(\frac{\partial^2 C}{\partial x^2} + \frac{\partial^2 C}{\partial y^2}\right)$ $+ \frac{E}{D}\left(\frac{\partial C}{\partial z}(z_s) - \frac{\partial C}{\partial z}(z_b)\right)$ $- \frac{1}{D}\left(CV_z(z_s) - CV_z(z_b)\right)$ $+ R_o$	Concentration of Species	
Salinity	$R_o = 0$	Salinity Concentration	Daily Avg. Seasonal
Coliform	$R_o = K_r$; where $K_r = f(\Theta)$	Coliform Bacteria Concentration	Monthly Avg. Seasonal

See Appendix A for Nomenclature

of the general three dimensional equation in the depth direction is a valid, simplifying assumption. Due to the specific nature of Mobile Bay, convective acceleration and the Coriolis force may make significant contributions in the momentum equations. Results can be generated for non-steady flow when boundary conditions are available as a function of time, or for quasi-steady flow when conditions are reasonably stable for a given time period.

The material transport model for Mobile Bay is based on the two dimensional form of the species continuity equation. This model is driven by tidal average velocities and dispersion coefficients generated by the hydrodynamic model. The results thus produced are average concentration distributions throughout the Bay. Modification of bottom boundary in areas where salt wedge (stratification or unmixed region) effects have been observed, has been used successfully to simulate three dimensional characteristics. Similarly, coliform die-off rate constants are introduced when these elements are being studied with the model.

These models have been used to study the influence of river discharge rate, wind direction and speed and tidal condition on bay circulation and material transport. A Univac 1110 computing system is used to numerically integrate the resulting differential equations.

Severe Condition Input Requirements

In order to describe the behavior of Mobile Bay water circulation under severe weather conditions, the particular forcing functions that apply must be known. With reference to Figure 2 this can be either in the form of river discharge data in which flooding conditions are exceeded (7×10^3 m^3/s) or in the case of severe storms, the hydrograph that exists at the Bay/Gulf exchange.

In the former case, sufficient data from the U. S. Army Corps of Engineers and U. S. Geological Survey can be used as primary input data to the hydrodynamic model. In the latter case, either field data or computed results from the storm surge model of Wanstrath (1978) can be used. Both methods were used as primary input data sources to the hydrodynamic model.

River Flooding Input Data

Flood stage river discharges are routinely reported by the U. S. Geological Survey for the Mobile River systems. These values are recorded at two stations located at Jackson and Claiborne on the Tombigbee and Alabama Rivers, respectively. Knowledge of the river flow rates at these locations, the general features of the

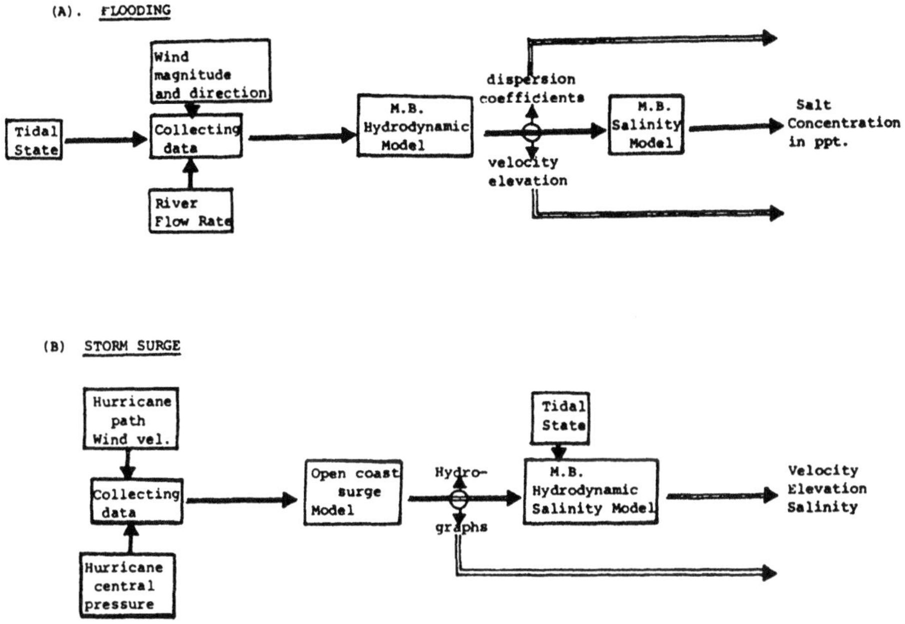

Figure 2. Severe Weather Input to the Mobile Bay Hydrodynamic Model.

river basin and estimated travel times provide the needed river discharge data to the Mobile Bay hydrodynamic model.

Calibration and Verification under Flood Stage Conditions

To achieve calibration and verification of the model (i.e., to make certain the model accurately describes bay behavior), data collected by William W. Schroeder (1977) during the flooding periods of March, April, and May 1973 are used. Eleven cruises were made by Schroeder aboard the University of Alabama System's 20 m research vessel, Aquarius (now the R/V G. A. Rounsefell). Data used in the model calibration phase are summarized in Table 2.

Using this information as input data, with knowledge of the tidal stage existing during Schroeder's data collection cruises, a comparison of actual water movement and elevation within the bay with model predicted results can be made. More importantly, using the resulting hydrodynamic results as input data to a previously

TABLE II CONDITIONS IN THE LOWER MOBILE BAY DURING THE 1973 FLOOD PERIOD

Cruise No.	Cruise Date	Time of Survey	Total Stations	Observed Tidal State	Discharge* m^3/s	Wind m/s
2	04/17/73	0900-1600	16	High→Ebbing	8410	4- 8 SE
5	04/27/73	0900-1400	13	Flooding	3060	6- 7 NW
6	04/30/73	0900-1500	18	High→Ebbing	3300	6- 8 SE
9	05/08/73	0830-1700	25	Flooding→High	7110	2-10 SW
11	05/15/73	0900-1430	22	High→Ebbing	4000	2- 5 NW

*This discharge is obtained as the average discharge from rivers in the Mobile River Basin at 5 days and 9 days prior to the bay survey dates.

verified salinity model permits a more direct and graphic comparison of the salt content within the lower reaches of the bay (i.e., salinity levels in the upper bay are too small as a result of the large fresh water river discharges). Because of the highly river dominant feature of the bay under these conditions, the comparison is restricted primarily to the lower Mobile Bay area. It should be pointed out that the model salinity values are averaged over the period specified by Schroeder for each cruise. River discharge rates were extrapolated from data gathered at Jackson and Claiborne 48 km up river from the bay. These values are corrected for time of travel (5-9 days) and additional runoff that accumulates in the basin (April and Raney 1979).

Cruise 2 by Schroeder was made near the maximum river discharge rate (8000 to 10000 m^3/s, during April 9-10, 1973) within the range of the first flooding period during March 30-April 18, 1973. The average river discharge rate was 8410 m^3/s, and the total volumetric discharge over this period was 15.2 x 10^9 m^3, that is 4.8 times the mean high water discharge volume into Mobile Bay. Results are shown in Figure 3 as model salinity data versus observed salt contents averaged in the depth direction.

There is good agreement between the observed depth averaged salinity profiles and the isohalines produced by the mathematical model. The values are well within experimental accuracy and the overall trend within the bay is maintained. Generally speaking under extremely high river discharge rates salinity intrusion within the bay from the Gulf through the main pass is restricted to a point 6 km north of the pass such that a 2 ppt salinity profile is obtained. The Bon Secour Bay area is likewise low in salt content ($\leqslant$5 ppt).

Cruise 5 and cruise 6 were made between the two flooding intervals during 1973. The river discharge rate was approximately 3200 m^3/s. Figures 3a, and b show the comparison of model results with field data obtained during cruise 6. Once again, the values are within field data accuracy and the overall trend is in agreement with real system behavior. It is also noted that the 2 ppt isohaline is shifted 1 km higher as compared with the results obtained at 8410 m^3/s. Salinity intrusion is restricted to the Main Pass area with a 2 ppt isohaline located 8 km or less from the pass. Most of Bon Secour Bay is at 5 ppt or less. The same general patterns were also observed during cruise 5, 9 and 11 which occurred when river flow rates were 4000 m^3/s.

Storm Surge Input Data

As in the case of river flooding, field data describing the

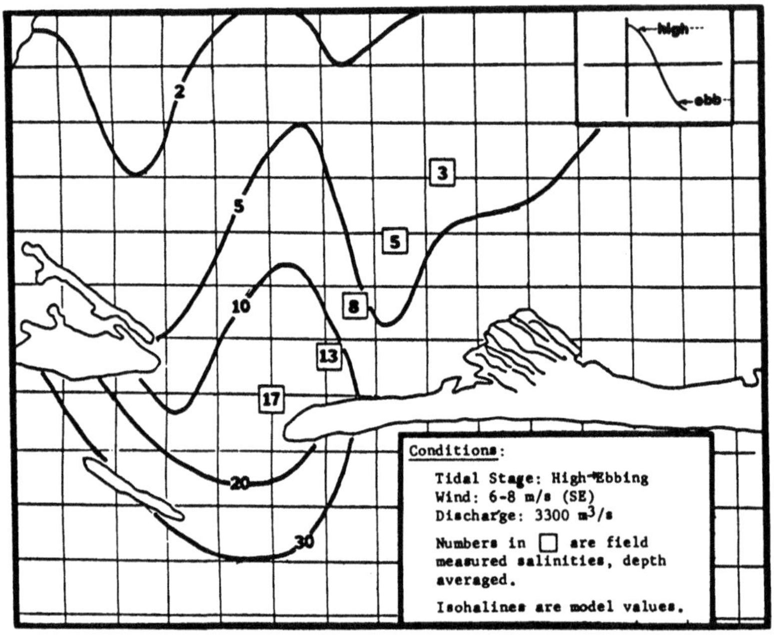

Figure 3a. Isohalines for the Lower Mobile Bay under River Flood Stage Conditions; (a) 3300 m^3/s.

storm surge hydrograph are also available in many cases. This information forms a valuable source for the calibration and verification of models designed to predict water movement and elevation. One such storm is used in this study for purposes of calibration and verification; Camille (1969). Since one of the objectives is the ability to predict the impact caused by storms on the coastal waters of Alabama, a model describing the hydrograph at the bay/gulf interface was adapted. The model is the open coast water model developed by Wanstrath, et al. (1976). Although the theoretical developments have been reported previously, application of Wanstrath's model to this segment of the Alabama coastline becomes an important part of this study.

Calibration and Verification under Storm Surge Conditions

Hurricane Camille traversed the extensive low-lying marsh area of the Mississippi Delta, moving over the shallow waters of Breton and Chandeleur Sounds for 4 hours prior to the landfall near Bay St. Louis, Mississippi at approximately 10:30 p.m. central standard time (Figure 4), August 17, 1969. The radius of maximum winds was

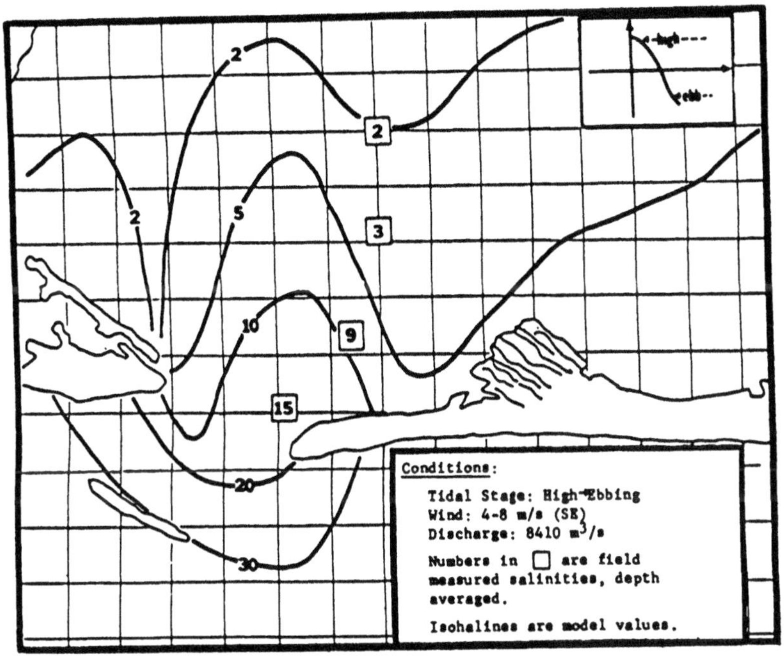

Figure 3b. Isohalines for the Lower Mobile Bay under River Flood State Conditions; (b) 8410 m^3/s.

about 37 km. The atmospheric pressure drop across the storm was 90 millibars. The maximum sustained gusts were estimated at 90 m/s.

The surge simulation extended over a 48-hour period beginning at 18:00 on August 16. A time step of 60 seconds was selected for numerical stability. The storm surge hydrograph at the Gulf/Bay interface is shown illustrated in Figure 5. There is a rapid buildup of water from 1.2 to 2.5 m in a 5 hour period. Initially a time shift occurred (computed elevation ahead of the observed elevation by 6 hours) between the observed and calculated hydrographs. Subsequent calculations diminished this time lag and increased the accuracy of the predicted hydrograph. In terms of storm surge models the results are reasonable; the frequency and amplitude of the hydrograph are in close agreement.

The maximum point on the computed elevation is 2.5 m, the lowest value is 0.2 m. The actual data (broken line) show a maximum elevation of 2.1 m and minimum elevation of 0.4 m. Inevitably with this computed data as input into the hydrodynamic model some deviation from the actual conditions will be introduced. However, the variations are such that the water movement and

Figure 4. Path of Hurricane Camille, August 16-18, 1969.

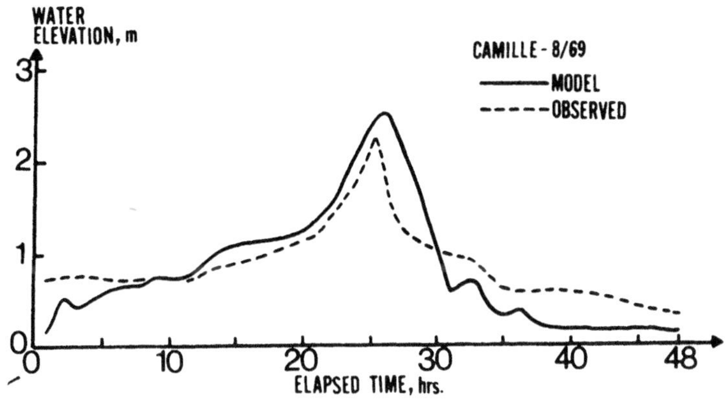

Figure 5. Storm Surge Hydrograph at Dauphin Island during Hurricane Camille.

elevation results predicted by the model are still within the accuracy associated with observed field data used in comparative studies contained in the next section.

CASE STUDIES

River Flooding

The influence of river discharge rate from 340 to 6960 m^3/s on Mobile Bay water movement, and salinity during non-flooding conditions has been studied in recent projects (April et al., 1976). The results presented in this study extend that analysis to higher flow rates and compares the results with those of the previous study. Trend charts will also be shown to illustrate critical conditions which result in rapid changes in water behavior within the bay.

Salinity Analysis

Figure 6 consists of tidal averaged isohalines that exist in Mobile Bay for four river discharge conditions. These discharges are 340, 1250, 6960, and 8520 m^3/s under a no wind condition (valid for wind speeds which are variable and less than 5-6 m/s) for comparison.

Figure 6. Tidal Cycle Averaged Salinity Profiles (in ppt) in Mobile Bay for Four River Flow Rates.

The progression of the high isohaline (15 ppt) at State Docks in Figure 6(a) at 340 m^3/s to the Main Pass at 8520 m^3/s in Figure 6(d), illustrates the impact that fresh water inflow to the bay has on salt water intrusion from the Gulf. A still greater compression of the isohalines occur in the Main Pass as the flow approaches flood stage conditions (8520 m^3/s).

Salt water penetration, as measured by the depth-averaged salinity profiles, shows noticeably large changes below values of 8500 m^3/s. Near this value and above, the nature of salt water intrusion seems stable.

This is perhaps best illustrated by studying the results shown in Figure 7. In these graphs the salinity (tidal cycle and depth averaged) is plotted against river discharge rate for three locations; Point Clear, Cedar Point and Bon Secour, as identified in Figure 3.

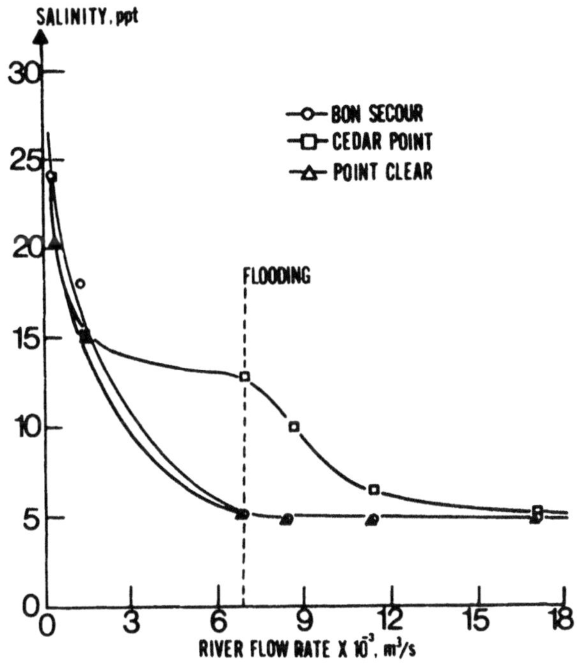

Figure 7. Salinity as a Function of River Discharge Rate for Three Mobile Bay Locations.

The idea of a critical flow rate condition is best illustrated in the Bon Secour Bay area. There is an approach to a constant level in the curve at 7000 m^3/s with nearly a constant, shallow slope beyond that point. The same general trend occurs at Point Clear.

At discharge values approaching the critical river flows, the bay becomes river dominant with large volumes of fresh water and runoff being introduced into the bay. Along with this condition comes potentially high levels of turbidity resulting from soil laden water and bay bottom scouring due to high flows. Also bacterial levels (measured as coliform) may increase affecting commercial shellfish operations in the area. Fortunately, the period of disturbance is short 3-5 days (except for very severe flooding conditions, 10-15 days), and recovery is rapid.

Water Elevation Analysis

The impact of flooding on water elevation is also important to analyze. This analysis will be made at three locations in the bay; State Docks, Point Clear, and Bon Secour Bay. The results are plotted in Figure 8. The water elevation increases rapidly with increasing river flow rate at State Docks. At Bon Secour Bay and Point Clear there is a less noticeable change in water elevation. This result can be explained by the bay geometry. In the broader lower bay area changes in volumetric throughput only slightly affect water elevation. Large volumes of water buildup are needed before substantial changes are noted. With the proximity of the passes, the water is rapidly discharged. That which accumulates during any given period is spread over the large expanse of available bay. As the water accumulates in the northern half of the bay, however, this condition changes. The broad area decreases by nearly a factor of 2 causing water elevation to increase. These increases are partially diminished by the open area between the rivers; however, there is often a time factor involved as a result of the rather narrow and restricted channels of the rivers.

Velocity Analysis

The velocity distributions within the bay during flooding conditions are illustrated in Table 3 for five locations.

At the State Docks, water flows only in the southern direction (270°, negative y-direction). The latitude component is nearly zero. This is due to the large amount of river flow into the bay. The velocity in each condition at the Main Pass shows a flow direction out of the bay, because of the large amount of river water inflow into the bay. These data also show that river water dominates the bay behavior while under a river flooding condition.

It should also be noted that the magnitude of the velocity does not greatly change as the river flow rate varies from 3×10^3 m^3/s to 8×10^3 m^3/s. This condition is also expected, since increases in water elevation cause the magnitude of the velocity to change only slightly as the river flow rate varies.

Table 4 shows the tidal cycle averaged velocity at river flow rates of 1.13×10^4 m^3/s and 1.7×10^4 m^3/s at the five locations previously specified. The flow directions are nearly constant. However, changes in the magnitude of the velocity as a result of river flow rates are noted. The greatest change occurs at Main Pass (highly restricted area). The discharge at Cedar Point is only slightly increased, indicating a majority of water movement leaving the bay via the Main Pass. Other locations (State Docks,

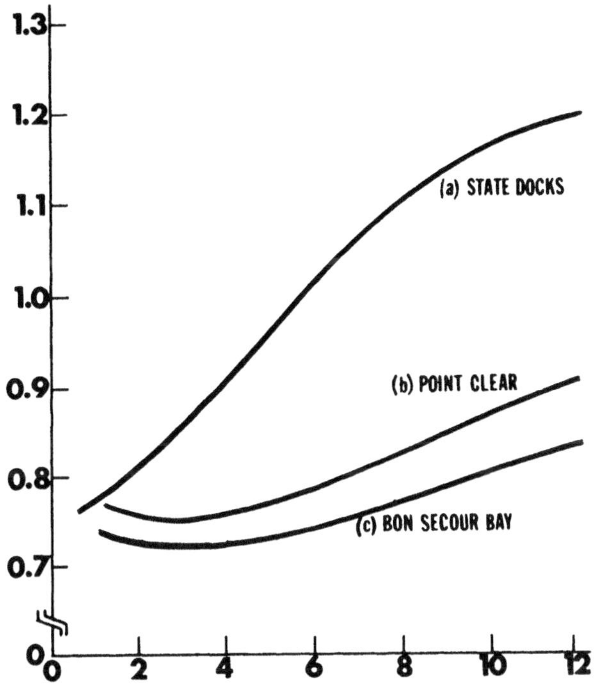

Figure 8. Maximum Water Elevations (metres) vs. River Flow Rate for (a) State Docks, (b) Point Clear and (c) Bon Secour Bay.

Point Clear, Bon Secour Bay) show only a slight increase.

Storm Surge

Similar comparisons of water elevation, velocity and salinity distribution can be shown under storm surge influence. A comparison of flooding results with those observed under storm conditions is also made to illustrate the severity and complexity of the disturbances caused by natural upsets.

Water Velocity Analysis

The magnitude and direction of the tidal cycle averaged velocity and the velocities over 4 hour time intervals within the surge period are shown in Figure 9. For each plot shown, lines 1, 2 and 3 represent a rising surge stage and lines 4, 5 and 6 represent a diminishing surge stage.

Table 3. Velocity Distribution of Mobile Bay Waters under River Flood Stage Conditions

Tidal Stage[b]	Flooding		High - Ebbing		Flooding - High		High - Ebbing	
River Flow Rates m^3/s	3000		5000		7000		8000	
Velocity Components[a]	Mag.	Dir.	Mag.	Dir.	Mag.	Dir.	Mag.	Dir.
Bay Location:								
Main Pass	1.33	267.8	1.36	269.0	1.41	269.0	1.46	268.0
Cedar Point	0.24	240.9	0.20	246.0	0.21	242.0	0.25	244.0
State Docks	0.29	270.0	0.39	270.0	0.51	270.0	0.60	270.0
Point Clear	0.06	310.0	0.10	324.0	0.10	324.0	0.12	322.0
Bon Secour Bay	0.06	216.0	0.09	199.5	0.07	189.0	0.07	220.0

[a] The velocity components include the magnitude (Mag.) measured in m/s and the direction (Dir.) measured in degrees where 270° represents flow to the south.

[b] Flood stages represent the periods over which field data were available for comparison purposes. The complete tidal cycle was broken down into fourths; flooding, high, ebbing and low, see Figure 9 for illustrations.

Table 4. Velocity Distribution (Tidal Cycle Averaged) at Two Flooding Conditions.

River flow rate m^3/s	1.13×10^4 (400,000 cfs)		1.7×10^4 (600,000 cfs)	
	Magnitude m/sec	Degrees*	Magnitude m/sec	Degrees*
Main Pass	0.45	268.	0.67	268.
Cedar Point	0.22	235.	0.24	240.
State Dock	0.08	270.	0.11	270.
Point Clear	0.09	315.	0.15	315.
Bon Secour Bay	0.03	268.	0.05	269.

*270° indicates a flow in the southerly direction.

As expected the directions of the velocity are good responses to the tidal height forcing function at each location. One exception is noted. The water movement at Cedar Point does not correspond to the surge hydrograph during the periods outlined. (See curves 2 and 4 in Figure 9 for Cedar Point). This deviation is thought to be caused by the interaction of the Main Pass with Cedar Point, especially during the critical periods just before and just after the surge crest. Such complex interactions have been observed under less severe conditions. As the velocity increases in magnitude through Main Pass from 0.2 to 0.4 m/s (period 1 to period 2), a corresponding shift from a westerly (period 1) to a southeasterly flow occurs at Cedar Point. As the surge crest is approached at the Main Pass (period 3), the flow at Cedar Point recovers to its original magnitude and direction. Note however, that the discharge is consistently from the Bay to the Sound during all surge periods blocking water inflow from the sound through Cedar Point. Under these conditions the bay behavior is dominated by the surge hydrograph. This is better illustrated by considering the tidal cycle averaged salinity profile for the bay (see next section). In this analysis, the wind condition is held constant at 76.8 km/hr. In actuality, the wind condition varies as the storm front moves over the coastal area. This variation in wind speed and direction could result in greater changes in behavior than those predicted by

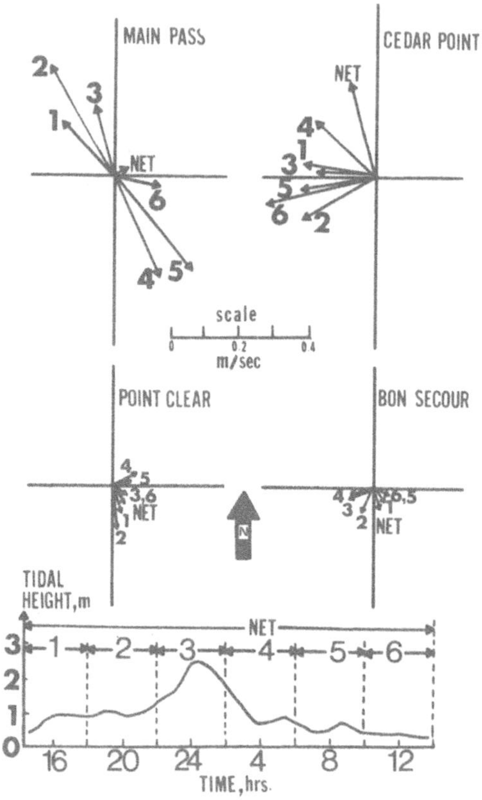

Figure 9. Four Hour Averaged Velocity Patterns during the Surge Hydrograph of Hurricane Camille.

considering the wind to be at a constant value. An improvement over the current model can be made by incorporating a wind field that is matched with the hydrograph. This would produce better results; however, the trend would be expected to show the same impact as that obtained in the simpler constant-wind case.

Salinity Analysis

Figure 10 shows the salinity profile within the bay under a storm surge condition. Comparison of the results with those in Figure 6 indicate a high level of salinity intrusion into the entire bay area. Since the river flow rate in this storm surge analysis was held constant at 1250 m^3/s, a comparison can be made with results obtained under normal conditions; and conditions describing

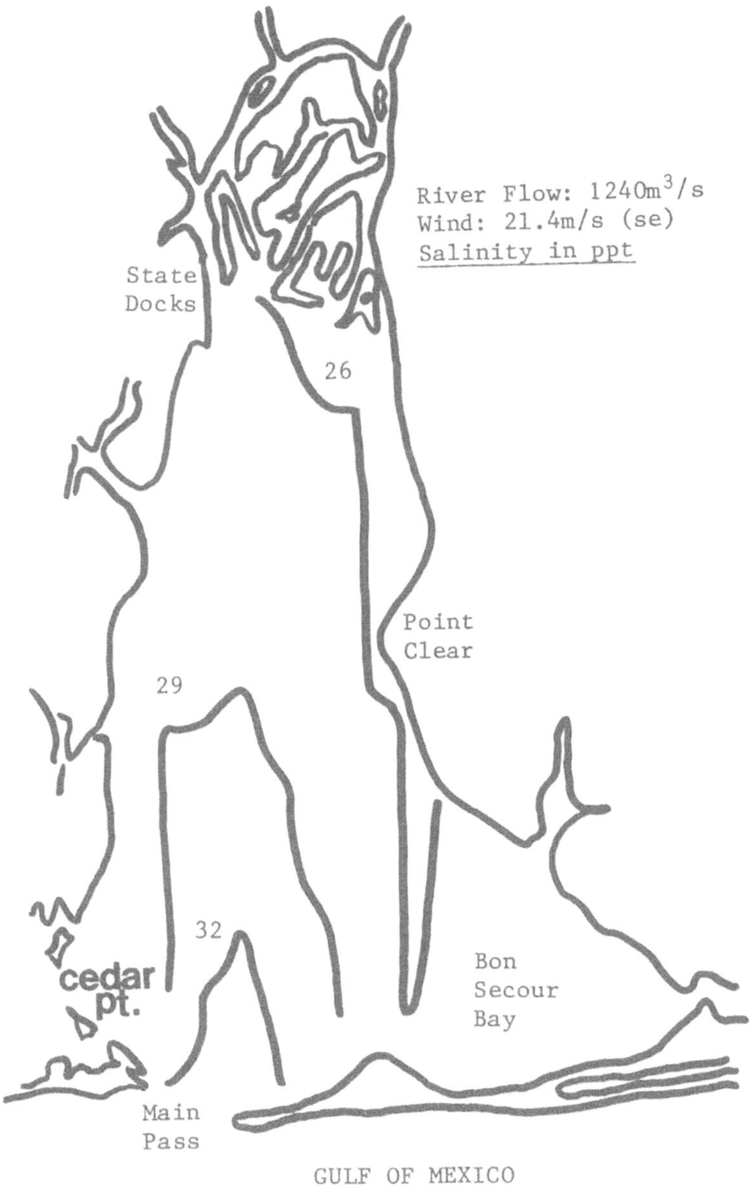

Figure 10. Salinity Profile in Mobile Bay under Storm Surge Conditions.

river flooding.

Table 5 shows the salt concentration for four different conditions, normal (with river flow rate 1250 m^3/s), river flooding (river flow rate of 1.13 x 10^4 and 1.7 x 10^4 m^3/s), and storm surge (river flow rate of 1250 m^3/s). The wind condition, except in the storm surge case (76.8 km/hr, SE), is calm. The first comparison is between normal and flooding conditions. The salt concentrations in all locations are lower in the river flooding case. The salinity levels also decrease as the river flow rate increases. A comparison between the normal condition with the storm surge case at the same river flow rate shows that the salinity levels increase, as would be expected.

In all cases representing extremely abnormal behavior, the bay water movement and properties are greatly altered. The ability of the system to return to normal conditions is severely tested when such events impact on the bay.

Water Elevation Analysis

Table 6 lists the water elevations for three different river flow rates and a storm surge upset. The elevations were increased everywhere. This is especially the case during the storm condition. For Bon Secour Bay the elevation under storm surge condition is nearly twice those of the other three conditions. The reason for this rise is due to the influx of the water entering the bay as a result of the storm hydrograph. The constant flow inward at higher than normal elevations causes the entire bay to increase in volume. Some dissipation occurs moving south to north as water overtops land areas. This is particularly true in the Bon Secour coastal areas and areas north of the bay between the rivers. Land affected in this manner can be clearly shown by using a SYMAP projection method. By comparison of the land masses at normal bay conditions with those at severe conditions, the extent of land flooding in each area can be identified.

These added facts support the observations that were made by considering the salinity and water movement patterns in the previous sections.

CONCLUSION

The salinity field data described by Schroeder (1977) provides a trend analysis of water movement in Mobile Bay during the river flooding stage of 1973. The results from this numerical simulation study are in agreement with these field data. The open coast surge

Table 5. Comparison of Salinity Values (in ppt) at Five Locations in the Bay as a Function of River Flow Rate and Storm Surge Condition

River Flow Rate m^3/s	1250	11300	17000	1250
Condition	Normal	River Flooding	River Flooding	Storm Surge
Bay Location:				
Main Pass	30	14	13	34
Cedar Point	18	5	3	23
State Docks	5	1	1	27
Point Clear	15	1	1	26
Bon Secour Bay	16	3	2	20

Table 6. Comparison of Maximum Water Elevations (in Meters) for Three Mobile Bay Locations at Varying River Flow Rates or under Storm Surge Conditions

River Flow Rate m^3/s	340	1250	6960	1250
Condition	Low Flow	Medium Flow	High Flow	Storm Surge
Bay Location:				
State Docks	0.78	0.79	1.10	1.19
Point Clear	0.77	0.76	0.78	1.48
Bon Secour Bay	0.75	0.74	0.75	1.65

model developed by Wanstrath (1978) also gives a reasonable computed water level at Dauphin Island. This surge hydrograph can be used to drive the general hydrodynamic model for Mobile Bay extending the capability to describe bay behavior under severe weather conditions.

Based on the investigations conducted in this study, the following conclusions can be reached:

River Flooding

The increase in water elevation, the lowering of the salinity levels in all bay locations, and the characteristic velocity profiles indicate that the bay water behavior is dominated by fresh water inflow from the rivers near flood stage levels.

(1) The magnitudes of velocity at State Docks increases from 0.29 m/s to 0.60 m/s as the river flow rate increases from a value of 3.0×10^3 m^3/s to 8.4×10^3 m^3/s. In each study case, the direction of flow is due South (270° at State Docks) as would be expected under river flood stage conditions. At Main Pass, the magnitude of velocity varies from 1.33 to 1.46 m/s, and the flow direction changes only slightly from 267° to 269° as the river flow changes from below flooding (3×10^3 m^3/s) to a point just above flooding (8.4×10^3 m^3/s). Variations in velocity magnitude and direction at points intermediate to these two locations follow the same general trend.

(2) The tidal cycle averaged 30 ppt isohaline is established about 10 km north of Main Pass under low river flow conditions; 340 m^3/s. As the river flow rate increases to flood stage and beyond (7000×10^4 m^3/s), the salinity level falls to 4 ppt at this location. The salt water intrusion is increasingly suppressed in the Lower Mobile Bay area as the river flow rate increases. When the river flow rate approaches flood stage (7000 m^3/s), the fresh river water begins to dominate and control the water behavior in the northern and central parts of Mobile Bay. When the river flow rate becomes greater than 8.5×10^3 m^3/s, no noticeable change in salinity levels appears in the lower Mobile Bay area. A stable condition exists between the fresh water and saline Gulf water in this part of the bay. Under these conditions, fresh water totally dominates the Mobile Bay system.

(3) At the State Docks, in the Northern Bay area, the elevation rises from 0.75 to 1.18 m as river flow rate varies from 340 to 11,300 m^3/s. For the same river flow rates, the elevation changes only slightly from 0.75 to 0.79 m in the Bon Secour Bay area. This rather small increase in elevation is attributed to the divergent nature of the bay in the lower portion, and the rapid discharge of waters through the Main Pass.

Storm Surge

During storm surge conditions, large amounts of saline water are introduced into the bay from the Gulf. The Mobile Bay water behavior is therefore said to be dominated by the intruding salt water.

(1) A tidal cycle averaged 27 ppt isohaline is observed as far as the State Docks under storm surge conditions typical of the hurricanes impacting the coast over the past century. This high intrusion of salt water can be compared with a 5 ppt isohaline at a normal condition with a river flow rate of 1.25×10^3 m^3/s. Similarly under a flooding condition with a flow rate of 1.70×10^4 m^3/s, salinities less than 1 ppt appear at State Docks. These comparisons illustrate the dominance of the Gulf water intrusion under storm conditions.

(2) The magnitudes and directions of the velocity are in good agreement with the storm surge hydrograph. The net velocity over the hydrograph period at Main Pass is 0.03 m/s at an average of 25.3 degrees (ENE). At Cedar Point the net velocity is 0.40 m/s at 105 degrees (NNW). The flow in these areas, which are the only two passes connecting the bay with the Gulf of Mexico (via Mississippi Sound at Cedar Point), show net flows into the bay over the surge period.

(3) The water elevation at State Docks is 0.79 m under a low river flow rate of 1,250 m^3/s, but increases to 1.19 m under storm surge conditions. The elevation at Bon Secour Bay varies from 0.75 to 1.65 m as conditions change from normal to storm surge status. The water elevation increases in the entire Bay, but is more pronounced in Bon Secour Bay due to the large and rapid inflow of water from the Gulf which is driven by the storm front as it approaches the coastline.

Comparison of Storm Surge Effects with River Flooding Effects

The impacts caused by storm surge and river flooding conditions upon the Mobile Bay area produce increases in water elevations, large variations in the salinity levels from those observed under normal river flow and wind field conditions. These large and rapid changes have caused, in the past, losses in human life, property and aquatic habitats within the area.

(1) At State Docks, the elevation under normal river flow rate conditions (1,250 m^3/s) is 0.79 m. When the river flow rate increases to 1.70×10^4 m^3/s, the elevation increases to 1.18 m. At Cedar Point the elevation under a normal flow rate condition, flooding condition (1.70×10^4 m^3/s) and storm surge condition are 0.76, 0.89 and 1.48, respectively.

The impacts under storm surge and river flooding stage will cause the water elevation to increase. The rise in water elevation caused by storm surge is more pronounced in the lower bay and results in a larger increase (1.65 m as compared to 0.74 m in

Bon Secour Bay). The greatest change in elevation under river flood stage conditions occurs at State Docks (1.19 m as compared with 0.78 m under normal flows).

(2) At State Docks, the salt concentration is 5 ppt at normal conditions (1.25×10^3 m^3/s river flow rate). This value decreases to less than 1 ppt under river flood stage conditions (1.70×10^4 m^3/s river flow rate). Conversely, the salinity increases to 27 ppt under storm surge conditions investigated in the study. The large variations in salinity can have a serious effect on Mobile Bay behavior.

RECOMMENDATIONS

(1) In order to test the predictive capability of models outlined in this study the open coast model used to produce the storm surge hydrograph at the Main Pass of Mobile Bay should be tested against, and compared with, field data from a second or third storm. The absence of accurate and complete data for Hurricane Betsy made it impossible to verify the hydrograph produced by Wanstrath's (1978) open coast model in this application.

(2) To insure greater accuracy in predicting water elevation, water movement and salinity within the bay, the use of a dynamic wind field matched to the hydrograph should be incorporated into the analysis. The work reported uses an average wind field condition over the surge period. Whereas, the overall trend behavior is useful from this analysis (i.e., comparing the effect on Mobile Bay of one storm with another), it does not permit the investigation of behavior in those areas such as Bon Secour Bay where a variable wind field could have an important effect on water behavior.

(3) Additional data are also needed to predict the corresponding hydrograph at Cedar Point (or at Petit Bois Pass translated to Cedar Point). Further study in the Mobile Bay area with similar numerical simulation models for the Mississippi Sound is needed to better describe the interactions that exist at the sound/bay exchange.

ACKNOWLEDGMENT

The research reported in this paper was accomplished under the sponsorship of a Water Resources Research Institute Contract (Project A-061-AL) at Auburn University, Research Grants Project No. 921 at The University of Alabama, and a Mississippi-Alabama Sea Grant Project (R/ES-4). In addition the authors would like to recognize the efforts of the following persons who contributed significantly to the projects; Donald O. Hill, Hua-An Liu, Samuel Ng, and Stephen Hu.

APPENDIX A

Nomenclature

C	Concentration of Species in the Water Column, M/L^3
D	Depth of Water in the Bay, L
E	Rate of Mass Transfer by Evaporation, L/T
	Dispersion Coefficient, L^2/T
f	Bay Bottom Friction Factor
g	Gravitational Acceleration, L/T^2
H	Height of Water above a Cell Datum, L
K	Constants
Q	Discharge Rate, L^2/T
R	Rate of Mass Transfer by Rainfall, L/T
	Rate of Disappearance or Appearance of Mass, M/L^3T
t	Time, T
V	Resultant of the Velocity Vector, L/T
v	Local Grid Velocity, L/T
W	Angular Velocity of Earth, L/T
x	Distance (East-West), L
y	Distance (North-South), L
z	Distance (Depth Direction), L
η	Wind Speed, L/T
ϕ	Wind Direction
ψ	Angle Measurement in the Coriolis Term

APPENDIX A (continued)

Subscripts

b	Bottom
o	Source or Sink Term
s	Surface
x	East-West Direction
y	North-South Direction

REFERENCES

Gary C. April, Donald O. Hill, et al., 1976. "Water Resources Planning for River Draining into Mobile Bay" National Aeronautics and Space Administration, George C. Marshall Space Flight Center.

Gary C. April, Samuel Ng and Stephen Hu, 1978. "Computer Simulation of Storm Surge and River Flooding in Mobile Bay," Interim Report, The University of Alabama, University, Alabama.

Gary C. April and Donald C. Raney, 1979. "Mathematical Modeling of Coastal Waters: A Tool for Managers and Researchers," Proceedings of the Symposium on the National Resources of the Mobile Estuary, Mobile, AL, May 1-2, 1979.

Donald O. Hill and April, Gary C., 1974. "A Hydrodynamic and Salinity Model for Mobile Bay," Report to NASA, The University of Alabama, BER Report No. 169-112, University, Alabama.

Philip E. LaMoreaux, R. L. Chermock, 1976. "Hurricanes and Tornadoes in Alabama," Geological Survey of Alabama, University, Alabama.

National Science Board, 1972. "Patterns and Perspectives in Environmental Science," Washington, Natl. Sci. Board, 426 p.

William W. Schroeder, 1977. "The Impace of the 1973 Flooding of the Mobile Bay River System on Mobile Bay and East Mississippi Sound," Northern Gulf Science, Vol. 1, No. 2, pp. 67-76.

John J. Wanstrath, 1978. "An Open-Coast Mathematical Storm Surge Model with Coastal Flooding for Louisiana," Report 1, U. S. Army Engineer Waterways Experiment Station, Vicksburg, Miss.

John J. Wanstrath, Robert E. Whitaker, Robert O. Reid, and Andrew C. Vastano, 1976. "Storm Surge Simulation in Transformed Coordinates," Technical Report No. 76-3, U. S. Army Corps Engineers, Coastal Engineering Research Center, Fort Belvoir, Va.

HYDRODYNAMIC-MASS TRANSFER MODEL OF DELTAIC SYSTEMS

L. M. Hauck and G. H. Ward

Espey, Huston and Associates, Inc.

3010 South Lamar, Austin, Texas 78704

ABSTRACT

A branching section-mean model has been developed for the simulation of the hydrodynamics and nutrient transport in estuarine deltaic systems, in which the momentum conservation, continuity and mass transfer equations are solved by the method of finite differences. Example executions of the hydrodynamic portion of the model are presented for the Lavaca, Guadalupe and Trinity deltaic systems. Computed water elevations and flows are compared to observations to evaluate model performance for a variety of conditions. The model is shown to obtain satisfactory results for conditions of variable freshwater inflow, such as associated with small-to-moderate floods, and meteorologically influenced tides, e.g., wind-induced setup or setdown, as well as low-flow, astronomical-tide regimes.

INTRODUCTION

The lowermost reaches of many river systems consist of deltaic systems with complex distributary networks and tidal flats with extensive marshes. These deltaic marsh systems are areas of low relief with narrow, interconnected channels, fed above by inflow from a river or river system and terminated below by a tidally forced, open-water area, either a coastal embayment or the sea. The head gradient is developed more from the slope in the water surface than the slope of the bed (the reverse of the situation in the upland reaches). Within the deltas, lateral areas contiguous to the channels are flooded and dewatered with rise and fall of water levels, which may arise from extreme tidal range, wind setup, floods in the river, storm surge, or a combination of these factors.

A branching section-mean model has been developed for the simulation of the hydrodynamics and constituent transport in estuarine deltaic systems of the coast of Texas. The basic hydraulic characteristic of these delta systems and the underlying postulate in the hydrodynamic model formulation is that the momentum of the flow pattern is concentrated in the longitudinal components of the channels. This characteristic prevails even at moderate levels of inundation, because the inundated areas function principally as volume storage and carry relatively little momentum. Various kinetic processes have been formulated in the companion mass-transport computation. At present this constituent transport model simulates concentrations of salinity, phosphorus, organic nitrogen, ammonia, nitrites, nitrates, total organic carbon, and two phytoplankton species as a function of time and space in the delta. (Clearly other constituents may be incorporated so long as the kinetic and boundary flux processes are capable of mathematical expression). The present paper discusses only the hydrodynamic aspects of the model.

MODEL FORMULATION

The formulation of the branching section-mean hydrodynamic portion of the model essentially follows the approach of Dronkers (1964) for calculation of tides in estuaries and tidal rivers and has been applied to estuaries in this country by Harleman and Lee (1969), among others. The mathematical model consists of the equations of momentum conservation and continuity, averaged across a section so that spatial variation is restricted to the longitudinal axis of the channel:

$$\frac{\partial Q}{\partial t} + \frac{\partial}{\partial x}\left(\frac{Q^2}{A}\right) + g\,A\,\frac{\partial H}{\partial x} + \frac{g n^2 Q|Q|}{2.22 A R^{4/3}} = 0 \qquad (1)$$

$$\frac{\partial H}{\partial t} + \frac{1}{B}\frac{\partial Q}{\partial x} - \frac{Qf}{SA} = 0 \qquad (2)$$

where:

Q = flow in conveyance channel,
A = cross-section area of conveyance channel,
H = water level (referenced to a standard datum),

R = hydraulic radius,
n = Manning's roughness parameter,
B = lateral width,
SA = surface area (includes lateral storage area if inundation is occurring),
Qf = discharge into channel (inflow or withdrawal),
g = gravitational acceleration,
x = distance, longitudinal direction, and
t = time.

Equations (1) and (2) constitute a set of two equations with two unknowns, Q and H, each a function of both x and t. Note that the momentum equation is employed in its full nonlinear form. Detailed derivation of the hydrodynamic and equations is presented in Ward (1973).

These equations are solved by the method of finite differences. In order to do this, the delta is segmented into discrete sections, and variables are staggered in both space and time in such a way as to maximize the accuracy of the finite difference approximations. The momentum equation is advanced in time by a leap-frog timestep, and the continuity equation by a second-order Adams-Bashforth. These explicit methods have associated numerical stability constraints but provide a satisfactory level of accuracy vis-a-vis intrinsically stabler implicit methods permitting larger time increments. A more detailed explanation of the finite differencing techniques employed may be found in Hauck (1977 and 1978).

Specification of boundary conditions is a necessary stipulation for the solution of the equations, i.e., for operation of the model. For the hydrodynamic solution, these may be formulated as either Q or H as a function of time at each boundary point. In practice, the lower boundary, i.e., toward the bay mouth, is taken to coincide with the location of a recording tide gauge so that H as a function of time is immediately available as recorded tide data. Though the upper condition may be either a specification of H(t) or Q(t); in practice, the simplest upstream conditions are specified by flows above the limit of tidal influences, i.e., the flows measured at U. S. Geological Survey (USGS) streamflow gauges.

An important feature of the model formulation is the accounting of flooding and dewatering of marshes and floodplains, a common occurrence in low lying marshes subject to variable water elevations through tidal and freshwater flow fluctuations. The model also incorporates the appropriate supplementary hydrodynamic equations and computational logic to incorporate the occurrence of intersecting channels with confluence or difluence of flow, locks (with open-close operating criteria), dams and levees (with overflow criteria), multiple tidal boundary conditions, and transient channels whose bed elevations preclude their carrying flow until the water level exceeds some threshhold value.

In general, the marsh and floodplain areas to either side of channels in deltaic areas are of uniform elevation and end abruptly at the beginning of upland areas, thus allowing a rather simplistic approach to incorporating the watering and dewatering of these regions adjacent to channels. Each section in the segmentation of the system is described by a streambank elevation and two surface areas, the area of the conveyance channel and the total area subject to inundation when the streambank elevation is exceeded. The surface area SA utilized in the continuity equation, equation (2), is time dependent on the water elevation in each section. At every time step, the water elevation of each section is compared to its section bank elevation, if the water elevation is less than the bank elevation, the conveyance channel surface area is used in equation (2), but if the water elevation exceeds the bank elevation, the surface area which includes the floodplain is used. However, the conveyance channel width B does not increase upon inundation, hence the previously mentioned assumption that the momentum of the flow pattern remains concentrated in the longitudinal components of the channels. Also, inundation of the entire floodplain adjacent to the channel is assumed to occur whenever bank elevations are exceeded, because of the assumed uniformly flat floodplain associated with each section. While certainly not true in the absolute sense, the low relief in deltaic regions indicates that a more sophisticated handling of this lateral storage term is unwarranted.

APPLICATION OF MODEL

The model has been applied to the deltaic system of several major Texas rivers which debouch into embayments along the coast, for the purpose of verifying the model performance against measured data. In this paper the results of implementation of the model to the Lavaca, Guadalupe and Trinity River Deltas will be presented.

In order to accurately utilize the model on a deltaic system, physiographic data is required to describe each section in the segmentation. Such information as section length, total surface area subject to inundation, and overbank elevation can normally be obtained from USGS topographic quadrangle maps. However, survey information of the area is required to obtain channel widths and depths and more accurate definition of overbank elevation, particularly the physiography of transient channels and swales that carry flow when water levels are sufficiently high.

In applications of the model to be presented, a two-step approach was utilized involving first model "calibration" and then model "verification." During the calibration process the data required as input to the model, including roughness coefficients, channel widths and depths (when not completely specified by survey data), and floodplain surface areas (when not completely specified by topographic data) were adjusted appropriately to achieve the optimum replication of historically gathered hydrodynamic data, e.g., water elevations from recording tide gauges in the system. Upon completion of this phase, the model was then tested, without any further adjustments, against other historical periods for which field data were available for comparison of model predictions and measurements. These model results are then considered as verified results and it is only these verified-noncalibrated-results that are presented in this paper.

During the application of the model to each deltaic system, the dearth of data for calibration and verification was a continual problem. The chart records from tide gauges maintained in the deltaic areas, of which many were temporarily placed gauges for the study period, were the most voluminous data source. Because of oil, gas and water subsurface production along the Texas coast, these deltaic regions are influenced by subsidence, which presents particular problems to maintaining accurate datums for tide gauges. As a result of the difficult terrain, the tide gauges utilized in each study area were not surveyed in on a closed circuit to a common benchmark, which resulted in what apparently were datum discrepancies between gauges.

The more preferred method of hydrodynamic verification than water-level records is velocity measurements or flows based on velocity and stream cross section. The available velocity data consist of either a single midstream measurement or the average of two midstream measurements at different depths. When such point-specific flows, which are subject to turbulent fluctuations, are compared to smoothed, section-mean simulated velocities considerable discrepancies can arise and are, in part, expected. This problem was aggravated by the remarkable occurrence that all field programs coincided with frontal passages and the very dynamic influences of such meteorological events on deltaic flows and water elevations from wind setdown and subsequent recovery.

Lavaca Delta

The Lavaca River Delta is a flat, low-lying marsh area located in upper Lavaca Bay, a segment of the Matagorda Bay system. The two major freshwater sources to the delta, the Lavaca and Navidad Rivers, merge near the head of the deltaic region. During high tides and/or high streamflow, large portions of the delta are subject to inundation. A map of the area with superimposed model segmentation is shown in Figure 1, on which are indicated the locations of two continuous recording tide gauges and several crest-stage gauges which provided data for the calibration and verification of the model. For each segment, Figure 1 also shows the associated floodplain area.

The upstream limits of the model segmentation were taken to coincide with two USGS gauging stations, one each on the Lavaca and Navidad Rivers, whose records were the source of the necessary flow information for the upstream boundary conditions. The lower segmentation limit is in Lavaca Bay, coinciding with the location of a U. S. Corps of Engineers (USCE) continuous recording tide gauge near Point Comfort, Texas. This gauge at the time of the simulation was operated by the USGS. (The deltaic region shown in Figure 1 does not include the southern portion of the bay, where the tide gauge is located.)

After calibration of the model, by adjusting the roughness coefficient for conditions of nonmeteorologically forced tides with constant streamflow, the model was applied to cases of variable inflow and/or meteorologically forced tides. One such verification case involved a period of high tide followed by a freshwater flood passage, from 5 June through 16 June 1974. The tide as recorded in Lavaca Bay during this time began as diurnal, phasing into dominately semidiurnal and back to diurnal on the last two days (Figure 2a). The principal meteorological feature of this period was the approach and passage of a slow-moving frontal system. Prior to the actual frontal passage on 10 June, southeasterly winds were augmented in advance of the front, producing elevated water levels as a result of wind setup in the Matagorda-Lavaca Bay system. The maximum of the daily-averaged combined discharge for both rivers during this period was 330 m^3/s occurring on 11 June.

The comparisons of simulated and measured water levels at the two continuous recording tide gauges at the Vanderbilt Tide Gauge (Section 31) and at the Lolita Tide Gauge (Section 37) demonstrate good agreement as indicated in Figures 2b and 2c, respectively. Both the high-tide-induced water elevation of 8 and 9 June and the floodwater-induced elevations of 11-13 June are adequately reproduced by the model. Some tidal phasing errors occur, but the errors never exceed a couple of hours. Simulated water elevations are generally within one-tenth of a metre of the recorded values, indicating good agreement, though occasional periods of larger discrepancies do occur.

Though both the flood waters and high tides resulted in the inundation of large portions of the deltaic region, only the

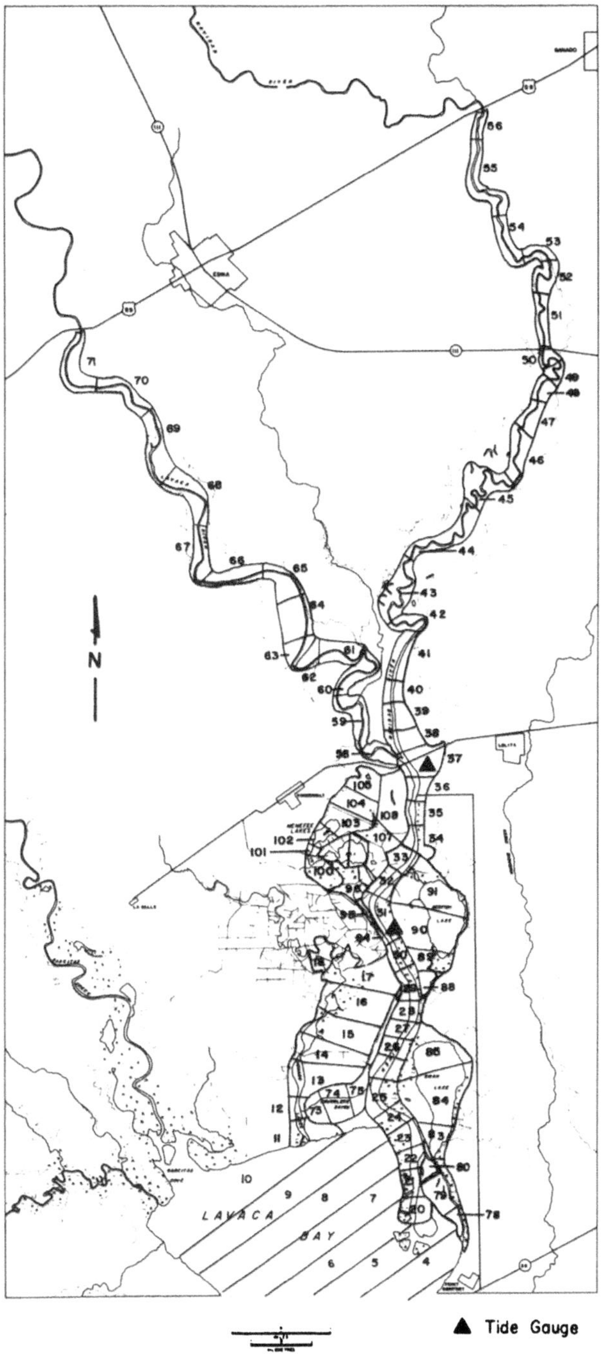

Figure 1. Lavaca Delta System showing floodplain area for each section.

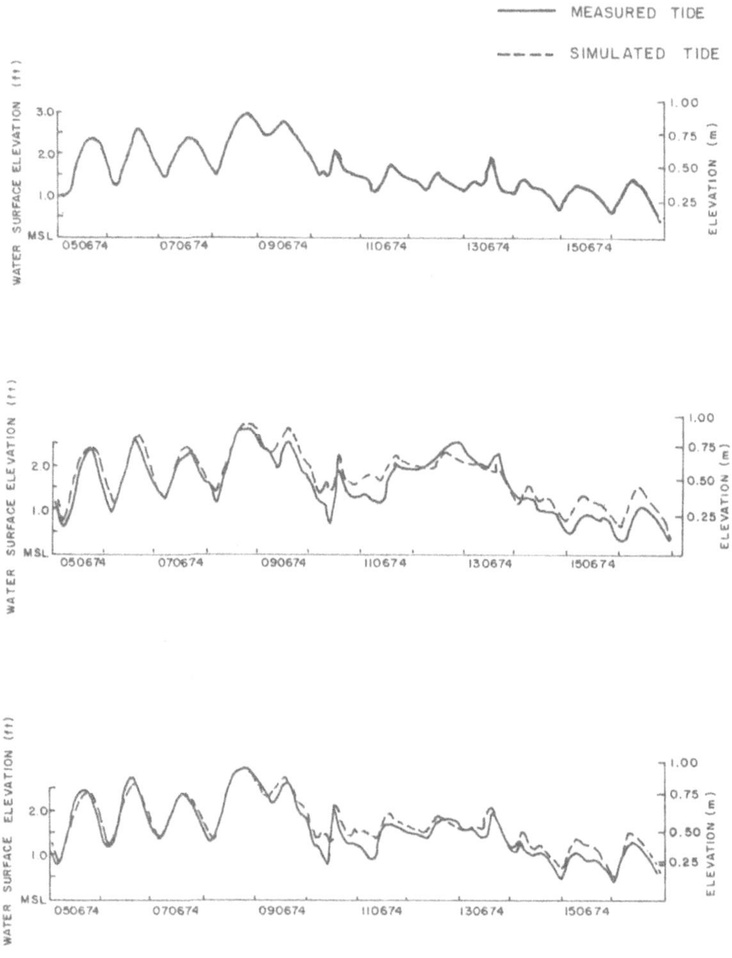

Figure 2. Tidal elevation record at (a) Section 2, gauge at Point Comfort; (b) Section 31, Vanderbilt Tide Gauge; (c) Section 37, Lolita Gauge.

high-tide event was recorded by the crest-stage gauges in the system. The comparison of recorded and measured water elevations at the crest-stage gauges is presented in Figure 3. At all gauge locations, with the exception of the Redfish Lake gauge, both water elevation and phasing were accurately replicated. The 0.2 metre discrepancy between measured and simulated elevations in Redfish Lake is difficult to explain. Since the measured elevation indicates water levels significantly higher than those recorded in the remainder of the system, it is suspected that the source of error is a gauge datum error.

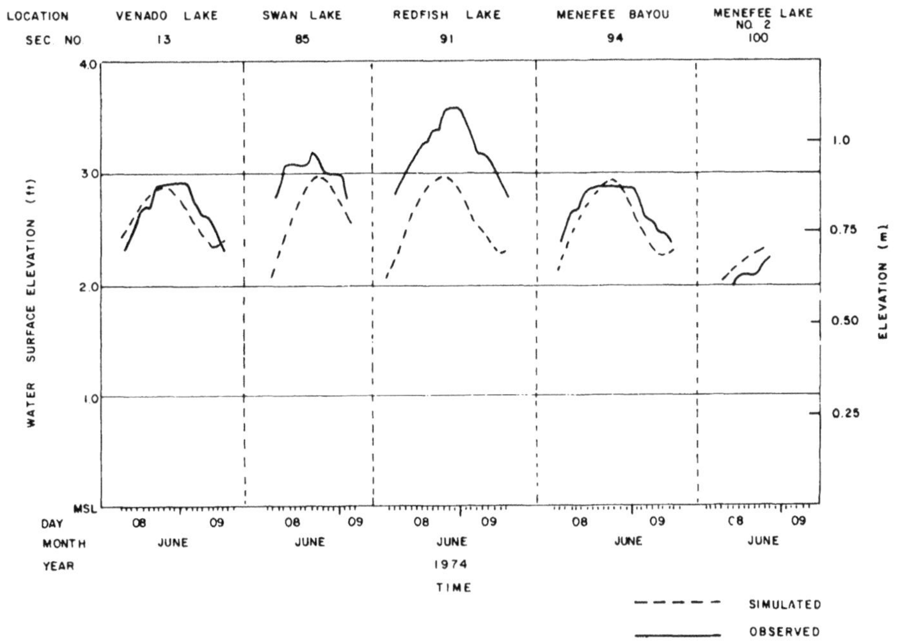

Figure 3. Tidal elevations at TWDB crest-stage gauges.

This example simulation is representative of other model operations in the Lavaca Delta which were verified by water elevation records from the gauges located within the system. Deltaic water elevation fluctuations resulting from pure astronomical tides (those without meteorological influences) and conditions of low to moderate inflow were accurately reproduced. As flow becomes variable and of sufficient magnitude to induce inundation of marsh areas, and as the tides become influenced by meteorological factors such as winds and barometric pressure, the response of water levels in the delta becomes very dynamic and discrepancies between simulated and measured water elevations become more prominent. However, even for the extremely dynamic series of events of high tides from wind setup followed by moderate flood flows, the model and segmentation of the Lavaca Delta reproduce measured water levels with reasonable accuracy. For a more complete presentation of the modeling of the Lavaca see Hauck et al. (1976).

Guadalupe Delta

Another model application was conducted on the Guadalupe River Delta, which is again a low-lying marsh area, but of greater complexity that the Lavaca River Delta. The delta is located in the upper reach of San Antonio Bay. A map of the delta with segmentation and recording tide gauge locations is shown in Figure 4. Two large marsh areas containing a series of interconnected channels, one area north and the other south of Mission Lake, comprise the majority of the delta. The southern marsh area receives little freshwater input other than from direct rainfall and consequently inundation of this area normally results from high tide levels in the bays. In contrast, the northern marsh receives considerable inflow from a series of bayous. The numerous tide gauges sited in the delta for this and related studies were a source of calibration and verification data for the model. In addition, a diurnal biological and hydrodynamic study conducted in November 1975 provided further data for model verification, including current measurements throughout the tidal cycle at several stations.

The lower boundary of the segmentation coincided with the USGS recording tide gauge near Seadrift in San Antonio Bay, which was the source of tidal data for the lower boundary contition of the model. (Operation of this gauge was discontinued in 1976.) The upper boundaries of the model were taken to coincide with the crossing of seven rivers and bayous by State Highway 35, and the flows for these streams was based on daily staff gauge readings by the Guadalupe-Blanco River Authority on the three major streams, the Guadalupe River, Hog Bayou and Goff Bayou.

The calibrated model was employed to simulate the period of 4 November through 12 November 1975, a period of low inflow. During this period flows were occurring at only two of the seven inflow points: the Guadalupe River at 50 m^3/s and Goff Bayou at 33 m^3/s. The passage of a front accompanied by strong northerly winds occurred in the early morning of 12 November and the resulting depression in deltaic water elevations from wind setdown is apparent in the downstream tide record (Figure 5a). Of the recording tide gauges positioned in the system, only those at Lucas Lake (Section 48), at Mamie Bayou (Section 22), and on Schwings Bayou (Section 137) were operating; a comparison of recorded and simulated elevations for these gauges is presented in Figures 5b through 5d, respectively. Tidal amplitudes and phases are correct at all three locations,

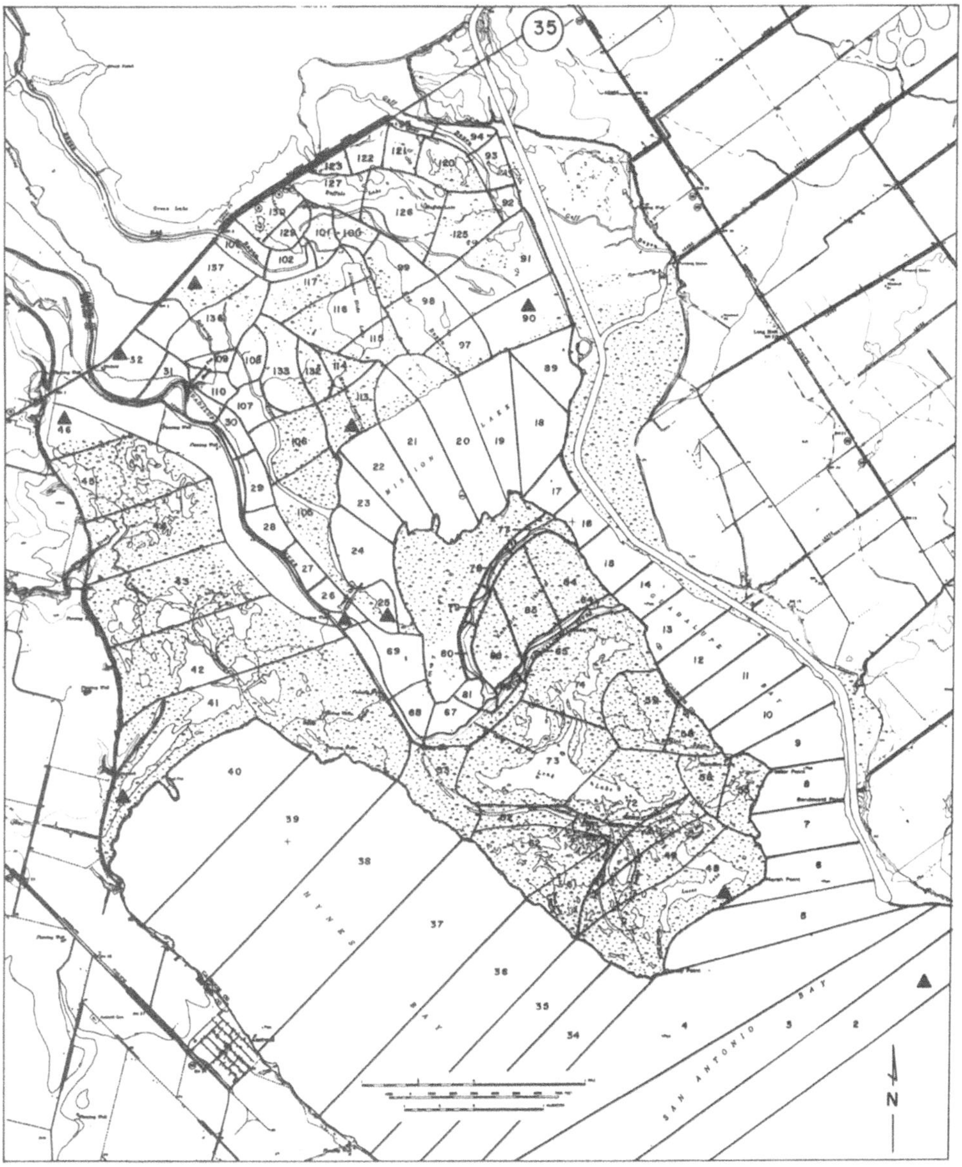

Figure 4. Guadalupe Delta System showing floodplain area for each section.

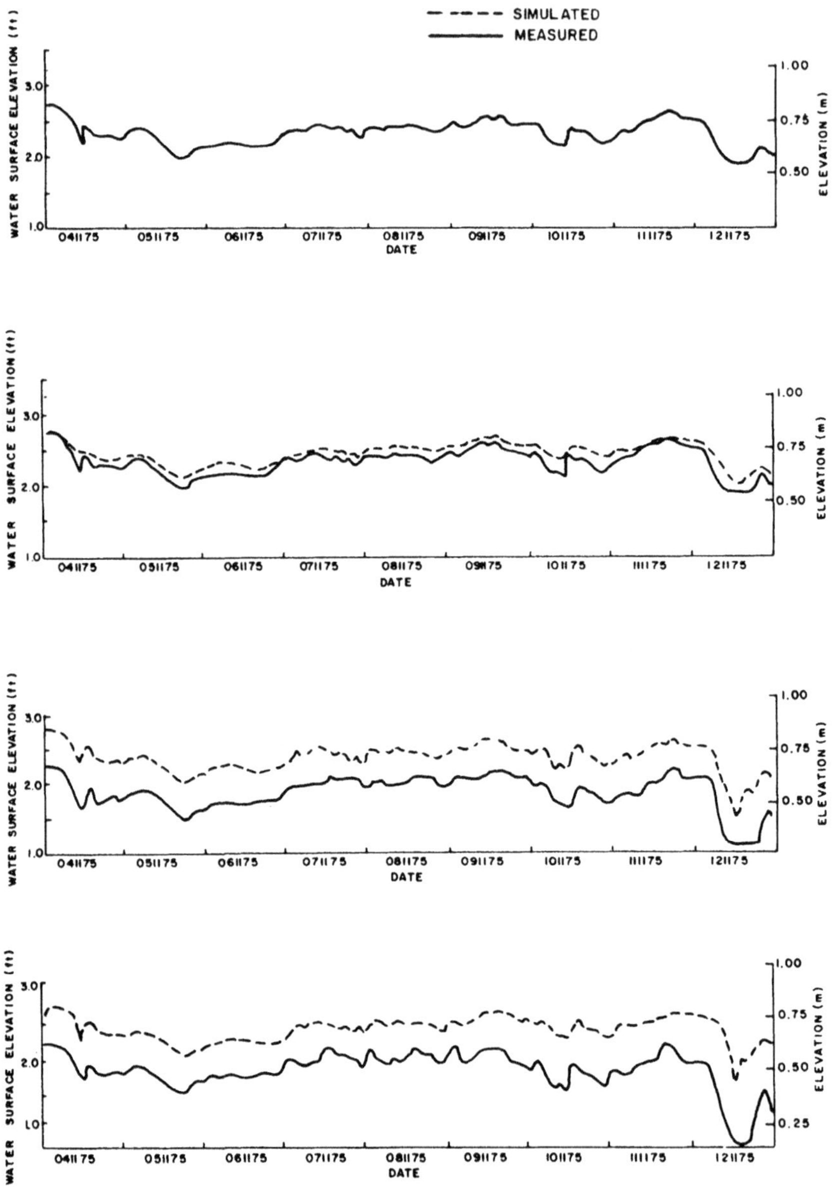

Figure 5. Tidal elevation record at (a) Section 2, gauge at Seadrift; (b) Section 48, gauge at Lucas Lake; (c) Section 22, Mamie Bayou Gauge; (d) Section 137, gauge on Schwings Bayou at Hwy. 35.

though an increase in simulation error occurs on 12 November with the increased dynamic conditions of the frontal passage and wind setdown. (The hydrodynamic model does not include the direct wind-stress or atmospheric pressure gradient, which may account for the discrepancy.) A 0.2-m displacement between measured and simulated water levels at both the Mamie Bayou and the Schwings Bayou gauges is apparent. This 0.2-m displacement was present in all simulations of the system, and the datum of the gauges appears to be the cause of the error. A strong indicator substantiating this view is obtained by a comparison of relative water elevations at the gauge locations and the bay. The driving tide in San Antonio Bay near Seadrift, Texas recorded a mean elevation of approximately 0.75 m above msl for this period, whereas a mean elevation of 0.6 m above msl was recorded at the two gauges in question. This would indicate a consistently lower water level in the delta region than in the bay, which is unlikely, certainly for such an extended period of time.

One feature of the Guadalupe Delta that should be especially noted is the feeble and distorted tidal influence in this system, in marked contrast to the regular stage variations in the Lavaca system. These irregular water level variations, e.g., Figure 5, are evidently of tidal scale and do propagate through the system. It is remarkable that the model replicates these irregularities as well as it does.

Since this period of simulation was selected to correspond to diurnal hydrodynamic measurements conducted at three-hour intervals from 1200 November 11 through 1200 November 12, additional sources of verification data exist in the form of current velocities. The comparison of measured and simulated velocities (speed and direction) at six sampling locations in Guadalupe Bay is presented in Table 1. In all six cases water direction is simulated correctly, though minor discrepancies are apparent, e.g., at Redfish Bayou from 0000 to 0300 November 12. Considering the dynamic influence of tides and winds on the flows in this area, compounded by the frontal passage during this period and the fact that even slight tidal phase errors can result in considerable error when comparing nearly instantaneous simulated and measured velocities, the magnitudes of the velocities compare favorably, with general patterns of increasing and decreasing velocity magnitude reasonably replicated.

This particular simulation period contains a meteorologically induced condition which is not completely accounted for by the present conservation of momentum equation in the model, i.e., water setdown resulting from wind stress acting on the water surface area. This wind stress, which normally results from the abrupt change in wind direction from the predominate southeasterlies to northerlies with the concurrent passage of a frontal system, produces dramatic

TABLE I

VELOCITY COMPARISON ON 11-12 NOVEMBER 1975

Time	Recorded		Simulated		Recorded		Simulated	
	Velocity (m/s)	Direction	Velocity (m/s)	Direction	Velocity (m/s)	Direction	Velocity (m/s)	Direction
	Guadalupe Bay (Section 8)				Redfish Bayou (Section 58)			
Nov 11 1200	.00	---	.02	Out	.13	In	.18	In
1500	.11	Out	.04	Out	.20	In	.11	In
1800	.09	Out	.04	Out	.13	In	.09	In
2100	.05	Out	.07	Out	.04	Out	.15	Out
Nov 12 0000	.05	Out	.05	Out	.05	In	.10	Out
0300	.09	Out	.06	Out	.05	Out	.02	In
0600	.26	Out	.18	Out	.40	Out	.13	Out
0900	.28	Out	.19	Out	.34	Out	.34	Out
1200	.17	Out	.04	Out	.11	Out	.52	Out
	Swan Lake Bayou (Section 61)				Schwings Bayou (Section 105)			
Nov 11 1200	.26	In	.19	In	.09	Out	.01	Out
1500	.22	In	.13	In	.02	In	.02	Out
1800	.19	In	.07	In	.06	Out	.01	Out
2100	.07	Out	.14	Out	.08	Out	.15	Out
Nov 12 0000	.05	Out	.10	Out	.08	Out	.09	Out
0300	.05	Out	.07	Out	.08	Out	.09	Out
0600	.73	Out	.23	Out	.13	Out	.14	Out
0900	.49	Out	.40	Out	.13	Out	.28	Out
1200	.55	Out	.40	Out	.13	Out	.15	Out
	Townsend Bayou (Section 40)				Varmum Bayou (Section 48)			
Nov 11 1200	.14	In	.30	In	.05	In	.17	In
1500	.34	In	.37	In	.06	In	.05	In
1800	.13	In	.30	In	.05	In	.05	In
2100	.10	Out	.04	In	.03	Out	.09	Out
Nov 12 0000	.05	Out	.11	Out	.01	In	.02	In
0300	.07	Out	.22	Out	.08	Out	.04	Out
0600	.24	Out	.49	Out	.27	Out	.37	Out
0900	.52	Out	.82	Out	.34	Out	.49	Out
1200	.28	Out	.91	Out	.67	Out	.49	Out

depressions in water elevation in the northern Gulf of Mexico which are transmitted to the adjacent bays as bay waters disgorge through the passes into the Gulf. This influence is rapidly felt even in the most inland bay regions, such as deltaic regions. That portion of the wind setdown recorded at the location of the driving tide, in this case the tide gauge in San Antonio Bay, is adequately accounted for in the model through the boundary condition. However, the additional setdown from wind action on the water expanse incorporated in the segmentation is not included in the model. As indicated in Figure 5, approximately 75% of the setdown for the frontal passage occurring on 12 November, is simulated in the present model.

The relatively limited verification data available for the Guadalupe Delta made it impossible to thoroughly verify the model for conditions of high streamflow. However, as indicated in the simulation presented, the model and segmentation adequately simulate water elevations and velocities for periods of low freshwater inflow. For further discussion of the modeling of the Guadalupe River Delta see Hauck et al. (1976).

Trinity Delta

The Trinity River is one of the two principal inflows to the Galveston Bay system. Its delta is located along the northern shore of Trinity Bay, the northern arm of the Galveston system. The lower reach of the Trinity River represents the largest application in areal extent of the model thusfar. A map of the deltaic region with superimposed segmentation and tide gauge locations is presented in Figure 6. A significant physiographic feature of this delta is the Wallisville levee, extending from Cotton Lake to the Trinity River, the original embankment for the Wallisville dam, whose construction was abandoned several years ago. A single freshwater source, the Trinity River, flows through an extensive marsh area, which may be conveniently separated into the east marsh, south marsh, and west marsh. The east marsh, located along the east bank of the Trinity River, is influenced by both river stage and tidal elevations. South of both the Trinity River and the Wallisville levee is the south marsh in which water levels are determined almost entirely by the tides. The west marsh is located west of the Trinity River and north of the levee. Tidal influence in this marsh is propagated by two channels traversing the south marsh and breaching the levee and also through a direct opening into the Trinity River near the eastern extremity of the levee. Water exchange between the west marsh and Trinity River occurs through this same opening as well as at channels at the north end of the marsh.

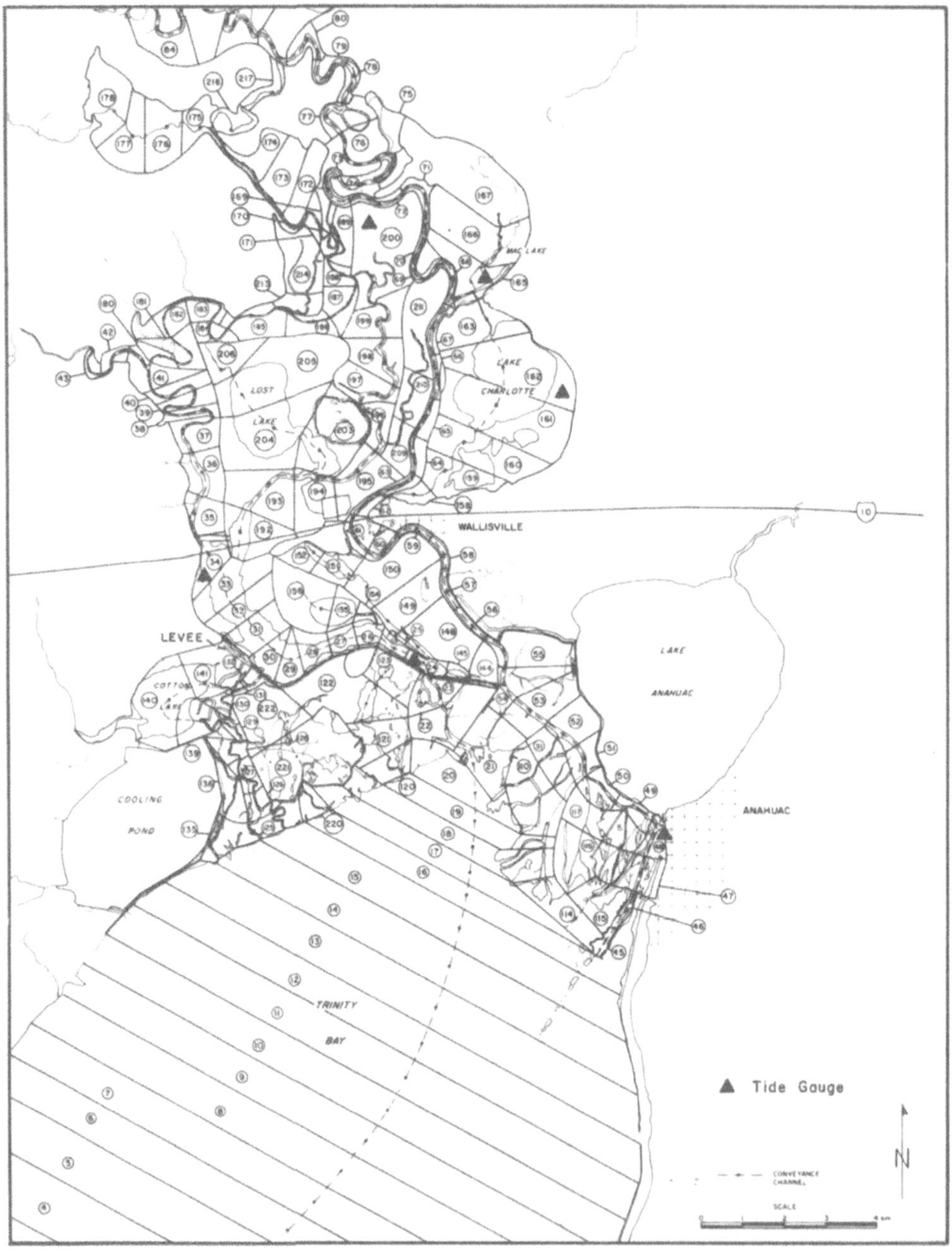

Figure 6. Trinity Delta System showing floodplain area for each section.

As indicated in Figure 6, several recording tide gauges are located within the delta region and the records from these gauges provided useful calibration and verification data. An intensive diurnal biological and hydrodynamic study conducted from 30 November through 3 December 1976 provided additional verification data. Boundary conditions to the model were supplied from the tide records of the USCE continuous recording gauge at Morgan Point, near the lower boundary of the model, and the flow record from the USGS streamflow gauge near Romayor, Texas, on the Trinity River was the source of the upstream model boundary condition.

The simulation presented here is for the period 25 November through 3 December 1976. For this simulation, the Trinity River flow was by necessity estimated because of the unfortunate circumstance that the USGS gauge near Romayor was inoperative during the time of interest. Streamflow was estimated at a nearly constant 70 m^3/s during this period at Liberty, a tidally influenced USGS streamflow gauge, and winds were light, except for 28 and 29 November when moderately strong north winds persisted, resulting in a substantial setdown on those days. The driving tide to the model as recorded at Morgans Point was diurnal with the wind setdown obvious on 28 and 29 November, see Figure 7a.

The hydrodynamic model was operated with the previously discussed boundary conditions and the model results are compared with all operative tide gauges in the delta, the Old River Cutoff Channel gauge (Section 24), the Anahuac Channel gauge (Section 48) and the Sulphur Barge Channel gauge (Section 165), in Figures 7b through 7d, respectively. The measured and simulated tides at the Old River Cutoff Channel and at Anahuac Channel compare favorably with tidal amplitude reproduced accurately and tide phasing within a couple of hours. At both gauges, a 0.1-m error between measured and simulated tides is evident, and was apparent in all simulations conducted on the Trinity Delta. As on the Guadalupe Delta, this discrepancy is expected to be the result of datum errors between the various gauges utilized in the simulations. If the 0.1 m datum difference is accounted for, the simulated tide during the wind setdown condition is approximately 0.3-m too low. No entirely satisfactory explanation of this error can be offered, though it appears that the tide gauges were established in a manner that prevented recording water elevations below -0.3-m msl. In fact, at these two gauge locations and at the Sulphur Barge Channel gauge, the tide records appear to have "pegged" during much of 28 and 29 November.

The Sulphur Barge Channel gauge record and simulated values also compare favorably, though a 0.1-m difference between simulated and recorded elevations also occurred at this gauge. However, this discrepancy was not apparent in the other simulations of the system and cannot therefore be attributed to gauge datum errors, as was

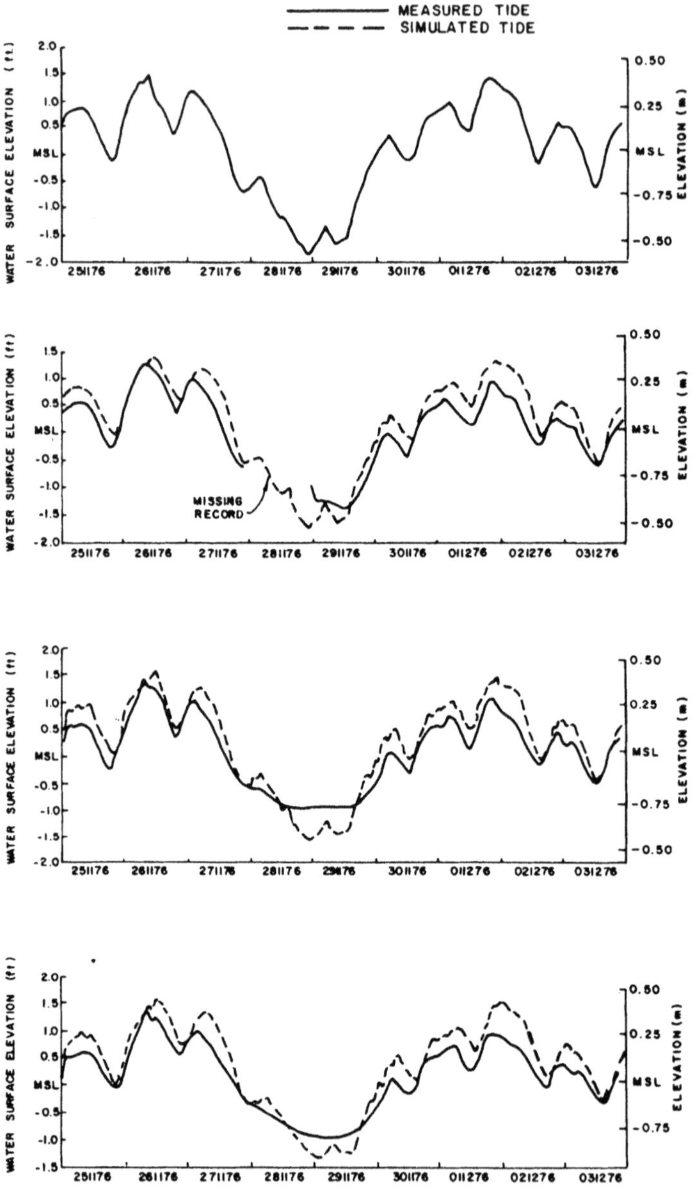

Figure 7. (a) Driving tide record at Section 2, Morgans Point Gauge; Tidal elevation record at (b) Section 24, Old River Cutoff Channel Gauge; (c) Section 48, Anahuac Channel Gauge; (d) Section 165, Sulphur Barge Channel Gauge.

the case at the other two gauges. A possible source of this error is the estimation of river flow, since the Sulphur Barge Channel is located in an area influenced by a combination of tides and river stage. As noted above, river flow had to be estimated because of the failure of the USGS gauge at Romayor. Possible overestimations of river flow could produce the simulation of water elevations that were too high. Because the other two gauges are located in the area referred to as the west marsh, and since that area is essentially beyond the influence of the river stage at the low river flows during this period, these gauges would not be influenced significantly by an error in river flow estimation.

Because the simulated period coincided with a period of intensive hydrographic study, direct comparisons of simulated and measured flows were produced at thirteen sampling points in the delta. Space does not permit the presentation of the entire comparison of flows; however, the results at three representative locations are presented in Table 2 for the Anahuac Channel (Section 47), Cross Bayou (Section 125) and Lost River (Section 192). As was the case in general, flow direction was accurately simulated at these sampling locations. An error that occurred occasionally is evidenced in the comparison at Anahuac Channel. The simulation at this location indicated a reversal in flow direction, whereas only a decrease in the flow magnitude, but not a reversal in direction, was measured. The results at the Lost River are exceptionally good. In all examples presented, flow magnitude was accurately reproduced, though in some of the smaller bayous (not presented) the error was somewhat larger. However, considering the dynamic event being simulated, the model produced good results at most points of comparison in the delta.

Additional results in the application of the model to the Trinity delta may be found in Hauck (1977), particularly involving the mass-transfer model to compute nutrient concentrations, with instream kinetics and bed exchange between different vegetational zones under conditions of transient inundation. (For this purpose, exchange rates measured by Armstrong and Gordon, 1977, were used.)

SUMMARY AND CONCLUSIONS

A branching section-mean model has been developed to simulate hydrodynamics and transport in estuarine deltaic systems, and has been applied to several systems along the coast of Texas. The model includes the ability to represent the complex, dendritic physiography of delta-marsh systems and to simulate the inundation and dewatering of the marsh areas adjacent to the distributary channels. The equations of longitudinal momentum conservation, continuity, and mass transfer conservation are solved by finite differences,

TABLE II

FLOW COMPARISON ON 30 NOVEMBER - 1 DECEMBER 1979

Time		Recorded Flow (m^3/s)	Recorded Direction	Simulated Flow (m^3/s)	Simulated Direction
		Anahuac Channel (Section 47)			
Nov 30	1045	66.3	Out*	53.9	Out
	1300	61.5	Out	38.2	Out
	1600	19.0	Out	1.8	Out
	1900	33.1	Out	17.8	In
	2200	47.3	Out	2.8	Out
Dec 1	0100	42.5	Out	22.3	Out
	0400	42.5	Out	18.4	Out
	0700	66.3	Out	30.2	Out
	1000	85.0	Out	54.9	Out
		Cross Bayou (Section 125)			
Dec 2	1130	12.7	Out**	17.4	Out
	1300	11.6	Out	16.0	Out
	1600	6.5	Out	4.4	Out
	1650	0.0	---	1.7	In
	1900	12.7	In	11.2	In
	2200	6.5	In	8.1	In
	2245	0.0	---	6.6	In
Dec 3	0100	9.3	Out	8.8	Out
	0400	6.5	Out	6.9	Out
	0700	8.7	Out	8.8	Out
	1000	10.6	Out	10.7	Out
		Lost River Near IH-10 (Section 192)			
Dec 2	1100	41.1	Out**	49.1	Out
	1230	43.9	Out	46.9	Out
	1300	42.5	Out	51.1	Out
	1415	46.7	Out	50.5	Out
	1500	36.8	Out	48.7	Out
	1625	24.1	Out	42.0	Out
	1700	21.2	Out	25.5	Out
	1800	4.2	In	3.4	Out
	1915	34.0	In	19.9	In
	2000	48.2	In	42.7	In
	2120	41.1	In	42.0	In
	2200	36.8	In	33.1	In
	2300	25.5	In	22.6	In
	2400	8.5	In	14.3	In
Dec 3	0210	19.8	Out	30.8	Out
	0325	11.3	Out	18.6	Out
	0400	7.1	Out	14.6	Out
	0500	7.1	Out	26.5	Out
	0600	21.2	Out	25.0	Out
	0715	31.2	Out	22.1	Out
	0800	36.8	Out	31.0	Out
	0915	32.5	Out	36.6	Out
	1000	29.7	Out	38.5	Out
	1105	28.3	Out	37.8	Out

*Out indicates down Anahuac Channel into Trinity Bay.
**Out indicates flow out of Cross Bayou into Trinity Bay.
***Out indicates flow direction towards Trinity Bay.

resulting in a time history of flows, water elevations, and constituent concentrations.

The hydrodynamic portion of the model has been extensively tested and verified for deltaic regions along the Texas Coast, including the Trinity, Guadalupe, and Lavaca Deltas. In all instances the model has been able to give good replication of water elevations for a variety of conditions including variable freshwater inflow, such as associated with small to moderate floods, and meteorologically influenced tides, e.g., wind induced setup or setdown, as well as low flow, astronomical tide regimes.

The model appears to offer a satisfactory tool for evaluating the complex behavior of these deltaic systems under even quite dynamic hydrometeorological regimes. The computer code (Hauck, 1978) is efficient and permits extended simulations of many days' duration for rather modest computer resources.

ACKNOWLEDGMENTS

Mr. R. J. Huston made considerable contributions to the development of this model, including the coding for the first computer solution. Freshwater inflow data and tide records were provided by the U. S. Geological Survey and the U. S. Corps of Engineers (Galveston). It is a pleasure to acknowledge the Texas Department of Water Resources, and its predecessor agency, the Texas Water Development Board, for its continued support of this work through its Bays and Estuaries Program. The TDWR coordinated the field studies in the deltas to obtain data for application and verification of the models, and furnished staff, facilities and equipment in the prosecution of the work.

REFERENCES

Armstrong, N. E. and V. N. Gordon. 1977. Exchange rates for carbon, nitrogen and phosphorus in the Trinity and Colorado River Delta marshes for the Texas Department of Water Resources. Center for Research in Water Resources, Univ. of Texas, Austin, Texas.

Dronkers, J. J. 1964. Tidal computations in rivers and coastal waters. Amsterdam: North-Holland Publishing Company.

Harleman, D. F., and C. H. Lee. 1969. The computation of tides and currents in estuaries and canals. Tech. Bul. No. 16, Committee on Tidal Hydraulics, U. S. Army Corps of Engineers.

Hauck, L. M. 1977. A study of the hydrological and nutrient exchange processes in the Trinity River Delta, Texas - Part III Development and application of hydrodynamic and mass transfer models of the Trinity deltaic system, Espey, Huston and Associates, Inc. Doc. No. 77101, Austin, Texas.

Hauck, L. M. 1978. User's manual for hydrodynamic mass transfer model of deltaic systems. Espey, Huston and Associates, Inc. Doc. No. 7815, Austin, Texas.

Hauck, L. M., G. H. Ward, and R. J. Huston. 1976. Development and application of a hydrodynamic model of the Lavaca and Guadalupe deltaic systems. Espey, Huston and Associates, Inc. Doc. No. 7663, Austin, Texas.

Ward, G. H. 1973. Hydrodynamics and temperature structure of the Neches estuary, Vol. II, theory and formulation of mathematical models. Doc. No. T73-AU-9510, Tracor, Inc., Austin, Texas.

CURRENT MEASUREMENTS AND MATHEMATICAL MODELING IN SOUTHERN PUGET SOUND

Philip J. W. Roberts

School of Civil Engineering, Georgia Institute of Technology, Atlanta, Georgia

ABSTRACT

Field observations and mathematical modeling were conducted in order to understand the circulation patterns in Nisqually Reach, Southern Puget Sound. Eight continuously recording current meters at four sites and a two-dimensional finite element model were used. Analysis of the current data showed the currents to consist of a first principal component which was essentially parallel to the channel walls. This component was primarily tidal, although both high and low frequency content was apparent. The high frequency content was attributed to fairly small-scale turbulence. The low frequency currents exhibited fluctuations on the order of several days, with the power spectra showing a secondary peak at 2.5 days. These low frequency fluctuations are probably due to wind effects, occurring both locally and non-locally. Typical circulation patterns predicted by the mathematical model are presented. The model reasonably reproduces the tidal currents but not the high and low frequency content. Other limitations of the model are discussed in light of the analysis of the current meter data.

INTRODUCTION

Estuarine hydrodynamics are so complex that the results of field observations, and both mathematical and physical modeling, should all be considered in yielding insight into their behavior. Each of these disciplines can give different information, but none alone gives a complete picture. A recent study in Nisqually Reach, Washington, in which extensive current measurements and mathematical modeling were combined illustrated some of these limitations. In this paper the results of these studies are presented, and the limitations and strengths of the different techniques discussed.

The area of study is known as Nisqually Reach; it is shown in the vicinity map in Figure 1, and in more detail in Figure 2. The area is complex; it has many bays and inlets, sharply sloping shoals, and large depth variability. The depths vary from about 4 to 110 metres, and the width of the channel is about 4 km. The Reach is connected to the Pacific Ocean by only one entrance, the Narrows at Tacoma. North of the Narrows lies Puget Sound. Freshwater inflows are primarily from the Nisqually River; 27-year records show discharges varying from 2.5 to 660 m^3/s with a mean of 52.5 m^3/s.

PREVIOUS WORK

Several field observations have been made in the vicinity of Nisqually Reach. These are reviewed by CH2M Hill (1978). The number of measurements of currents in the area are few, and are of only short duration. Recently, Cannon and Laird (1978) reported on extended measurements, including currents, in Puget Sound of one year duration. They found the mean flows to be out of the estuary at the surface and in at the bottom. This circulation appeared to be due to deep water intrusions caused by large tidal ranges. The prevailing Northerly winds in the area would aid this general circulation pattern. Based on the observed net flows and channel volumes, Cannon and Laird estimated a replacement time of about 9 days for the Sound.

A physical model of Puget Sound at the University of Washington has been used for current and dispersion studies. These studies, in addition to some performed as part of the present study, are reviewed in CH2M Hill (1978) and McGary and Lincoln (1977). The results indicate a complex flow pattern containing horizontal eddies and secondary circulations, particularly near slack water.

In the past few years, several studies of non-tidal circulation in estuaries have been conducted. These include Wang and Elliot (1978), Elliot and Wang (1978), Elliot (1978) in the Chesapeake Bay and Potomac Estuary; Weisberg and Sturges (1976) in Narragansett Bay; and Smith (1977) in Corpus Christi Bay, Texas. In these studies, some of currents, water levels, winds, and atmospheric pressures were measured. The results were numerically filtered to remove tidal and higher frequency effects. All of the studies found fluctuations in currents and/or water levels having periods of variability of several days. The wind was found to be the cause of these long period fluctuations through both local and non-local forcing. The local forcing, where the wind stress acts locally

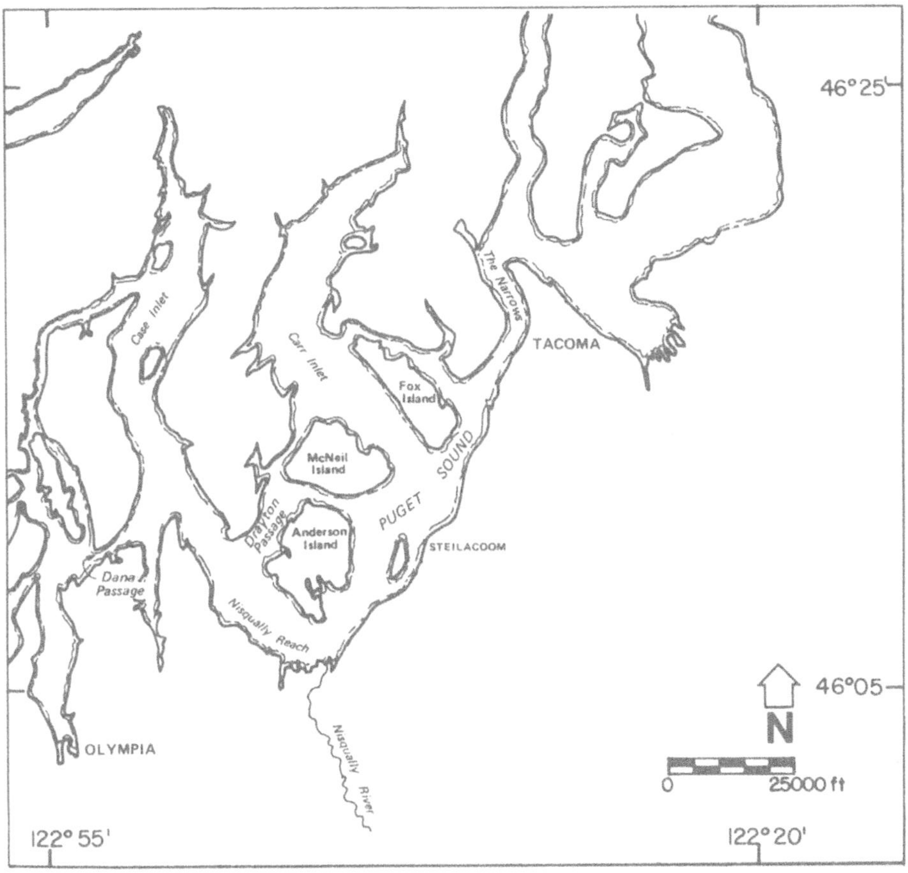

Figure 1. Study area vicinity

causing a water surface slope and associated flow, results in fluctuations of the order of 2-5 days. The non-local effect is caused by wind stress on the adjacent coastal waters which causes an Ekman drift resulting in a surface flow in or out at the surface near the estuary mouth. This non-local effect causes longer period fluctuations on the order of 20 days. These findings suggest that mean estuary replacement times can only be reliably estimated from long-period measurements of at least 30 days.

FIELD OBSERVATIONS

An extensive field monitoring program was conducted as part of the present study. Measurements included water level at three locations, salinity and temperature, various water quality parameters, and wind at four sites. Circulation patterns were measured by fixed current meters, portable meters, and drogue and surface drifters. Only the data obtained from the fixed current meters will be discussed here, the remainder of the data are discussed in CH2M Hill (1978).

The moored current meter locations are shown in Figure 2. Four sites, designated 1, 2, 3 and 4 were occupied; at each site were two meters, one at 6 m and the other at 33.5 m below mean lower low water. The surface meters are designated 1S, 2S, etc., and the bottom meters 1B, 2B, etc. The meters were Endeco Type 105. At each station, the two meters were attached to a common line anchored at the bottom by a 750 lb railroad wheel, and supported at the top by two buoys providing 80 lb of buoyancy. The meter integrates rotor revolutions over 30-minute intervals to measure current speeds, and employs a magnetic compass to reference magnetic north. The current-velocity range is 0 to 180.2 cm/s, with a threshhold of 2.57 cm/s, and accuracy of ±3% of full scale. Current direction is from 0 to 360°, with a sensitivity of ±5° at 2.57 cm/s and an accuracy of 2% above 2.57 cm/s. The analog data are recorded internally on 16 mm film and converted into digital form by the meter manufacturer. The data were retrieved at about 1 month intervals, with a total period of deployment from 6 May 1977 to 28 April 1978. In the present paper, only one month's worth of data will be analyzed and discussed in order to derive typical results. It is hoped to present the results of analyses of the entire data set in a later paper.

Analysis of Current Meter Data

The data were digitally analyzed in order to extract the information contained therein. The speed and direction values given were converted to orthogonal components, one in an Easterly direction and one in a Northerly direction. The current axes were then found which maximize and minimize the variance of the current speeds when projected onto those axes; these axes are orthogonal and are known as the principal axes. The approximate directions of the principal axes are shown in Figure 2, where the lengths of the arrows are proportional to the standard deviation of the components. The components of the currents in the direction of the principal axes are known as the principal components. A typical time series plot of the principal components at one station (2S) is shown in Figure 3. A summary of the properties of the principal

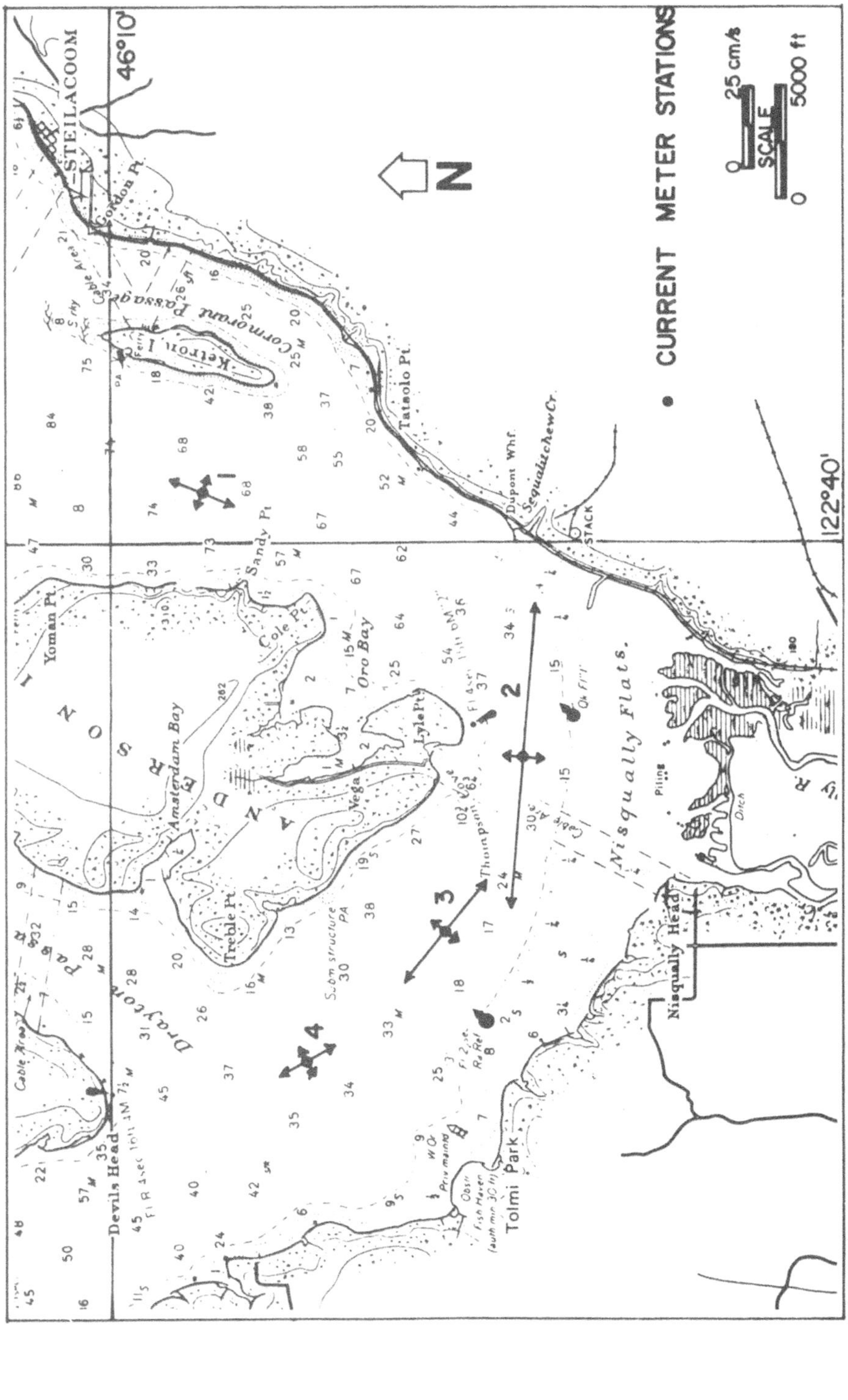

Figure 2. Approximate directions of principal components at each moored current meter station.

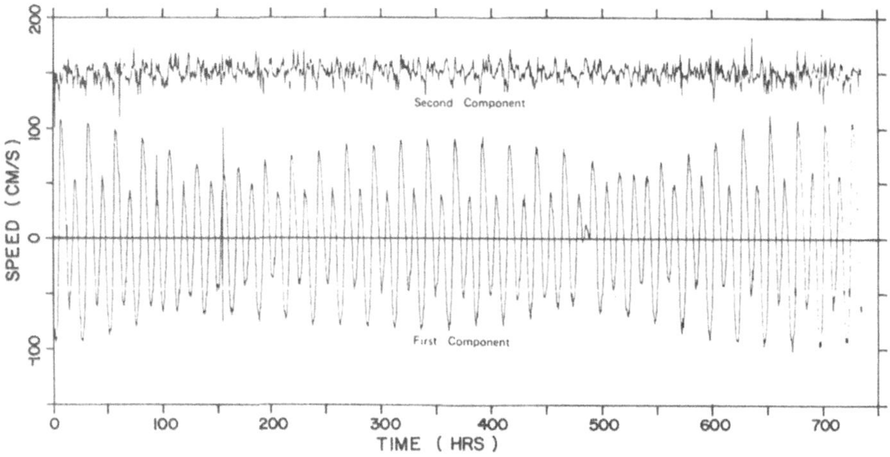

Figure 3. First and second principal current components, at Station 2S.

components is given in Table 1.

The first principal current component is approximately parallel to the local channel walls, and is obviously tidal. Although the first principal axes of the top and bottom meters are not exactly parallel, they deviate by less than 17 degrees. The second principal component, which is orthogonal to the first and hence approximately perpendicular to the channel walls, shows relatively more high frequency content and appears to be at least partially random.

The tidal nature of the first principal components was verified by the power spectra, which were sharply peaked at the diurnal and semidiurnal tidal frequencies. The first principal component also showed a secondary peak at about 2-5 days, and the spectrum continued to rise for periods longer than 5 days. Even the second principal component showed peaks in the power spectra at the tidal frequencies, although they were less pronounced than the high and low frequency content.

Phase relationships between the surface and bottom meters were examined in more detail. Figures 4 and 5 show superimposed plots on an expanded time scale for a portion of the principal components at stations 3B and 3S. The first principal components are in phase at tidal frequencies. Higher frequency fluctuations are apparent, however, which do not show any obvious phase relationship. The

Table 1. Properties of Principal Current Components[1]

Meter	Water Depth (metres)	Meter Depth (metres)	Direction, θ (See Figure 2) (Degrees)	Variance		Mean	
				First (cm^2/s^2)	Second (cm^2/s^2)	First (cm/s)	Second (cm/s)
1B	128	33.5	71.4	119.1	21.9	-2.82	1.71
1S	128	6	86.3	186.0	30.2	--	-0.47
2B[2]	52	33.5	176.7	2,465.2	25.7	1.43	4.60
2S[2]	52	6	179.5	2,492.7	108.0	-0.73	9.28
3B	46	33.5	141.5	373.8	23.3	8.45	-4.78
3S	46	6	133.3	669.0	35.3	4.98	-1.16
4B	63	33.5	128.8	270.0	20.8	--	-0.67
4S	63	6	111.1	422.6	26.2	-3.90	0.46

[1]Data period 10/11/77 at 1300 hours to 11/13/77 at 2200 hours

[2]From 10/11/77 at 1400 hours to 10/24/77 at 0200 hours

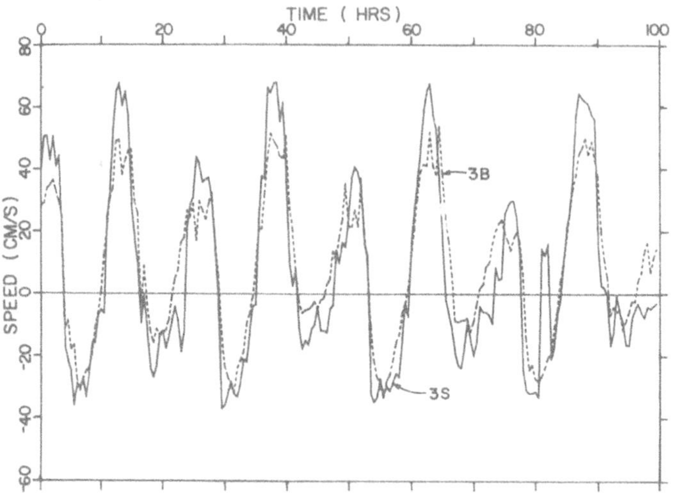

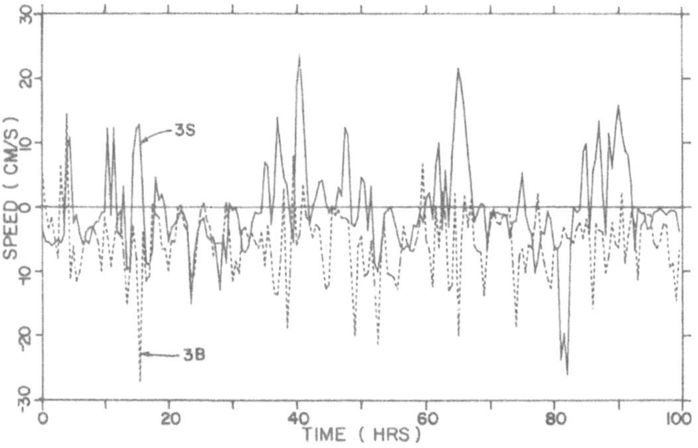

Figures 4 and 5. First and second principal current components, Stations 3B and 3S.

second principal components, consisting of relatively more high-frequency content, show no obvious phase relationships between the bottom and surface meters. The velocities are of smaller magnitude than the tidal component (note that Figure 5 has an expanded velocity scale).

The power spectra showed that current fluctuations, having periods of several days, exist. To investigate these fluctuations, the data were subjected to digital filtering by an exponential recursive filter in order to remove the effect of frequencies tidal and higher. The results are shown in Figure 6 for the first principal components, and Figure 7 for the second principal components.

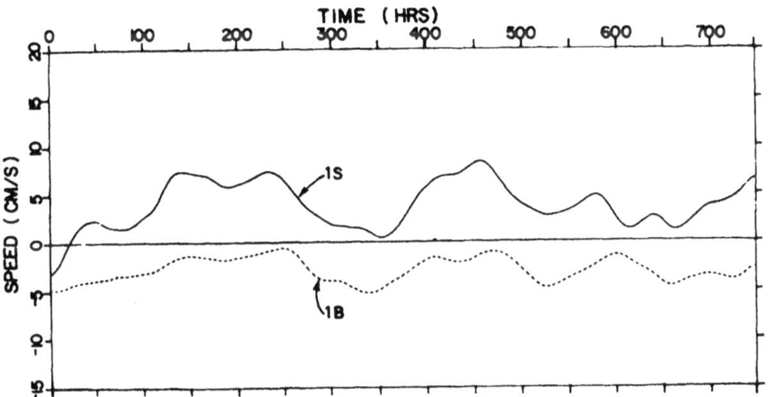

FIGURE 6. FIRST COMPONENTS

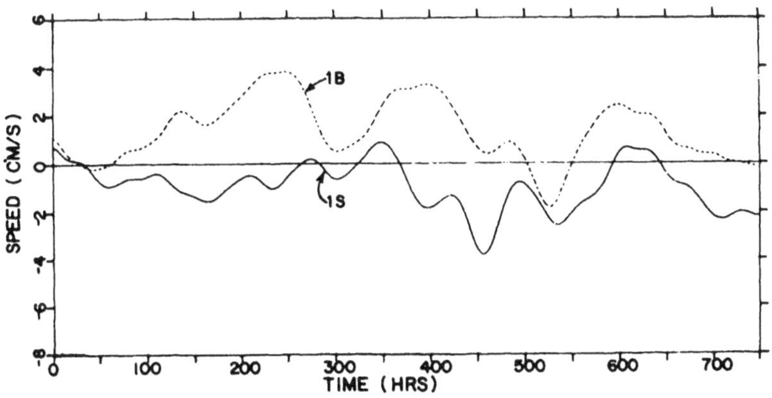

Figures 6 and 7. Filtered first and second principal current components, Meters 1S and 1B.

Correlations between the first principal current components and predicted tidal currents were investigated. The magnitude and times of occurrence of peak ebb and flood tidal currents and the times of slack water were obtained from tidal current tables (NOAA (1977)) for the Narrows (North End) of Puget Sound. The results were interpolated to values at half-hour intervals by means of a cubic spline fit. Figure 8 shows the first principal component at meter 3B and the interpolated tidal currents at The Narrows, plotted on an expanded time scale. In this graph the tidal data have been multiplied by 0.2 and displaced horizontally 2 hours to the right. The correlation coefficients between the first principal components at Stations 2B and 2S and the tidal current were both 0.86, similar results were obtained for the other stations.

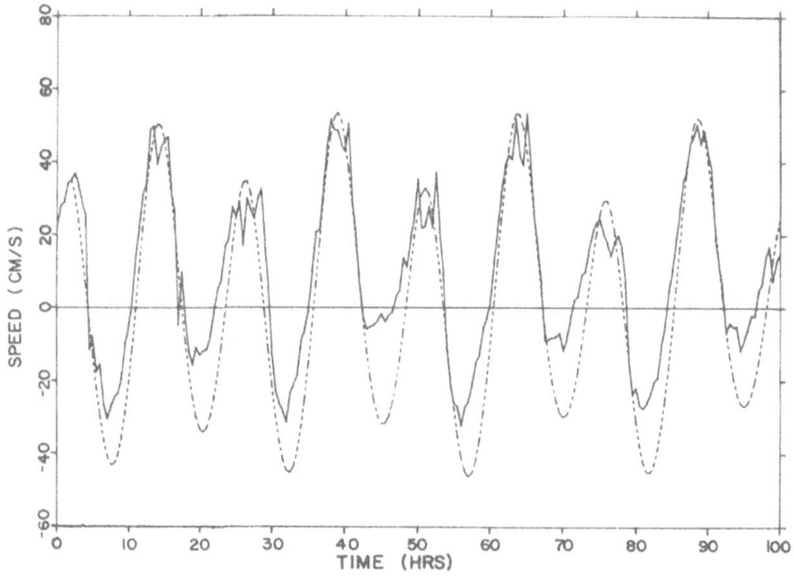

Figure 8. First principal component, meter 3B, and tidal currents at the Narrows.

Discussion

The principal component analysis showed the currents to consist of axial (first principal) and co-axial (second principal) components. The total energy of the axial components is much higher than that of the co-axial components (Figure 2), showing the predominance of the along-channel flow. The power spectra of the principal current components showed large peaks at the diurnal and semidiurnal tidal frequencies, and a secondary peak at about 2·5 days. At periods longer than about 5 days the energy content increased monotonically, although information in this range could not be reliably obtained from the short, one month, record investigated. The energy content at frequencies higher than tidal dropped off sharply. These spectra suggest a division of the currents by frequency categories: high, at frequencies shorter than 12·5 hours; tidal, at the diurnal and semidiurnal frequencies; and low, at periods longer than a day. Each of these is discussed separately below.

The high frequencies do not contain much energy. These fluctuations are probably due to random turbulent motions of relatively small length scale. This is suggested by the low correlation and lack of apparent phase relationships between the vertically separated meters (Figures 4 and 5).

The power spectra of the first principal components were sharply peaked at the diurnal and semidiurnal tidal frequencies. These currents were in phase at the tidal frequencies (Figure 4), and highly correlated with the predicted tidal currents at the Narrows (Figure 8). It was found that the axial currents could be reasonably predicted by a one-dimensional model. The currents are given by a constant multiple of the tidal current at the Narrows with an appropriate lag time. The constant of multiplication of any cross section is inversely proportional to the channel cross sectional area. For example, at Station 3 the constant is 0.2 and the lag time 2 hours. The comparisons between observed and predicted currents at Station 3B, Figure 8, show good agreement.

Low frequency current motions in estuaries have been experiencing increased attention in the past few years. These motions, termed residual or non-tidal, flows determine the residence times and hence, for example, the long term buildup of conservative contaminants continuously discharged into the estuary. The results obtained here are consistent with those obtained in other estuaries, discussed earlier. The mean currents contain fluctuations on the order of days and longer. The peak at about 2·5 days is probably due to local wind forcing. Although peaks at longer periods due to non-local forcing in the adjacent coastal waters may exist, they could not be discerned in the present data set, possibly due to its short duration.

The filtered data for the first principal components showed net transports generally less than 10 cm/s. A net shearing motion is indicated (Figure 6) with the flow out of the estuary at the surface and into the estuary near the bottom. These net motions appear to be in phase at the upper and lower meters. The second principal components also show net means, but of smaller magnitudes than those of the first principal components (Figure 7). The nature and reasons for these motions are unclear, but they could be due to secondary spiral circulations induced by channel curvature, non-uniformity, or Ekman flows, or eddies of a primarily horizontal nature.

MATHEMATICAL MODELING

Model Description

A two-dimensional finite element model was used to simulate circulation processes in the estuary. The equations of motion are:

$$\frac{\partial u}{\partial t} + u\frac{\partial u}{\partial x} + w\frac{\partial u}{\partial z} + g\frac{\partial h}{\partial x} + g\frac{\partial a_o}{\partial x} - \frac{\varepsilon_{xx}}{\rho}\frac{\partial^2 u}{\partial x^2} - \frac{\varepsilon_{xz}}{\rho}\frac{\partial^2 u}{\partial z^2} - 2\omega w \sin\phi$$

$$+ \frac{gu}{C^2 h}(u^2 + w^2)^{1/2} - \frac{\zeta}{h}V_a^2 \cos\psi = 0 \qquad (1)$$

$$\frac{\partial w}{\partial t} + u\frac{\partial w}{\partial x} + w\frac{\partial w}{\partial z} + g\frac{\partial h}{\partial z} + g\frac{\partial a_o}{\partial z} - \frac{\varepsilon_{zx}}{\rho}\frac{\partial^2 w}{\partial x^2} - \frac{\varepsilon_{zz}}{\rho}\frac{\partial^2 w}{\partial z^2} + 2\omega u \sin\phi$$

$$+ \frac{gw}{C^2 h}(u^2 + w^2)^{1/2} - \frac{\zeta}{h}V_a \sin\psi = 0 \qquad (2)$$

and the continuity equation:

$$\frac{\partial h}{\partial t} + \frac{\partial}{\partial x}(uh) + \frac{\partial}{\partial z}(wh) = 0 \qquad (3)$$

Here, u and w are the depth averaged velocities in the x and z coordinate directions, t is time, g is the acceleration due to gravity, a_o is the bottom elevation relative to a horizontal datum,

ρ is the fluid density, ε_{xx}, ε_{zx}, ε_{zz} and ε_{xz} are eddy viscosity coefficients, ϕ is the local latitude, C is the Chezy coefficient, h is the water depth, ζ is a windstress coefficient, V_a is the wind speed, and ψ the angle between the wind and the x-axis.

The model therefore includes the effects of free surface and bottom slopes, eddy viscosity, Coriolis forces, bottom friction, and wind stress. The boundary conditions which must be specified are velocity and water depth at an open boundary. The equations 1, 2, and 3 were solved by means of the finite element method on a fixed grid. For more details of the model see CH2M Hill (1978).

Results

The model was applied to predict currents in the region of Nisqually Reach extending from current meter stations 1 to 4 (Figure 2). The water velocity boundary conditions were supplied by the measured currents at these stations. Typical results from the model are shown in Figure 9. Comparisons between the predicted and measured currents at stations 2 and 3 were made.

Circulation characteristics were simulated with varying degrees of success by the model. At high current speeds, simulated currents were less than those measured. This is presumably due to the depth-averaging of the model. At low velocities the agreement was better. Observed predicted current directions corresponded closely during high velocity periods when the flow was predominantly parallel to the channel walls. The model could not, however, simulate the eddies which occur at times of low velocity or slack water. Also, the model will not generate the large horizontal eddies which would normally be expected to develop near corners, and so some of the flow patterns predicted may be suspect. The model also cannot predict the high frequency current fluctuations of small length scale, as these are propagated in the model from the boundary conditions, but in the field are locally generated.

DISCUSSION

The analysis of the current meter data shows the water motions in Nisqually Reach to be very complex. Motions at time scales of 1/2 hour, the sampling interval, up through many days were observed.

Both the mathematical modeling and field observations yielded insight into these complex hydrodynamic processes. The mathematical model can yield detailed information on the spatial variation of currents (see Figure 9, for example) which would be prohibitively expensive to obtain by field observations. Caution is necessary

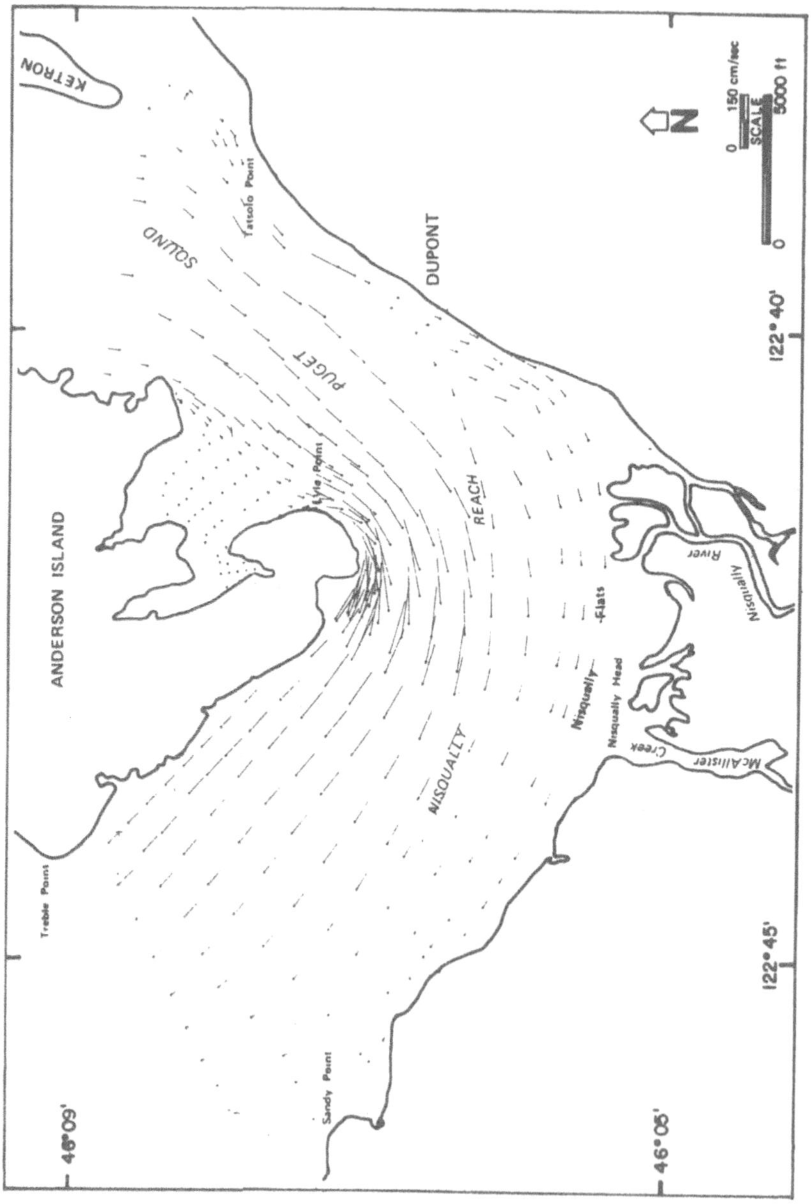

Figure 9. Typical current vector plot of mathematical model.

in interpreting the results, however, as the model may not predict large horizontal eddies, or gyres, which can occur. The model is also valuable in that the circulation response to arbitrarily specified boundary conditions, such as wind or tide, can be readily examined. On the other hand, the model, because it is depth averaged, cannot simulate the net shearing motions apparent in the flow. As these net motions govern long-term conservative pollutant buildup in the estuary, the mathematical model would be unsuitable as a means to predict them. The model may give reasonable estimates of pollutant dispersion over shorter time scales of a few days. The mathematical model also will not predict high frequency current fluctuations apparent. Thus, the model is most successful at predicting tidal currents. Unless a random current constituent were added, it cannot predict the high frequency current fluctuations. In order to predict the low frequency currents, coupling with the adjacent waters may have to be considered, as previous studies have shown these currents to be coupled with these waters. In fact, the model added little that could not be deducted from the interpretation of the current data.

REFERENCES

Cannon, G. A., and Laird, N. P. 1978. "Variability of Currents and Water Properties From Year-Long Observations in a Fjord Estuary," in Hydrodynamics of Estuaries and Fjords, J.C.J. Nichoul, Ed., Elservier.

CH2M Hill. 1978. "DuPont Site Hydrological and Modeling Studies." Ch2M Hill, Inc., Seattle, Washington.

Elliot, A. J. 1978. "Observations of the Meteorologically Induced Circulation in the Potomac Estuary," Estuarine and Coastal Marine Science, 6: 285-299.

Elliot, A. J. and Wang, D. P. 1978. "The Effect of Meteorological Forcing on the Chesapeake Bay: The Coupling Between an Estuarine System and its Adjacent Coastal Waters," Hydrodynamics of Estuaries and Fjords, J.C.J. Nichoul, Ed., Elservier.

McGary, N., and Lincoln, J. H. 1977. "Tide Prints: Surface Tidal Currents in Puget Sound," Washington Sea Grant Publication No. WSG 77-1. Seattle: University of Washington Press.

NOAA. 1977. "Tidal Current Tables, Pacific Coast of North America and Asia."

Smith, N. P. 1977. "Meteorological and Tidal Exchanges Between Corpus Christi Bay, Texas, and the Northwestern Gulf of Mexico," Estuarine and Coastal Marine Science, 5: 511-520.

Wang, D. P. and Elliot, A. J. 1978. "Non-Tidal Variability in the Chesapeake Bay and Potomac River: Evidence for Non-Local Forcing," J. Phys. Oceanography 8: 225-232.

Weisberg, R. H. and Sturges, W. 1976. "Velocity Observations in the West Passage of Narragansett Bay: A Partially Mixed Estuary," J. Phys. Oceanography. 6: 345-354.

THE ROLE OF PHYSICAL MODELING IN THE MATHEMATICAL MODELING OF THE SACRAMENTO-SAN JOAQUIN DELTA

Richard C. Kristof

Hydraulic Engineer, U. S. Bureau of Reclamation

Sacramento, California

INTRODUCTION

The Sacramento-San Joaquin Delta is an estuary located at the confluence of the Sacramento and San Joaquin Rivers. Its roughly triangular shape is defined by the principal tributary rivers flowing in at two corners of the triangle and out at the third corner towards San Francisco Bay and the Pacific Ocean. A map of the Delta is shown in Figure 1. The delta, which includes over 700,000 acres (283,290 hectares) of land interlaced with approximately 700 miles (1130 kilometers) of meandering waterways, originally consisted of swampland covered primarily with tules, willows, and cottonwoods. During the past century, the lands have been reclaimed for agricultural use by the construction of levees to form approximately 50 islands and tracts. In addition to providing water for agricultural purposes, the Delta channels support municipal and industrial demands, commercial navigation, recreation, and provide spawning, nursery and/or habitat areas for a variety of anadromous and freshwater fish.

The Delta is located between the water-sufficient northern part of the state and the water-deficient southern part of the state. Federal and state agencies have developed water projects that store water in the northern basin, release it according to demand to the Delta where it is diverted for agricultural, municipal, and industrial uses in the San Francisco Bay Area, the San Joaquin Valley, and Southern California.

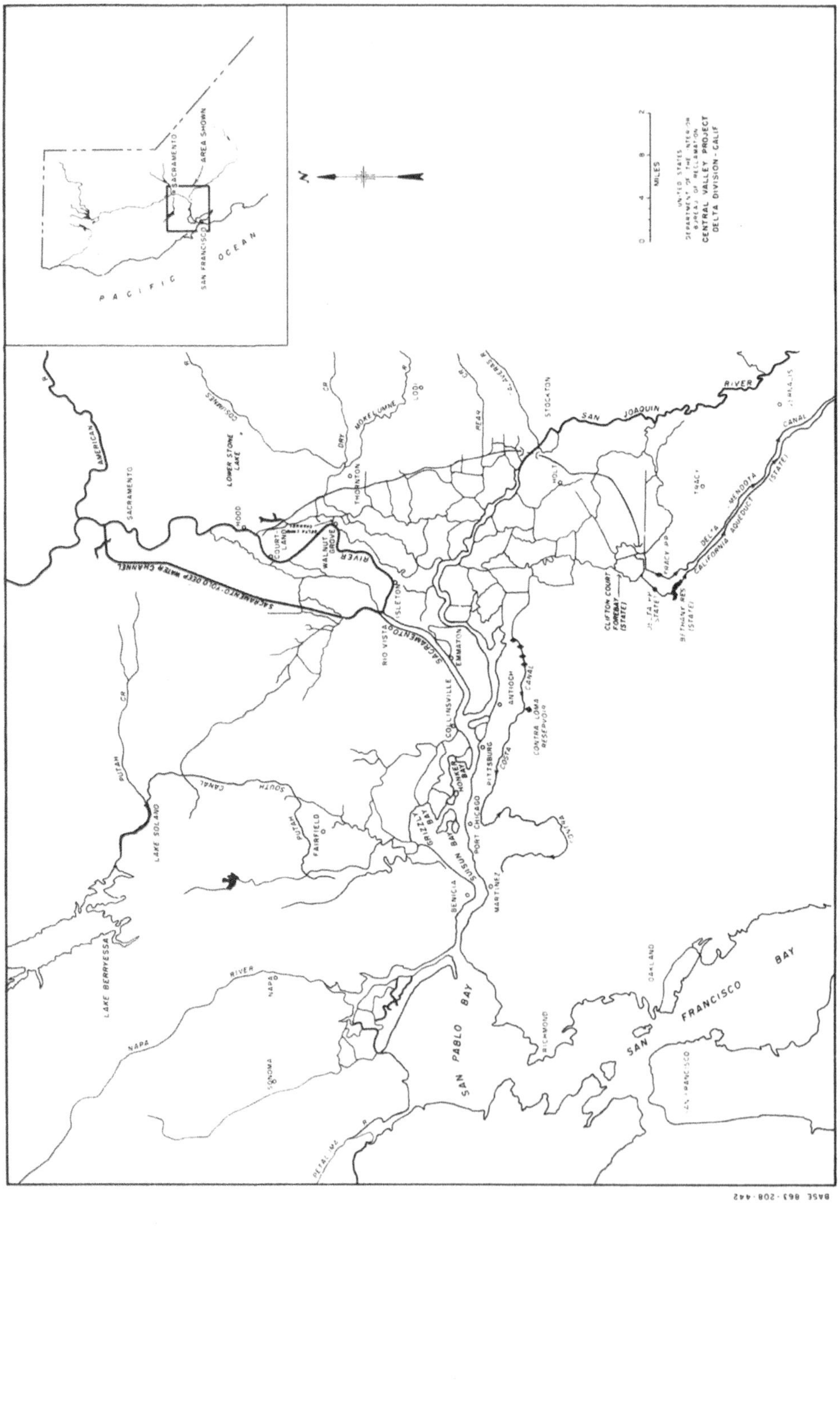
PACIFIC
OCEAN
SACRAMENTO
AREA SHOWN
SAN FRANCISCO
MILES
CENTRAL VALLEY PROJECT
DELTA DIVISION - CALIF
AMERICAN R
SACRAMENTO
SACRAMENTO-YOLO DEEP WATER CHANNEL
COSUMNES R
LOWER STONE LAKE
DRY CR
MOKELUMNE R
LODI
HOOD
THORNTON
COURTLAND
WALNUT GROVE
RIVER
ISLETON
RIO VISTA
SACRAMENTO
EMMATON
COLLINSVILLE
STOCKTON
SAN JOAQUIN RIVER
HOLT
TRACY
CLIFTON COURT FOREBAY (STATE)
TRACY PP
BETHANY RES (STATE)
DELTA-MENDOTA CANAL
CALIFORNIA AQUEDUCT (STATE)
ANTIOCH
CONTRA LOMA RESERVOIR
PITTSBURG
COSTA
CANAL
HONKER BAY
GRIZZLY BAY
SUISUN BAY
PORT CHICAGO
BENICIA
MARTINEZ
FAIRFIELD
PUTAH SOUTH CANAL
PUTAH CR
LAKE SOLANO
LAKE BERRYESSA
NAPA RIVER
NAPA
SONOMA
SAN PABLO BAY
RICHMOND
OAKLAND
SAN FRANCISCO BAY
BASE 863-208-442

Figure 1.

SALINITY INTRUSION

The Delta is subject to ocean salinity intrusion as a natural consequence of its connection to the Pacific Ocean via Suisun and San Francisco Bays and of its arid region hydrology characterized by long periods of low freshwater inflow. This problem has worsened over the years due to increased upstream diversion for agriculture. A key element of the Federal and State Water Projects is the maintenance of a "hydraulic barrier" which requires enough freshwater outflow from the Delta to the ocean to prevent salinity intrusion from damaging beneficial water uses within the Delta as well as beneficial uses of the water diverted from the Delta. The amount of freshwater required to maintain the hydraulic barrier has eluded precise definition because of the difficulties involved in measuring or computing Delta outflow. The difficulty in measurement of outflow is illustrated by the fact that the minimum net freshwater flow of 2,000-3,000 cfs (56-85 m^3/s) is only a small fraction of the tidal flow which peaks at 250,000-300,000 cfs (7,080-8,495 m^3/s). Thus, the magnitude of error in measurement of the gross flow is greater than the size of the flow to be measured. Calculation of the net freshwater outflow by summation of the various flows into and out of the Delta is further complicated by uncertainty in the amount of water depleted by users in the Delta.

Salinity intrusion is affected by export as well as outflows. For high levels of export, reversal of net tidal flow occurs in some channels resulting in increased salinity in some places and decreased salinity in others.

Alternative salinity control facilities have been examined, and some have been constructed in an effort to minimize the amount of freshwater outflow required to prevent salinity intrusion. Among the facilities evaluated is the proposed Peripheral Canal, which would transport freshwater flow around the Delta in a channel hydraulically isolated from the rest of the estuary. This plan would maximize the amount of water that could be exported while minimizing the adverse effects of export upon the Delta itself.

In order to define the salinity-outflow relationship, efficiently operate existing facilities, and evaluate the benefits of proposed facilities, a number of hydraulic and water quality models have been developed, including two physical models, three analog models, and several mathematical models. The number of models that have been developed is in itself a statement of the complexity of the system.

Controversy often arises over whether it is best to use a mathematical model or the physical model. Each approach has certain advantages over the other and the selection of the appropriate method is entirely dependent on the circumstances of the situation to be modeled. However, there are also situations in which the best approach may be to use a synergistic combination of both mathematical and physical models. The purpose of this paper is to illustrate this technique.

USE OF MATHEMATICAL MODELS

Because of the complexity of the Delta system, all models have suffered from some difficulties in defining various system coefficients and boundary conditions. For mathematical models, these unknowns have usually been determined during calibration by modeling a historical sequence of inflows and adjusting the unknowns to force the model output to match the prototype data collected during the historical sequence. While this is an accepted procedure, it tends to result in a model that performs well under conditions similar to those used in the calibration process, but is of uncertain reliability when applied to significantly different conditions, which are often the most important application of the model. Modelers often deal with this uncertainty by noting that although the actual values predicted by the model are subject to some uncertainty, the incremental change between a model run with and without the test condition should be valid.

While emphasis on the incremental changes between model tests is well placed, there is often no real evidence that the incremental changes are representative of the incremental changes that would occur in the prototype, simply because there is no experience with the same conditions in the prototype. This is the case in the evaluation of the Peripheral Canal since no prototype data for verification can exist unless the canal is constructed.

An example of this dilemma concerns a particular mathematical model developed for simulation of hydraulics and water quality in the Sacramento-San Joaquin Delta. Verification was achieved using historical prototype data from several years of record. Based on the verification results, the model was assumed to be valid for application to the Delta with a Peripheral Canal. An opportunity to test this assumption was afforded by a levee break that occurred in June 1972 on Andrus Island in the northern part of the Delta. The levee break and the Peripheral Canal have similar impacts in that they both result in major (although different) variations in channel flow distribution. It was hypothesized that the degree of success with which the levee break salinity could be reproduced would be an indication of the degree of confidence that could be placed in the model results for prediction of salinity with the Peripheral Canal. Figure 3 shows some results of the levee break simulation.

OREGON
CALIFORNIA
NEVADA
N
CLAIR ENGLE LAKE
LEWISTON LAKE
SHASTA LAKE
KESWICK
WHISKEYTOWN LAKE
COW CREEK UNIT
CLEAR CREEK SOUTH UNIT
RED BLUFF DIVERSION
LAKE ALMANOR
CORNING CANAL
SACRAMENTO
STONY CANAL
LAKE OROVILLE (STATE)
TEHAMA-COLUSA CANAL
LAKE TAHOE
SUGAR PINE
AUBURN
JENKINSON LAKE
FOLSOM LAKE
LAKE NATOMA
LAKE BERRYESSA
FOLSOM SOUTH CANAL
DELTA CROSS CHANNEL
CONTRA COSTA CANAL
NEW MELONES LAKE (CE & USBR)
TRACY PUMPING PLANT
DON PEDRO
LAKE McCLURE
DELTA-MENDOTA CANAL
SAN LUIS (FEDERAL-STATE)
SAN FELIPE DIVISION
MADERA CANAL
MILLERTON LAKE
JOAQUIN
DOS AMIGOS PUMPING PLANT (FEDERAL-STATE)
PINE FLAT LAKE (CE)
FRIANT-KERN CANAL
SAN LUIS DRAIN
SAN LUIS CANAL
COALINGA CANAL
CENTRAL VALLEY BASIN BOUNDARY

Figure 2.

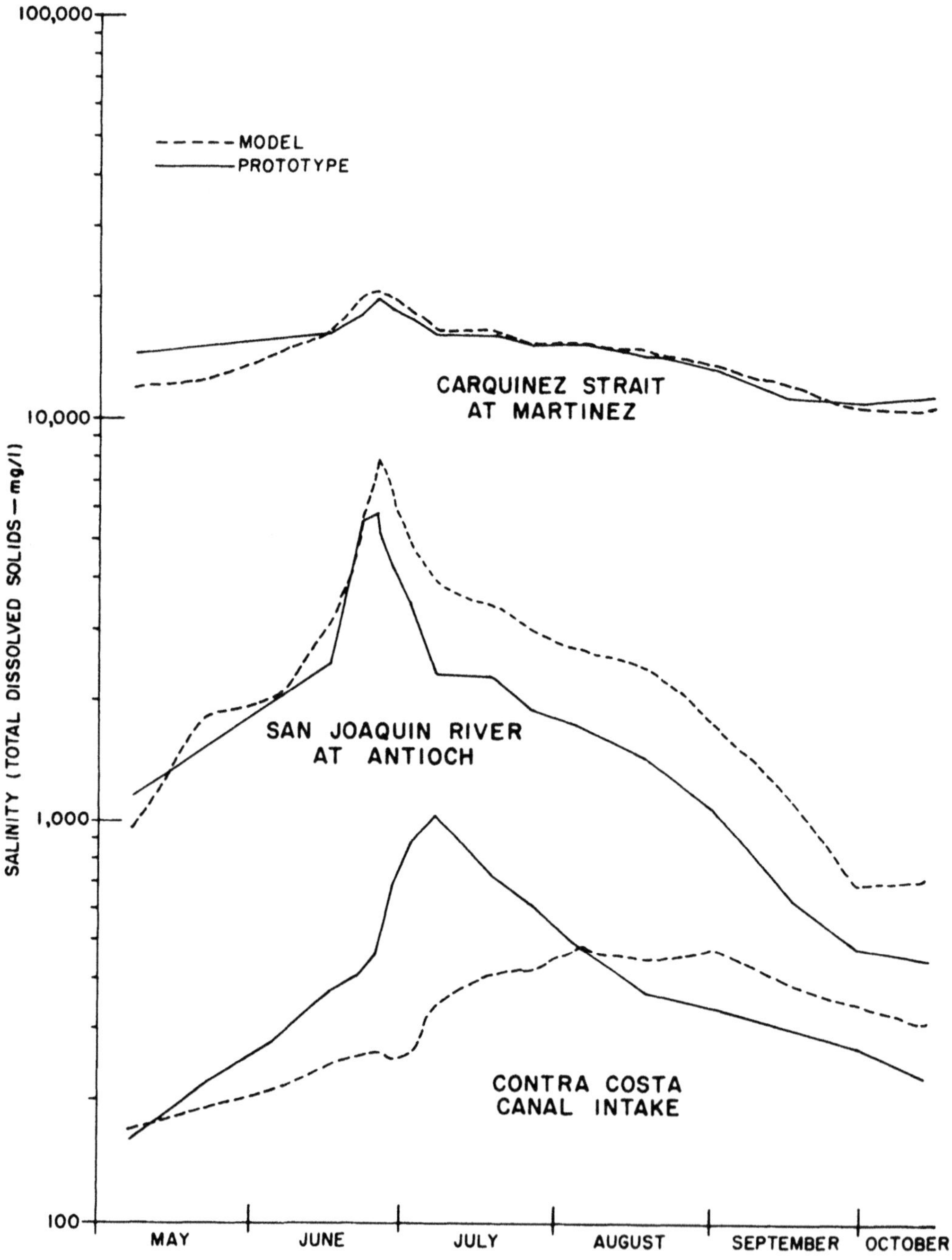

Figure 3. Comparison of model results with prototype data for 1972 Andrus Island levee break.

For most salinity stations the model simulation was remarkably good. The results for the Martinez and Antioch salinity stations shown in Figure 3 are typical. However, in the case of the salinity station at the Contra Costa Canal intake (see Figure 1), the model fails to give a reasonable approximation of prototype behavior. This failure was characteristic of model performance at all other stations along Old River. Old River is, perhaps, the most critical channel in the Delta since all major exports for agricultural, municipal, and industrial use originate in this channel. While the model could certainly be recalibrated after the fact to remedy the deficient performance, it is clear that had it been used as a predictive tool for anticipating the consequences of the levee break, it would have failed to predict the water quality crisis that in fact did occur in the Old River portion of the prototype. The results of this study caused understandable concern as to the validity of model runs evaluating the Peripheral Canal.

There are probably a number of reasons why a model may simulate historical conditions well and fail to predict hypothetical or future conditions. The most obvious reason is that empirical quantities (roughness and dispersion coefficients) are usually chosen to force a match between historical data and model output and are not necessarily representative of the actual physical process. As mentioned earlier, in the case of the Delta, considerable uncertainty exists as to the quantity of water flowing from the estuary to the ocean (Delta outflow). If the flows used in the verification process are in error, the advective transport of salt is also in error and must be compensated for by adjustment of dispersion coefficients that do not correctly account for the dispersive transport. This situation will provide adequate predictive capability only so long as the error in dispersion coefficients compensates for the incorrect advective transport. In the case of a condition which causes major changes in channel flows (advection), this will probably not be true.

Another reason the traditional verification process may be inadequate relates to the change in the nature of system hydrology with increased development of water resources. Data taken in the early stages of water resources development will likely reflect the lack of control of flows in the system. These data are characterized by rapidly changing hydrology. It is data of this nature that will most probably be used in model verification. On the other hand, the prime use of the model may be for prediction of conditions of "ultimate development" under which storage reservoirs and diversion facilities will have reduced the variability of hydrologic inputs. Empirical coefficients developed under the former situation are not actually representative of a single given flow condition but rather are a "best fit" for a range of flow conditions. Thus, when the model is applied to a markedly different situation, like that of future operation, these coefficients may not provide a reliable simulation.

ROLE OF PHYSICAL MODEL IN MATHEMATICAL MODELING

Calibration of a physical model involves a similar approach. In the case of the San Francisco Bay and Delta model constructed by the Corps of Engineers in Sausalito, California, copper strips were added to the channel cross section and adjusted in order to match model velocities and tide heights with prototype measurements. However, model salinities are not usually matched to prototype data, because of the difficulties involved in adjusting dispersion in a physical model. Thus, the water quality behavior of a physical model must usually be accepted regardless of how well it reproduces the prototype.

There does not seem to be any reason to believe that calibration and verification of a physical model is any less subject to difficulty than mathematical models. However, one aspect of a physical model that is commonly overlooked is that apart from how well it simulates the prototype, it is a real hydraulic system in its own right wherein the real processes are simulated. If it is scaled such that the relationship between gravitational and inertial forces is correctly simulated (Froude number) and the relationship between viscous and inertial forces (Reynolds number) falls within a suitably narrow range, then the model should be described by the same mathematical formulations as the prototype. The importance of this is that it allows testing of mathematical abstractions on the physical model where geometry, boundary conditions, and system parameters are known or can be determined. Successful application of a mathematical technique to prediction of an unverified condition on the physical model provides a basis for application of that same mathematical technique to the same unverified condition in the prototype.

As an example, a simple mass balance equation was used in an attempt to determine the true Delta depletion: (Fischer, 1974)

$$QC_s + \frac{\delta c}{\delta t}\, dv = K\,\frac{\delta c}{\delta x} + \Sigma q_i c_i$$

Where Q is net outflow from the Delta, C_s is mean tidally averaged salinity at a control station at the western edge of the Delta, v is Delta Channel volume, K is a mixing coefficient, x is distance along the axis of the channel, and $q_i c_i$ are net discharges and salinities of various inflows to and exports from the Delta.

This equation states that the salinity carried out of the Delta by the net advective flow, plus the salinity accumulated in the Delta by change of storage equals the salinity entering the Delta through the sum of diffusion across the control section and input from the upland discharge.

Q can be considered to be the difference between the total Delta inflow, Q_{DI}, and the net Delta Channel depletion, Q_{DCD}.

$$Q = Q_{DI} - Q_{DCD}$$

Making this substitution into the mass balance equation, dropping the integral term under the assumption of steady state conditions, and rearranging the terms, the equation becomes:

$$Q_{DI} - \frac{\Sigma q_i c_i}{C_s} = \frac{K}{C_s} \frac{\delta C}{\delta x} + Q_{DCD}$$

The unknowns in this equation are Q_{DCD} and K. If the diffusion coefficient K is considered to be constant over the range of flows evaluated, a plot of $Q_{DI} - \Sigma q_i c_i$ versus $\frac{1}{C_s} \frac{\delta c}{\delta x}$ will yield a straight line of slope K and intercept Q_{DCD}. (The linearity of the resulting line will provide a measure of the validity of the constant K assumption.)

In the course of using this approach, it was found that the results were very sensitive to the value used for C_s, the salinity at the control section. The method of choosing the C_s to correctly represent the diffusion term was found to be critical. An analytical procedure was used to determine the correct value of C_s to use given the salinity at both ends of the control section.

$$C_s = \frac{Cx_1 - Cx_2}{-\ln (Cx_2/Cx_1)}$$

Where Cx_1 and Cx_2 are the salinities at either end of the control section across which diffusion into the Delta is evaluated.

In order to test the validity of this approach, it was hypothesized that if this approach was valid for the prototype, it should also be valid for the physical model. That is, mass is conserved in the model as well as the prototype. The channel depletion in the physical model is a known value. Thus, if the mass balance equation could be used with data from the physical model to correctly compute the channel depletion in the physical model, there would be a basis for the assumption that the approach would also correctly compute the prototype Delta channel depletion.

Applying the equation to a model situation using steady state hydrology for which the net model Delta Channel depletion was 3,400 cfs (96.3 m^3/s) resulted in a computation of Q_{DCD} of 3,000 cfs (85.0 m^3/s) which is in error by 400 cfs (11.3 m^3/s). (See Figure 4.) Of course, this discrepancy may result to some degree from inaccuracies in flow measurements in the physical model, as well as from deficiencies in the mathematical model. Depending on the circumstances, this estimate may or may not be considered to be of adequate precision for use in making prototype estimates. The point is that it does provide a basis for making such an evaluation.

It is seen that this procedure makes use of analytically developed relationships, makes use of the physical model to evaluate the validity of the mathematical model, and can be applied to the prototype using actual prototype data. It is suggested that this procedure may be used to avoid some of the pitfalls common to the usual calibration/verification procedures and is believed to provide a better approach than using either the mathematical or the physical models separately.

Further research is being conducted in order to develop a more comprehensive approach to the conjunctive use of mathematical and physical models. Dr. William Yeh, UCLA, is currently conducting research which involves mathematical modeling of the portion of the physical model between the western edge of the Delta, located near the city of Pittsburg, and the western edge of Grizzly Bay, located near the city of Martinez. (See Figure 1.)

The ultimate objective of this research is to develop a two-dimensional unsteady model of the hydrodynamics and salinity variation of the entire Bay and Delta physical model. The purpose of trying to model the model is that the geometry and boundary conditions are well defined or relatively easily measured; whereas there are a number of unknowns and uncertainties in the prototype values. Thus, this research will not be concerned with how well the physical model simulates the prototype, but will actually

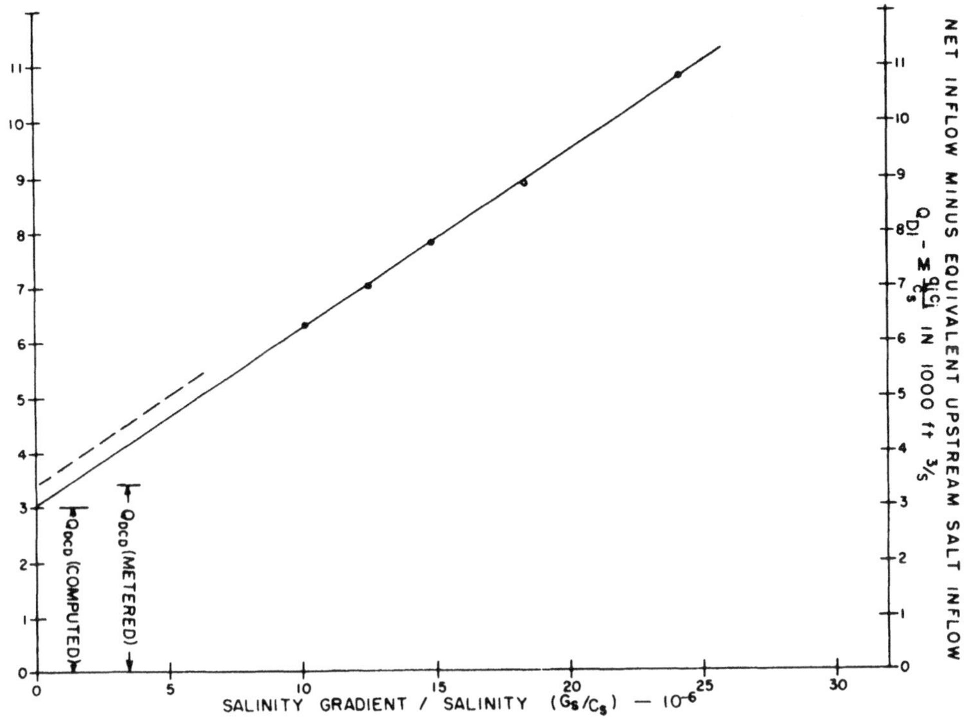

Figure 4. Projected Delta channel depletions (Bay and Delta model). July. Export - 5,000 ft.^{3} /s. Gradient: Port Chicago to Pittsburg.

consider the physical model as the prototype. It is expected that this approach will allow the emphasis to be placed on the mathematics and provide a distinction between inadequacies of the mathematics and inadequacies in knowledge of the prototype. When completed, the model will provide a much needed tool that can be used to supplement the physical model when time or money constraints make use of the physical model impractical or to enhance the physical model testing by providing a relatively inexpensive means of testing the hypothesis to be examined. It is also expected to provide a means by which the behavior of the prototype, under conditions for which no verification data are available, can be inferred. For example, use of a mathematical model to evaluate prototype channel modifications is complicated by the uncertainty of how channel modifications may alter preselected roughness and dispersion coefficients. Determining what variation in coefficients is required to adapt a mathematical model to physical model data, with and without channel modification, may provide some insight as to how coefficients in a mathematical model of the prototype should be varied.

CONCLUSION

It is not suggested that all modeling situations are adaptable to this approach nor even that it is necessarily always the best approach for situations in which it is adaptable. An obvious limitation is the availability of a physical model. In cases where a physical model does not already exist, the complexity of the system and the importance of accurate modeling must be weighed against the cost of building a physical model. Each case should be examined on an individual basis for other possible limitations. While the concept of combining the best aspects of both mathematical and physical models in order to evaluate or anticipate the behavior of the prototype was arrived at independently as a result of dealing with a complex hydraulic system, it is presumed that the procedure is neither unusual nor uncommon. However, experience indicates it is often overlooked and its value perhaps not completely appreciated. Aside from the rather simple example presented here, there are a number of other applications which have been made or are being considered in relation to the Sacramento-San Joaquin Delta. It is suggested that there are probably systems aside from the Delta as well as other phenomena besides hydraulics and salinity that may well benefit from the judicious blending of mathematical and physical models.

ACKNOWLEDGMENT

The author gratefully acknowledges the input and assistance received in the preparation of this paper from Donald J. Herbert, Regional Research Coordinator, and Earll D. Dudley, Civil Engineer, of the Bureal of Reclamation, Sacramento, California

REFERENCES

Fischer, H. B., 1974, "A Reanalysis of the Outflow vs. Salinity Relationship for the Sacramento-San Joaquin Delta, California."

Fischer, H. B., 1976, "A Review and Comparison of Salinity Models of the San Francisco Bay-Delta System."

Dudley, E. D. and R. C. Kristof, 1977, "Modeling of Salinity Intrusion in the Sacramento-San Joaquin Delta," ASCE Fall Convention Preprint 3055.

Yeh, W. W-G., 1977, "Mathematical Modeling of Salinity Variation in the Hydraulic Model of the San Francisco Bay and Delta Estuary," Research Proposal to the University of California Water Resources Center.

MODELING SEDIMENT TRANSPORT IN A SHALLOW LAKE

Y. Peter Sheng

Aeronautical Research Associates of Princeton, Inc.

50 Washington Road, P. O. Box 2229, Princeton, N. J. 08540

ABSTRACT

A systematic approach to study the sediment transport in shallow waters is presented. The approach combines mathematical modeling, remote sensing by satellite and aircraft, and laboratory and field experiments. Two- and three-dimensional hydrodynamic models are utilized and combined with significant wave models to account for the important mechanisms of sediment dispersion: convection, turbulent diffusion, gravitational settling, and resuspension and deposition at sediment-water interface. Two examples of model application are presented: (1) a feasibility study of direct pipeline discharge of dissolved solids into the Central Basin of Lake Erie; and (2) a realistic sediment dispersion event of 3 days in the Western Basin of Lake Erie. For the sediment dispersion event, the model successfully simulates the observed general sediment dispersion pattern as well as the significant concentration gradient in the horizontal direction. The bottom sediments underwent appreciable resuspension and deposition during the 3-day event, while the net change in sediment thickness over much of the basin was quite small. The basic approach can be extended to study the sediment transport in estuarine and coastal regions.

INTRODUCTION

The transport of sediments in shallow coastal and inland waters is a complex yet important problem. The ever-increasing activities of energy exploration and production and dredging operations demand better understanding of the pollutant transport and quantitative assessment of the impact of these activities on the aquatic ecosystems. As a step toward the goal of an accurate predictive tool, this study presents a systematic approach to model the transport of sediments in shallow waters with a combination of a mathematical modeling, remote sensing, field study and laboratory experiments.

As shown in Figure 1, the river-supplied sediments in a lake or estuary are convected by the wind-driven currents and mixed vertically as well as horizontally by turbulence. Settling of sediments takes place due to gravitation. The combined action of mean currents and wind waves can resuspend the sediments deposited at the shallow bottom. The basic modeling approach encompassing the above-mentioned mechanisms is shown in Figure 2. In the following, attempts are made to summarize the present understanding on each of the important processes and to briefly describe the calibration and verification of respective models. To illustrate the use of a three-dimensional dispersion model, a feasibility study of direct pipeline discharge of dissolved solids into Central Basin of Lake Erie is presented. Application of the overall sediment dispersion model to realistic events in Western Basin of Lake Erie is then illustrated. Although one-dimensional and two-dimensional (one horizontal and one vertical) models have been used extensively in this study, the present work emphasizes the three-dimensional model and its results. Differences among various models of the same kind are discussed. The present state-of-the-art modeling capability of the complex phenomena of sediment transport are assessed and recommendations for future research given.

HYDRODYNAMIC MODEL

Hydrodynamic modeling of shallow lakes or estuaries have been advanced during the last decade (Nihoul, 1975; Cheng et al., 1976; Edinger and Buchak, 1980). Models have been developed to simulate the steady-state or time-dependent circulations including one or two or three spatial dimensions. Although many of the simpler models may be quite useful and give reasonable results, their limitations have to be understood, particularly when used to provide the required input for dispersion models. For instance, accurate resolution of the vertical dimension is especially important for water quality modeling in shallow waters. Vertical shear of horizontal velocities generally result in appreciable turbulent mixing in the vertical direction. However, many parameters such as nutrients and sediments are primarily supplied from the bottom yet others such as oxygen and heat are primarily supplied from the surface, hence appreciable vertical gradients often exist. Accurate representation of the boundary conditions and turbulent

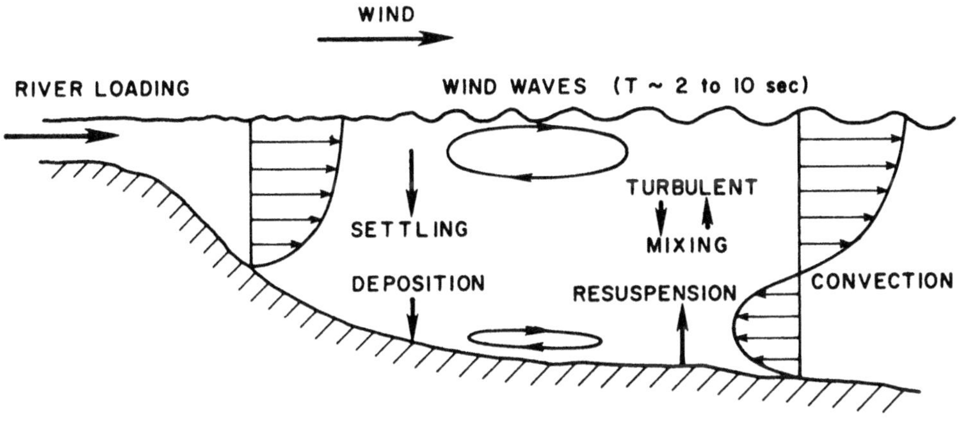

Figure 1. Schematics of dominant mechanisms of sediment dispersion in shallow waters.

mixing are essential to an accurate dispersion model. Some of these aspects will be briefly discussed in the following.

Vertically-Averaged Model Vs. Three-Dimensional Model

Vertically-averaged models have been used extensively to predict the water level in lakes, estuaries, and ocean (e.g., Leendertse, 1967; Reid and Bodine, 1968; Haq et al., 1974; Sheng, 1975; Butler, 1980). The bottom shear stress in these models, for shallow waters, is generally assumed as

$$\underset{\sim}{\tau}_B = \rho A_v G \frac{\underset{\sim}{U}}{h^2} \qquad (1)$$

where $\underset{\sim}{\tau}_B$ is the bottom stress, ρ is density, A_v is the eddy viscosity, U is the vertically-averaged velocity, h is the depth, and G is a function of the velocity profile. In general, A_v is

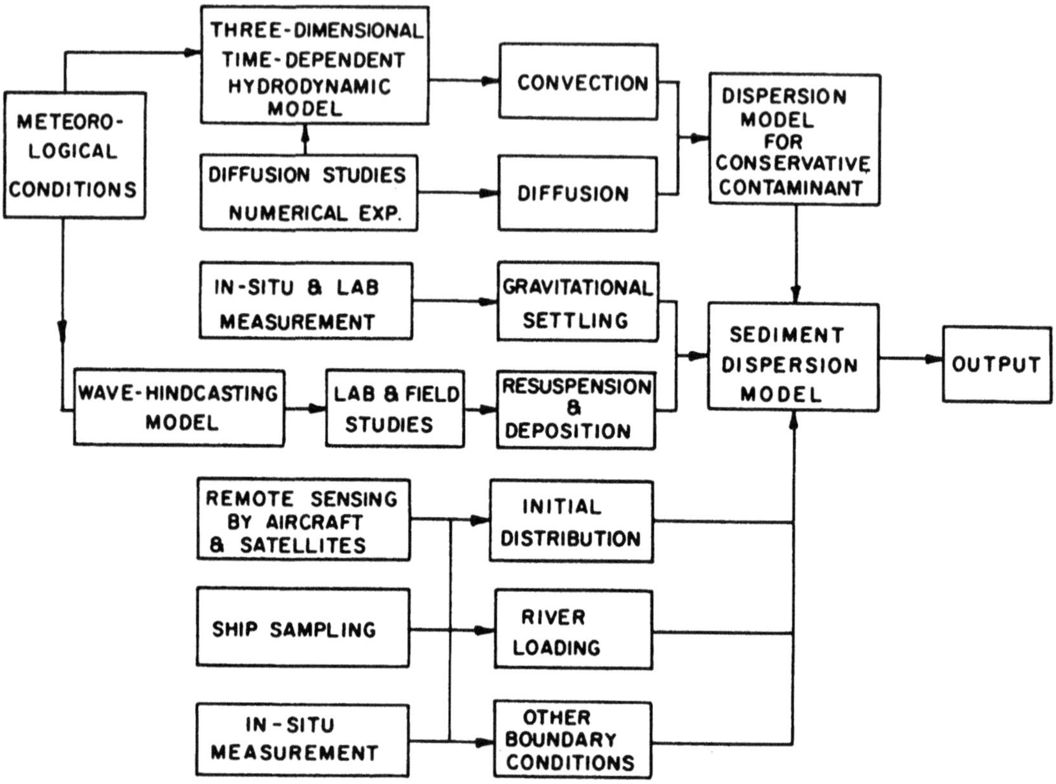

Figure 2. Hierarchical structure of the overall model for sediment dispersion.

assumed to be proportional to U. The vertically-averaged model, when applied to Lake Erie, predicted water levels qualitatively similar to the results of three-dimensional model. But the predicted water level at any instant of time, as well as in the steady state, may be as much as 35 percent off the three-dimensional results, due to the errors in specifying the bottom stress. This can be understood as follows. Figure 3 shows the steady-state wind-driven currents in the Cleveland near-shore region of Lake Erie (Sheng, 1975). In the deeper mid-lake region, significant return currents exist which are opposite to the surface wind and currents, resulting in small vertically-averaged currents and hence small bottom stress according to Equation (1). In the shallow near-shore region, however, the currents are more or less in the same direction and the vertically-averaged currents are quite appreciable. Comparing with the three-dimensional model in which the bottom shear stress is related to the near-bottom currents,

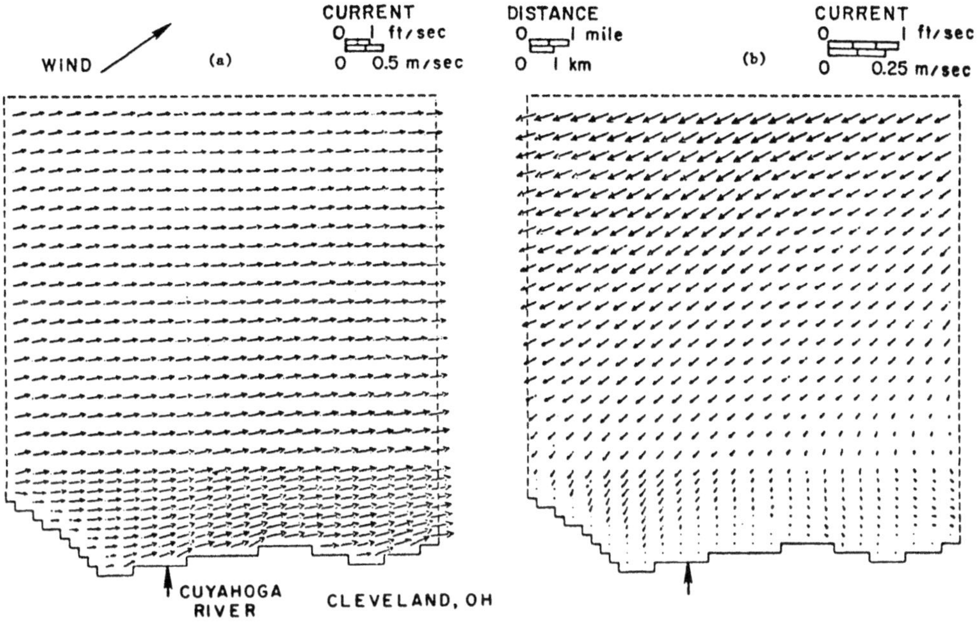

Figure 3. Horizontal velocities in the Cleveland near-shore region caused by a 7.6 m/sec wind from SSW: (a) at the surface; (b) at the 9.14 m from the surface.

Equation (1) tends to underestimate the bottom stress in deeper mid-lake region but overestimate that in the shallow near-shore region. Considering the steady-state force balance (the sum of wind stress and bottom stress balances the pressure gradient due to surface slope) for simplicity, it is apparent that using Equation (1) would overestimate the water level in the shallow region but underestimate it in the deeper region. In addition, when applied to compute the dispersion contaminants, vertically-averaged models will not be able to predict the vertical gradient of contaminants resulting from horizontal convection by surface currents and bottom currents in opposite directions.

Rigid-Lid Model Vs. Free-Surface Model

The important time scales of time-dependent circulation in a shallow lake such as Lake Erie are: t_1, a viscous diffusion time, t_2, a period of oscillation of surface gravity waves, t_3, inertial period and t_4, a flux time (Haq et al., 1974). For Lake Erie, the dominant t_2 is about 14 hours and t_1 and t_3 are about 18 hours.

Rigid-lid models eliminate the gravity waves from the problem and allow larger time steps for numerical integration. Free-surface models which are comparable in efficiency have also been developed. However, the two models may give significantly different results of the wind-driven currents within a few cycles of the seiche (or tidal) oscillation. A detailed comparison of the two models can be found in Sheng et al., (1978).

Bottom Boundary Condition

The bottom shear stress in shallow waters may be comparable to the magnitude of the wind stress at the water surface. Hence, the specification of the bottom boundary condition can significantly affect the accuracy of the vertical profile of the velocities and other parameters. Existing hydrodynamic models use either the no-slip condition

$$u = v = w = 0 \quad @ \quad z = 0 \tag{2}$$

or the quadratic-stress law:

$$\underset{\sim}{\tau}_B \propto \underset{\sim}{u}_+ |\underset{\sim}{u}_+| = \rho C_D \underset{\sim}{u}_+ |\underset{\sim}{u}_+| \tag{3}$$

where ρ is the density, C_D is a drag coefficient and $\underset{\sim}{u}_+$ is the velocity at a height z_+ above the bottom.

The no-slip condition is strictly only valid in the limiting case of laminar flow or when there is sufficient number of grid points in the laminar sublayer of turbulent flow. On the other hand, the validity of the quadratic stress law has been tested in both field and laboratory experiments. It has been shown that adjacent to the bottom of a planetary or benthic boundary layer, a constant-flux layer generally exists within which the vertical gradient of mean velocities, $\partial u/\partial z$, is related to the momentum fluxes to the bottom, u_* (friction velocity = $\sqrt{\tau_B/\rho}$):

$$\frac{\partial u_+}{\partial z} = \frac{u_*}{K z_+} f_1 \left(\frac{z_+}{L}\right) \tag{4}$$

where K is the von-Karman constant, z_+ is the height above the bottom, L the Monin-Obukhov length, and f_1 the Monin-Obukhov similarity function which represents the effect of stability. Similar relationships exist for temperature and species (Lewellen and Teske, 1973; Businger et al., 1971).

Equation (4) can be integrated vertically

$$u_+ = \frac{u_*}{K} \left[\ln \frac{z_+}{z_o} + \phi \left(\frac{z_+}{L}\right) \right] \tag{5}$$

where z_o is the roughness length at which all velocities go to zero and ϕ is a function of stability.

Equation (3) is equivalent to Equation (5) if one assumes:

$$C_D = \left[\frac{K}{\ln \frac{z_+}{z_o} + \phi} \right]^2 \tag{6}$$

Thus, the drag coefficient in the quadratic stress law is a function of the bottom roughness, the stability, and the height at which u_+ is measured or computed. Sternberg (1972) measured the steady-state flow over a variety of bottom conditions and found C_D to be generally in the neighborhood of 0.004 when the mean velocity at the edge of the logarithmic layer is used. The use of Equation (2), rather than Equation (3), in a three-dimensional model was found to result in generally smaller vertical gradients of velocities, thus reducing the near-bottom velocities by more than 50 percent (Sheng, 1978).

The bottom boundary of the atmospheric or aquatic environments are frequently covered with obstructions such as canopy of vegetations or topographical features. The representation of the surface features in terms of an effective roughness z_o is generally adequate so long as the primary interest is in the flow above the canopy. However, for flow within and adjacent to the canopy, a more detailed representation is required. Lewellen and Sheng (1979a, b) included the effect of canopy on the mean flow and turbulence in a higher-order turbulence model and achieved good agreement with field data on momentum and heat transfer over a variety of surface conditions. Such a canopy model may be used in studying highly vegetated wetlands.

Turbulence Parameterization

The hydrodynamics and dispersion of contaminants in a shallow lake or estuary, particularly under stratified conditions, depends strongly on the turbulent transport processes. Most of existing hydrodynamic models for lakes and estuaries have parameterized turbulence with the eddy viscosity (or first-order closure) concept (Bowden, 1977). Such models are quite efficient and do give reasonable results when the eddy viscosity for a particular environment has been sufficiently calibrated with data. Richardson-number-dependent eddy viscosities have been used to model the effect of buoyancy on turbulence (Sundaram and Rehm, 1970; Sheng, 1978). When flow measurements in an aquatic environment are not sufficient enough to provide calibration of the empirical model constants, indirect calibrations and verifications are then generally performed to warrant confidence on the model. The alternative approach is to use some of the more recent invariant models (Donaldson, 1973; Lumley and Khajeh-Nouri, 1974; Lewellen and Sheng, 1979a, b) which do not contain model constants that require adjustment for each application. These models contain the proper dynamics of the mean flow variables as well as the second-order turbulent correlations and have been sufficiently verified for laboratory flows and atmospheric applications. Recently, measurements of mean flow variables and turbulent fluxes in the aquatic environment have been quite advanced (Gordon and Dohne, 1973; Gardner et al., 1980) such that these data may be useful for verification of both the eddy viscosity models and the second-order closure models.

In the present study of sediment transport in the Western Basin of Lake Erie, a three-dimensional, free-surface hydrodynamic model with quadratic stress law was used. The hydrodynamic model includes the effects of time-dependence, nonlinear inertia, Coriolis effect, horizontal and vertical turbulence. The turbulence was parameterized by eddy coefficients which, in the vertical direction, depend on the wind stress and have been calibrated with measured currents under several different conditions. Strictly conservative numerical schemes were used to solve for the hydrodynamic equations. To properly include the bottom topography, the vertical coordinate is stretched such that equal number of grid points exist at deeper as well as shallow parts of the basin. Details of numerical procedure can be found in Sheng (1978).

WAVE MODEL

Hydrodynamic models generally have not considered the effects of waves which become increasingly important as the depth of water decreases. In order to include the direct effect of waves in causing sediment resuspension and contaminant transport, the present study tested six wave hindcasting models against data from a station in the Western Basin of Lake Erie (Sheng and Lick, 1979). The results indicated the SMB (Sverdrup-Munk-Bretschneider) method (U. S. Army CERC, 1973) for shallow water gave the best correlation with data on wave period and wave height. Other methods overestimated the wave parameters in the following ascending order: deep-water SMB, SMC (Bokuniewicz et al., 1977), Liu (Liu, 1977), Mitsuyasu (Mitsuyasu, 1971), and JONSWAP (Hasselmann, et al., 1973) (Figure 4). The wave model has been extended to compute the bottom shear stress and has been applied to the entire Western Basin of Lake Erie. Results indicated strong dependence on the wind fetch as effected by the wind direction. Figure 5 indicates the contours of bottom shear stress under different wind direction and wind speed. It is interesting to note that along the Ohio shore, substantial bottom stresses were generated by the moderate Easterly wind. For simplicity, detailed interaction of waves with islands have not been included in the model.

The same wave model was also employed by the present author to compute the bottom stresses along a transect in Long Island Sound between New Haven, Connecticut and Port Jefferson, New York. The results indicated that under Easterly wind (longest fetch), very little resuspension of sediments would occur below the 20m-depth (Figure 6). Following Hurricane Belle in 1976, X-ray radiographs of box cores taken from the 13m, 20m, and 27m stations (Station A, B, C in Figure 6) showed evidence of sediment resuspension of approximately 2cm, 0.2cm and 0cm, respectively (McCall, 1978). Previous workers applied the SMC method (Bokuniewicz et al., 1977) and deep-water SMB method (Lam and Jaquet, 1976) to shallow water environments, which may have resulted in possible overestimation of the wave effect in causing sediment resuspension. Close examination of the six tested methods indicated only the shallow water SMB method contains the explicit effect of water depth and bottom friction. The other methods contain empirically determined constants based on wave data from deep lakes or oceans where either the physical environment or the wave spectrum is quite

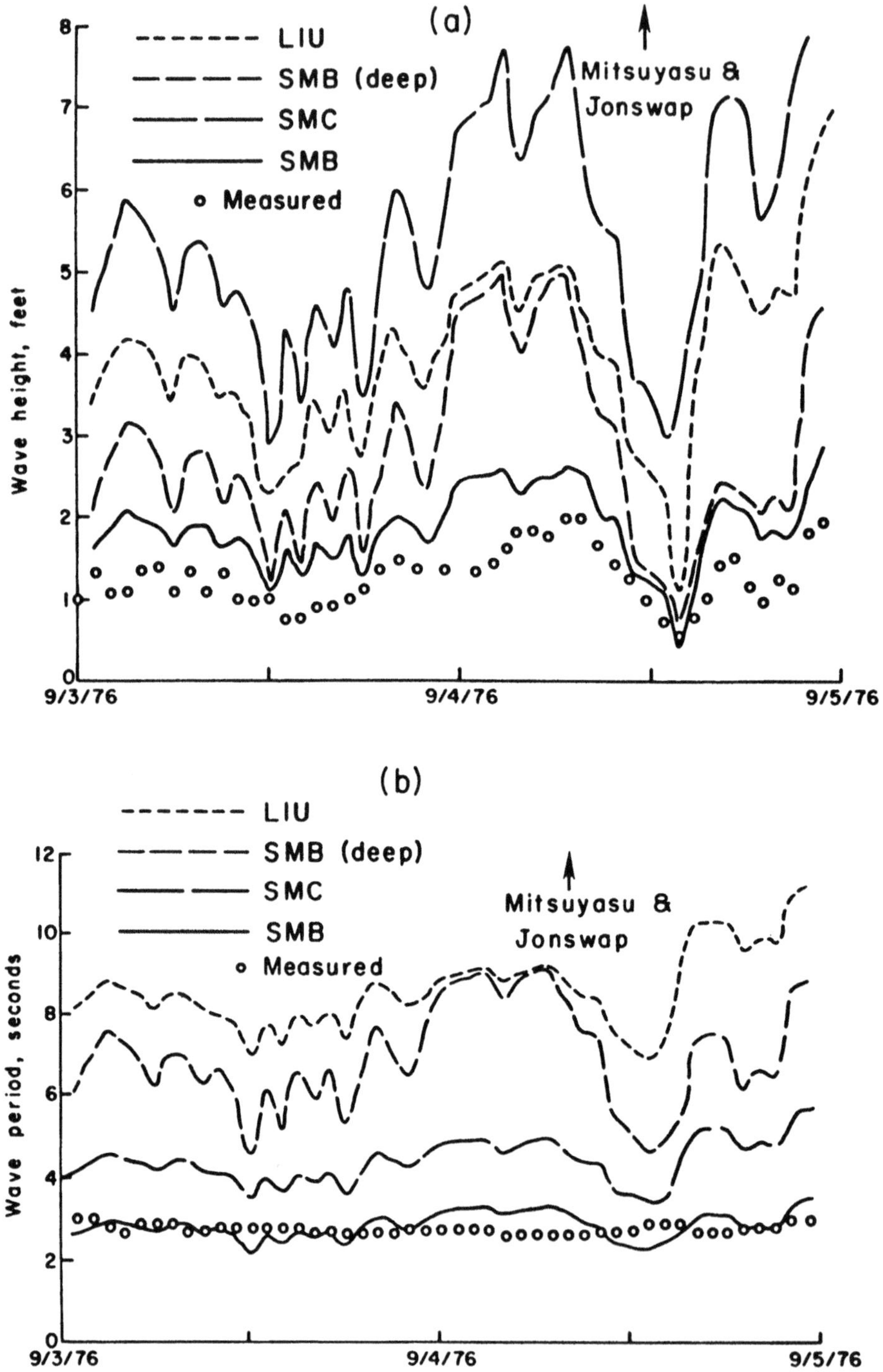

Figure 4. Significant wave parameters during September 3-5, 1976, computed by various methods: (a) wave height; (b) wave period.

different from the shallow Lake Erie.

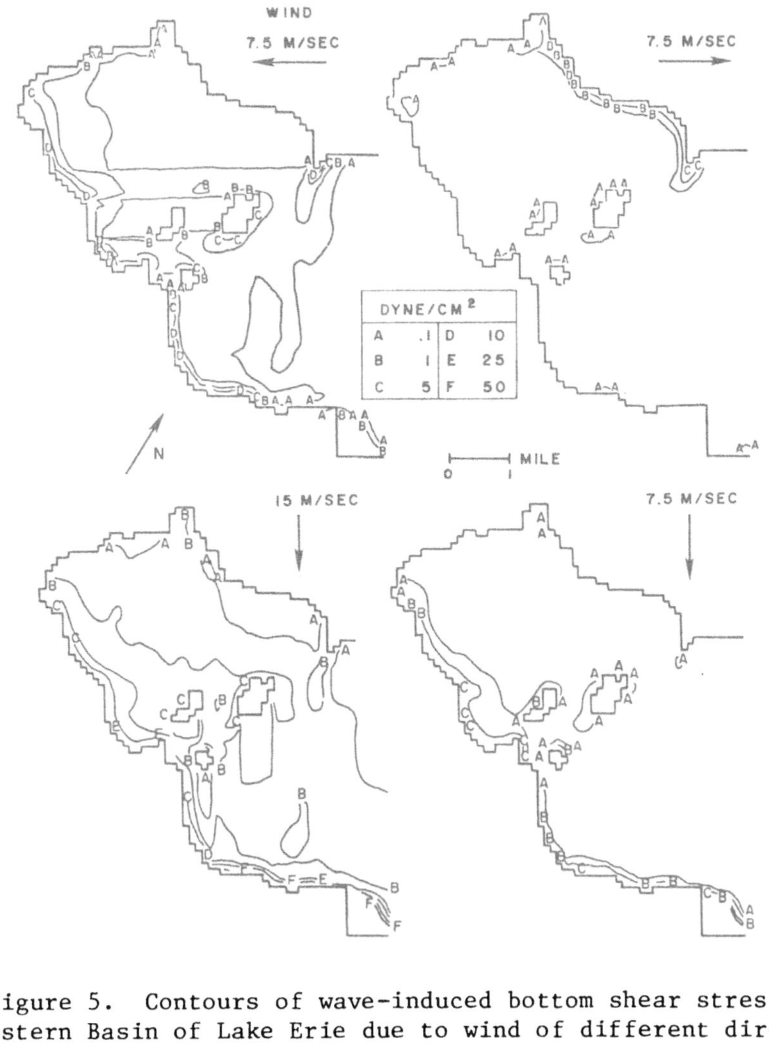

Figure 5. Contours of wave-induced bottom shear stress in the Western Basin of Lake Erie due to wind of different directions and different speeds.

Recently, Resio and Vincent (1977a) modified the numerical model for wave generation by Barnett (1968) and applied it to the five Great Lakes. Remarkable agreement between computed and observed wave data was achieved at several stations in the lakes, including two stations in the relatively deeper Central Basin of

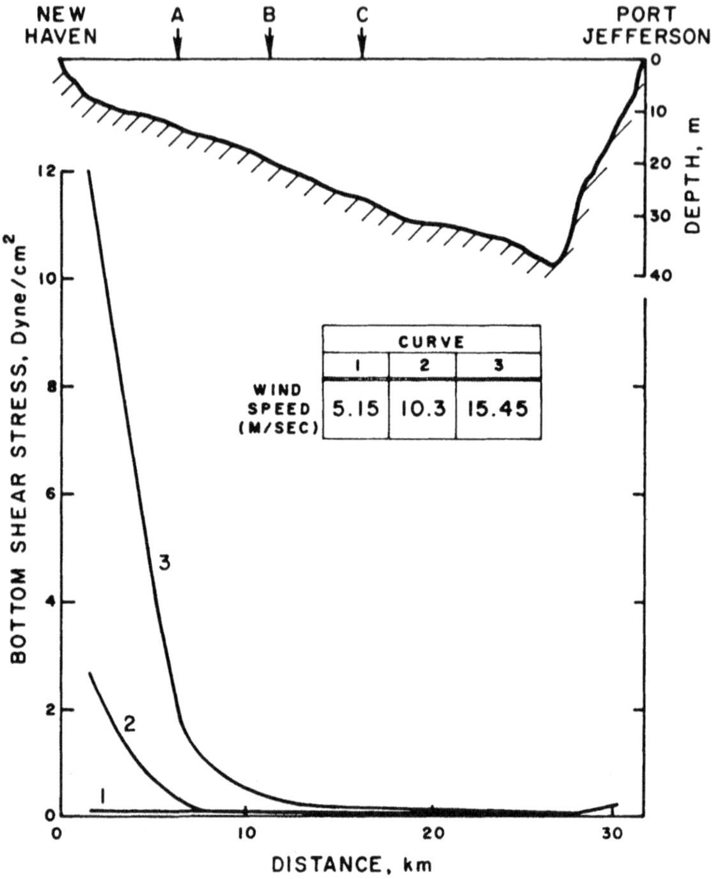

Figure 6. Wave-induced bottom shear stresses along a transect in Long Island Sound between New Haven and Port Jefferson.

Lake Erie (Resio and Vincent 1977b). Their model appears to be by far the most extensively verified model for wave generation in deep waters. For shallow waters, the effects of bottom friction and wave breaking should be included. Resio-Vincent model computes the entire wave spectrum, in addition to the significant wave parameters required for this study, and hence is generally quite time-consuming compared to other methods such as the shallow water SMB method. Consequently, it was not used in the present study.

DISPERSION MODEL

The basic equation describing the dispersion of a contaminant is

$$\frac{\partial C}{\partial t} + \frac{\partial (Cu)}{\partial x} + \frac{\partial (Cv)}{\partial y} + \frac{\partial [C(w+w_s)]}{\partial z} = \frac{\partial}{\partial x}\left(D_H \frac{\partial C}{\partial x}\right)$$

$$+ \frac{\partial}{\partial y}\left(D_H \frac{\partial C}{\partial y}\right) + \frac{\partial}{\partial z}\left(D_v \frac{\partial C}{\partial z}\right) + S \tag{7}$$

where C is concentration (or temperature), x and y are the horizontal coordinates, z is the vertical coordinate, t is the time, (u,v,w) are the three-dimensional fluid velocities in (x,y,z) directions, w_s is the vertical settling velocity of the contaminant relative to the fluid, D_H and D_V are the turbulent eddy diffusivity in the horizontal and vertical direction, and S is a source term. The inherent assumptions and the validity of the above equation have been discussed in Sheng (1975).

The net vertical flux of the contaminant is specified at both the surface and the bottom. The conditions for sediment particles are:

$$-w_s C + D_v \frac{\partial C}{\partial z} = 0 \qquad @ \qquad z = 0 \tag{8}$$

$$-w_s C + D_v \frac{\partial C}{\partial z} = \beta C - E = \beta(C - C_{eq}) \qquad @ \qquad z = -h(x,y) \tag{9}$$

where h(x,y) is the bottom, βC is the deposition, E is the resuspension, and C_{eq} is defined as E/β. β and E are generally functions of the sediment properties and bottom shear stresses. β and E for this study were constructed from experimental data obtained in a laboratory annular flume. The flume can generate bottom shear stresses up to 10 dyne/cm^2 with the rotating lid at reasonable speed (Fukuda, 1978). The depth of the flume is sufficiently shallow such that the sediment concentration in the flume is approximately uniform vertically and the flume depth can be considered to represent the first bottom numerical grid in the dispersion model. A bottom of sediments is prepared and the flow is gradually set up to a desired bottom shear stress.

Initially, there is little deposition to the bottom and resuspension causes the concentration to increase in time. Deposition gradually increases with concentration and the suspended sediment reaches equilibrium concentration eventually when the deposition balances the resuspension at the bottom. The time history of suspended sediment concentration allows the determination of β and E. It can be shown that E is equal to the asymptotic rate of change of concentration. The rate of resuspension as a function of bottom stress for two types of Lake Erie sediments were constructed and are shown in Figure 7. Curve A corresponds to sediments from the Western Basin off the Ohio shore which contain a large proportion of silt and clay. Curve B corresponds to sediments from the Central Basin which contains less clay and silt but more fine sand. β in both cases was found to be relatively constant at 0.008 cm/sec. Notice the interest of the laboratory experiment is not only in the threshhold of sediment transport (Komar and Miller, 1973) but also in the resuspension rate at various bottom stresses.

The bottom shear stress that appears in the resuspension relationships in Figure 7 have to be computed from the hydrodynamic and the wave models. In general, a much thinner wave boundary layer exists within the thicker current boundary layer, and turbulent fluxes due to one mechanism tend to enhance that generated by the other. In the present study, due to the lack of data in Lake Erie and lack of resolution within the bottom boundary layer, a simplifying procedure to combine the wave-generated and current-generated stresses (Sheng and Lick, 1979) were used.

The overall procedure for determining the rate of resuspension of a particular type of sediment is summarized in Figure 8.

PIPELINE DISCHARGE OF DISSOLVED SOLIDS INTO A LAKE

The problem considered here is the dispersion of dissolved solids in the Painsville near-shore area of Lake Erie (Figure 9) after entering the lake from a discharge pipeline. The Environmental Protection Agency proposed to the Diamond Shamrock-Painsville Chemical Plant which discharged about 6,000,000 lbs/day dissolved solids (chlorides) into Grand River, to consider direct discharge into the lake with a one-mile or two-mile pipeline. The following is a preliminary feasibility study to investigate the possible impact on the water quality at the two Painsville water intakes and beaches. As shown in Figure 9, the dissolved solids were to be discharged at either location 1 (8.08m depth) or location 2 (11.13m depth), within 1.5 to 3m from the bottom. Physical parameters of the discharge waste were:

Total discharge	12.1×10^6 gal/day
Density of waste stream	8.85 lb/gal
Amount of solid discharge	6.16×10^6 lb/day
Discharge Temperature	80°F

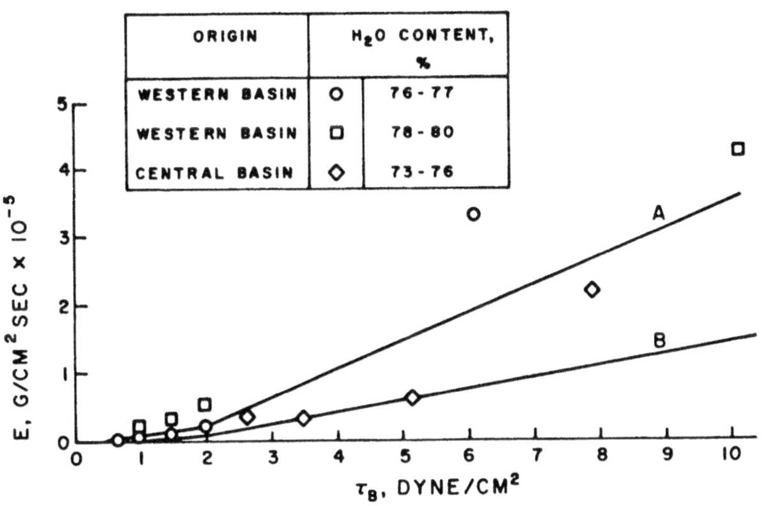

Figure 7. Rate of resuspension as a function of bottom shear stress for two types of Lake Erie sediments. Curves A and B are used for the model computation.

A steady-state hydrodynamic model (Sheng, 1975) was used to calculate the three-dimensional flow field in the near-shore region with a one-quarter mile grid in the horizontal direction and seven grid points in the vertical direction. The steady-state wind-driven currents in the near-shore region caused by uniform wind from South, Southwest, North, and Northwest were calculated. Moderate wind stress of 0.875 dyne/cm^2 and vertical eddy viscosity of 16.9 cm^2/sec were assumed. Once the currents are known, the time dependent dispersion of contaminants can be calculated. It is assumed that (1) the dynamics of the flow are not affected by the contaminants, (2) the contaminants are dissolved solids which move with the fluid, (3) eddy diffusivity is the same as eddy viscosity and is constant, and (4) the net flux of dissolved solids is zero at the bottom as well as the surface.

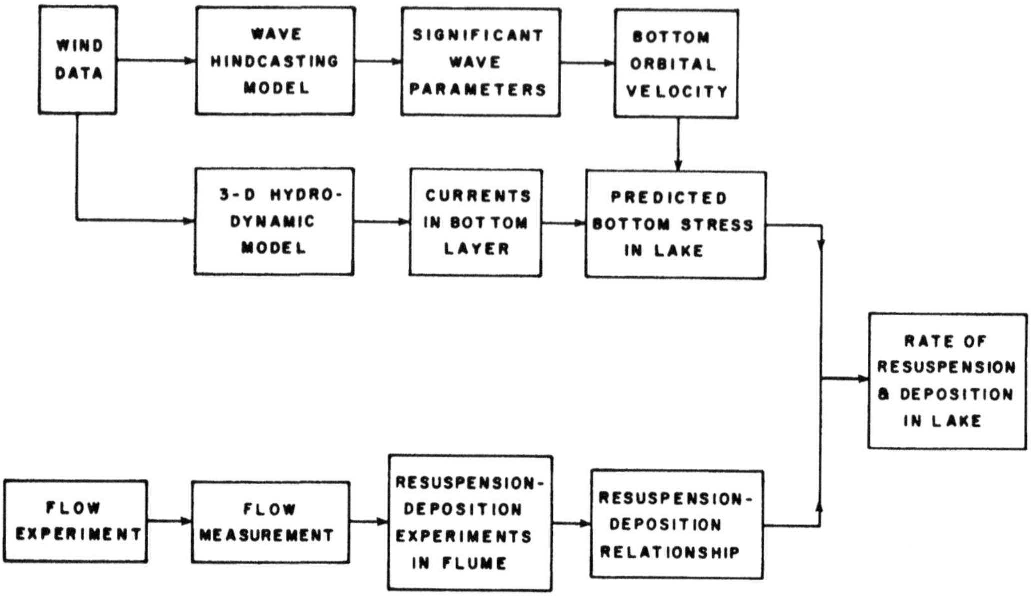

Figure 8. Schematics for computation of rate of resuspension and deposition for a given type of sediment in Lake Erie.

The waste was assumed to discharge at a constant rate starting at some initial time. Although the effects of inertia and buoyancy may be important locally near the discharge points, gravitational settling and turbulent mixing tend to mix the waste more uniformly in the surrounding waters thereby decreasing its density to near the water density. It is assumed that the discharge system will allow the waste to be uniformly mixed within the discharge grid cell within a time step (20 min). With each of the calculated four velocity fields, the dispersion of dissolved solids discharged from the one-mile and two-mile discharges were calculated for a period of 24 hours. The results for each case were compared in terms of predicted concentrations of dissolved solids at water intakes and along lake shore (Table 1). The maximum concentration at water intake 1 is attained with the combination of Southwesterly wind, one-mile discharge at both the surface level and the discharge level (SW-1-S and SW-1-D), while that at water intake 2 is attained with Northerly wind, one-mile discharge (N-1-S and N-1-D). Under Southwesterly wind, strong near-shore currents exist at the discharge level as well as the surface level (Figures 10a, b). The dissolved solids were initially convected downstream from the discharge level while gradually being transported upward by turbulent diffusion. Far downstream of the discharge point, the dissolved solids appear to be well mixed in the vertical direction. For the two-mile

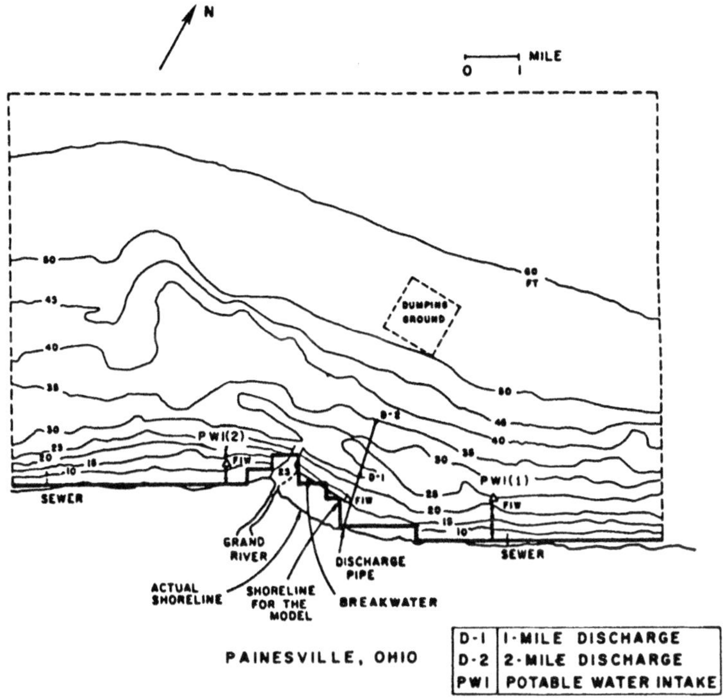

Figure 9. The Painsville near-shore region of Lake Erie.

discharge, the currents at the discharge level are directed away from the shore while the surface currents are along the shore. The results indicated that the difference between concentrations at the surface level and the discharge level are more pronounced than in the one-mile discharge case (Figure 11).

Under Northerly wind, the currents at discharge level are much weaker than the surface currents (Figures 12a, b). The dissolved solids were first transported upward by turbulent diffusion, then convected by the strong surface currents toward the West, followed by vertical diffusion to the discharge level (Figure 13). Available measurements by EPA near Water Intake 2 indicated that high concentrations of chlorides always occurred on days with moderate Northerly wind (Amendola, 1976). From the stand-point of causing less pollution nearshore, the two-mile discharge appears to be more desirable than the one-mile discharge. This example illustrated the importance of three-dimensionality in which vertical diffusion and horizontal convection are the major transporting mechanisms. Their combined transport corresponds to the horizontal turbulent diffusion generally referred to in dye diffusion experiments.

Table 1. Model Predicted Dissolved Solids Concentration in Lake Erie after 24 Hours of Pipeline Discharge

	Max. Conc. mg/1	Water Intake (1)	Water Intake (2)	Harbour Mouth	Eastern Breakwall Exit	(miles) Shoreline over which conc. > 5 mg/1
SW-1-S	25.6	18.69	0	0	0	8
SW-1-D	207.2	18.69	0	0	0	5.5
SW-2-S	8.06	.8	0	0	0	0
SW-2-D	182	.775	0	0	0	0
S-1-S	82.2	16.02	0	0	0	2.6
S-1-D	325.7	10.68	0	0	0	2.6
S-2-S	31	3.5	0	0	0	0
S-2-D	294.5	3.7	0	0	0	0
NW-1-S	113.2	3	0	0	0	2.1
NW-1-D	380	3	0	0	0	2.1
NW-2-S	46.5	.2	0	0	0	.25
NW-2-D	279	.1	0	0	0	0
N-1-S	85.4	0	32	50	.5	2.5
N-1-D	320.4	0	32	50	.5	2.4
N-2-S	31	0	7.8	2	0	0
N-2-D	302.3	0	6.8	1.0	0	0

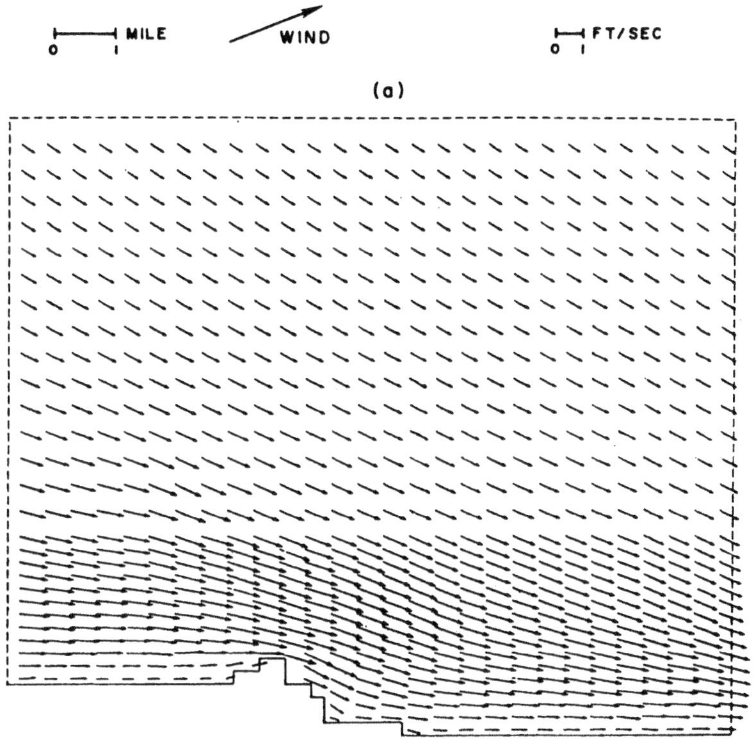

Figure 10a. Horizontal velocities in the Painsville near-shore region caused by a 0.875 dyne/cm^2 wind stress from SW: (a) at the surface.

SIMULATION OF SEDIMENT DISPERSION IN THE WESTERN BASIN OF LAKE ERIE

Annually, about 13 million tons of fine grained solids are transported into Lake Erie. About 5 million tons are from river loading, particularly the Maumee and the Detroit Rivers, which contribute about 85 percent of the sediment load to the Western Basin. Much of the river loading takes place during the heavy run-off events following the melting of ice in the spring. The large amount of sediment loading and the shallow depth of the Basin result in generally high surface sediment concentrations and fast response to meteorological changes, making the Western Basin an ideal and unique environment to study the dispersion of sediments by a combination of remote sensing, field measurements, and mathematical modeling.

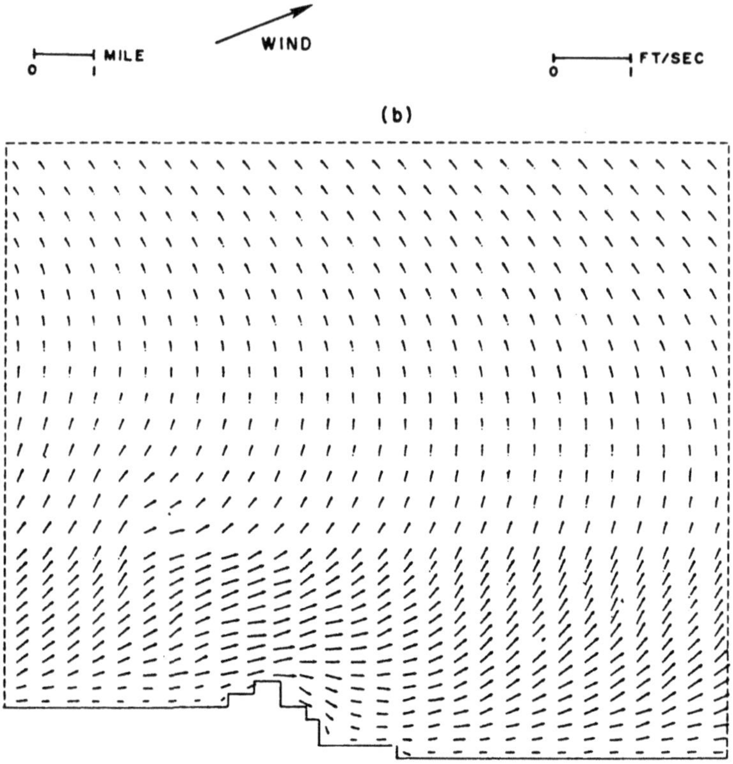

Figure 10b. Horizontal velocities in the Painsville near-shore region caused by a 0.875 dyne/cm^2 wind stress from SW: (b) at 1/6 depth from bottom.

Intensive surveys of the Western Basin of Lake Erie took place in the spring of 1976. From March 8 through March 11, NASA aircrafts equipped with the 10-channel Ocean Color Scanner (OCS) were flown over the entire Western Basin several times. Concurrent ship surveys were also conducted at several stations to provide the ground truth data necessary for calibration of the OCS data. Specifications of the OCS (Gedney et al., 1977) are shown in Table 2.

Single channel radiance data from Channel 8 was found to give the best correlation with measured suspended solids concentration near the lake surface and is independent of water types. The OCS provided data over much of the Western Basin on March 8 and March 11. Starting from condition on March 8, the present study attempted to simulate the sediment dispersion for three days. However, to eliminate any transient set-up, the hydrodynamics computation was carried out for six days starting March 6.

Table 2. Specifications of Ocean Color Scanner Used in Aircraft Remote Sensing Surveys (from Gedney et al., 1979)

	OCEAN COLOR SCANNER (OCS)									
Channel	1	2	3	4	5	6	7	8	9	10
Center wavelength λ_c(nm)	428	466	508	549	592	632	674	714	756	794
Bandwidth $\Delta\lambda$(nm)	40	40	40	40	40	40	40	40	40	40

Ground Speed = 834 km/hr; Scan Rate = 4./scans/sec;
Sample Rate = 350 samples/90° scan;
Ground Resolution = 60m x 60m

The time-dependent wind-driven currents in the Western Basin were computed by means of a three-dimensional free-surface model (Sheng, 1978). A one-mile numerical grid was used in the horizontal direction throughout the Western Basin. An 8-mile grid was used in the Central and Eastern Basin. The flows in the 1-mile grid and the 8-mile grid are dynamically coupled through an overlapping region (Figure 14). Either the surface elevation or the horizontal velocity is specified along the two boundaries of the overlapping region. The coupling procedure ensures that (1) the mass, momentum, and energy are conserved from one region to another, (2) no discontinuity occurs across the two regions, and (3) any effect in one region is felt through the other region. Details of the basic procedure can be found in Sheng (1975). Such a coupling procedure can be applied to an imbedded grid system in modeling coastal hydrodynamics. In the vertical direction, the coordinate was stretched such that the depth corresponds to 1 in the new coordinate and five grid points were included. This gives a total of 62 x 72 x 5 grid points. Notice that the grid points are staggered in both the horizontal and vertical directions. In the vertical direction, the shear stresses and vertical velocity are defined at full grid points including the bottom and the surface while the horizontal velocities are defined at half grid points. Such a grid system allows straightforward derivation of conservative finite-difference equations and specification of boundary conditions. A time step of 75 sec. was used for computing the surface elevation and a step of 15 min. was used to compute the three-dimensional velocities. Hourly wind data at nine weather stations around the lake were used to generate the necessary wind stress field over the lake surface.

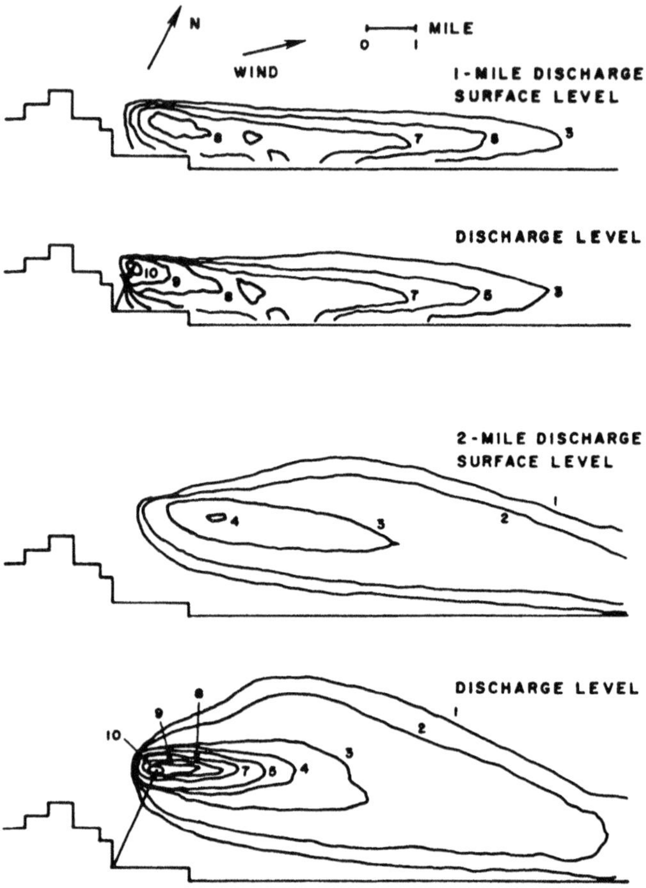

CONTOUR	1	2	3	4	5	6	7	8	9	10
MG/L	.53	1.07	5.3	8	10.7	13.4	16	21.4	32	107

Figure 11. Contours of dissolved solids concentration in the Painsville near-shore region at 24 hours after initial discharge, under SW wind. Background concentration is zero everywhere.

Table 3 summarizes the representative meteorological data during the six-day period and schedules of remote sensing, field survey and mathematical simulation carried out for the event.

Measured lake level data at six stations (four in the Western Basin, one in the Central Basin, one in the Eastern Basin) were compared with computed level at nearby grid points. As shown in

Table 3.

	3/6	3/7	3/8	3/9	3/10	3/11
Maumee River Flow Rate (m^3/sec)	1470	1320	930	700	520	400
Detroit River Flow Rate (m^3/sec)	6650	750	6450	5960	6250	5940
Toledo Wind Speed at Toledo (m/sec)	7.8	7.0	2.8	3.5	4.3	4.7
Resultant Wind Direction at Toledo (day)	260	280	360	80	250	70
Resultant Wind Speed at Windsor (m/sec)	10.3	11.3	4.7	4.4	4.6	5.9
Resultant Wind Direction at Windsor (day)	270	270	337.5	90	292.5	337.5
Landsat Satellite Data						X
Ocean Color Scanner Data (NASA LeRC)			X			X
Ship Data (OSU)			X	X	X	
3-D Hydrodynamic Simulation	X	X	X	X	X	X
3-D Dispersion Simulation			X	X	X	X

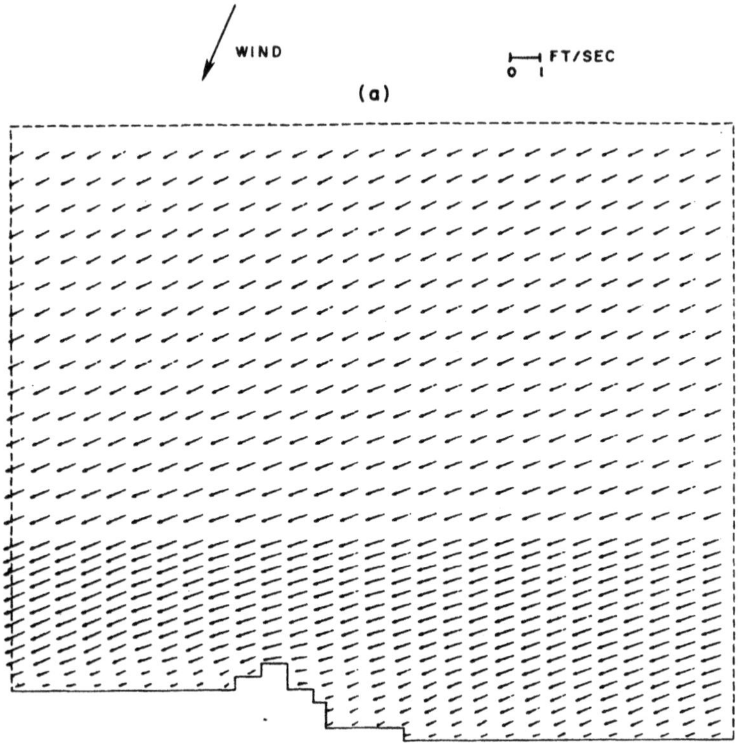

Figure 12a. Horizontal velocities in the Painsville near-shore region caused by a 0.875 dyne/cm^2 wind stress from N: (a) at the surface.

Figure 15, generally good agreement exists at all stations. The period of dominant seiche oscillation (∿14 hours) is apparent. Typical results of the horizontal velocities computed by the three-dimensional hydrodynamic model are shown in Figure 16. It is interesting to note that the currents near the bottom (σ = -0.9) are opposite to the direction of those near the surface (σ = -0.1). The bottom currents may generate appreciable shear stress to re-suspend the bottom sediments.

Using the three-dimensional currents computed from the hydrodynamic model, the dispersion of sediment between noon March 8 and noon March 11 were then computed. The initial condition of sediment concentration as produced from remote sensing and ship survey is shown in Figure 17. For simplicity, the sediments in the entire basin were assumed to be uniform in the vertical direction. Based on laboratory studies of lake sediments, a value of 0.01 cm/sec

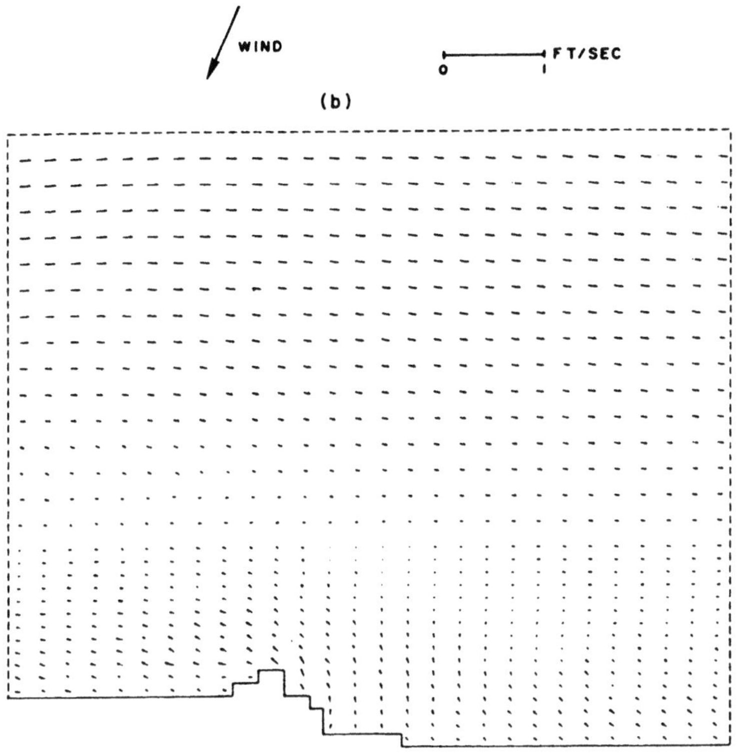

Figure 12b. Horizontal velocities in the Painsville near-shore region caused by a 0.875 dyne/cm^2 wind stress from N: (b) at 1/6 depth from bottom.

was assumed as the settling speed of the sediments. The fine-grained sediments contain a large fraction of clay or silt, the individual particles tend to aggregate and increase the effective size of particles. As a first example, the resuspension rate-shear stress relationship as shown by curve A in Figure 7 was used in the model simulation.

The winds in the Western Basin were not strong but appreciable change in wind direction occurred during the three-day period. For example, the wind data at Toledo indicated a shift fron NE to SE and then to NE again. Consequently, the wave-generated bottom shear stress showed significant variation with time and space. For example, high resuspension prevailed along the Ohio shore around noon March 9, while moderate resuspension occurred along the Canadian shore around noon March 10. On March 11, moderate resuspension occurred along the Ohio shore again. The repeated cycles of resuspension, deposition and transport resulted in the appreciable

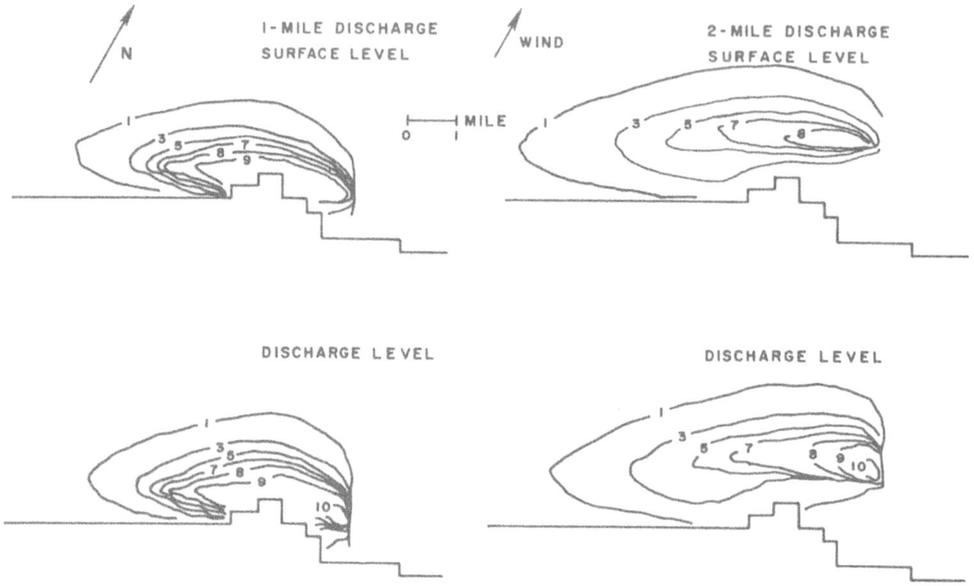

Figure 13. Contours of dissolved solids concentration in the Painsville near-shore region at 24 hours after initial discharge, under N wind. Background concentration is zero everywhere.

temporal and spatial variation in sediment concentration (Figure 18). The computed results at noon March 11 clearly exhibited the strong concentration gradient off the Ohio shore as observed by both the Landsat and OCS (Figures 19a, b).

It is of interest to examine the net amount of sediments eroded from the lake bottom during the three-day period (Figure 20). The largest amount of erosion occurred along the Ohio shore where significant bottom shear stress occurred frequently and the sediments contain a large amount of silt and clay. Appreciable erosion apparent in the vicinity of the Detroit River and around the islands were responsible for causing the computed concentration in these areas to be higher than the observed data. Recent studies indicated sediments in these areas actually contain a larger fraction of coarser material such as fine sands. The Detroit River sediments are composed primarily of coarse sediments suspended from Lake St. Clair. The strong river currents at the river entrance generally do not allow the sediments to be deposited in its vicinity until farther into the lake. Thus, the resuspension in these areas were probably overestimated.

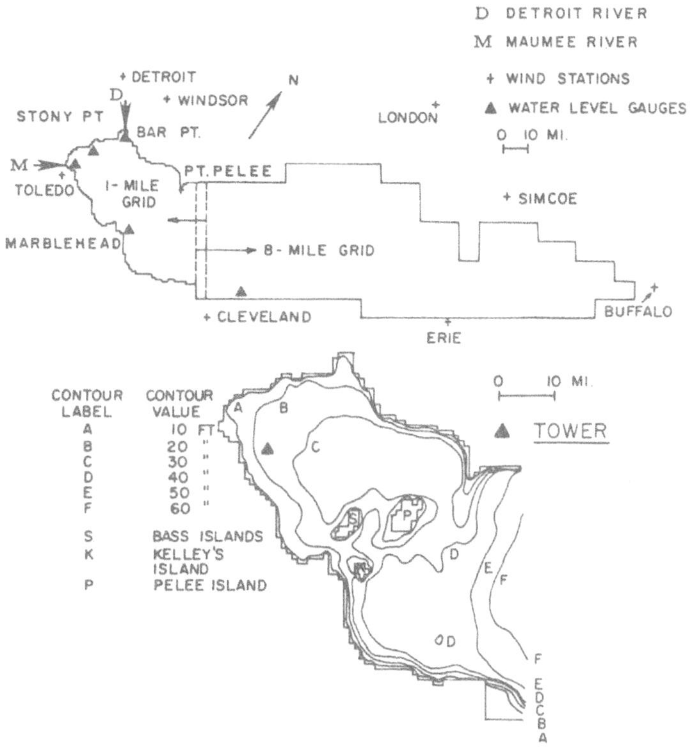

Figure 14. (a) Numerical grid structure of Lake Erie and locations of wind stations and water level gauges, (b) Bottom topography of Western Lake Erie.

Assuming the sediments in the Western Basin are uniformly coarser and follow the resuspension relationship shown by curve B in Figure 6, the sediment dispersion were again computed. The computed results at noon March 11 (Figure 21) agree well with observations in the vicinity of Detroit River and around islands but underestimated the concentration along the Ohio shore where the sediments are finer. The results suggest that in the case of appreciable spatial inhomogeneity in sediment composition, the lake bottom should probably be classified into different regions and more than one resuspension relationship be used in the model simulation.

The remote sensing effort over the Western Basin continued in April, May and June 1976. The OCS data taken after April 29 did not show any appreciable surface concentration even when winds

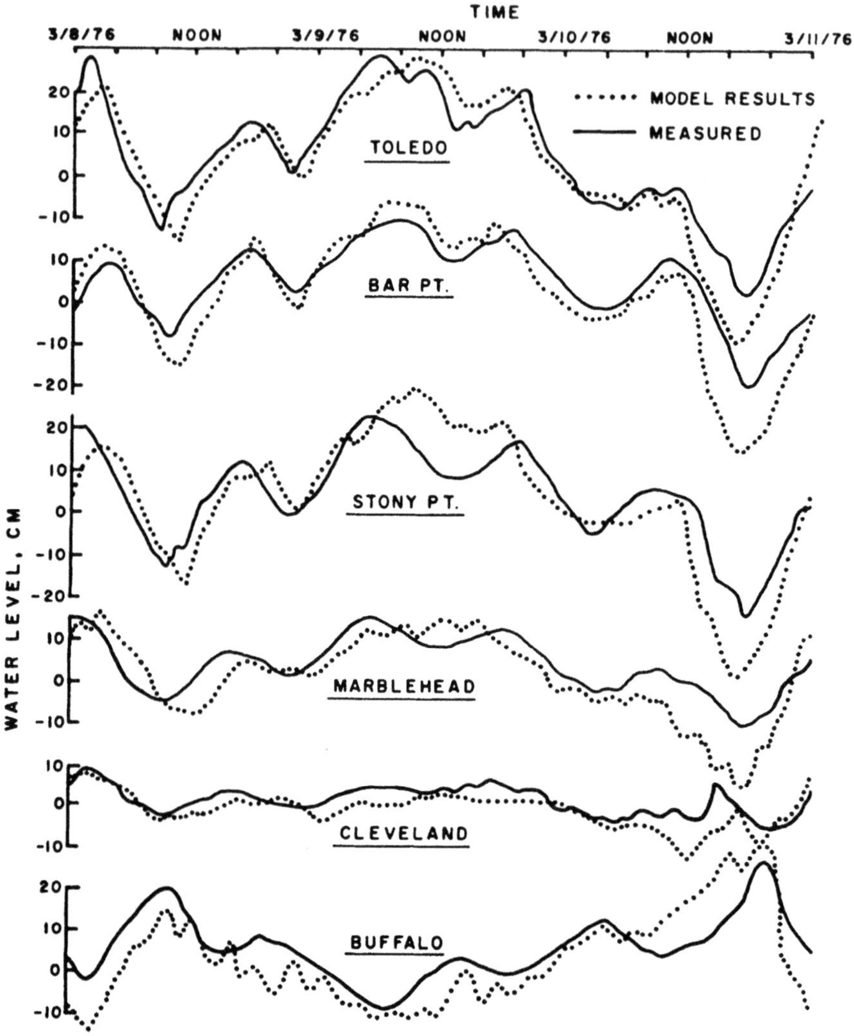

Figure 15. Computed and measured water level at stations in the Western Basin (Toledo, Bar Point, Stony Point, and Marblehead), Central Basin (Cleveland) and Eastern Basin (Buffalo) of Lake Erie.

were comparable to those in March 1976. This suggests that most of the sediments supplied from rivers in spring had probably been transported into the deeper Central Basin by this time.

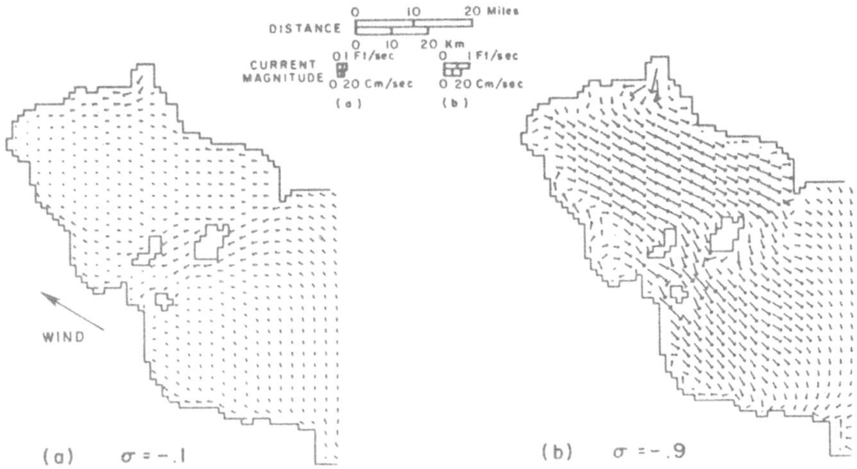

Figure 16. Horizontal velocities at a constant: (a) 0.1 depth; and (b) 0.9 depth from lake surface at noon, March 9, 1976.

SUMMARY

A systematic approach has been undertaken to model the dispersion of sediments in shallow waters. Applications to Lake Erie have demonstrated the basic soundness and feasibility of the modeling approach.

The study indicated the dominant effect of waves and currents in causing sediment resuspension. By properly including the mechanisms of resuspension and deposition, in addition to convection, diffusion and settling, the model was able to successfully reproduce the observed general sediment dispersion pattern, particularly the sharp spatial concentration gradient. The model is capable of keeping track of the change in bottom sediment thickness. Although the bottom sediments may undergo appreciable resuspension and deposition during an event, the net change in sediment thickness may be quite small (only a fraction of 1 cm in this study). The study indicates that most of the river-supplied sediments probably reaches the central Basin within several months. It appears that the consolidated sediments at depth do not become resuspended in general, hence it is not necessary to consider the variation of sediment properties in different vertical layers. In modeling such a complex problem as sediment dispersion, the success hinges on understanding and proper modeling of each of the important processes. Although the accurate boundary condition on resuspension and deposition still requires further study, the accuracy of the

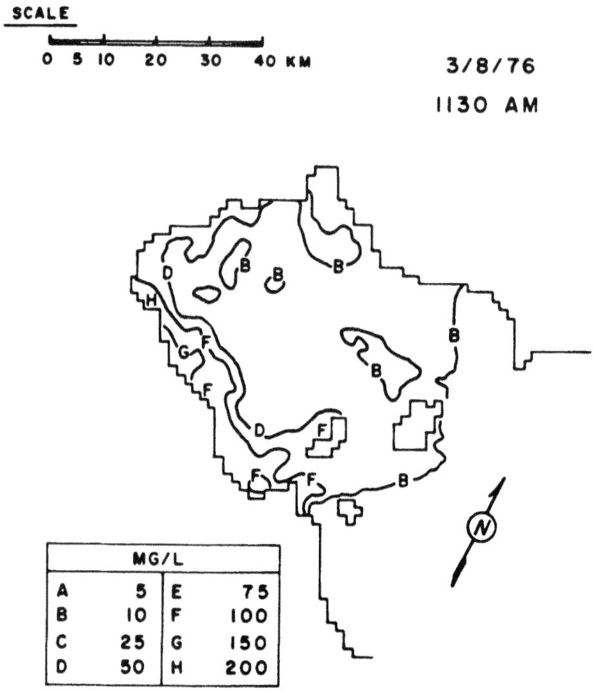

Figure 17. Contours of total suspended solids concentration in the Western Basin at noon, March 8, 1976 as obtained from NASA OCS and calibrated with ship data.

hydrodynamic model and the wave model is very essential. The hydrodynamic model and wave model used in this study have both been verified with available data.

The processes of sediment transport are quite general and there exist striking similarities between those in lakes and those in estuaries. The sediments in both environments are primarily composed of fine-grained cohesive sediments. The currents in both environments are primarily wind-driven, while seiche oscillation (in a lake) and tidal oscillation (in an estuary) can also induce circulation. In addition, physical dimensions of some lakes, say, the Western Basin of Lake Erie, are similar to an estuary such as Long Island Sound or Chesapeake Bay. Turbulent mixing and wave effect can be quite significant in the water column of both environments. The models used in this study may be extended for application in the estuarine and coastal environments.

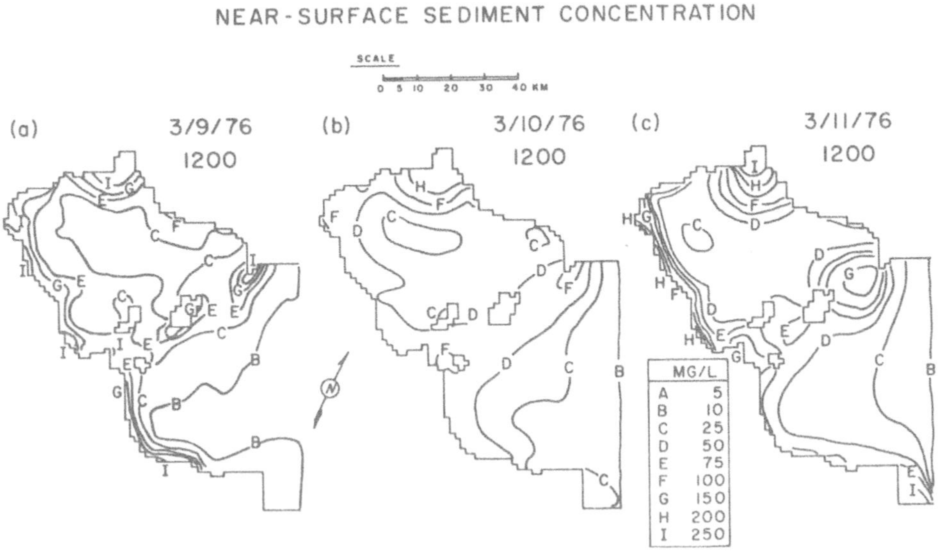

Figure 18. Computed near-surface sediment concentration at noon: (a) March 9, (b) March 10, and (c) March 11, 1976.

Future Research

The basic models outlined in this study are sufficiently general, to be used as bases to study a variety of problems such as sediment transport, dredging problems, phytoplankton and zooplankton growth, community succession, etc., in aquatic environments. With the inclusion of the salinity equation and tidal condition at the open boundary, the models can be applied to an estuary or coastal region. Further numerical experiments with the models should be performed to better understand the important model parameters under a variety of conditions. Efforts should be made to compare various models and determine the limiting conditions under which simpler models can be used.

Further studies are needed to better understand the resuspensio and deposition of sediments under oscillatory flow conditions and stratified conditions. Models containing the proper dynamics of turbulence may be utilized in conjunction with laboratory and field experiments.

The effects of wind waves have been included in this study. Existing wave models can be improved by further validation of the model constants with measured wave data under various wind conditions

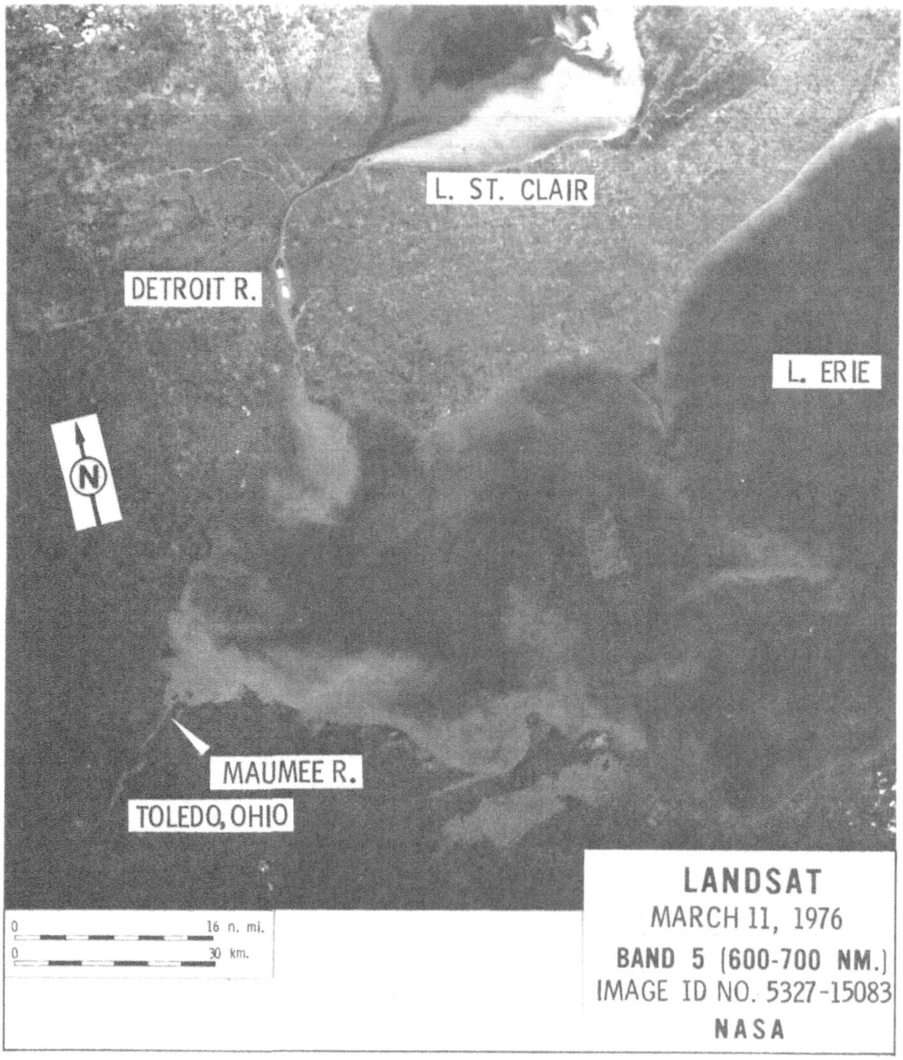

Figure 19a. Imagery of total suspended solids concentration at noon, March 11, 1976: (a) LANDSAT satellite.

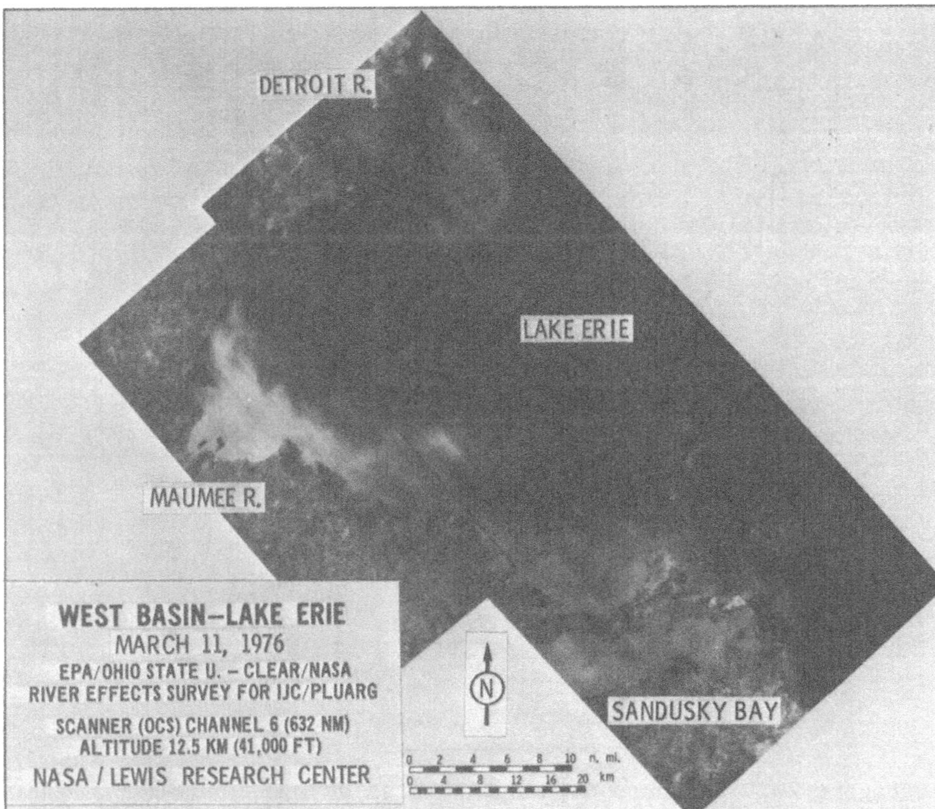

Figure 19b. Imagery of total suspended solids concentration at noon, March 11, 1976: (b) NASA Ocean Color Scanner.

in different physical environments. The interaction of waves with wind-driven currents and tidal currents may be included by parameterizing the eddy coefficients in the hydrodynamic models.

As a step toward better presentation of transport of momentum, heat, and contaminant in shallow waters, higher-order closure models as well as eddy-viscosity models may be applied to the aquatic environments and their results compared. Recently, the higher-order closure model of A.R.A.P. was used to predict the development of buoyancy flux in the experiments of Kotsovinos and List (1977) for transition from a plane jet to a plane plume. As shown in Figure 22, remarkable agreement with experimental data

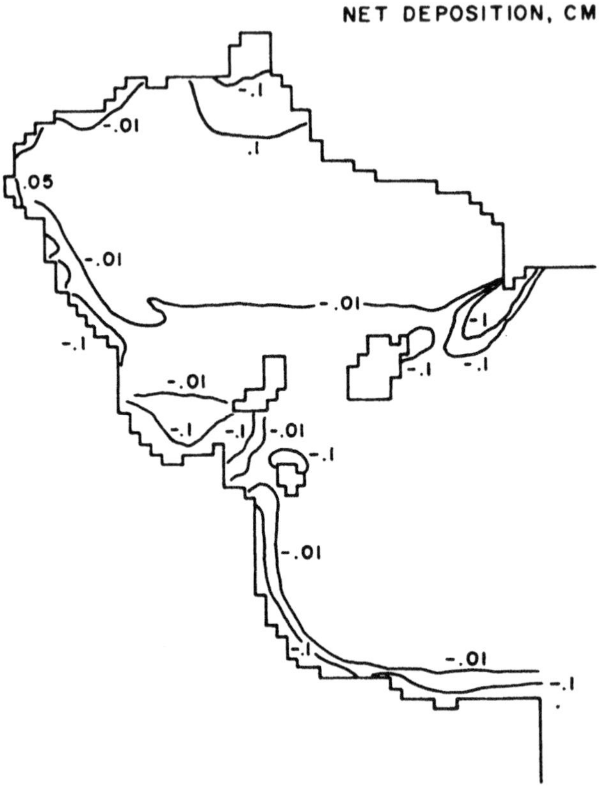

Figure 20. Contours of net thickness of sediments deposited to the lake bottom during the 3-day period from March 8 to March 11.

was achieved without any adjustment of the model constants.

To remove empiricism from the remote-sensing procedure, a radiative model should be developed and further verification with mathematical modeling and field studies should be planned.

ACKNOWLEDGMENT

I am grateful to Drs. W. Lick and R. T. Gedney for many helpful suggestions. Dr. R. T. Gedney and Mr. F. B. Molls of NASA Lewis Research Center provided the valuable remote-sensing data and generous assistance on computer resources. I would also like to thank Drs. P. L. McCall, P. C. Liu, and W. S. Lewellen for useful discussions and V. Anderson for typing the manuscript.

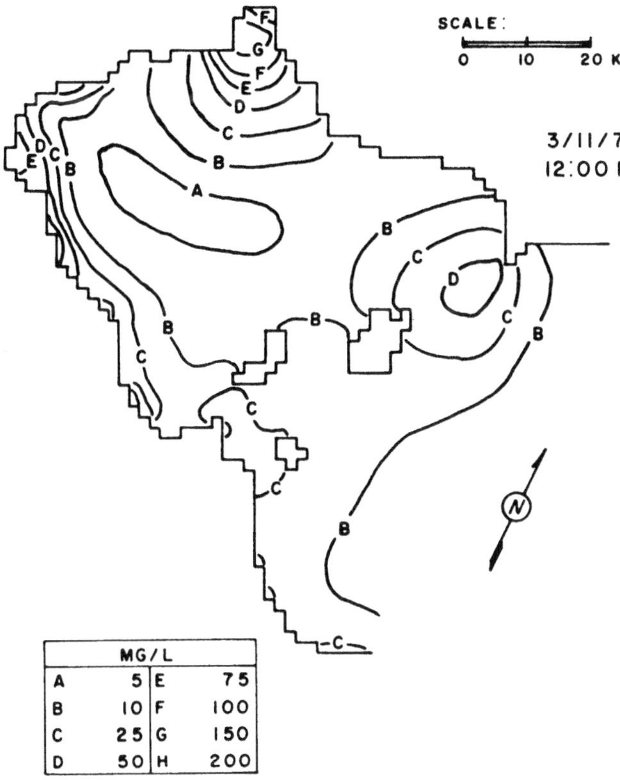

Figure 21. Computed near-surface sediment concentration at noon, March 11, 1976. Based on the resuspension rate-bottom stress relationship for coarser sediments (curve B in Fig. 7).

The Environmental Protection Agency through Case Western Reserve University provided partial funding for this work. Dr. David Dolan served as project officer.

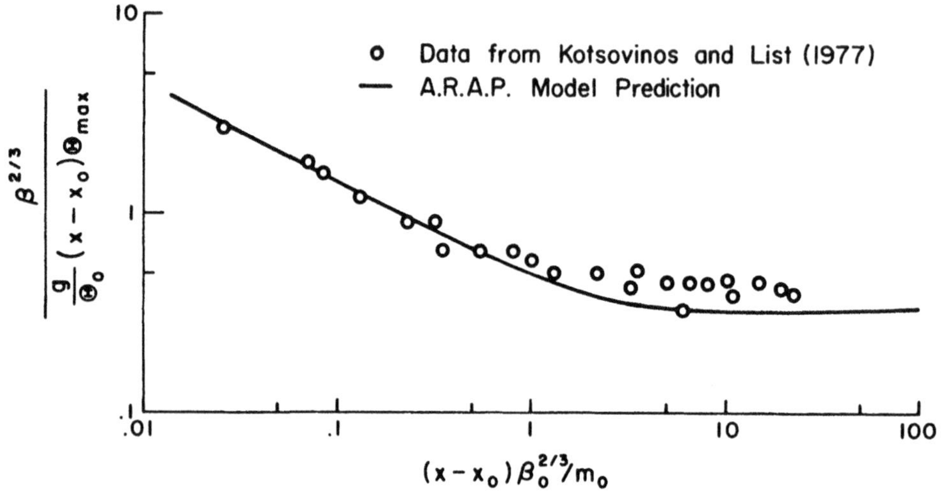

Figure 22. Comparison of A.R.A.P. model prediction with data from Kotsovinos and List (1977) on the decay of normalized buoyancy flux as a function of downstream distances for a jet transition to a plume. β is the buoyancy flux, Θ is the mean temperature, m is the momentum flux, and x is the downstream distance. Subscript o denotes initial value.

REFERENCES

Amendola, G. 1976. Private communications.

Barnett, T. P. 1968. On the generation, dissipation, and prediction of wind waves, J. Geophys. Res., 73, 513-529.

Bokuniewicz, H., J. Gebert, R. Gordon, P. Kaminsky, C. Pilbeam, M. Reed, and C. Tuttle. 1977. Field study of the effects of storms on the stability and fate of dredged material in subaqueous areas. Tech. Rep. C-77-22, U. S. Army Eng. Waterways Exp. Sta., Vicksburg, Miss.

Bowden, K. F. 1977. Turbulent processes in estuaries, in geophysics and the environment. ed. Geophys. of Estuaries Panel, National Academy of Sciences, 46-56.

Businger, J. A., J. C. Wyngaard, Y. Izumi, and E. F. Bradley. 1971. Flux-profile relationships in the atmospheric surface layer. J. Atmos. Sci. 28, 181-189.

Butler, H. L. 1980. Evolution of a numerical model for simulating long-period wave behavior in ocean-estuarine systems. This volume.

Cheng, R. T., T. M. Powell and T. M. Dillon. 1976. Numerical models of wind-driven circulation in lakes. Appl. Math. Modeling 1, 141-159.

Donaldson, C. duPont. 1973. Atmospheric turbulence and the dispersal of atmospheric pollutants. in AMS Workshop on Micrometeorology, ed. D. A. Hangen, Science Press, Boston, 319-390.

Edinger, J. E. and E. M. Buchak. 1980. Numerical hydrodynamics of estuaries. This volume.

Fukuda, M. K. 1978. The entrainment of cohesive sediments in fresh water. Ph.D. dissertation, Case Western Reserve University.

Gardner, G. B., A. R. M. Nowell, and J. D. Smith. 1980. Turbulence in estuaries. This volume.

Gedney, R., C. A. Raquet, J. A. Saltzman, T. A. Coney, R. V. Svehla, and D. Shook. Coordinates aircraft/ship surveys for determining the impact of river inputs on Great Lakes waters-Remote sensing results, report for International Joint Commission, NASA Lewis Res. Center, Cleveland, Ohio, 1977.

Gordon, C. M., and C. F. Dohme. 1973. Some observations of turbulent flow in a tidal estuary. J. Geophys. Res., 78, 1971-1978.

Haq, A., W. Lick and Y. P. Sheng. 1974. Time-dependent flows in large lakes with application to Lake Erie, Technical Report, Case Western Reserve University.

Hasselmann, K., T. P. Barnett, E. Bouws, H. Carlson, D. E. Cartwright, K. Enke, J. A. Ewing, H. Gienapp, D. E. Hasselmann, P. Kruseman, A. Meerburg, P. Muller, D. J. Olbers, K. Richter, W. Sell, and H. Walden. 1973. Measurements of windwave growth and swell decay during the Joint North Sea Wave Project (JONSWAP), Deutsche Hydrog. Z. Suppl. A(8), 12, 95 p.

Komar, P. and M. Miller. 1973. The threshold of sediment movement under oscillatory water waves. Journal of Sedimentary Petrology, 43, 1101-1110.

Kotsovinos, N. E. and E. J. List. 1977. Plane turbulent bouyant jets. Part 1, Integral Properties. J. of Fluid Mech., 81, 25-44.

Lam, D. C. L., and J-M Jaquet. 1976. Computation of physical transport and regeneration of phosphorus in Lake Erie, Fall 1970, J. Fish. Res. Bd. Can., 33, 550-563.

Leendertse, J. J. 1967. Aspects of a computational model for long-period water-wave propagation. Rand Corporation, Santa Barbara, Calif., RM-5294-PR, 165 p.

Lewellen, W. S. and Y. P. Sheng. 1979a. Modeling for dry deposition of SO_2 and sulfate aerosols, A.R.A.P. Report prepared for EPRI.

Lewellen, W. S. and Y. P. Sheng. 1979b. Influence of surface conditions on tornado wind distributions, presented at AMS 11th Conf. on severe local storms, Kansas City, Mo.

Lewellen, W. S. and M. E. Teske. 1973. Prediction of the Monin-Obukhov similarity functions from an invarient model of turbulence. J. Atmos. Sci., 30, 1340-1345.

Liu, P. C. 1977. A hindcast of Lake Superior waves during the disastrous storm of 10 November 1975, GLERL Contribution No. 98, GLERL, National Oceanic and Atmospheric Administration, Ann Arbor, Michigan.

Lumley, J. L. and B. Khajeh-Nouri. 1973. Computational modeling of turbulent transport. Paper presented at Second IUIAM-IUGG symposium on Turbulent Diffusion in Environmental Pollution, Charlottesville, VA, 8-14 April.

McCall, P. L. 1978. Spatial-temporal distributions of Long Island Sound infauna: the role of bottom disturbance in a nearshore marine hatitat, in Estuarine Interactions, Academic Press, N. Y., 191-219.

Mitsuyasu, H. 1971. On the form of fetch-limited wave spectrum, Coastal Eng. in Japan, 14, 7-14.

Nihoul, J. C. 1975. Hydrodynamic models in modeling of marine systems, ed. Nihoul, J. C., Elsevier, Amsterdam, 41-68.

Reid, R. O. and Bodine, B. R. 1968. Numerical model for storm surges in Galveston Bay. ASCE J. Waterways and Harbor Div., 33-57.

Resio, D. T. and C. L. Vincent. 1977a. A Numerical hindcast model for wave spectra on water bodies with irregular shoreline geometry, H-77-9, Rept. 1, Waterways Experiment Station, Vicksburg, Miss.

Resio, D. T. and C. L. Vincent. 1977b. A Numerical Hindcast Model for Wave Spectra on Water Bodies with Irregular Shoreline Geometry, H-77-9, Rept. 2, Waterways Experiment Station, Vicksburg, Miss.

Sheng, Y. P., 1975. The wind-driven currents and contaminant dispersion in the near-shore of large lakes, Report H-75-1, U. S. Army Engr. Waterways Experiment Sta., Vicksburg, Miss. (NTIS AD-A017694).

Sheng, Y. P., W. Lick, R. T. Gedney and F. B. Molls. 1978. Numerical computation of three-dimensional circulation in Lake Erie; A comparison of free-surface and rigid-lid model. J. Phys. Oceanogr., 8, 713-727.

Sheng, Y. P. 1978. Three-dimension hydrodynamic models of large lakes and estuaries, A.R.A.P. Report, 200 pp.

Sheng, Y. P. and W. Lick. 1979. The transport and resuspension of sediments in a shallow lake. J. Geophys. Res., 84, 1809-1826.

Sternberg, R. W. 1972. Predicting initial motion and bedload transport of sediment particles in the shallow marine environment, in Shelf Sediment Transport, ed. D. P. Swift, Dowden, Hutchinson, Ross, Stroudsburg, PA.

Sudaram, T. R. and R. G. Rehm. 1970. Formation and maintenance of thermocline in stratified lakes, AIAA paper 70-238.

A NUMERICAL SIMULATION OF THE DISPERSION OF SEDIMENTS SUSPENDED BY ESTUARINE DREDGING OPERATIONS

Donald F. Cundy and W. F. Bohlen

Marine Sciences Institute, The University of Connecticut

Groton, Connecticut, 06340

ABSTRACT

A predictive numerical model designed to simulate the dispersion of sediments suspended by estuarine clam-shell dredging operations is described. The model evaluates the downstream distribution of the column of materials introduced by each vertical pass of the dredge bucket using a modified conservation of mass approach in which a horizontal moment term is used to represent the spatial distribution of the suspended mass concentrations. Solution of the resultant equation in finite difference form provides a time history of the 0th to 4th moments of the dispersing mass introduced by each bucket pass. A representation of the sum total effect of these discrete injections forming the downstream plume is then developed through linear superposition. This scheme provides a description of the gross characteristics of the dispersing mass without requiring large amounts of computer time and storage.

Required inputs to the model include specification of the local mean velocity characteristics, sediment settling velocities and turbulent mass diffusion coefficients. Field data obtained under a variety of conditions are used to supply these inputs and to test the accuracy of the computational scheme.

Preliminary comparisons suggest that this model provides a reasonable analogue of observed field conditions. Accuracy appears to be primarily dependent on the specification of settling velocity and mass diffusivity representing second-order influences.

INTRODUCTION

In most coastal areas, the dredging of estuarine sediments is a common occurrence. The materials removed in these operations are often contaminated by a variety of organic and inorganic compounds. During the dredging process, some fraction of these sediments is introduced into the ambient waters potentially impacting local biota and perturbing overall water quality. Evaluation of these impacts requires an understanding of spatial distributions of the entrained sediments. Numerous models have been developed to predict suspended materials distributions resulting from a point source, but none meet the initial and boundary conditions imposed by a clam shell dredge, the most common technique used in the northeastern United States.

During the past two years several detailed surveys of clam shell dredge-induced suspended material plumes have been conducted in the Thames River estuary near New London, Connecticut. The results of these surveys have been used to evaluate the spatial distributions of suspended materials under a range of flow conditions (Bohlen et al., 1979) and to examine the major processes controlling the dispersion of dredge-induced sediment plumes (Bohlen, 1979). This combination provides an initial framework for the development of a predictive numerical model.

The following paper outlines the characteristics of this model and presents the results of initial comparisons with selected field data. Model details can be found in Cundy (1979.)

STUDY AREA

The Thames River estuary is one of the major harbors of Connecticut. A submarine base and a variety of recreational and commercial marine activities are located along its shores. Demands for larger draft vessels have required frequent dredging and deepening of the main navigation channel. Since 1974 over 3 million cubic meters of silt and clay have been removed from the estuary by clam shell dredge for disposal at an open water site.

Hydraulically, the Thames is representative of a typical medium-scale New England estuary. Annual average streamflow of this system over the past five years has varied between 70 and 76 m^3/s. Flows through the river display a regular daily variability dominated by the semi-diurnal tide. Mean tidal range varies progressively from 0.78 m at New London to 0.9 m at Norwich, Connecticut, approximately 25 km upstream. This combination of tidal and stream-flow characteristics generally permits extensive intrusion of saline water, producing a density field displaying persistent vertical stratification.

Field observations obtained during mid-ebb tidal cycles indicate that the local hydraulic conditions favor development of a narrow, well-defined plume of the materials suspended by the dredging operation (Bohlen, et al., 1979). Typically, this plume extends downstream for less than 1000 m and the particles remain in suspension for times that are short compared to the half tidal period. This plume can be divided into three primary regions: an initial mixing zone, immediately adjacent to the dredge, where concentration levels and mixing are dominated by the vertical motion of the dredge bucket; beyond this area, a secondary zone where concentration levels rapidly decay under the influence of gravitational settling; and a final zone, within which concentrations gradually approach background levels due to the combined effects of settling and turbulent diffusion. This combination of temporal and spatial characteristics is used in formulating the numerical modeling scheme.

PREVIOUS WORK

The dispersion of mass in turbulent shear flows has been intensively studied both experimentally and theoretically for many years. Descriptions of this process require an ability to specify the mixing characteristics of three dimensional turbulent flows and the behavior of the dispersing mass. Neither of these is easily defined. As a result, attempts to model dispersion have usually been based on simplifications of the actual problem. Moreover, the majority of these efforts have been concerned solely with dispersion in single phase systems and have seldom considered the problem of dispersion of solid particulates in fluid flows. These two-phase systems are typically difficult to treat using a simple phenomenological approach and are best analyzed by solution of the advection-diffusion equation. Until digital computers became available this represented a time-consuming computational task.

Yotsukura and Fiering (1964) were among the first to numerically solve the advection-diffusion equation for two-dimensional flows containing neutrally buoyant particles. Although the results of their work are not directly applicable to the problem of negatively buoyant dispersants, the numerical techniques developed in their study provided a framework for more detailed future work.

Sayre (1968) was the first to employ finite difference techniques to predict the dispersion of sediments in open channel flows. His approach incorporated the moment technique developed by Aris (1956) in which the actual concentration distribution is represented by the moments of its longitudinal distribution. From the moment description, he calculated the mean, variance and skewness of the dispersed materials and used a Pearson Type III distribution to estimate the suspended material concentrations from the statistical parameters. The advantage of this technique is that it is as accurate as previous numerical simulation and takes an order of magnitude less computation time due to the elimination of the longitudinal variable.

Utilizing essentially the same technique as Sayre, and with the initial conditions developed by Chatwin (1968), Atesman (1975) simulated dispersion from a point source for a variety of neutrally buoyant materials. By studying the effect of source position in the vertical on the dispersant distribution, he determined that the mixing characteristics are strongly dependent on the location of the instantaneous point source. These results imply that the introduction of materials entrained during the vertical migration of the dredge bucket is more properly treated as a vertical cylindrical source rather than either a point or plane source.

Koh and Chang (1973) used the moment technique to simulate the long term dispersion of suspended particulates in three dimensions. They were interested in predicting the fate of materials dumped from a barge in open water. During the descent through the water column the transport of these wastes proceeds through a series of well-defined phases. During the final phase of this process material distributions are governed primarily by turbulent dispersion and concentrations vary in response to the combination of fluid flow characteristics and waste properties. The assumption was made that the distribution of the materials during this phase was Gaussian and could be sufficiently well represented by the 0th through the 2nd moments of the concentration distribution.

For the case of sediments introduced into the water column during clam shell dredging, the techniques used by Koh and Chang are applicable but require some modification in order to more accurately represent the character of the concentration field in the immediate vicinity of the dredge (i.e., the initial mixing zone). In particular, within this area there is little reason to believe that material distributions are simply Gaussian. A variety of experimental and analytical investigations have shown this to be true for both neutrally and negatively buoyant materials (Fischer, 1967; Sayre, 1968). To accurately simulate non-Gaussian distributions the Koh and Chang solution must be extended to include the 3rd and 4th moments of the concentration distribution. In addition, the initial conditions must be modified to permit replacement of the computational methods used to specify the inputs to the dispersion phase by a simple definition of the character of the sediment source produced by the vertical migration of the dredge bucket.

METHODS AND PROCEDURES

Model - The model to be applied to the problem of dispersion of dredge-induced sediments is based on the Aris (1956) moment solution to the dispersion equation as applied by Koh and Chang (1973). It numerically describes in three dimensions the mixing of a single instantaneous vertical cylindrical source of materials, introduced by a clam shell dredge, with the surrounding water. The model assumes that the character of the entrained material and the hydrodynamic regime do not appreciably change over the period of interest.

A conservation of mass approach within a control volume is used to develop the general equation for dispersion. The equation utilizes the turbulent mass diffusivity concept to describe the mixing, resulting from the turbulent flow field. With the exception of settling, which is represented by a fall velocity, the suspended particulate material is assumed to behave in the same manner as water particles. Since we are dealing with a straight, dredged channel of constant depth over periods of time of less than 1 hour, the lateral and longitudinal components of velocity and turbulent mass diffusivity coefficients are assumed to be only a function of depth. These assumptions impose several restrictions on the utilization of this model in other estuarine areas. Although the mean component of the vertical velocity is assumed to be negligible, a vertical diffusion term is included since this is an important factor which serves to modify the vertical distribution of the particulate matter. The governing equation is:

$$\frac{\partial C}{\partial t} + U \frac{\partial C}{\partial x} + W \frac{\partial C}{\partial z} = \frac{\partial}{\partial y} (K_y \frac{\partial C}{\partial y} - W_s C) + \frac{\partial}{\partial x} (K_x \frac{\partial C}{\partial x}) + \frac{\partial}{\partial z} (K_z \frac{\partial C}{\partial z}) \qquad (1)$$

in which x, y, and z are spatial coordinates in a right handed system having its origin at the surface adjacent to the dredge. x is the longstream coordinate, y the vertical coordinate. K_x, K_y, K_z are the mass diffusivity coefficients; U and W are the ambient time averaged velocities; C is the concentration averaged over a short time period; and W_s is the settling velocity.

Field observations (Bohlen, 1979) have shown that the initial mixing zone is defined by the cylindrical wake produced by the dredge bucket passing through the water column. Sediment is entrained into this zone by the impact with the bottom and leakage from the ascending bucket. This produces an initial concentration field which decreases progressively from the bottom to the surface. This column is divided into a series of discrete horizontal laminae. Material distributions within each lamina are assumed to be uniform resulting in a single concentration value for each layer.

Conditions are imposed on equation 1 within both the near surface and near bottom layers. In the surface layer, the effects of settling and vertical diffusion on the suspended sediments are equal and opposite and as a result no material may pass through the water surface. Within the bottom layer, exchange processes result in net deposition along the lower boundary adjoining the sediment-water interface. These deposited materials are represented by a mathematical array D. The information in this array may be used to describe the distribution of the deposited materials. The equation governing the bottom layer also assumes that these materials once deposited are not resuspended. The equations describing the above conditions are as follows:

$$K_y \frac{\partial C}{\partial y} - W_s C = 0 \tag{2}$$

for the surface layer, and

$$K_y \frac{\partial C}{\partial y} - W_s C = D \tag{3}$$

in the bottom layer.

Prior to numerical solution, equation (1) is transformed from an equation for concentration distribution to an equation for the moment

$$C_{k,\ell} (y, t) \equiv \int_{-\infty}^{\infty} \int_{-\infty}^{\infty} x^k z^\ell C (x, y, z, t) \, dxdz \tag{4}$$

Following Aris (1956) and Koh and Chang (1973), the transformation is accomplished by performing the operation defined in equation (4) on each term in equation (1).

The moment characteristics describing the total mass of suspended material may be found by integration of (4) over the depth of the water column.

Following the moment transformation (1) becomes:

$$\frac{\partial C_{k,\ell}}{\partial t} - kU\, C_{k-1,\ell} - \ell W C_{k,\ell-1} =$$

$$k(k-1)K_x\, C_{k-2,\ell} + \ell(\ell-1)\, K_z\, C_{k,\ell-2} +$$

$$\frac{\partial}{\partial y}\left(K_y \frac{\partial C_{k,\ell}}{\partial y} - W_s\, C_{k,\ell}\right) \qquad (5)$$

The moment transformation is also applied to the boundary conditions.

Using a modified forward time, center space scheme, the equations are converted to finite difference form. The velocity profiles, diffusion coefficients and settling velocity are substituted into the difference equation resulting in a set of simultaneous equations. The equations are then solved using Thomas' algorithm (Ames, 1977). The solution provides a time history of the 0th to 4th moments of the longitudinal and/or transverse distributions of the dispersing mass. These moments are then used to compute the total mass, mean position, variance, skewness and kurtosis of the suspended material.

In order to facilitate comparisons between the moment descriptions and available field data, a method involving Hermite polynomials (Atesman, 1970) is used to convert the moment statistics into mass distribution. This conversion initially assumes that the material distributions must be Gaussian. Distributions developed using this assumption are subsequently corrected for the contribution of the 3rd and 4th moments. As expected (Sayre, 1969) the distributions progressively approach normality as time increases.

Inputs - The model requires inputs to describe the ambient hydrodynamic regime and the characteristics of the dispersing material, in particular its settling velocity. Since this is a dispersion model and not a hydrodynamic model, the time-averaged characteristics of the flow field and parameters describing vertical, transverse and longitudinal turbulent mass diffusion must be specified.

Longitudinal (U) velocity conditions are set using field data obtained during high and normal stream flows. These data shown in Figures 1 and 2 indicate that the primary differences between these flow conditions are confined to the top three meters of the water column. In the high flow case, the peak velocity in this region reaches 68 cm/sec. For normal flows, maximum velocities average less than 20 cm/sec. Below 3 meters the profiles under both conditions are quite similar with a mean velocity of approximately 15 cm/sec. For all cases the transverse velocity W is assumed to be zero.

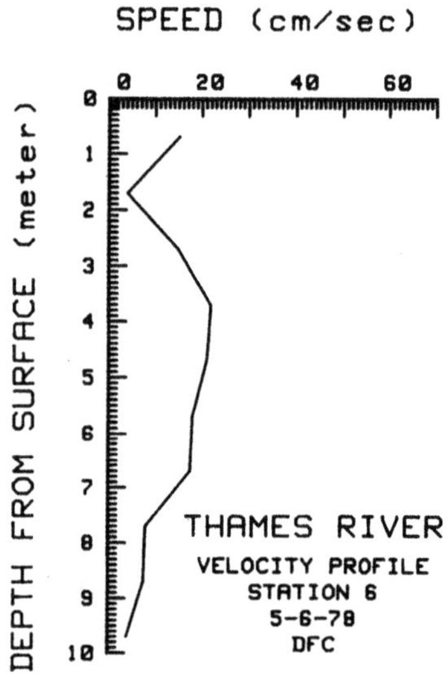

Figure 1. Mean Velocity Profile - Normal Flow Conditions

Turbulent mass diffusivity values are specified using one procedure for the horizontal directions and a second for the vertical. The transverse and longitudinal coefficients are computed using:

$$K_i = AL_i^{4/3} \quad \text{(Orlob, 1959)}$$

$$i = x,\ z \qquad (6)$$

where L is a characteristic length scale of the dispersing materials and A is a constant which must be specified as an input to the program. Since the length scale is constantly changing, these coefficients are recomputed before each time step. For the vertical diffusivity K_y, a range of values was selected from the literature for conditions approximating those found in the Thames River. The values of K_y tested in the model ranged from 50 to 500 cm^2/sec.

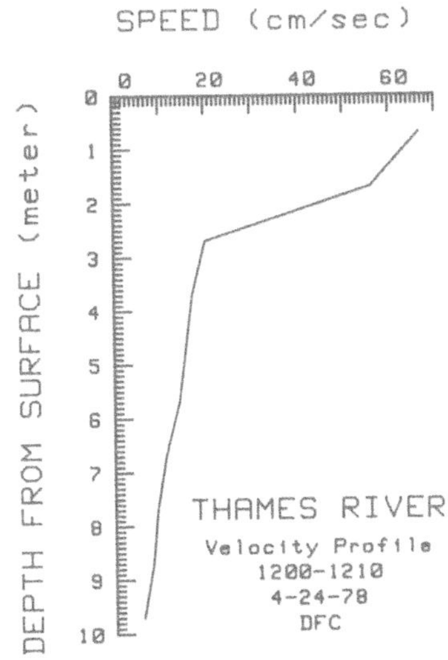

Figure 2. Mean Velocity Profile - High Flow Conditions

Settling velocities for the model were obtained from analysis of selected field data. Bohlen (1979) calculated that 90% of the materials suspended by a dredge settled with a fall velocity of 4.8 cm/sec within the first 100 meters downstream. A review of these data indicates that the remaining materials settle with a substantially lower velocity. Applying these observations, the total suspended mass is assumed to consist of two populations, each with a distinct settling velocity. The model computes the dispersion of each population for a defined number of time steps and then prints an output. At that time, the distribution for the distinct particulate populations are summed, resulting in a distribution for the assemblage of suspended matter. The dual dispersion computations continue until less than 1 percent of the population with the higher settling velocity remains in suspension. The simulation then continues, using only the lower settling velocity until 99 percent of the remaining suspended sediment is deposited.

RESULTS AND CONCLUSIONS

All runs of the model were made for a water depth of 10 meters, and a vertical grid spacing of 0.25 meters. The initial suspended material distribution applied in the model was similar to that observed in the field study with concentrations ranging from 500 mg/l at the surface to 1000 mg/l at the bottom. For simplicity a linear vertical gradient was assumed. Materials were distributed throughout a vertical column 20 meters in diameter. 80% of the entrained material was assumed to have a settling velocity of 5.0 cm/sec while the remaining 20 percent settled at 0.9 cm/sec. The time steps used were 10 sec. for the high settling velocity material and 2 sec. for the low velocity material. The constant A used in the horizontal mass diffusivity computations was set at 0.01. Runs were made with vertical mass diffusion coefficients of 50 cm^2/sec and 200 cm^2/sec.

Figures 3 and 4 show representative outputs for the normal flow condition and a vertical mass diffusion coefficient of 200 cm^2/sec.

Figure 3 provides a time history of the distribution of a single entrainment of particulate matter at 100 second intervals. The sharp decrease in the magnitude of the peak values within the first 100 meters is a result of the rapid settling of the 5 cm/sec particulates. Further downstream, the suspended material is composed primarily of the 0.9 cm/sec population which results in a slower decay of the distributions due to the significant decrease in the rate of deposition. Although this figure indicates that the plume extends downstream for approximately 250 meters, review of the raw output shows significant material remaining in suspension at 350 m. This discrepancy is the result of the limited resolution of the plot.

For the case of the operating dredge, field observations indicate that materials are introduced into the water column as a series of discrete injections each separated in time by approximately 100 sec. These columnar sources are dispersed downstream and

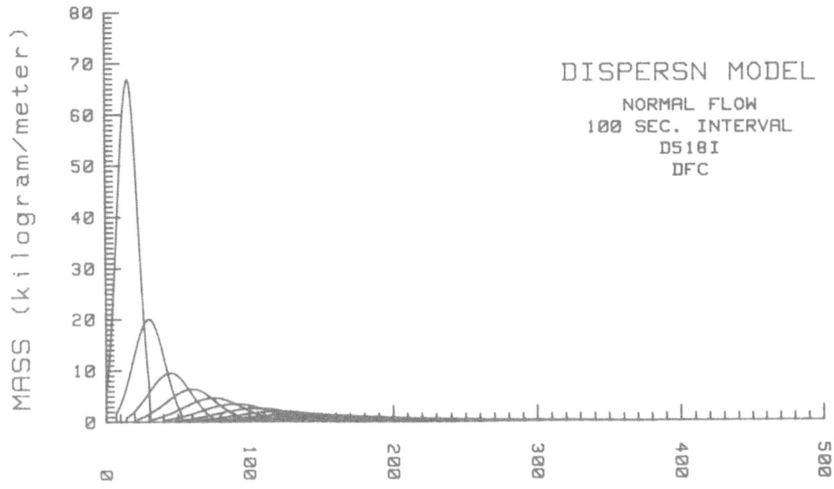

Figure 3. Model Output - Predicted dispersion of a single injection of suspended materials.

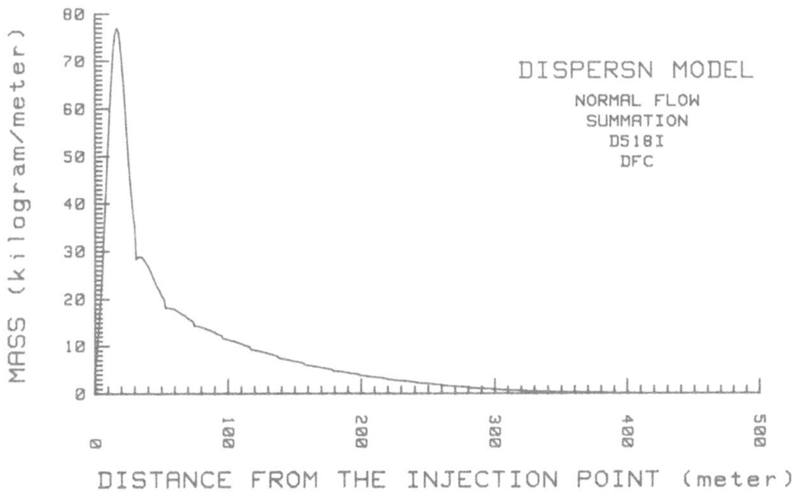

Figure 4. Model Output - Predicted suspended material distribution downstream of an operating dredge.

progressively coalesce, forming a well-defined plume. Within the model the linearity of the governing conservation of mass equation permits representation of this process by simple superposition. The effects of i discrete injections, each separated by a specified time interval, are summed to provide a simulation of the downstream suspended natural distribution. Figure 4 shows the results of such a calculation for 19 injections, each separated by 100 sec. Again, the mass of materials in suspension displays a high initial rate of decay followed by a progressive decrease in decay rates with increasing distance downstream.

Comparing the results of the simulation shown in Figure 4 to selected field data obtained in the Thames River (Figure 5) indicates that the numerical schemes provide a reasonable representation of the actual field distributions. Qualitatively, the two curves are essentially similar. Quantitatively, direct comparisons are difficult since the centerline point sampling routine used in the field surveys (Bohlen, et al., 1979) precludes calculation of the total mass of suspended materials across a given transverse section. Surveys using a white light transmissometer show a regular increase in plume width as a function of distance downstream from the dredge (Table 1).

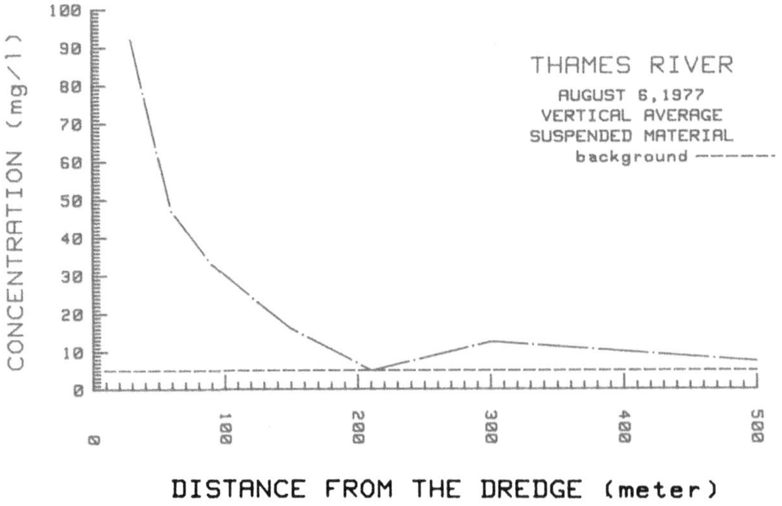

Figure 5. Measured centerline suspended material distribution downstream of an operating dredge.

Table 1. Thames River Suspended Material Plume Width - August 1977

Distance Downstream from Dredge m	Plume Width m
30	40
100	60
200	80

On the assumption that the dredge sediments are uniformly distributed across each transverse section at concentrations equalling those observed on the center-line, these widths suggest that approximately 36 kg of material is in suspension at the 30 m section, 15 kg at the 100 m section and 5 kg at the 200 m section. Comparison of these values to those shown in Figure 4 indicates that the mass estimates provided by the model differ from the field data by less than 10%.

The close agreement between the field data and model outputs suggests that the numerical scheme may be relatively insensitive to a wide range of values used in the specification of the input parameters. To evaluate this sensitivity a series of model outputs were obtained for a range of mass diffusion, settling velocity and mean velocity conditions. For each test, runs were conducted in which one parameter was sequentially varied over a selected range while the remaining inputs were held constant. The results of these tests indicate that the simulation is most sensitive to changes in settling velocity. Reducing this parameter from 5 cm/sec to 1 cm/sec increases the time materials remain in suspension from 300 sec to 2000 sec. Equivalent reductions in vertical mass diffusivity result in, at most, a 50% increase in suspension time. Mean velocity variations are even less significant within the range of conditions tested (Figures 1 and 2) producing no evident modification in suspension time. These results suggest that the accuracy and general applicability of this model will in large part depend on the ability to specify particulate settling velocities. The field data indicate that this parameter cannot be simply established using laboratory analysis of grain size characteristics (Bohlen, 1979) and that a variety of in situ measurements are required before its value can be routinely specified. The present inability to accurately define settling velocity appears to represent the major unknown affecting the modeling of the dispersion of dredge entrained suspended sediments.

SUMMARY

Dredging represents a common and continuing source of suspended materials to estuarine areas.

A numerical model has been developed to simulate the dispersion of suspended materials induced by a clam shell dredging operation. This dispersion model uses the moment technique to mathematically describe the dispersion of the sediments. The method offers a reduction in computer time without sacrificing necessary detail. The inputs to the model include specification of the ambient hydrodynamic conditions and the character of the dredge-resuspended sediments.

Initial comparison of the model output with field data yields reasonable quantitative agreement.

The sensitivity of the model to input parameters was tested. Outputs are most sensitive to the selected range of settling velocities. Changes in the mean flow and mass diffusion produce second order variations in suspension time when compared to settling velocity. Accurate definition of particulate settling velocities appears to be the major unknown affecting the modeling of the dispersion of dredge-resuspended sediments.

ACKNOWLEDGMENTS

This research has been supported by the U. S. Navy through Contract N00140-77-C-6536 and the University of Connecticut Research Foundation. Raymond Sosnowski and John Tramontano assisted in the field work and laboratory analysis. The authors gratefully acknowledge this support and assistance.

REFERENCES

Ames, W. F., 1977. Numerical methods for partial differential equations. Academic Press, New York. 365 pps.

Aris, R., 1956. On the dispersion of a solute in a fluid flowing through a tube. Proc. Roy. Soc. London, Vol. 235A:67-77.

Atesman, K. M., 1970. The dispersion of matter in turbulent shear flows. Unpubl. Ph.D. Thesis, University of Colorado at Fort Collins, 158 pps.

Atesman, K. M., 1975. Point source dispersion in turbulent open channels. Proc. Amer. Soc. Civ. Eng., J. of Hyd. Div., Vol. 101, No. HY7:789-799.

Bohlen, W. F., D. F. Cundy and J. M. Tramontano, 1979. Suspended material distributions in the wake of estuarine channel dredging operation. Estuarine and Coastal Marine Science (in press).

Bohlen, W. F., 1979. Factors governing the distribution of dredge-resuspended sediments. Proc. of Sixteenth Coastal Eng. Conf., Hamburg, Germany (in press).

Chatwin, P. C., 1968. The dispersion of a puff of passive contaminant in the constant stress region. Quart. J. Roy. Met. Soc., London, Vol. 94:350-360.

Cundy, D. F., 1979. A finite difference model of the dispersion of dredge entrained sediments. MS Thesis, University of Connecticut, Storrs, Ct. (in preparation).

Fischer, H. B., 1967. The mechanics of dispersion in natural streams. Proc. Amer. Soc. Civ. Eng., J. of Hyd. Div., Vol. 93, No. HY6:187-216.

Koh, R. C. Y. and Y. C. Chang, 1973. Mathematical model for barged ocean wastes. Prepared for the United States Environmental Protection Agency, National Environmental Research Center, Corvallis, Oregon, 178 pps + Apps.

Orlob, G. T., 1959. Eddy diffusion in homogeneous turbulence. Proc. Amer. Soc. Civ. Eng.- J. of Hyd. Div., Vol. 85, No. HY9:75-101.

Sayre, W. W., 1968. Dispersion of mass in open-channel flow. Hydrology Papers, Colorado State University at Fort Collins, Colorado, No. 75, 73 pps.

Sayre, W. W., 1969. Dispersion of silt particles in open channel flow. Proc. Amer. Soc. Civ. Eng., J. of Hyd. Div., Vol. 90, No. HY3:1009-1038.

Yotsukura, N. and M. B. Fiering, 1964. Numerical solution to dispersion equation. Amer. Soc. Civ. Eng., J. of Hyd. Div., Vol. 90, No. HY5:83-104.

SHORT-TERM FLUXES THROUGH MAJOR OUTLETS OF THE NORTH INLET MARSH IN TERMS OF ADENOSINE 5'-TRIPHOSPHATE[a]

L. Harold Stevenson,[1] Thomas H. Chrzanowski,[1] and Bjorn Kjerfve[2]

Departments of Biology[1] and Geology,[2] and the Belle W. Baruch Institute for Marine Biology and Coastal Research, University of South Carolina, Columbia, South Carolina 29208

ABSTRACT

Transects across three major creeks joining the North Inlet marsh system to the neighboring ocean and bay environments were characterized in terms of the temporal fluctuations, distribution, and short-term transport of total microbial biomass (measured as adenosine 5'-triphosphate [ATP]). The mean ATP density ranged from 0.865 to 1.357 mg per m^3. Highest densities were recovered during flood tides. The distribution of mean ATP densities as well as net flux through each interface proved to be complex with both vertical and horizontal stratification apparent at some locations. A net import of ATP at a rate of about 40 mg per s was noted at the two creeks that interfaced directly with the oceanic environment. A net export was noted through the creek that emptied into the bay. The results indicate that the characterization of a tidal creek interface in terms of ATP, or similar parameters, requires the simultaneous measurement of both the component of interest and directional velocity.

---

[a]Contribution No. 318 of the Belle W. Baruch Library in Marine Sciences

INTRODUCTION

Interaction between a salt marsh and the adjoining oceanic environment can take place by several mechanisms. One possible route for interaction between the two systems is via the tidal water that alternately floods and drains the marsh. As oceanic water covers the marsh at high tide, the activity of the system could enrich the water in certain components resulting in an export or outwelling of the components during the ensuing ebb tide. Alternately, if the activity resulted in a depletion of the overlying water of the substance in question, then the marsh would be inwelling or importing in regard to the material in question. The net movement of several classes of substances into or out of several marshes has been reported (Heinle and Flemer, 1976; Valiela et al., 1978). The movements of carbon (van Es, 1977; Woodwell et al., 1977), suspended solids and inorganic nutrients (Duedall et al., 1977), chlorophyll a (Duedall et al., 1977; Erkenbrecher and Stevenson, 1978), detritus (Hanes, 1977), and microorganisms (Axelrad et al., 1976; Chrzanowski et al., 1979) have been investigated.

The movement of microorganisms in tidal waters can be monitored through the quantification of a major metabolite found in all living cells--adenosine 5'-triphosphate (ATP). The extraction and quantification of the nucleotide present in aquatic microflora is indiscriminate in that ATP will be removed from any viable microbe; consequently, the movement of ATP into or out of a marsh would be indicative of the movement of all viable microorganisms, (i.e., phytoplankton, fungi, bacteria) regardless of type of microbe involved.

In an effort to determine if a marsh is a source or sink for total microbial biomass, sampling stations were established across creeks that form major tributaries to the North Inlet marsh system and the movement of ATP into and out of the system was measured. This paper reports on the initial characterization of those transects. The levels and fluctuations of ATP at each interface, as well as short-term transport through each boundary, are reported to enhance understanding of microbial biomass at marsh-nonmarsh boundaries. The characterization was done prior to the initiation of long-term sampling designed to answer the flux question.

Portions of the material included in this communication have been either published previously (Chrzanowski et al., 1979) or are in the process of being published (Chrzanowski and Stevenson, 1980; Kjerfve et al., in manuscript form). The information is being included in this symposium volume so that the research may be summarized and brought to the attention of a group of investigators who might not normally peruse the microbiology literature.

Materials and Methods

The North Inlet system, located near Georgetown, South Carolina, has an area of approximately 3500 ha and is typical of many of the inlet-marsh enclosures along the southeastern Atlantic coast. The marsh interfaces with the Atlantic Ocean through an inlet (North Inlet) along the northeastern boundary and with a river embayment (Winyah Bay) along the southern fringes (Figure 1). The interface with the Atlantic was sampled at the two creeks, Town and Jones, that form the inlet. The transect in Jones Creek at this location is referred to as North Jones Creek. The southern boundary was sampled at the south end of Jones Creek, referred to as the South Jones Creek (Figure 1). Town Creek was sampled from 10 positions and both North and South Jones were each sampled from 3 locations (Figure 2).

The details of station positioning, velocity measurements, sample collection, and ATP quantification are given in Chrzanowski et al., (1979). A synopsis of the methodology is given herein. Water was pumped from 3 depths, 0.2 m below the surface, 0.2 m above the bottom, and a point equidistant between the surface and bottom (Figure 2). Samples were collected synoptically from each of the sampling points (9 or 30 depending on the transect) for 25 h comprising two tidal cycles. Ten or 20 ml aliquots were filtered using Whatman GF/F glass-fiber filters for determination of microbial ATP levels. Extraction of the retained organisms and quantification of the ATP were done using the method of Holm-Hansen (1973) employing an SAI photometer (model 3000) operating in the peak height mode.

The cross-sectional bathymetry was determined using a Raytheon fathometer. In addition, the water depth was measured at each sampling time using a lead-weighted line. Tidal curves were constructed by subtracting instantaneous depth values from the net depth and computing a mean across the transect. Velocity measurements, obtained using lead-weighted, biplane current crosses (Pritchard and Burt, 1951), were made from each boat comprising the transect, first at the surface and then at meter intervals to within 0.5 m of the bottom. Matching vertical profiles of both ATP concentration and velocity at each station were interpolated from actual measurements using the method of Kjerfve (1975) (Chrzanowski et al., 1979). Mass flux calculations, statistical analysis, and computer mapping were done as outlined in Chrzanowski et al., (1979).

Results

The transect at Town Creek (Figure 2a) was 320 m across and the primary channel had a mean depth of about 7.5 m. The water velocity through the interface ranged from 0 to 230 cm per s. The

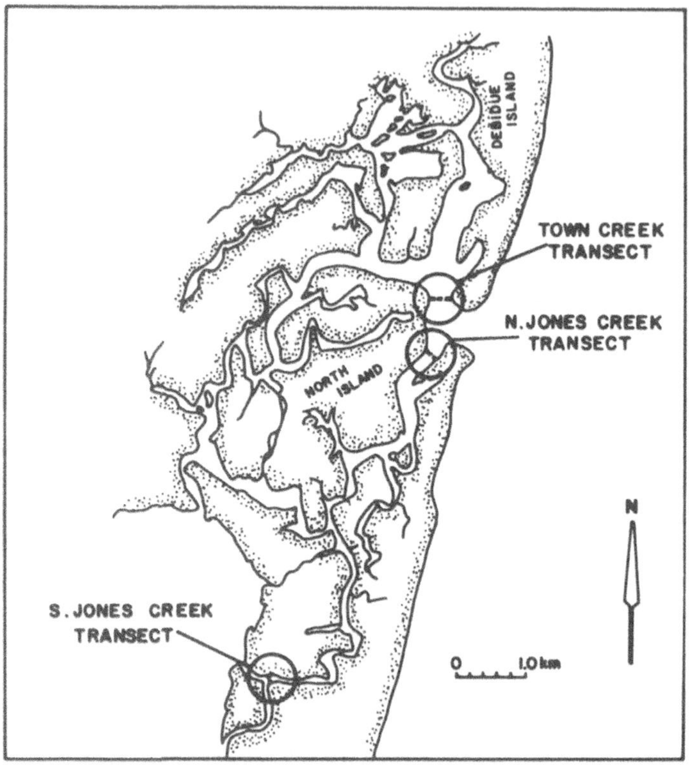

Figure 1. Map of the North Inlet marsh system illustrating the locations of the transects from which samples were collected.

net velocities, which relate not only to the speed of the current but also to the direction of flow, recorded across the transect are illustrated in Figure 3a. Dominant ebb-directed velocities were observed in the primary channel; an area of flood-directed movement was observed in the western or shallow portion of the interface.

The bathymetry at North Jones Creek (Figure 2b) differed from the Town Creek location in that it was about one-fourth as wide and had a less well defined primary channel. The mean depth at the deepest point was about 5.8 m and the eastern shore had a relatively shallow slope. The maximum water velocity observed in the transect was 270 cm per s. Preparation of a net velocity profile indicated that the center section of the transect (station B02) was characterized by a net flood-directed current and the two sides of the interface experienced net ebb-directed flow. A strong, ebb-directed velocity core was noted in the center of the primary channel (Figure 3b).

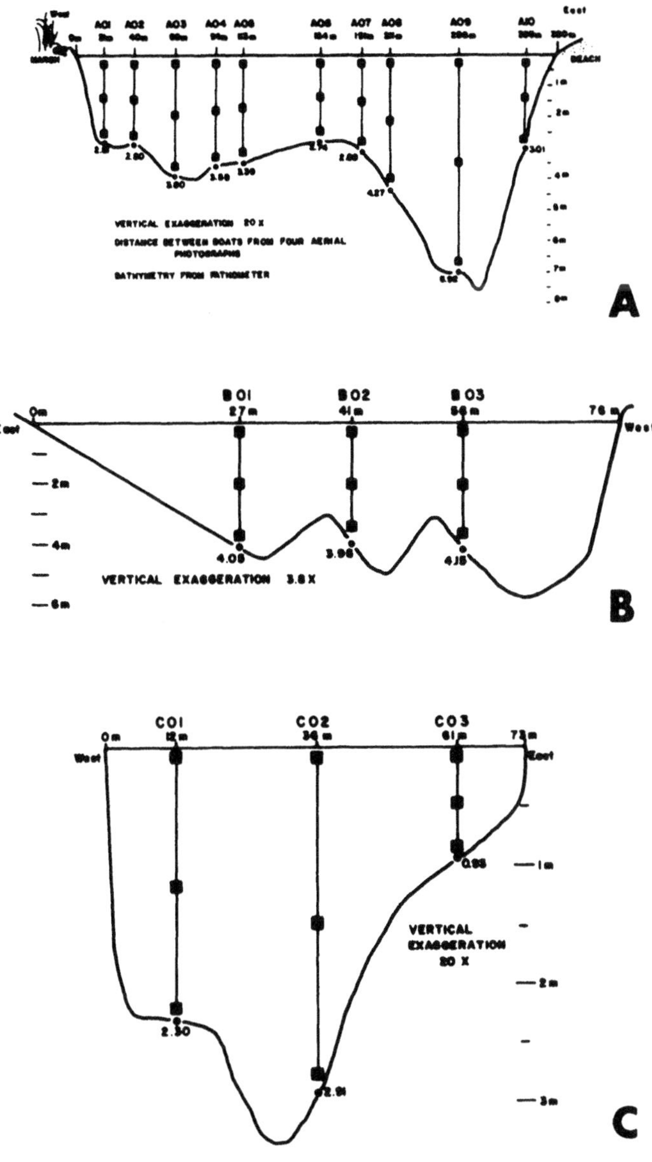

Figure 2. Bathymetry profile of each interface indicating the position of sampling sites, distance between sites, and the approximate locations from which samples were collected (represented by squares), (A) Town Creek, (B) North Jones Creek, (C) South Jones Creek. Drawings in this and following figures were reconstructed from figures previously published in Chrzanowski et al. 1979, and Chrzanowski and Stevenson 1980.

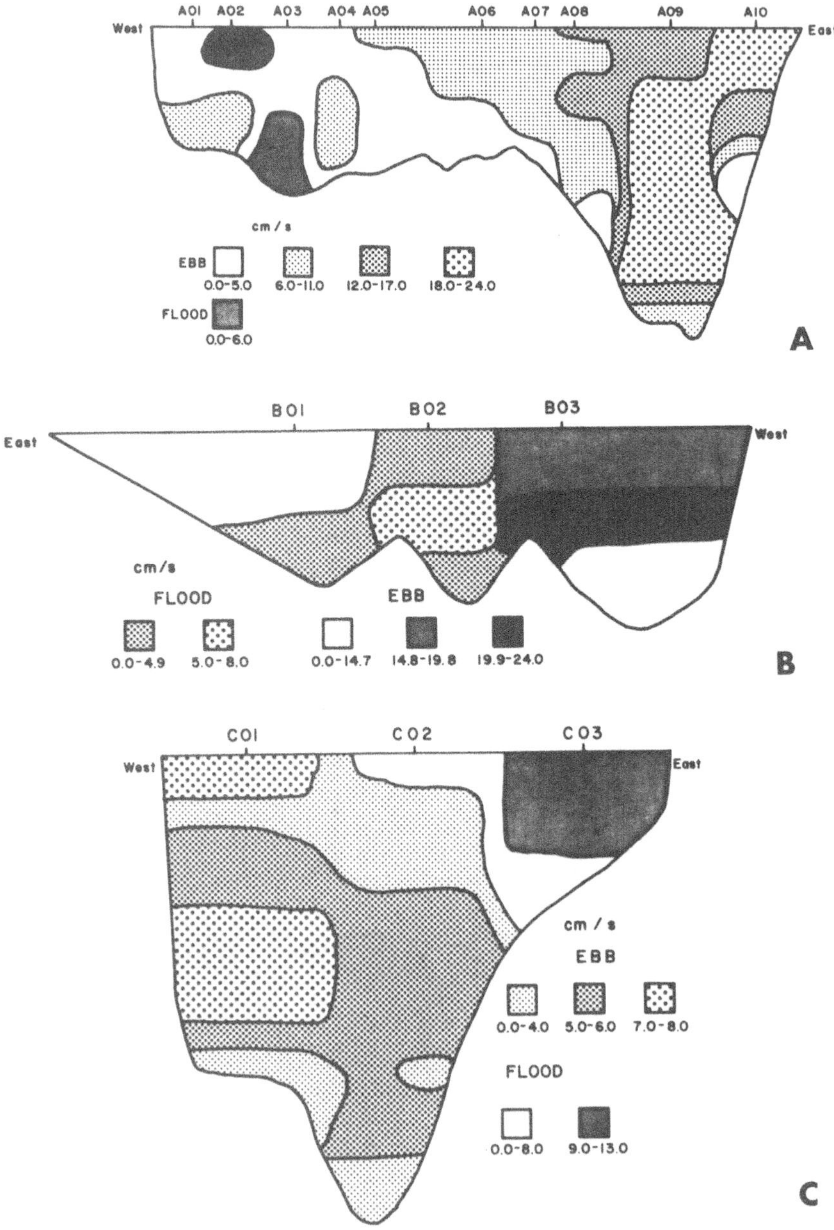

Figure 3. Isopleth of net velocities, corrected for direction of movement, observed during each of the 25-h study periods at the Town Creek (A), North Jones Creek (B), and South Jones Creek (C).

The transect at South Jones Creek differed substantially from the oceanic interfaces (Figure 2c). The creek had very steep banks and was about 3.4 m deep. The primary channel was located in the approximate center of the creek. The shallow portion of the creek experienced net flood-directed velocities while net ebb-directed currents were observed in the remainder of the transect (Figure 3c). A significant circulation anomaly was observed at site C03. Flood-directed currents were experienced for approximately 3 h during the second ebb tide of the study; concurrently, ebb-directed movements were recorded at the other two sites in the deeper portions of the interface.

The means and ranges of ATP concentrations measured in samples collected at each of the interfaces are shown in Table 1. The highest mean value was observed during the expedition to North Jones Creek, and the lower mean was recovered from samples collected at Town Creek. The wide ranges in ATP concentrations observed at each location were due, in part, to fluctuations associated with tidal movements (see Chrzanowski et al., 1979; Chrzanowski and Stevenson, 1980). Nucleotide levels were rhythmical with elevated levels of ATP concentration generally observed at periods of high tide and lowest values of ATP density recovered from samples collected during periods of low tide; however, these fluctuations in biomass levels were not precisely in phase with variations in water level. Additionally, the fluctuations in biomass levels were more erratic in samples collected at South Jones Creek (Chrzanowski and Stevenson, 1980). The day-night cycle had no apparent influence on biomass concentrations recovered from any of the locations.

Salinity determinations indicated that the interfaces were "well mixed" with respect to density (Kjerfve and Proehl, 1979); however, the cross-sections were manifestly not "well mixed" with respect to the concentrations of particulate, viable microbiota (Figures 4, 5, and 6). The spatial distribution of ATP, illustrated by the mean concentrations computed from values obtained during the 25-h study, show that the Town Creek transect was heterogeneous with respect to ATP density; however, no vertical stratification across the interface was evident (Figure 4a). Horizontal divisions were apparent with marked discontinuities in mean values between sites A05 and A06 as well as between locations A08 and A09. The heterogeneity of the interface was even more pronounced when the cross products of individual ATP concentrations and instantaneous directional velocities were examined (Figure 4b). The eastern half was a predominantly exporting system with a dominant region in the major channel. The western section was predominantly importing with a marked import occurring along the surface at stations A01 to A07. A computation of net flux of ATP for the entire cross-sectional area for the complete sampling revealed that ATP was imported at a rate of 39.8 mg per s which represents about 3.6 kg of ATP imported during the 25-h study.

Table 1. Descriptive statistics for adenosine 5'-triphosphate (ATP) data recovered from samples collected at each transect reported as mg ATP per m^3.

Location	Date	N	Mean concentration	Range
Town Creek	November 1977	2750	0.865	0.020 - 5.578
South Jones Creek	March 1978	825	0.893	0.367 - 2.837
North Jones Creek	June 1978	858	1.357	0.610 - 3.390

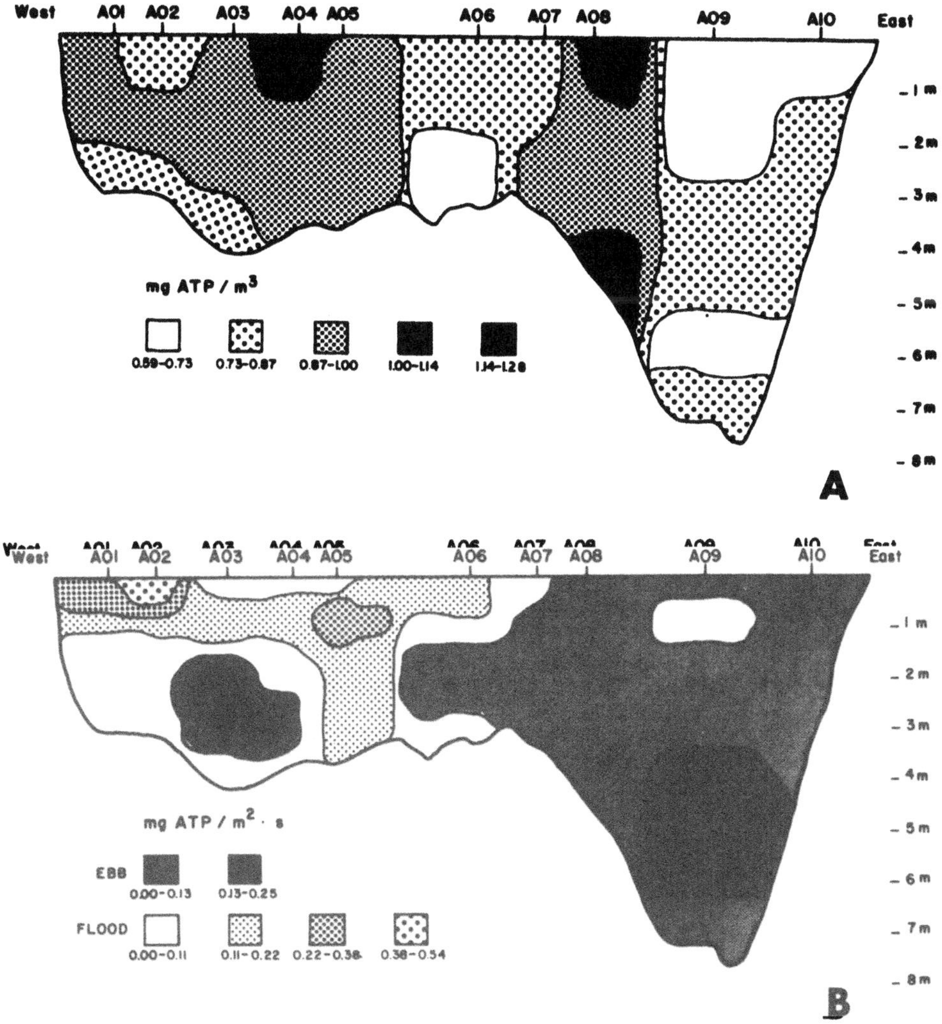

Figure 4. Distribution of the mean ATP concentrations for the complete 25-h sampling period at Town Creek (A) and two-dimensional isopleth illustrating the net flux of ATP through the Town Creek interface during the study (B).

The heterogeneity across an interface was even more pronounced at the North Jones location (Figure 5). The distribution of mean concentrations computed from values obtained from all samples collected during the 25-h study indicated that this interface was divided both vertically and horizontally (Figure 5a). Generally higher concentrations were found on the surface and lower values

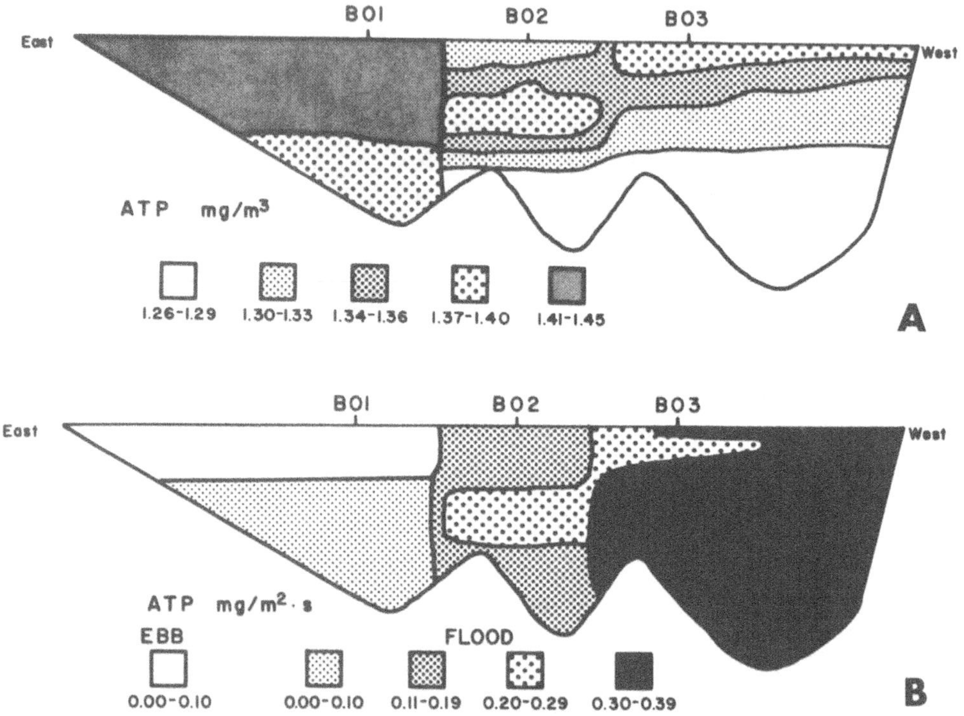

Figure 5. Distribution of the mean ATP concentrations for the complete 25-h sampling period at North Jones Creek (A) and two-dimensional isopleth illustrating the net flux of ATP through the North Jones interface during the study (B).

recovered from samples collected in bottom areas and higher mean values were observed in the eastern side of the interface. In addition, both a horizontal and vertical partitioning of North Jones Creek was evident when the movement of microbiota across the interface was computed (Figure 5b). A net import of ATP occurred across most of the interface. The primary channel on the western side (station B03) was the strongest inwelling (flood-directed) region and the surface on the eastern side (station B01) was the location of the only net export. In total, ATP was imported across the interface at a rate of 40 mg per s.

The isopleth of the mean concentrations of ATP recovered during the sampling of the South Jones Creek location revealed that stations C01 and C02 were homogenous with respect to ATP from top to bottom (Figure 6a). Some stratification was evident at the shallow station (C03) with the lower concentrations found on the bottom portion. The highest mean concentrations were recovered from samples obtained

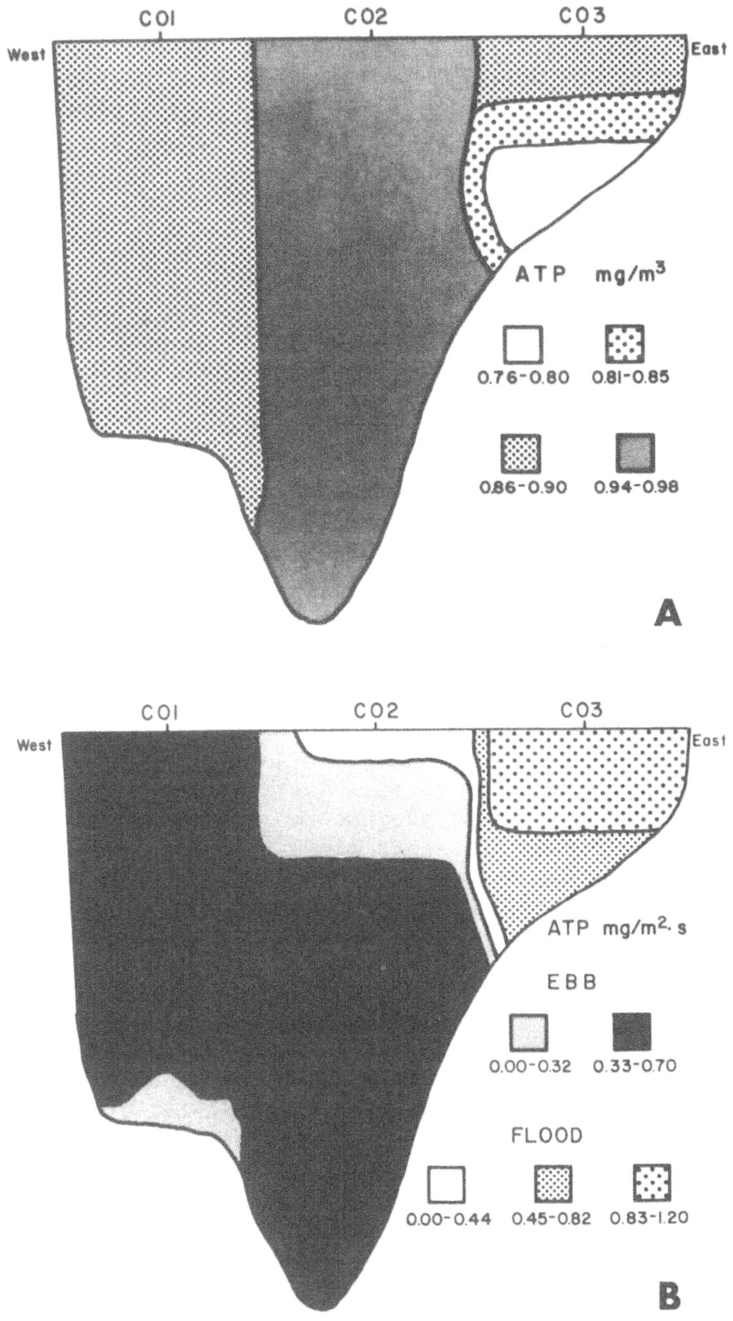

Figure 6. Distribution of mean ATP concentrations of the complete 25-h sampling period at South Jones Creek (A) and two-dimensional isopleth illustrating the net flux of ATP through the South Jones interface during the study (B).

in the center of the interface. The major cross-sectional area of the interface was revealed to be a homogenous exporting system; however net import was noted along the surface of station C02 and the shallow eastern side of the creek (Figure 6b). Biomass was exported through this interface at a rate of 3.6 mg per s during the 25-h study.

DISCUSSION

The sampling duration during each of the studies described in this communication was relatively short and the analysis of results obtained from two tidal cycles is not sufficient for the resolution of the long-term source-sink question for this marsh relative to the oceanic environment. However, the studies do characterize the major tributaries of this marsh in regard to the movement of microbial biomass. The intensive sampling regimes employed allow for the formulation of strategies for expeditions of longer duration. Even though the three locations differed significantly in size and bathymetry, even though the hydrography of each interface varied substantially, and even though the transects were sampled at different seasons of the year (Table 1) there are similarities among the comments that could be made following a scrutiny of the information obtained at each location.

The detailed pictures of biomass distribution within, and flux across, each interface caution against the random selection of sampling locations across transects similar to the ones described. The collection of samples from only one depth at such locations is similarly counter indicated. Neither velocity nor concentration data alone were sufficient for the characterization of the interfaces. Characterization of a tidal creek interface in terms of ATP or similar parameter requires the simultaneous measurements of both the component of interest and velocity and the calculation of cross products. The computation of the mean concentration of some material in ebbing and flooding waters coupled with the assumption that the mass of water moving in each direction is equal will result in erroneous conclusions about the inwell-outwelling nature of a system. This was well illustrated by Kjerfve et al., (1980) during a critical analysis of spatial measurement density and errors of the data collected during the course of this study. They computed both the net discharge and ATP flux across the Town Creek interface by employing data from selected stations and comparing those values to the complete data set. Net movements were computed using data generated from thirty different combinations of stations (using from one to ten locations). A net export of water from the system and a simultaneous import of microbial biomass was calculated in almost every case.

The primary channel in each creek was the location of maximum movement of ATP even though the density of the biomass was not always highest in that location. In fact, the mean concentration of ATP in the Town Creek interface was lowest in the deep channel. The net flux of biomass through the primary channel was in one direction while there was a concurrent net flux in the opposite direction on the other side of the interface, an observation that demonstrates the importance of both the simultaneous measurement of water movement and density of the component in question and the judicious placement of stations. The selection of representative sampling stations that will yield representative flux data is possible through a careful examination of the data collected at each station. For example, Kjerfve et al. (1980) have shown that flux values computed employing the data collected at three Town Creek stations (A01, A05, and A09) are statistically identical to values computed using all ten stations.

Several additional conclusions are warranted from the data obtained during this study. One point relates to the dissimilarities between consecutive tidal cycles. The temporal fluctuations of ATP density reported at each interface exhibited a different pattern for each tidal cycle. These observations together with net velocity patterns for each cycle indicate that the monitoring of a single tidal cycle and the extrapolation of the results to a longer time interval, i.e., a month, may be misleading. A clarification of that possibility must await the results of extended studies currently in progress. Another aspect that deserves mention is the preliminary evidence showing the apparent importing of microbial biomass into the marsh from the oceanic environment.

ACKNOWLEDGMENTS

This work was supported by National Science Foundation Grant DEB 76-83010. We thank the students, faculty, and technicians of the Marine Science Program and Department of Biology for their assistance in collecting the field data. We are expecially grateful to Steven Knoche and Jeffrey Proehl for their capable technical assistance.

REFERENCES

Axelrad, D. M., K. A. Moore, and M. E. Bender. 1976. Nitrogen, phosphorus and carbon flux in Chesapeake Bay marshes. Bulletin 79, Virginia Water Resources Center.

Chrzanowski, T. H., L. H. Stevenson, and B. Kjerfve. 1979. Adenosine 5'-triphosphate flux through the North Inlet marsh system. Applied and Environmental Microbiology 37:841-848.

Chrzanowski, T. H., and L. H. Stevenson. 1980. Microbial biomass variability in salt-marsh creek cross-sections. Marine Geology (submitted).

Duedall, I. W., H. B. O'Connors, J. H. Parker, R. E. Wilson, and A. S. Robbins. 1977. The abundances, distribution and flux of nutrients and chlorophyll a in the New York Bight Apex. Estuarine and Coastal Marine Science 5:81-105.

Erkenbrecher, C. W. and L. H. Stevenson. 1978. The transport of microbial biomass and suspended material in a high-marsh creek. Canadian Journal of Microbiology 24:839-846.

Haines, E. B. 1977. The origins of detritus in Georgia salt marsh estuaries. Oikos 29:254-260.

Heinle, D. R. and D. A. Flemer. 1976. Flows of material between poorly flooded tidal marshes and an estuary. Marine Biology 35:359-373.

Holm-Hansen, O. 1973. Determination of total microbial biomass by measurement of adenosine triphosphate, p. 73-89. In Estuarine Microbial Ecology (L. H. Stevenson and R. R. Colwell, eds.) University of South Carolina Press, Columbia.

Kjerfve, B. 1975. Velocity averaging in estuaries characterized by a large tidal range to depth ratio. Estuarine and Coastal Marine Science 3:311-323.

Kjerfve, B., and J. A. Proehl. 1979. Velocity variability in a cross-section of a well-mixed estuary. Journal of Marine Research 37:409-418.

Kjerfve, B., L. H. Stevenson, J. A. Proehl, T. H. Chrzanowski, J. D. Spurrier, and W. M. Kitchens. Estimation of material fluxes in an estuarine cross-section; a critical analysis of spatial measurement density and errors. Limnology and Oceanography (submitted).

Pritchard, D. W. and W. V. Burt. 1951. An inexpensive and rapid technique for obtaining current profiles in estuarine waters. Journal of Marine Research 10:180-189.

Valiela, I., J. M. Teal, S. Volkmann, O. Shafer, and E. J. Carpenter. 1978. Nutrient and particulate fluxes in a salt marsh ecosystem: tidal exchanges and inputs by precipitation and groundwater. Limnology and Oceanography 23:798-812.

van Es, F. B. 1977. A preliminary carbon budget for a part of the Ems estuary: The Dollard. Helgol. wiss. Meeresunters. 30: 283-294.

Woodwell, G. M., D. E. Whitney, C. A. S. Hall and R. A. Houghton. 1977. The flax pond ecosystem study: Exchanges of carbon in water between a salt marsh and Long Island Sound. Limnology and Oceanography 22:833-838.

SOURCES AND VARIABILITY OF SUSPENDED PARTICULATES AND ORGANIC CARBON IN A SALT MARSH ESTUARY

Robert C. Harriss[1], Benny W. Ribelin[2], and C. Dreyer[3]

[1]NASA Langley Research Center, Mail Stop 270, Hampton, Virginia 23665, [2]Sea Farms de Honduras, Choluteca, Honduras, [3]Dept. of Oceanography, Florida State University, Tallahassee, Florida

ABSTRACT

Juncus roemerianus salt marsh ecosystems bordering the Northeast Gulf of Mexico are an apparent source of suspended particulates to adjacent coastal waters. More than 98 percent of the detrital particulates collected from ebb tide waters are comprised of amorphous aggregates, derived primarily from organic films produced by benthic microflora. Vascular plant fragments from the predominant macrophyte in the marshes, *Juncus roemerianus*, are not an important source of detritus to the estuarine water column. Tidal cycle, light levels, and weather-related episodic phenomena all influence the production and distribution of suspended particulates and organic carbon in estuarine waters.

The transport of dissolved organic carbon from low salinity marsh source areas to relatively high salinity offshore waters exhibits linear dilution characteristics. Particulate organic carbon exhibits a nonlinear relationship to salinity in estuarine waters, primarily due to the influence of sediment resuspension by water column turbulence.

The data from this study offer an opportunity to explore the relative importance of components of variability in the suspended particulate distribution through water-quality simulation modeling.

INTRODUCTION

Salt marshes exchange inorganic and organic materials, particulate and dissolved, with adjacent coastal waters. Fluxes of these materials between marsh and estuary can play an important role in both the nutrient status and water quality of coastal marine environments. This paper reports on factors which determine the variability of total particulate and dissolved organic carbon concentrations in Juncus roemerianus (needlerush) marshes and adjacent estuarine waters of the North Florida Gulf coast. We place particular significance on the implications of the measured variability for water-quality monitoring and modeling. What frequency of sampling is required to quantify concentrations and fluxes of particulates in salt marsh tidal creeks? What sources of energy drive quantitatively important fluxes of suspended particulate and organic materials between marsh and estuary? How should the time scale of a coastal water-quality model be structured to simulate important characteristics of particulate flux dynamics?

The extensive literature on particulate materials and organic carbon in salt marsh and estuarine environments has been received in several recently published papers (Woodwell et al., 1973; Woodwell et al., 1977; Haines, 1977; Valiela et al., 1978; Happ et al., 1977; Correll, 1978; Smith, 1979); the reader is referred to these papers for background. One of the more controversial and significant issues which remains unanswered in this problem area is the question--Are salt marshes a source or sink for estuarine organic-rich particulate materials? The answer seems to depend on a variety of geographical, temporal, and environmental considerations. For example, in South Florida's mangrove estuaries or the salt marsh dominated estuaries of the Mississippi Delta the relatively high ratio of marsh area to tidal creek area is clearly a dominate influence on the particulate organic content of estuarine waters (Heald, 1971; Happ et al., 1977). In other coastal regions the influence of salt marshes on the particulate organic content is not a predominant factor, with other primary producer systems such as phytoplankton and/or seagrasses having a more important role (Correll, 1978). However a careful assessment of existing data on sources and fluxes of suspended particulates in estuarine waters suggests that, in part, at least, the issue of what role salt marshes play in water quality will not be resolved until improved techniques and regional sampling schemes are applied to the problem. Important contributions by Pickral and Odum (1976), Happ et al. (1977), Woodwell et al. (1977) and Valiela et al. (1978) have brought attention to the complex variability in suspended particulate and organic carbon concentrations in marsh and estuarine waters. The variability appears to have important temporal and spatial components related to a variety of diurnal, seasonal, and episodic factors.

In this paper, we report on analysis of data on suspended particulate and organic carbon concentrations associated with J. roemerianus salt marsh ecosystems. Most of the existing scientific literature on salt marshes along the Atlantic and Gulf coasts of the United States pertains to Spartina alterniflora marshes (e.g., Haines, 1977). We also focus on an assessment of the temporal characteristics of variations in suspended particulate and organic matter concentrations as a contribution to defining improved strategies for water-quality sampling in future studies of coastal environments.

Environmental Setting

This study focused on marsh and tidal creek systems typical of the North Florida Gulf coast. J. roemerianus is the most common macrophyte in the marshes of this region. Juncus grows in the upper intertidal environment and is an important component of salt marsh floras elsewhere in the world (Chapman, 1974). The detailed physiographic and vegetative features of North Florida marshes have been described by Kurz and Wagner (1957). Studies related to the standing crop, production, and decay of J. roemerianus have been made in Virginia (Wass and Wright, 1969), North Carolina (Foster, 1968; Stroud and Cooper, 1968; Waits, 1967; Williams and Murdoch, 1972), Florida (Heald, 1971), Mississippi (de la Cruz, 1973; Hackney, 1977) and Louisiana (White et al., 1978).

Two small Juncus marsh ecosystems were sampled for the studies reported in this paper. Both marshes are approximately 14 to 16 hectare systems, located in Franklin County, Florida, J. roemerianus covers more than 95 percent of the total area of each marsh. Daily tidal amplitude averages of 0.73 metre, with maximum high tides in August and September (+1 metre above mean low water).

Methods

An intensive sampling of tidal creek waters was accomplished in the study marshes by a combination of manual and automated sampling techniques. The sampling strategy was designed to characterize the total particulate and organic carbon concentrations over a wide range of time scales from hourly to seasonal.

A battery-powered automatic water sampler allowed the regular hourly collection of 500 ml water samples throughout a 24-hour period. The water sampler was placed on a floating platform anchored in the middle of a tidal creek, approximately 75 metres from the discharge point to the Gulf of Mexico. The sampling location was sheltered from direct wind and surf. A water intake tube was placed at a depth of 10 centimetres below the water surface.

All water samples were treated with 2 ml of 2 percent $HgCl_2$ immediately upon collection. Samples were analyzed within 12-24 hours after collection.

Quantities of total suspended particulates were determined by dry weight measurements of material retained by prewashed, pre-weighed 0.45 μm filters.

Organic carbon determinations were made using a Total Carbon Analyzer (Oceanography International, Inc.) that operated, with minor modifications, on the principles described by Menzel and Vaccaro (1964). Samples for dissolved organic carbon (DOC) concentration were filtered through precombusted Gelman Type A glass fiber filters. Five millilitres of the filtrate were placed in a precombusted 10 ml glass ampoule containing 200 mg of potassium peroxydisulfate and then acidified with 0.1 ml of concentrated phosphoric acid. The ampoules were purged of inorganic CO_2, sealed, autoclaved for 4-6 hours to oxidize organic compounds, and analyzed for CO_2. Particulate organic carbon (POC) was determined with the same analytical procedure just described. Filters containing particulate matter were folded and placed in ampoules. Five millilitres of distilled water were added to each POC ampoule.

Standards were developed with sodium carbonate-phosphoric acid solutions. Reagent blank values were determined for every set of samples for both DOC and POC. The average precision at the 95 percent confidence limit was ± 0.39 mg/l for DOC and ± 0.20 mg/l for POC.

Variability in Suspended Particulate Concentrations

An almost continuous record of suspended particulates, collected hourly for 96 hours in June 1977, in the middle reaches of a tidal creek at a constant depth of 10 centimetres is presented in Figure 1. The few missing data points are periods of very low tidal levels where the intake for the automated sampler did not function. This data set illustrates several interesting and commonly observed characteristics of the patterns of variability in suspended particulates in the tidal creeks studied. These characteristics include: (1) The concentrations of suspended particulates in ebb tide waters are typically greater than concentrations measured in flood tide waters. The *Juncus* marshes studied are apparently a source of suspended particulates to adjacent estuarine waters. However, a quantitative particulate mass flux on a marsh ecosystem is extremely difficult to measure due to high temporal variability

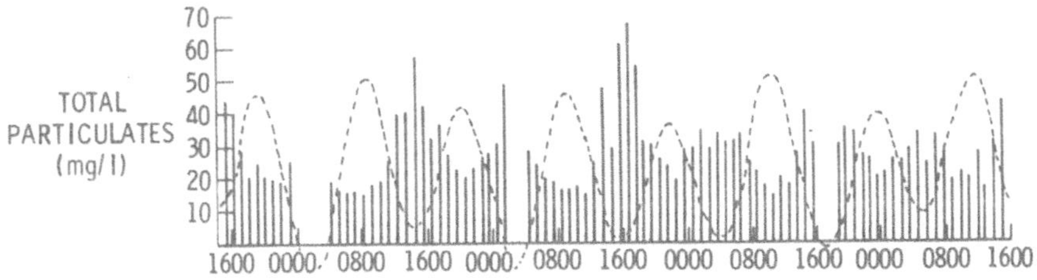

Figure 1. Total suspended particulate concentrations in ebb and flood tide waters from a Juncus marsh tidal creek. The dashed line represents the relative tidal stage. A few data points were lost when the sampler was grounded at low tide.

in the flux patterns. (2) The concentration of suspended particulates is commonly higher in daytime ebb tide waters than in nighttime ebb tide waters. (3) The rate of change in suspended particulate concentrations reaches values up to approximately 9 mg/l hr^{-1} during the transition from flood to ebb or vice versa when other factors (e.g., wind, benthic animal activity) are at a minimum.

Maximum suspended particulate concentrations were observed in August and September. Figure 2 illustrates a 72-hour series of samples taken in August 1977, at the same station and depth as in the case discussed above. Several important characteristics of this data set are: (1) the suspended particulate concentrations associated with ebb and low tide waters are approximately four times higher in August than in June when similar environmental conditions are compared. (2) On August 9, an intense rainfall passed over the study site at 1900 hours producing considerable surface runoff. The stormwater runoff increased both the peak concentrations and the duration of continuous high concentration levels in the estuarine water column by approximately a factor of two. These data illustrate the important role episodic storm events can play in particulate concentrations, fluxes, and

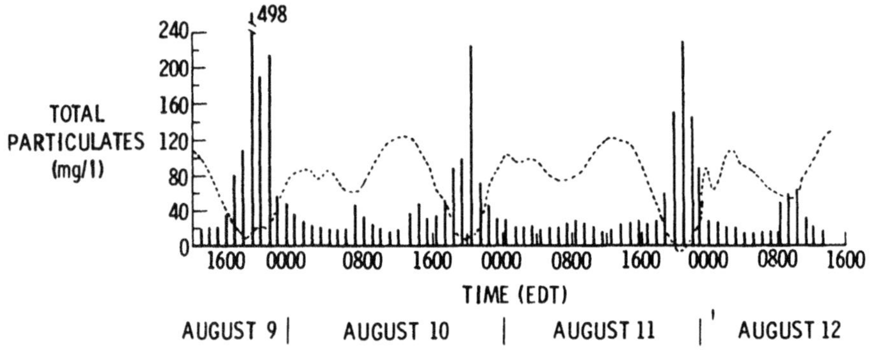

Figure 2. Total suspended particulate concentrations in ebb and flood tide waters from a <u>Juncus</u> marsh tidal creek. The dashed line represents the relative tidal stage. Note the above-scale particulate concentration (498 mg/l) collected immediately following a rain storm at 1900 hours, August 9, 1977.

consequently in the chemical budgets of the marsh and adjacent estuarine waters.

Seasonal variation in maximum concentrations (exclusive of storm-related concentrations) observed in the <u>Juncus</u> marsh tidal creek are illustrated in Figure 3. These data indicate lower concentrations in the winter months, increasing through the spring and early summer to maximum values in August and September. The scale of the seasonal variability in suspended particulate concentrations is similar to the variability observed on an hourly to daily basis during intensive sampling periods. Suspended particulate concentrations in flood tide water typically do not vary by more than a factor of two on a daily or seasonal basis. The suspended particulate concentrations in ebb tide waters vary by an order of magnitude on a seasonal basis. Qualitatively these data suggest that the <u>Juncus</u> marsh probably exports particulates to estuarine waters during approximately 9 months of the year.

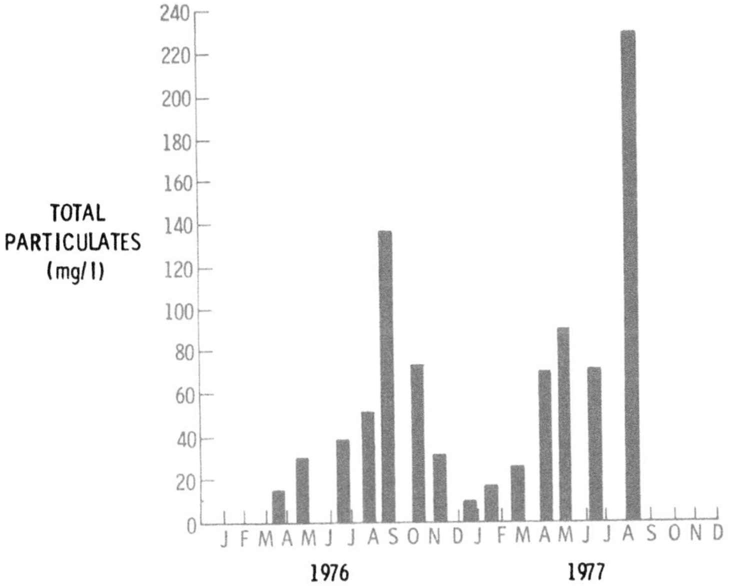

Figure 3. Seasonal variation in maximum ebb tide total suspended particulate concentrations in a Juncus marsh tidal creek. Suspended particulate concentrations associated with the August 9, 1977 storm event were not included in the data base for this graph.

Origin of Suspended Particulates Exported from Juncus roemerianus Marshes

An exhaustive study of the origin of suspended particulates in the tidal creek waters of the Juncus marsh has been conducted by Ribelin (1978). Both field and laboratory studies support the hypothesis that suspended particulates are complex, organic-rich, detrital aggregates which are produced by the benthic microflora of the marsh. Field observations provide qualitative evidence on the mechanism of particulate formation. Flood tides lift films of aggregate material from the dense community of benthic algae that carpet the sediment surface of the marsh. Ebbing tides transport the floating films into tidal creeks where any mild disturbance of the water surface results in dispersal of the films, which sink into the estuarine water column. Suspended particulates produced by the disruption of the floating films typically range from less than 1 μm to over 100 μm in diameter, with the majority in the

25 µm to 50 µm range.

Examination of suspended particulates with both light microscopy and scanning electron microscopy further confirms their origin in the benthic microflora habitat. The ebb tide particles are aggregates of diatom frustule fragments, with species of the benthic genera Nitzschia and Navicula being most common, and many small (<1 µm) unidentifiable inclusions, all bound together by an adhesive, gelatinous substance that is yellowish-brown in color. Visual observations and cellulose staining techniques indicate that vascular plant fragments comprise less than 1 percent of the total suspended load.

Particles collected from flood tide waters are quite different in appearance from those in ebb tide samples. Detrital aggregates are smaller and fewer in number than those already described from ebb tide collections. Whole centric diatoms are often present in flood tide samples represented by Chaetocerus spp. and Rhizosolenia spp. which characteristically exhibit unbroken spines. When these planktonic species are observed in ebb tide waters, they generally exhibit considerable superficial damage and are covered with small aggregates and gelatinous substance.

An examination of floating surface films collected in quiet tidal creek ebb waters and of surface films on the sediment surface of the marsh reveal a mixture of materials essentially identical to the components of the suspended particulates, with Nitzschia and Navicula being the predominant genera of the benthic microflora in the Juncus marsh and the major identifiable component of the surface films. Thus, visual observations provide strong evidence for a casual linkage between the organic-rich films on the marsh sediment surface, floating organic-rich films on ebb tide waters in tidal creeks, and suspended particulate materials being exported from the Juncus marsh to estuarine waters.

Temporal Variability in the Source of Suspended Particulates

Having established a qualitative model for the origin of suspended particulates in ebb tide waters, we now explore how well the model accounts for measured variations in concentrations described in earlier sections of this paper. Three identified components of variability were a day/night difference, seasonal variations, and storm-related episodic increases in particulate concentrations. The storm-water component of the variability is easily explained by increased erosion of the marsh sediment surface. Particulates collected from ebbing waters associated with storm conditions exhibit the same general characteristics described above for normal conditions, except for a higher total concentration and a slightly higher inorganic content. The potential importance of storm events in the suspended particulate budgets of small watersheds and salt marshes has been recognized in a number of studies (e.g., Turner et al., 1975; Pickral and Odum, 1976). Both this study and the study by Pickral and Odum (1976) emphasize that storms may be particularly important in flushing areas of high marsh (i.e., Juncus). The timing of the storm in relation to tides and the character of wind and rain intensity are important variables which can determine erosion rates.

The seasonal variability in ebb tide particulate concentrations (Figure 3) correlates qualitatively with the abundance of microflora on the marsh sediment surface. The carpet of filamentus algae and benthic diatoms embedded in mucilaginous material is much more abundant in warm months (i.e., April-October) when particulate concentrations are also higher. During summer months, the diatoms become so abundant at the sediment surface that they can completely cover the filamentous green algae. The magnitude of average high-tide above mean sea level is also approximately 0.3m higher during July to October, which results in more extensive flooding of the Juncus marsh and consequently more erosion of the marsh surface. This combination of higher benthic biomass and more extensive flooding during April to October appears to best explain the seasonal character in suspended particulates concentrations in ebb tide waters associated with Juncus roemerianus marsh.

The higher concentrations of particulates observed in daytime ebb waters (Figure 1) correlate qualitatively with the frequently observed phenomenon of migration and phototaxis in benthic diatoms (Aleem, 1950). Motile members of the diatom community are most abundant at the marsh surface at relatively high levels leading to potentially higher loss rates to erosional processes occurring during daylight hours.

Fate of Organic Carbon in Salt Marsh and Estuarine Waters

So far we have only considered the concentrations and sources of total suspended particulate materials in the Juncus marsh environment. In addition to these studies, we have conducted considerable research on the transport and fate of the organic carbon fraction of the suspended and dissolved materials in Juncus marsh tidal creeks and adjacent nearshore estuarine waters of the northeastern Gulf. Particular emphasis was placed on organic carbon as a tracer for studying the hypothesis that salt marshes produce and export organic-rich materials which are important to estuarine productivity.

One method for determining whether organic-rich materials are exported from the marsh environment to coastal waters is to examine the distribution of dissolved organic carbon (DOC) and particulate organic carbon (POC) over salinity gradients from low salinity source areas to higher salinity coastal waters. Any significant input or removal processes in the estuarine environment should perturb the transport process and result in non-ideal dilution behavior. The data in Table 1 summarize the results of a statistical analysis of salinity-DOC relationships for 165 samples collected in the tidal creek of a Juncus marsh and in adjacent nearshore estuarine waters. A wide range of environmental conditions were sampled over a 7-month study period, with an observed salinity range of less than 0.5 to 33.4 parts per thousand and a range in DOC of 3.1 to 25.9 mg/l. These data indicate a very strong negative correlation between salinity and DOC for most environmental conditions sampled. Only in cases where extreme high tides occurred, resulting in limited salinity gradients, did the linear, inverse salinity-DOC relationship break down.

The DOC in Juncus tidal creeks can be derived from several sources including upland freshwater runoff, groundwater seepage, and marsh vegetation. Both the salinity-carbon relationships measured and mass balance considerations indicate that freshwater runoff from adjacent pine forests are the primary source of DOC in the marsh studied.

POC concentrations showed no strong statistical relation with salinity. There is a qualitative indication in the data that winds and high tides produce sufficient turbulence in these shallow waters to generate frequent suspension of off-shore bottom sediment which produces a significant impact on the salinity-POC relationship.

The results of these studies on carbon-salinity relationships should not be generalized beyond the North Florida Juncus marsh ecosystem. Flocculation and coagulation processes which would enhance deposition of carbon in a salinity gradient are affected by variables such as the source and characteristics of the DOC, degree of estuarine turbulent mixing, time required for transport through the salinity gradient, and the amount and type of biological activity in the estuary. The Juncus ecosystems discussed in this paper are small estuaries with relatively rapid mixing and transport of materials through the salinity gradient, minimizing the effectiveness of removal mechanisms.

Implications for Wetland and Estuarine Water Quality Modeling

The results of our studies provide a detailed data set on particulates and organic carbon in a Juncus roemerianus salt marsh ecosystem. However, even this relatively large data base leaves

Table 1. Dissolved organic carbon-salinity correlation in tidal creek and estuarine waters adjacent to *Juncus* marsh.

Date	Number of Samples	Salinity Range (o/oo)	DOC Range (mg/l)	Correlation Coefficient
January 13	8	0.0-20.4	12.6-25.9	-0.98
January 27	9	0.0-26.6	5.0-20.2	-0.95
February 7	9	4.0-27.7	4.4-20.5	-0.95
March 6	13	8.4-19.0	14.8-21.5	-0.94
March 15	13	6.3-26.3	6.0-18.6	-0.78
April 18	18	24.0-28.3	3.4- 9.0	+0.11
April 25	16	8.2-27.6	3.1- 6.6	-0.93
May 5	16	12.9-30.1	3.6- 8.9	-0.92
June 2	14	26.0-31.7	3.5- 8.0	+0.11
June 9	14	23.3-33.6	3.3- 6.5	-0.64
June 21	9	6.2-28.5	6.6-22.4	-0.98
July 3	16	19.0-26.9	3.4- 8.2	-0.22

us far short of our goal of a quantitative understanding of the role of *Juncus* marshes in regional biogeochemical cycles and productivity. Were the two small marshes we studied representative of the entire *Juncus* dominated coastline in the Northeastern Gulf of Mexico? What is the significance of high energy, infrequent, episodic events (e.g., hurricanes) to long-term water quality and productivity relationships? Can quantitative monitoring of dissolved and particulate fluxes from small marsh systems be accomplished without disturbing the ecosystem itself (i.e., are we trapped by the Uncertainty Principle)? We offer the following propositions as a summary of our results and conclusions:

1. The results of our intensive monitoring studies of total suspended particulates in ebb and flood tide waters in a Juncus roemerianus marsh support the hypothesis that these marsh ecosystems are a net source of particulate materials to adjacent estuarine and nearshore waters, at least during the months of April to October.

2. Attempts to quantitatively monitor fluxes of DOC and POC across the salinity gradient from relatively low salinity marsh source areas to higher salinity offshore waters were complicated by the complex circulation dynamics in these shallow waters. The results suggest that DOC is transported from marsh to offshore waters by ideal dilution. Particulate organic carbon transport processes could not be resolved with the data collected due to the effects of sediment resuspension. While more intensive field monitoring, particularly with in situ, continuous sampling could possibly contribute to a better understanding of POC fluxes from Juncus marshes to adjacent coastal waters, the recommendation of Pickral and Odum (1976) for laboratory flume studies to quantify bedload and suspended particulate transport in different hydrologic regimes is supported by our experience.

3. The results of this study identify three important components of variability in the distribution of suspended particulates--day/night, seasonal, and episodic events related to storms. Our data offer a semi-quantitative estimate of the effects of these variations on suspended particulate concentrations available for export from the Juncus marsh. An interesting and useful water quality simulation modeling exercise could be developed, based in part on the data in this paper, to examine the relative importance of each of the identified components of variability in the potential mechanisms for long-term chemical coupling of Juncus marshes to adjacent waters of the Gulf of Mexico.

REFERENCES

Aleem, A. A. 1950. The diatom community inhabiting the mud-flats at Whitside. New Phytol. 49, 174-182.

Chapman, V. J. 1974. Salt Marshes and Salt Deserts in the World. Interscience Publishers, New York.

Correll, D. L. 1978. Estuarine productivity. BioScience. 28, 646-650.

de la Cruz, A. A. 1973. The role of tidal marshes in the productivity of coastal waters. Assoc. S. E. Biol. Bull. 20, 147-156.

Foster, W. A. 1968. Studies on the distribution and growth of Juncus roemerianus in southeastern Brunswick County, North Carolina. M. S. Thesis, North Carolina State University, Raleigh.

Hackney, C. T. 1977. Energy flux in a tidal creek draining an irregularly flooded Juncus marsh. Ph.D. Thesis, Mississippi State University.

Haines, E. B. 1977. The origins of detritus in Georgia salt marsh estuaries. Oikos. 29, 254-260.

Happ, G., Gosselink, J. G., and Day, J. W. 1977. The seasonal distribution of organic carbon in a Louisiana estuary. Estuarine Coastal Mar. Sci. 5, 695-705.

Heald, E. J. 1971. The production of organic detritus in a South Florida estuary. Sea Grant Tech. Bull. No. 6, University of Miami.

Kurz, H. and Wagner, K. 1957. Marshes of the Gulf and Atlantic coasts of Northern Florida and Charleston, South Carolina. Florida State University Studies No. 24, Tallahassee.

Menzel, D. W. and Vaccaro, R. F. 1964. The measurement of dissolved organic carbon and particulate organic carbon in seawater. Limnol. Oceanogr. 9, 138-142.

Pickral, J. C. and Odum, W. E. 1976. Benthic detritus in a salt marsh tidal creek. In Estuarine Processes, Vol. II, pp. 280-292 (Wiley, M., ed). Academic Press, New York.

Ribelin, B. W. 1978. Salt marsh detrital aggregates: A key to trophic relationships. Ph.D. Thesis, Florida State University, Tallahassee.

Smith, T. J. 1979. Estuarine productivity revisited. BioScience. 29, 149-151.

Stroud, L. M. and Cooper, A. W. 1968. Color-infrared aerial photographic interpretation and net primary productivity of regularly-flooded North Carolina salt marsh. University of North Carolina, Water Resources Research Inst. Rept. No. 14.

Turner, R. R., Harriss, R. C. and Burton, T. M. 1975. The effect of urban land use on nutrient and suspended-solids export from North Florida watersheds. In Mineral Cycling in Southeastern Ecosystems. ERDA Symposium Series 740513, pp. 686-708.

Valiela, I., Teal, J. M., Volkmann, S., Shafer, D. and Carpenter, E. J. 1978. Nutrient and particulate fluxes in a salt marsh ecosystem: Tidal exchanges and inputs by precipitation and groundwater. Limnol. Oceanogr. 23, 798-812.

Waits, E. D. 1967. Net primary productivity of an irregularly-flooded North Carolina salt marsh. Ph.D. Thesis, North Carolina State University, Raleigh.

Wass, M. L. and Wright, T. D. 1969. Coastal wetlands of Virginia. Virginia Inst. Mar. Sci., Spec. Rept. Appl. Mar. Sci. Ocean Eng. No. 10.

White, D. A., Weiss, T. E., Trapani, J. M., and Thien, L. B. 1978. Productivity and decomposition of the dominant salt marsh plants in Louisiana. Ecology. 59, 751-759.

Williams, R. B. and Murdoch, M. G. 1972. Compartmental analysis of the production of *Juncus roemerianus* in a North Carolina salt marsh. Chesapeake Sci. 13, 69-79.

Woodwell, G. M., Rich, P. H., and Hall, C. A. 1973. Carbon in estuaries. Brookhaven Symp. Biol. 24, 221-240.

Woodwell, G. M., Whitney, D. E., Hall, C. A., and Houghton, R. A. 1977. The Flax Pond ecosystem study: Exchanges of carbon in water between a salt marsh and Long Island Sound. Limnol. Oceanogr. 22, 833-838.

TIDAL WETLANDS AND ESTUARINE COLIFORM BACTERIA

Paul Jensen,[1] Andrew Rola[2] and John Tyrawski[3]

[1]Espey, Huston and Assoc., formerly Delaware Sea Grant Marine Advisory Service. [2]Graduate Student, Department of Civil Engineering, University of Delaware. [3]Senior Environmental Specialist, State of New Jersey, formerly Research Associate, College of Marine Studies University of Delaware.

ABSTRACT

High concentrations of indicator bacteria (total and fecal coliforms) are common in Delaware estuarine waters which have large areas of adjacent tidal wetlands. The relation between tidal wetlands and these high coliform bacteria levels is explored through direct observation and statistical analysis of possible causative factors. Statistical analyses are performed on two representative tidal rivers using data collected by the state as part of it's water quality monitoring program. Statistical results are used to suggest possible mechanisms for wetland/coliform bacteria interactions and to identify those parameters which are most important in a predictive model of estuarine coliform bacteria concentrations.

INTRODUCTION

The national goal of fishable/swimmable waters has produced a great deal of research and analysis of water quality and the principal factors (point and non-point sources) that affect it. Of all the parameters typically considered in water quality analysis, coliform or indicator bacteria are perhaps one of the most significant. Indicator bacteria are used in regulations for primary contact recreation and the harvesting of shellfish. When indicator bacteria levels exceed very specific standards, uses of the water body (some of which have direct economic significance) are affected. The same statement is much more difficult to make when, for example, a dissolved oxygen standard is not met.

Delaware's tidal river estuaries traditionally have supported substantial stocks of oysters, clams and mussels (Maurer et al., 1971). However, since shellfish-harvesting regulations based on indicator bacteria have been instituted, the vast majority of these tidal-river, estuarine areas have been closed to shellfishing. Figure 1, compiled from State Division of Public Health maps, shows the shellfish closure areas in Delaware. Shellfishing is allowed in most of Delaware Bay and in the open portions of the smaller bays in the south, but essentially all of the tidal-river estuaries in the state are closed to shellfishing.

Figures 2 and 3 show the median total and fecal coliform levels (plus or minus one quartile, 25% of the observations exceed the upper bound and 25% are below the lower bound) for two representative tidal rivers, the Broadkill and Mispillion. These data are surface grab samples collected by the State Division of Environmental Control on a monthly basis from April through October, 1976 through 1978. Coliform concentrations were measured with a 3-tube, 3-dilution MPN technique (Standard Methods, 1971). The coliform standards used to regulate shellfish harvesting are also shown on the figures.

These high coliform bacteria levels may be related, in a complex fashion, to the presence of large areas of adjacent tidal wetlands. This possibility is suggested by a combination of several factors including direct observation, and the process of elimination.

While conducting water quality modeling studies (Jensen and Tyrawski, 1977) on the Broadkill River, it was found impossible to calibrate a total coliform model without either an unrealistically low coliform die-off rate or unrealistically inflated coliform loadings. Point sources were eliminated both quantitatively and by analogy-the Mispillion River has no point sources yet has similarly high coliform levels. Other potential sources eliminated included urban runoff (urban area <5% of watershed), tributary inputs (Ritter and Scheffler, 1977), and waterfront septic systems (Jensen et al., 1977). Attention was turned to the possible effects of wetlands on stream coliform levels.

A limited number of direct observations of coliform bacteria exchange between marshes and their adjacent estuaries were carried out. Figure 4 shows total and fecal coliform levels collected at approximately hourly intervals over a 24-hour period at the mouth

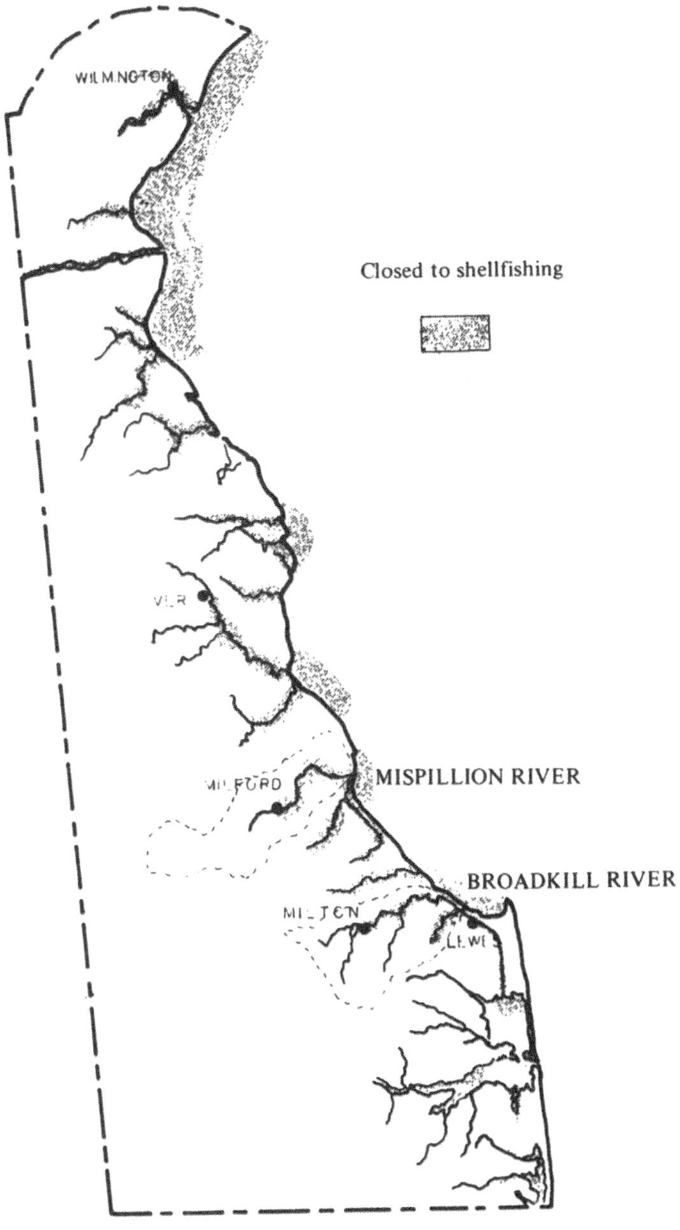

Figure 1. Delaware shellfish closure areas

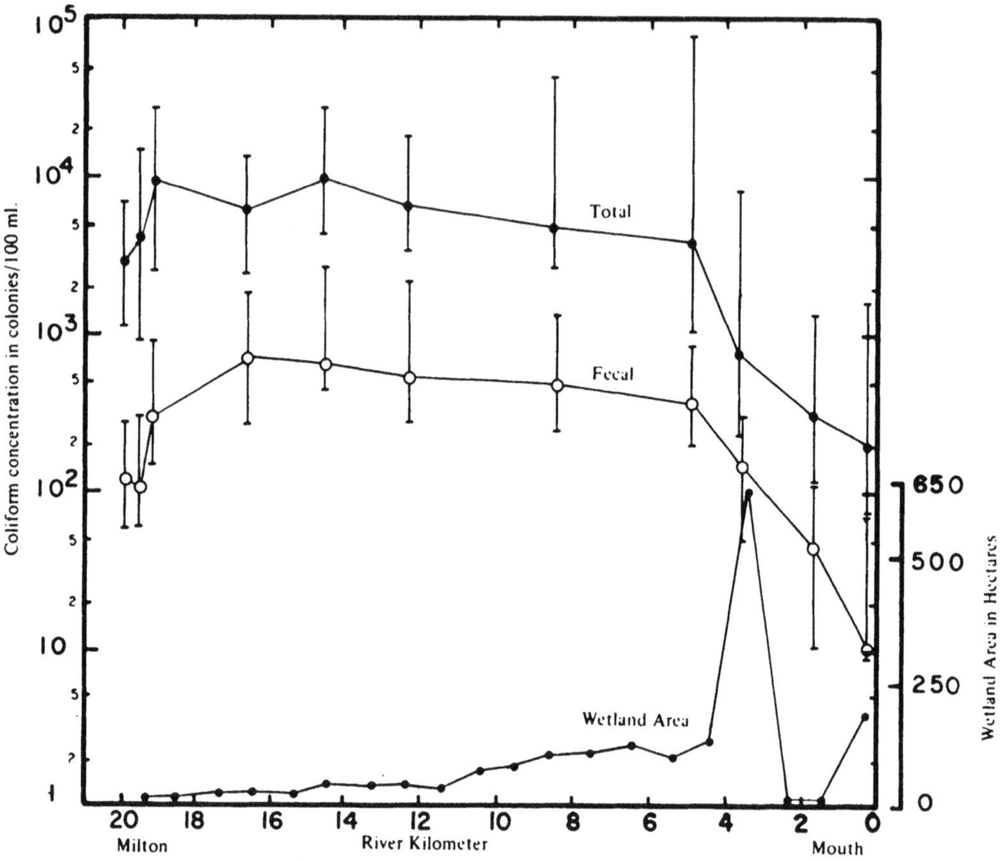

Figure 2. Broadkill River coliform concentrations and wetland areas

of Canary Creek marsh near the mouth of the Broadkill River. The Canary Creek marsh is approximately 186 hectares and has a mean tidal range of 1.1 meters. The results indicate a very strong correlation between tide height and direction and coliform bacteria concentrations. When the net transport over the 24-hour period was summed, it resulted in approximately 9 x 10^{12} total and fecal colonies transported into the Broadkill River for that day (Jensen et al., 1977). In sampling two other marshes for two separate tidal cycles each, the net transport results were inconclusive, although high coliform levels (>1000 col/100 ml) were observed at low tide in the morning and much lower levels at high tide.

It was deduced that although wetlands may have a strong effect on the tidal river coliform concentrations, there was not a simple relation between adjacent wetland area and coliform bacteria

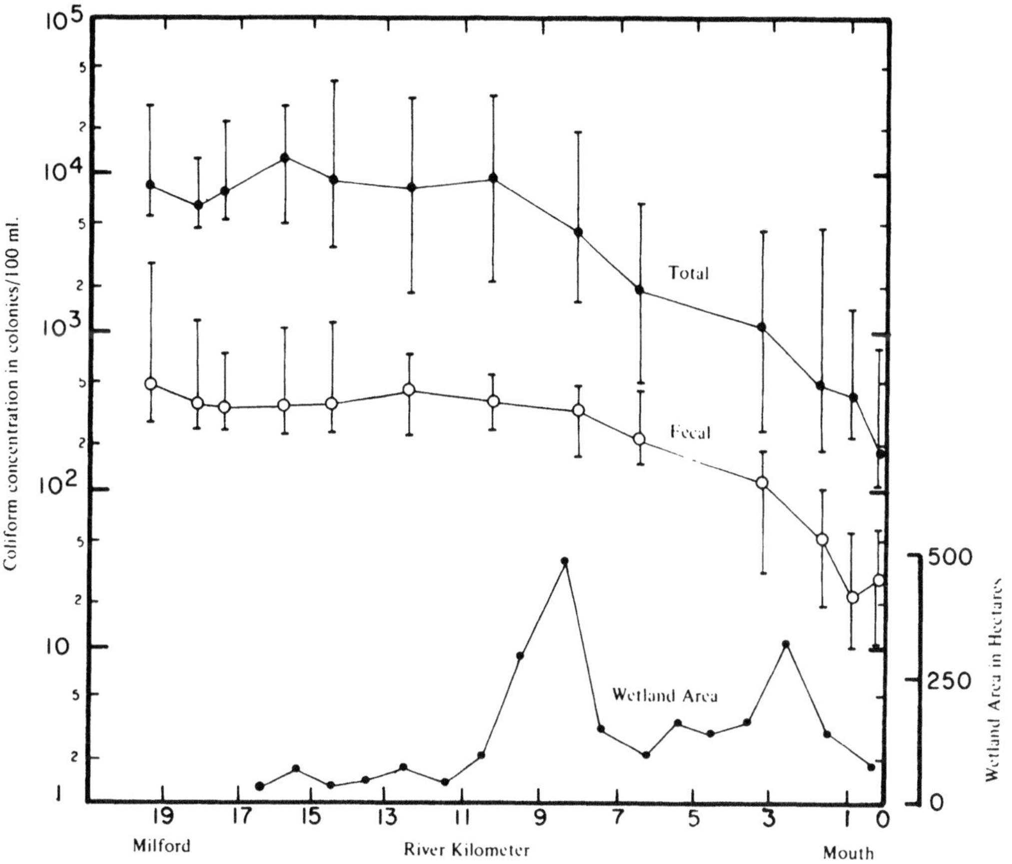

Figure 3. Mispillion River coliform concentrations and wetland areas

exported. Figures 2 and 3 illustrate this point.

The coliform organisms, particularly fecal, should, in theory, be of mammalian or avian origin. However, the biomass of these forms on a tidal marsh is small relative to areas traditionally considered as sources of fecal contamination (e.g., pasture lands). Because of this it is not likely that wetlands are acting as a simple direct source of bacterial contamination. It is hypothesized that tidal wetlands create an environment suitable for the increased survival or even growth of coliform organisms. To understand possible wetland effects, it is useful to review factors that may affect coliform bacteria survival. Several researchers (Carlucci and Pramer, 1960; Orlob, 1956; Rittenberg et al., 1958; Vaccaro et al., 1950; Won and Ross, 1973) have observed that addition of organic substrates and/or inorganic nutrients to enteric bacterial

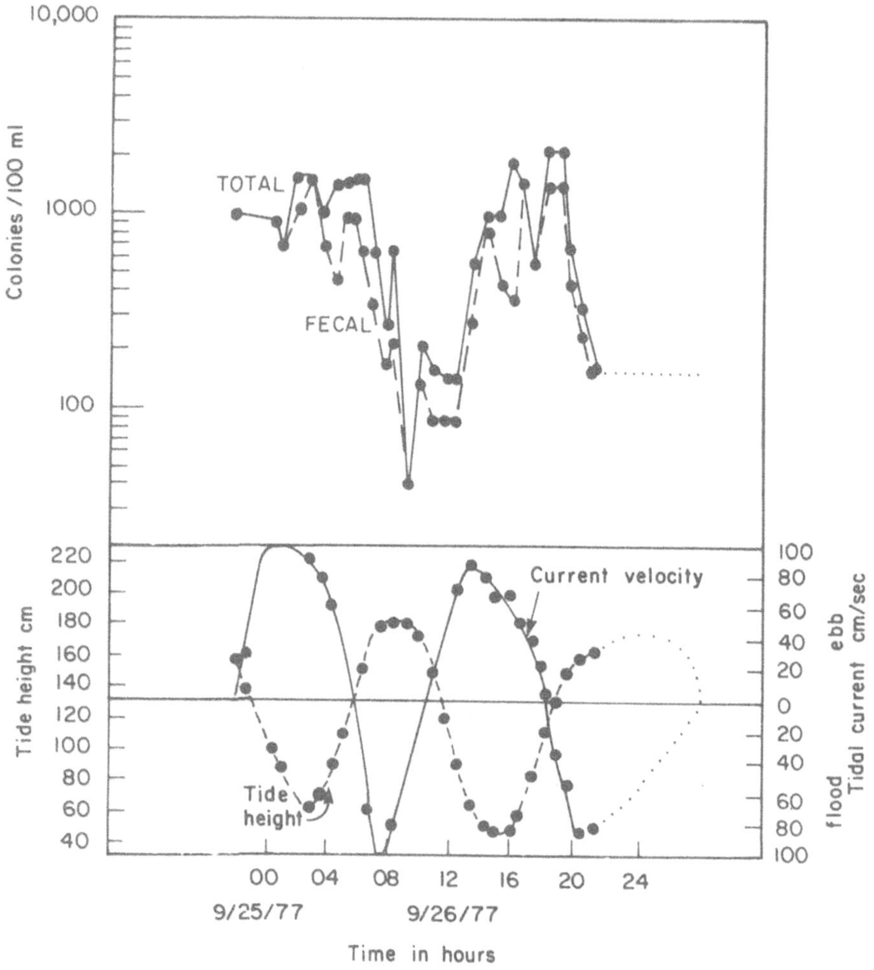

Figure 4. Coliform concentrations at the mouth of Canary Creek Marsh

populations in seawater have led to decreases in the die-off rate, and under certain circumstances the growth, of coliform organisms. Sediment resuspension is also known to cause bacterial increases in the water column in shallow waterbodies (Carney et al., 1975; Gerba et al., 1976). It is well documented (Matson et al., 1978; Rittenberg, 1958; Weiss, 1951; Faust et al., 1975) that coliform organisms tend to absorb onto and persist in the sediments. Levels in the sediments are often 1,000 times higher than in the water column. In shallow waterbodies, sediment resuspension can be a significant factor, especially in light of the findings by Van Donsel and Geldrich cited by Goyal et al. (1977) that coliforms tend to accumulate in the upper few centimetres of sediment. Another important factor affecting bacterial survival or die-off rate is light intensity. Gammeson and Gould (1974) conducted coliform (E. coli.) die-off experiments for several ocean outfalls in the United Kingdom. In addition to the expected scatter in their data, they found a strong relationship between average light intensity and the time required for a 90% reduction in population. Bellaire et al. (1977) also found a similar relation in Australian ocean outfall studies using fecal coliform procedures. Other factors that may affect coliform survival are salinity (Faust, 1975; Carlucci and Pramer, 1959; Goyal et al., 1977; Mitchell and Chamberline, 1974; and Orlob, 1956) and temperature (Canale, 1973; Faust, 1976). Finally, tide stage and amplitude might be expected to affect a coliform observation at a location where the longitudinal concentration gradient is strong.

Tidal marshes affect several of these parameters. The ability of tidal wetlands to export dissolved and particulate organic matter as well as inorganic nutrients to their adjacent waterways has been well documented (Axelrad, 1974; Daiber et al., 1976; Heinle and Flemer, 1976) and, to an extent, quantified (Jensen and Tyrawski, 1978). In addition, sediments in marsh creeks tend to be highly organic and easily resuspended, resulting in high turbidity of marsh waters relative to open bay waters. This relatively high turbidity (Secchi disc readings are typically 15-50 cm in both tidal rivers) significantly reduces depth-averaged light intensity over what might be expected in an open estuary or coastal waters.

To test the hypothesis that tidal wetlands affect coliform survival concentrations through the previously mentioned mechanisms, a straightforward statistical analysis was performed. Additional objectives were to determine which factors would be most useful in predicting coliform levels and whether the independent variables employed could account for a significant component of the total variance observed.

PROCEDURE

A statistical analysis of coliform data collected by the State Division of Environmental Control was performed using the following independent variables: Tide stage, tidal amplitude, chloride concentration, freshwater flow, time since last rain over 0.5 inches in a day, organic nitrogen concentration, five-day biochemical oxygen demand (BOD_5), turbidity, water temperature, and light intensity. The data consisted of 18 (Broadkill River) and 19 (Mispillion River) observations at monthly intervals (April-October) for three years. Due to the large data set, only 5 stations spaced along each river were analyzed statistically.

Tide stage and amplitude were obtained from tide table predictions. Tide stage was entered as a number between 0 and 6, with 6 being low water slack and 0 high water slack. Tide stage for upriver stations was computed using the tide wave velocity of 1.1 m/s calculated by DeWitt (1968). Tide amplitude was taken as the highest predicted water level height within 12 hours preceding sample collection. Freshwater flow was entered directly as the flow at USGS gauging station in each basin. In the Broadkill River, Sowbridge Branch near Milton (01484300) was used, while Beaverdam Branch at Houston (01484100) was employed on the Mispillion River. Chloride and organic nitrogen concentration, turbidity and water temperature were taken directly from the state laboratory analysis sheets. Average light intensity for the day prior to sample collection was obtained from Atlantic City, New Jersey percent possible sunshine data, adjusted for latitude, and entered as Langleys per day using an empirical method developed by Hamon et al. (1954). The day before collection was used because with morning sample collection, light on the day of observation would have relatively little effect. Time (days) since last rain was counted from rainfall observations at Lewes, near the mouth of the Broadkill River.

Each station's data was analyzed using the University of Delaware's DEC-10 computer and a statistical analysis package called MINITAB. Pearson product-moment correlation coefficients were computed for all variables, and a multiple regression analysis was performed.

RESULTS

Correlations between total and fecal coliform and the independent variables are presented in Table 1. A great deal of variability exists between stations and even between results obtained for total and fecal at the same station. There are also systematic differences between rivers. Because of the noisy data, few conclusions can be reached with statistical confidence. To aid in the interpretation of the correlation coefficient, a 95% confidence of significance requires an R value of 0.485 (0.47 for the Broadkill with 19 observations). An 80% confidence requires an R value of 0.315 (0.315 for the Broadkill).

Organic nitrogen exhibited the expected reasonably high positive correlations with the notable exception of a few stations. BOD_5 proved to be a less useful indicator of organic substrate than organic nitrogen. Light intensity exhibited a negative correlation at most stations, but the correlation was significant in a smaller number of stations. Water temperature has the same seasonal pattern

TABLE I

CORRELATION COEFFICIENTS[1] BETWEEN TOTAL AND FECAL COLIFORM AND INDEPENDENT VARIABLES

Broadkill River

	STATION				
	0	A	B	C	D
River Kilometer	.2	1.8	3.6	5.0	8.5
Parameter					
Fecal Coliform	.81	.96	.78	.72	.77
Tide Stage	.05/-.09	-.08/-.02	.12/-.05	.35/.25	.62/.32
Tide Amplitude	.21/.29	-.20/-.21	-.23/-.24	-.30/-.29	-.44/-.31
Chloride	-.00/2.1	.04/.01	-.31/-.01	-.41/-.28	-.38/-.23
Freshwater Flow	.03/-.19	.41/.45	.46/.47	.45/.26	.30/.18
Time Since Rain	-.36/-.04	-.33/-.36	-.53/-.40	-.51/-.49	-.51/-.46
Org-Nitrogen	.17/.28	-.45/-.37	-.16/-.13	.60/.61	.61/.69
BOD_5	.21/.33	-.04/.02	-.18/.07	.54/.29	.32/.13
Turbidity	-.25/-.16	.03/-.00	-.08/-.04	.08/-.21	.13/.13
Temperature	.21/.24	-.27/-.32	-.08/-.33	.43/.06	.24/-.08
Light	-.10/.21	-.57/-.59	-.42/-.60	-.01/-.29	.00/-.24

Mispillion River

	STATION				
	0	A	C	E	G
River Kilometer	.2	1.8	6.4	10.3	14.5
Parameter					
Fecal Coliform	.22	.16	.80	.60	.64
Tide Stage	.11/-.09	.15/.02	-.10/-.23	-.01/-.07	-.00/.25
Tide Amplitude	.06/.42	.22/.25	-.22/-.22	-.02/-.21	-.04/.03
Chlorides	-.53/.17	-.52/-.02	.51/.01	.10/.19	-.12/.06
Freshwater Flow	.31/-.05	.44/-.26	-.39/-.25	-.32/-.32	-.05/-.25
Time Since Rain	-.34/-.25	-.32/-.21	.04/.07	..09/.09	.13/-.11
Org-Nitrogen	.05/.54	.22/-.08	.18/.27	.12/.45	.50/.62
BOD_5	-.38/-.25	-.09/-.50	.11/.16	-.17/.15	.30/.44
Turbidity	-.22/-.17	-.08/-.23	-.19/-.14	-.38/-.18	-.22/.17
Temperature	-.34/.09	-.23/-.13	.01/.29	-.04/.36	.27/.46
Light	-.08/.09	.08/-.06	-.25/-.16	-.30/-.15	-.11/.14

[1]R total coliform/R fecal coliform

as light intensity, and although correlated with light intensity, was relatively poorly correlated with coliform concentrations.

It was expected that turbidity would be positively correlated with coliform concentration. This was not the case. Although turbidity does not correlate well with coliform concentration, as one would expect from the discussion on sediment resuspension, the results of this study are in accord with those of Goyal et al. (1977). Goyal also found little positive relation between turbidity and coliform levels on finger canals near Galveston, Texas. In both Delaware rivers, turbidity levels were 3 to 5 times higher than in nearby open bay areas (Delaware and Rehoboth) where coliform levels are quite low. There would appear to be a relation between turbidity and coliform levels from different geographic areas but not for the data from an individual station.

A linear multiple regression analysis also was conducted to obtain an estimate of the variance that could be explained with the available data. Table 2 presents the highest percentage variance (R^2) explained and the independent parameters found significant at an α level of 0.1 and 0.2. The α level is the probability of the coefficient in the regression equation being insignificant. Care was taken to avoid colinearity in the regression equations, which occurs when two independent variables are strongly correlated to each other. Colinear variables occurring in this study were tide stage and chlorides, temperature and light, and freshwater flow and time since last rain. BOD was not used in the regressions.

Table 2 reveals that, at best, only approximately half of the variance can be explained with a linear model of the independent parameters. Several logrithmic transforms of the data were performed but produced no significant improvement. However, when one considers the variance contribution of the MPN test itself, the percentage explained becomes more acceptable. Woodward (1957) concludes that the standard deviation of a log MPN observation can be estimated by $0.55/N^{1/2}$. For the 3 tube MPN's in these data, the standard deviation is 0.318 and variance 0.10. Since the actual variance of the log MPN observations at all stations ranged from 0.18 to 1.36, the component of the variance due to the MPN test itself ranges from 7.3% to 55%. Furthermore, errors in each of the independent variables must be considered. It is reasonable to expect, for example, that if light intensity were measured at the point of sample collection, and if actual tide records rather than predicted tides were employed, a significant improvement would result.

CONCLUSIONS

Although it is impossible to draw conclusions with certainty

TABLE II

REGRESSION ANALYSIS RESULTS

PARAMETERS SIGNIFICANT AT A GIVEN ALPHA LEVEL

Broadkill River

	Total Coliform				Fecal Coliform			
Station	α=.2	R^2	α=.1	R^2	α=.2	R^2	α=.1	R^2
0	Rain	12.9						
A	org-N light rain	53.6	org-N light rain	53.6	org-N light rain	51.2	org-N light rain	51.2
B	chloride light rain	43.2	rain light	39.7	fr. flow light	54.7	fr. flow light	54.7
C	org-N temp tide st rain	52.5	fr. flow temp	37.5	org-N tide ht light rain	53.9	org-N light	47.3
D	org-N tide st rain	64.2	org-N tide ht rain	56.0	org-N tide st rain	53.9	org-N	47.2

Mispillion River

	Total Coliform				Fecal Coliform			
Station	α=.2	R^2	α=.1	R^2	α=.2	R^2	α=.1	R^2
0	chlorides rain	35.6	chlorides	28.0	org-N tide ht	37.4	04g-N	28.9
A	chlorides rain	30.8	chlorides	26.8	rain	10.2		
C	chlorides fr. flow	30.9	chlorides	25.5				
E	turbidity fr. flow	21.7	turbidity	14.6	org-N fr. flow	33.1	org-N	20.3
G	org-N	25.3	org-N	25.3	org-N	37.8	org-N	37.8

from these data, some trends are suggested where laboratory studies and field data are in agreement. Light intensity and organic nitrogen both appear to be definite factors in the survival and ultimate concentration of both total and fecal coliform bacteria. Because tidal wetlands affect both of these parameters, it is a reasonable tentative conclusion that this mechanism is at least a partial explanation for the high observed coliform levels. Another potential mechanism for tidal wetlands influencing coliform levels is by sediment resuspension. However, the lack of correlation in these data between turbidity and stream coliform levels does not reinforce this possibility.

A statistical analysis in conjunction with an appreciation for analytical and measurement errors in the input data suggest that a dynamic model of total or fecal coliform concentration may be possible. Such a model would be complex, as a number of factors significantly impact coliform concentration. Also, the data requirements may be more rigorous than with typical models used for water quality management and waste load allocation. However, there are at least two areas in Delaware alone where the water quality issue is not low dissolved oxygen, but rather use of the waters for shellfishing. A "typical" model would obviously be inappropriate. The most significant factors in a dynamic coliform model would appear to be light intensity and penetration (needed to compute depth-averaged light intensity), and organic material concentration. Organic nitrogen would appear to be a good candidate but other parameters such as organic carbon, BOD or chemical oxygen demand may also serve adequately. Physical factors such as tide stage, salinity (chlorides) and rainfall would ordinarily be incorporated in a water quality model.

In view of the direct economic significance of coliform bacteria levels, two avenues of research should prove fruitful. One is a reexamination of total or fecal coliform standards used to regulate shellfish harvesting. These standards were developed well before 1950, based largely on empirical data developed from areas with significant inputs of poorly treated human wastes. While the standards used by the states under the National Shellfish Sanitation Program have been generally successful -- as judged by the extremely low incidence of disease from eating shellfish -- the standards may be unnecessarily rigorous where the bacterial source would appear to be largely of natural origin. Another research direction should be to continue to improve understanding of bacterial dynamics in natural waterways. Through this should come the ability to make better management decisions on an important aspect of water quality.

REFERENCES

Axelrad, D. M. 1974. Nutrient flux through the salt marsh ecosystem. Ph.D. Dissertation, College of William and Mary. 133 pp.

Bellaire, J. T., G. A. Parr-Smith and I. G. Wallis. 1977. Significance of diurnal variations in fecal coliform die-off rates in the design of ocean outfalls. Journal of Water Pollution Control Federation, 49, pp. 2022-2030.

Canale, R. P., R. L. Patterson, J. J. Gannon, and W. F. Powers. 1973. Water quality models for total coliform. Journal Water Pollution Control Federation 8. pp. 325-336.

Carlucci, A. F. and D. Pramer. 1960. An evaluation of factors affecting the survival of Escherichia coli in sea water. II - Salinity, pH, and nutrients. Applied Microbiology, 8, pp. 247-250.

Carney, J. F., C. E. Carty and R. R. Colwell. 1975. Seasonal occurrence and distribution of microbial indicators and pathogens in the Rhode River of the Chesapeake Bay. Applied Microbiology 30, pp. 771-780.

Daiber, F. C., V. A. Lotrich, and L. E. Hurd. 1976. Unpublished data from NOAA Project #04-5-158-22. Nutrient flux, energy flow production in salt marsh ecosystems in Delaware.

deWitt, W., III. 1968. The Hydrography of the Broadkill River Estuary. Master's Thesis, U. of Delaware.

Faust, M. A. 1976. Coliform bacteria from diffuse sources as a factor in estuarine pollution. Water Research 10, pp. 619-627.

Faust, M. A., A. E. Aotaky and M. T. Hargardon. 1975. Studies on the survival of Escherichia coli MC-6 in diffusion chambers in an estuarine environment. Unpublished manuscript. Smithsonian institution, Edgewater, Maryland.

Gameson, A. L. H. and D. J. Gould. 1974. Effects of solar radiation on the mortality of some terrestrial bacteria in sea water. Proceedings of the International Symposium on the Discharge of Sewage from Sea Outfalls. Aug. 127-Sept. 2. Pergamon Press. NY., pp. 209-219.

Gerba, C. P. and J. S. McLeod. 1976. Effect of sediments on the survival of Escherichia coli in marine waters. Applied Environmental Microbiology, 32, pp. 114-120.

Goyal, S. M., C. P. Gerba, and J. L. Melnick. 1977. Occurrence and distribution of bacterial indicators and pathogens in canal communities along the Texas coast. Applied and Environmental Microbiology, 34. pp. 139-149.

Hamon, R. W., L. L. Weiss and W. T. Wilson. 1954. Insolation as an empirical function of daily sunshine duration. Monthly Weather Review, 82.

Heinle, D. R. and D. A. Flemer. 1976. Flows of materials from poorly flooded tidal marshes and an estuary. Marine Biology, 35(4), pp. 359-373.

Jensen, P. A. and J. M. Tyrawski. 1978. Wetlands and water quality. Presented at the March 14-16, 1978 Coastal Zone 78 Conference held at San Francisco, California.

Jensen, P. A., and J. M. Tyrawski. 1977. Water quality modeling and analysis. Report on Task 2359. For the Coastal Sussex Water Quality Program, by the College of Marine Studies, and College of Agricultural Science, Univ. of Delaware.

Jensen, P. A., W. Ritter, and J. M. Tyrawski. 1977. Coliform bacteria loadings and dynamics. Report on 23596. The College of Marine Studies and Agricultural Sciences, Univ. of Delaware.

Karpas, R. M. and P. A. Jensen. 1977. Hydrodynamics of Coastal Sussex County Estuaries. Report on Task 2331, by the College of Marine Studies and Agricultural Sciences, Univ. of Delaware.

Matson, E. A., S. G. Horner, and J. D. Buck. 1978. Pollution indicators and other micro-organisms in river sediment. Journal Water Pollution Control Federation, 50, pp. 13-19.

Maurer, D., L. Watling, and R. Keck. 1971. The Delaware oyster industry: a reality? Transactions of the American Fisheries Society 100, pp. 100-111.

Mitchell, R. and C. Chamberlain. 1974. Factors influencing the survival of enteric micro-organisms in the sea: an overview. Proc. of the Int. Symp. on the Discharge of Sewage from Sea Outfalls. 27 Aug - 2 Sept. 1974. Pergamon Press, pp. 237-251.

Orlob, T. G. 1956. Viability of sewage bacteria in seawater. Sewage and Industrial Wastes, 28. pp. 1147-1167.

Rittenberg, S. C., T. Mittwer and D. Ivler. 1958. Coliform bacteria in sediments around three marine sewage outfalls. Limnology and Oceanography, 3, pp. 1010-108.

Ritter, W. F. and G. Sheffler. 1977. Monitoring of Non-Point Source Pollution in Coastal Sussex County. Report on Task 2332 of the Coastal Sussex Water Quality Program. College of Marine Studies and College of Agricultural Sciences, University of Delaware.

Standard Methods for the examination of Water and Wastewater. 1971. 13th Edition, APHA, Washington, D. C.

Vaccaro, R. F., M. P. Briggs, C. L. Carey, and B. H. Ketchum. 1950. Viability of Escherichia coli in sea water. American Journal of Public Health, 40, pp. 1257-1266.

Weiss, C. M. 1951. Adsorption of E. coli river and estuarine silts. Sewage and Industrial Wastes, 12, pp. 227-237.

Won, W. D. and H. Ross. 1973. Persistence of virus and bacteria in sea water. Journ. of Env. Eng. Div., ASCE. 99(EE3), Proc. Paper 9781. pp. 205-211.

Woodward, R. L. 1957. How probable is the most probable number? Journal American Water Works Association. (August), pp. 1060-1068.

RATE OF SEDIMENTATION AND ITS ROLE IN NUTRIENT CYCLING IN A LOUISIANA SALT MARSH

R. D. DeLaune and W. H. Patrick, Jr.

Laboratory for Wetland Soils and Sediments, Center for Wetland Resources, Louisiana State University, Baton Rouge, Louisiana, 70803

ABSTRACT

The Gulf Coast salt marshes in the deltaic plain of the Mississippi River are in a rapidly subsiding zone where accretion processes are important for maintenance of the marsh surface within the intertidal range. Incoming sediment is essential for maintaining the marsh surface and for supplying nutrients for plant growth. In an area that is apparently maintaining its surface with respect to sea level, ^{137}Cs dating shows an accretion rate of 1.35 cm/yr. In an adjacent deteriorating marsh the sedimentation rate is 0.75 cm/yr, not enough to compensate for subsidence. The incoming sediment also is a major source of plant nutrients for *Spartina alterniflora*, with inputs as great as 231, 23.1 and 991 kg/ha of nitrogen, phosphorus and potassium, respectively. Mineralization of these nutrients in the sediment provides a significant portion of the plant's requirements, but growth of salt marsh plants is still limited by available nitrogen, as addition of nitrogen fertilizer confirms.

INTRODUCTION

Louisiana contains a vast and very productive wetland ecosystem. Features of its coastal wetland are related to the geological history of the Mississippi River. Over the past several thousand years frequent flooding of the distributaries over their natural levees resulted in the formation of broad expanses of marshland near the coast. The tidal range within these salt marshes are lower than their counterparts on the Atlantic coast and marsh inundation is generally influenced more by winds than tides. Marsh relief is extremely low and is characterized by slightly elevated natural levees adjacent to stream and water-bodies, which gradually slope into inland depressions where the dominant marsh grass Spartina alterniflora is sparse or completely absent.

Over historic time, land building exceeded erosion and subsidence, resulting in the formation of a large deltaic plain. However, artificial leveeing of the Mississippi River over the past century has prevented distributary switching and annual overbank flooding. Consequently, Louisiana salt marshes are not presently receiving a direct supply of fluvial sediment sufficient to counteract rapid subsidence. Louisiana is currently losing approximately 16 square miles of its coastal wetlands per year primarily to subsidence, (Gagliano and Van Beck, 1970). The marsh surface maintains an approximate elevation with respect to sea level datum by continual accumulation of dead plant material and by entrapment and stabilization of organic detritus and some nutrient enriched mineral sediment. Nutrient cycling in these salt marshes is closely interrelated with these geological processes. Nitrogen cycling is of special interest since nitrogen has been found to be the limiting nutrient for growth of Spartina alterniflora (Patrick and DeLaune et al., 1976; Buresh et al., 1979).

This paper is a review of several previous papers by the authors dealing with sedimentation and nutrient cycling in salt marsh ecosystems in Louisiana's Barataria Basin.

Rates of Sedimentation and Vertical Marsh Accretion

^{137}Cs dating has recently been used to document sedimentation rates in Louisiana's rapidly accreting salt marshes (DeLaune et al., 1978). In this study duplicate cores were taken from streamside and inland Spartina alterniflora salt marsh and from an adjoining shallow water lake (Airplane Lake) in Louisiana's Barataria Basin (29°13'N, 90°7'W). ^{137}Cs activity in each core was determined from oven dried, 3 cm sections of the profile using a lithium drifted germanium detector and multichannel analyzer. ^{137}Cs is a fallout product of nuclear testing and does not occur naturally in the environment. Accretion rates were calculated from peak ^{137}Cs concentrations found in the marsh profile. These concentrations can be correlated to 1963, the year of peak ^{137}Cs fallout and 1954, the first year of significant ^{137}Cs fallout (Pennington et al., 1973). Profile distribution of ^{137}Cs showed rapid vertical marsh accretion (Figure 1). Parts of the marsh nearer the stream were accreting at a rate of 1.35 cm/yr compared to 0.75 cm/yr for an adjoining inland marsh. The inland marsh is beginning to deteriorate into small open water

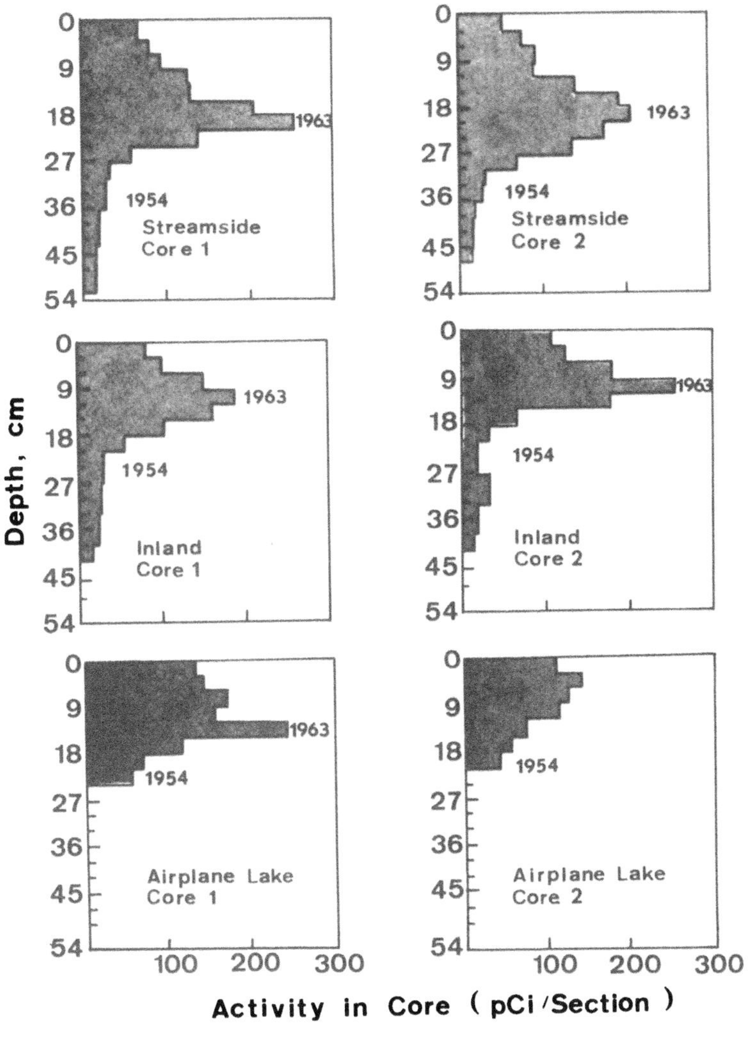

Figure 1. Profile distribution of ^{137}Cs

areas, presumably because its vertical accretion rate is not able to compensate for subsidence and eustatic sea level rise. The distribution of ^{137}Cs in the lake cores indicates accretion rates of 1.1 cm/yr. Subsidence in this area based on tide gauge measurements is estimated to be greater than 1 cm/yr (Swanson and Thurlow, 1973).

Density and carbon analyses of the sediment cores used in the ^{137}Cs study shows that the marsh vertically accretes through organic detritus accumulation and sediment input (DeLaune et al., 1978). Density of marsh soils ranged between 0.10 g/cm^3 and 0.30 g/cm^3 depending on depth and location. As would be expected, a large portion of the volume in these low density, rapidly accreting marsh soils is occupied by water and entrapped gases. On a dry weight basis, organic matter generally accounts for less than 20% of total marsh soil solids. However, organic matter plays a very important role in providing structural support to these soils, for volume-wise organic matter may nearly equal the mineral fractions, which consist mainly of silt and fine clays material. (Figure 2). A characteristic of these soils is that they shrink considerably upon drying, but when rewetted do not expand back to their original volume.

Amount of Nutrients Supplied through Sedimentation Processes

Nitrogen, phosphorus, and potassium content of sediment retained in sediment traps placed on the marsh at each location where ^{137}Cs dating cores were taken, show that sedimentation can be an important source of nutrients (DeLaune et al., 1980). Shallow porcelain pans (37 cm length x 23 cm width x 6.5 cm depth) were used for capturing sediment. The pans were pushed into the marsh allowing the sides of the pan to extend 3 cm above the surface. The traps were left on the marsh for 3 months (March through May). The nitrogen, phosphorus, and potassium concentration in the sediment retained by the traps was 6880 μg/g, 680 μg/g and 3040 μg/g, respectively. (There was little difference between nutrient content of sediment trapped at streamside and inland sites.) Marsh accretion rates, soil parameters and nutrient concentrations were used to calculate the possible contribution of nutrients to the marsh through sedimentation processes.

The equivalent of 231 kg/ha/yr of nitrogen and 23.1 kg/ha/yr of phosphorus was being supplied to the streamside marsh. Most of the nitrogen was organic nitrogen, which must be mineralized before it can be taken up by plants. Due to low mineralization of organic nitrogen under anaerobic marsh soil conditions probably less than 10 percent would be made available during the first year. Each succeeding year additional nitrogen would be mineralized, however. Sedimentation was also supplying a large amount of nutrients to the inland marsh. The phosphorus concentration of incoming sediment at both the streamside and inland marsh was greater than the average phosphorus concentration in the local sediment profile. Regression analysis of phosphorus in the soil profile showed a decrease in phosphorus with depth, interpreted as resulting from removal through plant uptake.

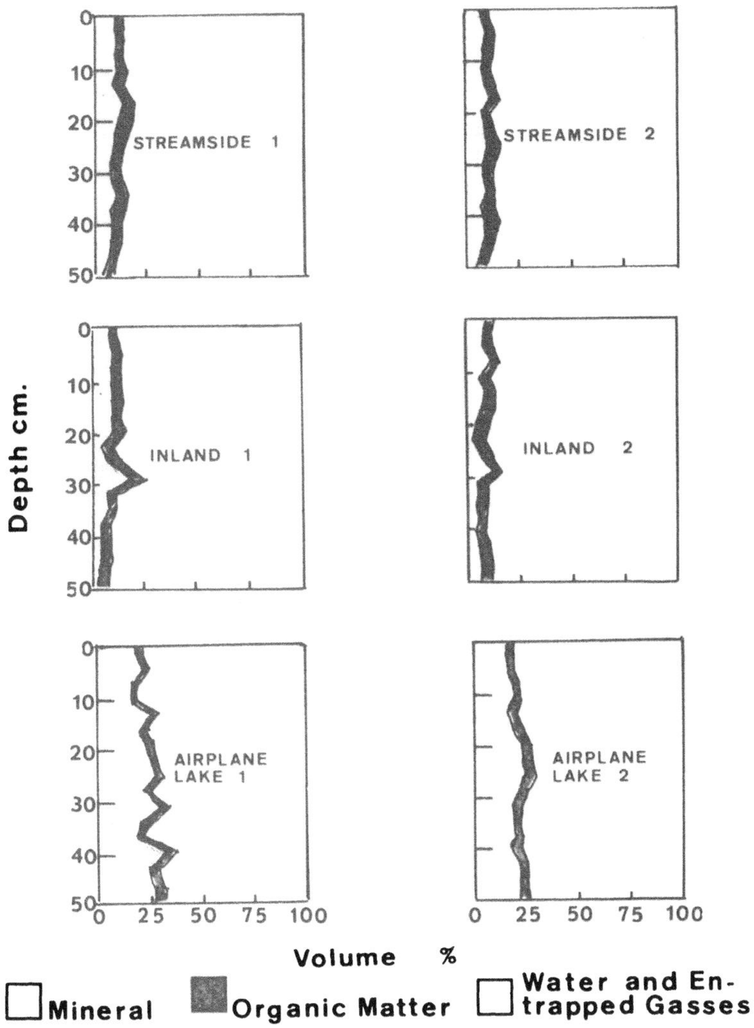

Figure 2. Volume percent of mineral and organic matter in soil and sediment profiles

Sediment deposited on the surface of streamside and inland marshes also contained more manganese, cadmium, copper, and zinc than did sediment in lower sections of the soil profile (DeLaune et al., 1980). Possibly these elements are also being taken up by S. alterniflora and released into the ecological food web. Even though the metal content of incoming sediment was greater than that in the soil profile, the concentrations were indicative of non-polluted sediment.

Sedimentation and Plant Growth

The nutrient content of the *Spartina alterniflora* salt marsh sediment may be quite high on a total dry mass basis, but rather low on a volume basis, especially in soils containing small quantities of inorganic sediment. Data expressed on a volume basis was found to be more meaningful in relating plant growth to soil nutrient concentrations (DeLaune et al., 1979). In this study, neither extractable phosphorus, potassium, magnesium, calcium, sodium, or total nitrogen expressed on a dry weight basis (µg/g) was significantly related to growth of *S. alterniflora* as measured by standing crop biomass. However, each of these constituents when converted to a volume expression (µg/cc) was positively correlated with plant growth.

The amount of inorganic sediment deposited on the marsh, as reflected by a greater soil density, was shown to be directly related to increased *S. alterniflora* biomass (Figure 3). Soil densities of less than 0.20 g/cm^3 in these salt marshes will not support growth of *S. alterniflora*. Soil density decreases with increasing distance from the stream and is paralleled by a corresponding decrease in plant biomass (DeLaune et al., 1979). The taller grass in the streamside region produced dry weight yields slightly above 2000 g/cm^2. Soil density values in this study were quite low and ranged from approximately 0.10 g/cm^3 to 0.49 g/cm^3. The content of soil organic matter does not account for the variations in density; rather, the density of these marsh soils is directly related to the quality of mineral material per volume of soil. The higher soil density observed in the streamside marsh reflects a greater input of nutrient rich suspended sediments to the natural levee adjacent to the stream, than to the marshes further inland.

Accumulation of Nutrients in the Soil Profile

Accumulation rates for selected elements (Table 1) were determined from sedimentation rates and plant nutrient and heavy metal concentrations in the sediment profile (DeLaune et al., 1980). The values reported, represent sediment chemical composition <u>after</u> the sediment has been exploited by plant roots and subjected to possible loss through biological and chemical transformations.

The marsh is undoubtedly a sink for nitrogen. Nitrogen was shown to be accumulating at rates of 210 kg/ha/yr, 134 kg/ha/yr, and 153 kg/ha/yr in the streamside marsh, inland marsh, and Airplane Lake, respectively. The original source of most of this accumulated nitrogen is from deposition of nitrogen-enriched organic sediment. Some of the nitrogen is mineralized to ammonium and utilized by

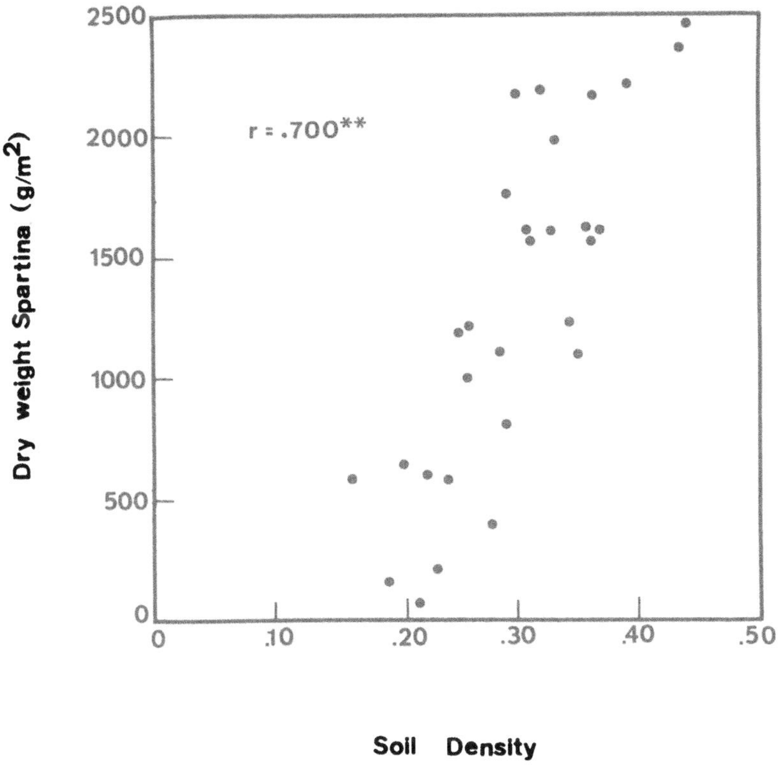

Figure 3. Correlation of soil density with aboveground biomass of Spartina alterniflora.

S. alterniflora in the streamside and inland marshes. Nitrogen fixation undoubtedly also contributes to nitrogen accumulation, although Casselman (1979) found fixation to be a less important process than sedimentation. A large portion of the inorganic nitrogen taken up by S. alterniflora is returned to the marsh in organic form. Decomposition of dead plant material is slow in these soils because they remain anaerobic throughout the year.

Based on accretion rates and carbon measurements, organic carbon is accumulating at rates of 3930 kg/ha/yr, 2370 kg/ha/yr, and 2310 kg/ha/yr in the streamside marsh, inland marsh, and Airplane Lake, respectively. Using the factor 1.724 to convert organic carbon to organic matter (Wilson and Staker, 1932), the equivalent of 6780 kg/ha/yr, 4090 kg/ha/yr, and 3980 kg/ha/yr of organic matter is accumulating. Most of the organic matter exists as an approximately one meter thick layer of peat overlying recent Mississippi alluvial mineral sediment.

Table 1. Accumulation of selected elements and mineral matter

	kg/ha/yr[1]		
	Streamside	Inland	Airplane Lake
N	210.00	134.00	153.00
P	16.50	7.50	32.20
K	991.00	324.00	1930.00
Fe	598.00	199.00	1327.00
Mn	3.70	1.30	15.00
Cd	0.06	0.03	0.12
Pb	0.90	0.35	1.70
Cu	0.60	0.20	0.81
Zn	2.10	0.70	4.00
C	3930.00	2370.00	2310.00
Mineral matter	27250.00	11110.00	58560.00

Net production of *S. alterniflora* in the marsh surrounding Airplane Lake was estimated to be equivalent to 5060 kg carbon/ha/yr and productivity by phytoplankton in the lake itself was 1980 kg carbon/ha/yr (Stowe et al., 1971). This productivity estimate for the marsh is conservative because productivity of belowground plant biomass was not included. When plant production is compared with the amount of carbon accumulating in the system, it is obvious that a significant portion of the carbon from primary production is remaining on the marsh or is being exported and accumulating in the lake sediment.

Phosphorus is accumulating at rates of 16.5, 7.5, 32.2 kg/ha/yr in the streamside marsh, inland marsh, and Airplane Lake, respectively. Sedimentation processes are supplying adequate amounts of phosphorus for plant growth. A portion of the phosphorus associated with the incoming sediment subsequently becomes available to *S. alterniflora*.

Large amounts of potassium are accumulating in the system. Levels of potassium present in the marsh soil are adequate for plant growth. The streamside marsh is accumulating more potassium than the inland marsh because of a faster accretion rate and greater deposition of mineral material (potassium is associated with the mineral component of the sediments). Sediments in Airplane Lake, with greater densities and more mineral material than the marsh soil, are accumulating potassium at a rate of 1930 kg/ha/yr.

Greater iron and manganese accumulation occurred in locations receiving larger mineral sediment inputs and correspondingly greater plant growth. In addition to supplying nutrients, added sediment can possibly interact with other factors to influence plant growth. For example, iron in the soil mineral component can help neutralize plant toxins such as sulfide and modify the nature of the root rhizosphere, thus influencing plant growth. Total sulfide levels on the order of several hundred μg/g of sediment have been measured in these soils (DeLaune et al., 1976). Only small amounts of manganese were accumulating in the streamside and inland marsh. Manganese content of the sediment in these organic marsh soils was low compared to concentrations in recent Mississippi alluvial sediment, the original source of sediment in the salt marsh. There is evidence that manganese may be unstable in these wetland systems due to its redox and solubility properties.

Effect of Added Nutrients

Even though mineralization of nutrients supplied by sedimentation provides a significant portion of plant growth requirements, nitrogen fertilization experiments indicate that growth is still limited by nitrogen availability. Supplemental labelled ammonium nitrogen applied in the spring increased aboveground biomass of *Spartina alterniflora* at a streamside marsh by 15% (Patrick and DeLaune, 1976). The added nitrogen caused a yield increase equivalent to 2500 kg/ha. Practically all of the inorganic nitrogen in a reduced soil is found in the ammonium form. Nitrate is ineffective as a nutrient because it is subject to rapid denitrification under reducing soil conditions. Not only was productivity increased by the added nitrogen, but the content of nitrogen in the plant material was also increased. The use of labelled nitrogen made it possible to distinguish between plant nitrogen derived from the sediment and the added fertilizer. The amount of plant nitrogen derived from the sediment was about 59% during June and July and increased to about 69% by September. The added nitrogen supply was depleted toward the end of the growing season, probably through nitrification-denitrification reactions as well as prior plant uptake. Only 29% of the 200 kg/ha of added nitrogen was recovered in the aboveground portion of the plants in September.

A similar experiment in an inland marsh receiving less mineral sediment showed a greater response to added nitrogen (Buresh, 1978). The addition of 200 kg/ha of labelled nitrogen in May significantly increased total aboveground plant biomass and plant height by 28% and 25%, respectively during the growing season. The increase in plant biomass was almost twice the increase observed from the addition of an equal amount of added nitrogen at the streamside location. The inland marsh sediment supplied approximately 50% of the nitrogen taken up by the plants over the growing season as compared to 69% for sediments from the streamside marsh. These fertilization experiments confirm that inland marsh soils, receiving minimal nitrogen enriched sediment, supply less nitrogen for plant growth than streamside marsh soils which receive larger amounts of nitrogen through sedimentation.

There was no increased growth of S. alterniflora in response to the addition of phosphorus at the rate of 200 kg/ha/yr at either the streamside or inland marsh. However, the concentration of phosphorus in the plant tissue increased about 20% from these additions of inorganic phosphorus. The increase in phosphorus uptake represented a very small fraction (about 1%) of the added phosphate.

CONCLUSION

The entrapment and stabilization of suspended sediment is a very important process in Louisiana's salt marshes. Sediment input contributes to vertical marsh accretion which is important in keeping the marsh intertidal. This process compensates for rapid subsidence due to compaction of recent Mississippi River alluvial sediment underlying the marsh surface. Incoming sediment also supplies essential nutrients for plant growth, which in turn enhance further sediment entrapment and stabilization. The increased primary production contributes to the organic pool of these peaty soils. Maintenance of a viable marsh is thus affected through aggradation process of plant growth, organic detritus accumulation and inorganic deposition.

Most water management projects--hurricane and flood control levees, spoil embankment, dredging of canals--reduce the sediment load to affected marshes. These practices can be expected to decrease vertical marsh accretion, soil density, and plant growth, and subsequently lead to more rapid marsh deterioration.

ACKNOWLEDGMENT

This paper is a result of research sponsored by NOAA Office of Sea Grant, U. S. Department of Commerce.

REFERENCES

Buresh, R. J. 1978. Nitrogen transformation and utilization by Spartina alterniflora in a Louisiana salt marsh. Louisiana State University, Baton Rouge, La.

Casselman, M. E. 1979. Biological nitrogen-fixation in a Louisiana Spartina alterniflora salt marsh. M.S. Thesis, Louisiana State University, Baton Rouge, Louisiana.

DeLaune, R. D., W. H. Patrick, Jr., and R. J. Buresh. 1978a. Sedimentation rates determined by ^{137}Cs dating. Nature 275:532-533.

DeLaune, R. D., W. H. Patrick, Jr., and J. M. Brannon. 1976. Nutrient transformation in Louisiana salt marsh soils. Sea Grant Publication No. LSU-T-76-009. Louisiana State University, Baton Rouge, Louisiana.

DeLaune, R. D., R. J. Buresh and W. H. Patrick, Jr. 1979. Relationship of soil properties to standing crop biomass of Spartina alterniflora in a Louisiana marsh. Est. Coastal Mar. Sci. 8:477-487.

DeLaune, R. D., C. N. Reddy and W. H. Patrick, Jr. 1980. Accumulation of plant nutrients and heavy metals through sedimentation processes and accretion in a Louisiana salt marsh (accepted for publication in Estuaries).

Gagliano, S. M. and J. L. Van Beck. 1970. Hydrologic and geologic studies of coastal Louisiana. Rep. No. 1 (Center for Wetland Resources, Louisiana State University).

Patrick, W. H., Jr., and R. D. DeLaune. 1976. Nitrogen and phosphorus utilization by Spartina alterniflora in a salt marsh in Barataria Bay, Louisiana. Est. Coastal. Mar. Sci. 3:59-64.

Pennington, W., R. S. Cambray and E. M. Fisher. 1973. Observation on lake sediment using fallout ^{137}Cs as a tracer. Nature 242:324.

Smith, W. D. 1943. Density of soil solids and their genetic relations. Soil Sci. 56:263-272.

Stowe, W. C., C. Kirby, S. Brkich, J. G. Gosselink. 1971. Primary production in a small saline lake in Barataria Bay, Louisiana. Louisiana State University, Coastal Studies Bulletin No. 6: 27-37.

Swanson, R. L. and C. I. Thurlow. 1973. Recent subsidence rates along the Texas and Louisiana coast as determined by tide measurements. J. Geophys. Res. 78:2665.

Wilson, B. D. and E. V. Staker. 1932. Relation of organic matter to organic carbon in the peat soils of New York. Journal of the American Society of Agronomy 24:477-481.

AN INFILTROMETER TO MEASURE SEEPAGE IN SALT MARSH SOILS

Roger Burke, Harry Hemond and Keith Stolzenbach

Ralph M. Parsons Laboratory for Water Resources and Hydrodynamics, Department of Civil Engineering, Massachusetts Institute of Technology, Cambridge, Massachusetts 02139

ABSTRACT

This study considers the design, construction, and operation of a device for measuring how much water seeps across the surface of salt marsh soil as it is inundated by flooding tides. The device was used in Great Sippiwisset Marsh, Cape Cod. Preliminary results confirm that infiltration during flood tide is followed by widespread exfiltration, of comparable magnitude, during ebb. Total water flux decreased with increasing distance from adjacent creeks.

The infiltrometer quantifies water flux across the sediment interface under relatively undisturbed conditions and represents a definite step forward in studies of the seepage component of salt marsh water budgets.

INTRODUCTION

A knowledge of how water moves across the peat surface of salt marsh soils is valuable to several areas of research, such as (1) development of a tidal circulation model for wetlands, (2) plant zonation, and (3) exchange of dissolved nutrients with outlying areas. A review of the literature reveals that surprisingly little work has been done on the seepage component of water flow through marsh systems.

It was thus necessary to find a new approach for this study. An in-situ device was desired, one that would measure infiltration during the entire time the flooding tidal water covered the marsh flats. The infiltrometer's operation was to be based on the following general principle: isolation of a column of water, duplicating the outside head conditions inside the isolated column, while bookkeeping on two quantities: namely (1) the water volumes required to maintain equal head conditions, and (2) the changes in water height of the isolated column. From these two quantities the flux across the section of marsh underneath the water column can be determined from the following equation.

$$Q_f = Q_p - H \cdot A \qquad (1)$$

where Q_f = water flux across the peat surface

Q_p = water volume pumped

H = change in height of water column

A = cross-sectional area of water column

In contrast to conventional ring infiltrometers used in terrestrial systems, (Rodda, 1976), which confine a column of soil, the device developed here confines only water. The underlying plot of soil remains more or less undisturbed; as part of the marsh continuum, it is subjected to the hydrologic factors actually controlling seepage. This approach has the advantage that the seepage measurements - having been collected under natural circumstances - correspond to the conditions to which they are to be applied.

DESCRIPTION OF INFILTROMETER APPARATUS

The infiltrometer which was developed is shown schematically in Figure 1. A column of water is isolated by a circular cylinder pushed into the peat a small distance. Any difference between inside and outside water levels is sensed by two weights, which control the operation of a bi-directional pump - the combination of the weight triggering mechanism and pump maintain equal head conditions. A flow recorder does the bookkeeping on water volumes transferred by the pump, while two stage recorders keep track of water height.

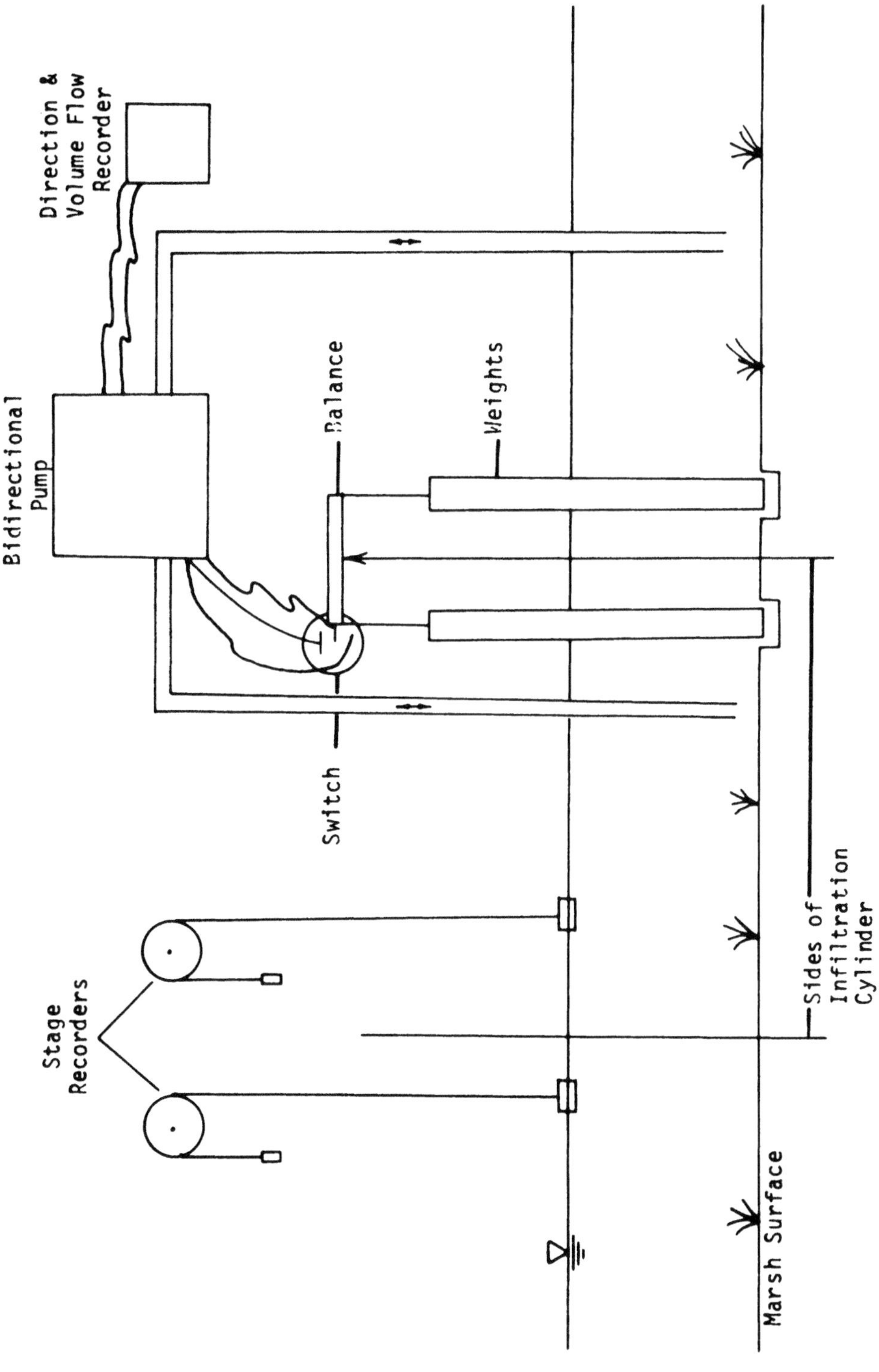

Figure 1. Schematic Representation of Infiltrometer

Mechanical Construction

The infiltrometer consists of five major parts: two equal weights hung from opposite ends of a lever arm that pivots about it's midpoint; a cylindrical barrel; two Stevens Type F Water Level recorders; a Master-flex peristaltic pump; and a 4-channel Rustrak event recorder. These components interact with one another as follows (Figure 2): a change in water height establishes a head difference across the infiltration cylinder, creating a torque that tips the lever arm from it's horizontal balance position. As the lever arm tips, one of the mercury switches strapped to either side of the lever arm's midpoint closes, thus turning on the peristaltic pump. Depending on the direction of tip, water is transferred either into or out of the cylinder until equal head conditions are again established. Rotation of the pump shaft causes a microswitch to alternately open and close, thus activating one pen of the Rustrak event recorder. Two water level recorders use floats to keep track of changes in water level inside and outside the infiltration cylinder (only one water level recorder is shown in Figure 2).

A PRELIMINARY SALT MARSH INFILTRATION STUDY

The infiltrometer was used to study seepage in Great Sippewissett Salt Marsh, located outside of Falmouth in Cape Cod. Great Sippewissett has an area of about .25 km^2, with a single entrance for all tidal water. The tide range is about 1.6 m. The marsh has an input of fresh groundwater, as evidenced by the lower salinity of the flooding seawater at ebbtide (Valiela et al., 1978).

Infiltration Measurements

Three different areas of the marsh were investigated (see Figure 3), each having it's own soil and vegetation characteristics. The relative positions of the sites within each area are shown in Figure 4. In the first area, the soil at site #1 contained a higher percentage of organic material (as estimated on the basis of texture) than the second site, and supported a denser crop of dwarf form *Spartina alterniflora*. The second area was considerably sandier than the first, with fewer *Spartina* plants per square foot. Measurements were made at varying distances from a tidal pool. In the third study area, measurements were taken at various points along a 110 meter stretch of creek bank. Sites 7 and 8 had soil and vegetation characteristics similar to those of area 2; site 9 was sandier, with only scant vegetation. Figure 5 summarizes the seepage measurements obtained from Great Sippewissett Marsh. From the data, three general patterns emerge. They include: (1)

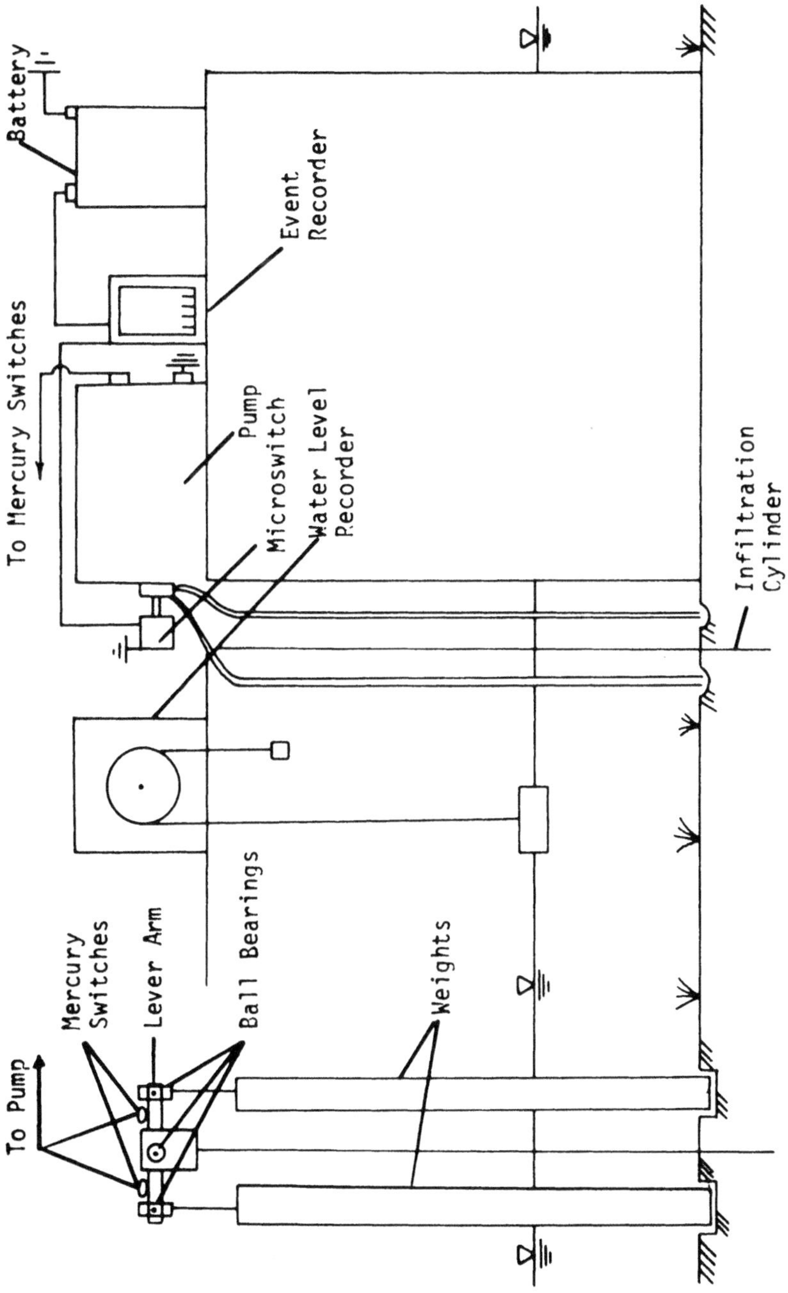

Figure 2. Infiltrometer

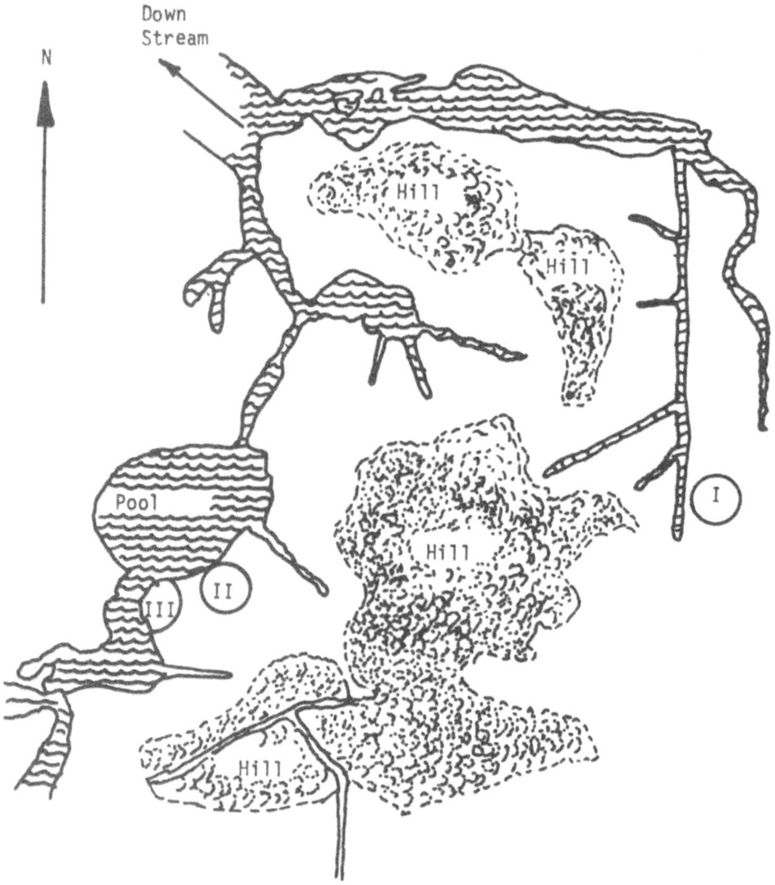

Figure 3. Experimental Areas in Great Sippewissett Marsh. Circles labeled I, II, III Indicate Regions of Seepage Measurement. See Figure 4 for Location of Sites in Each Area.

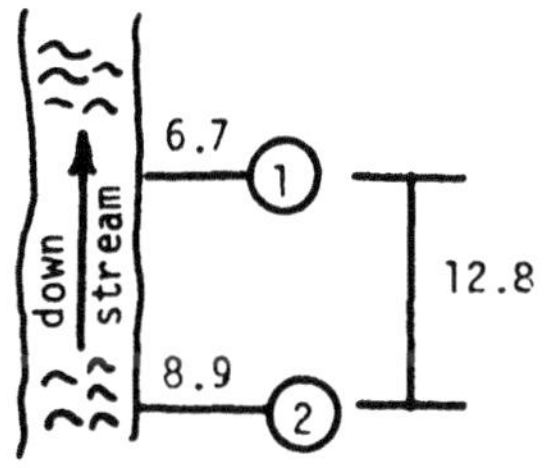

Figure 4A. Relative Positions of Sites in Area I. (Distances in Metres) See Figure 3 for Overall Map of Sippewissett Marsh.

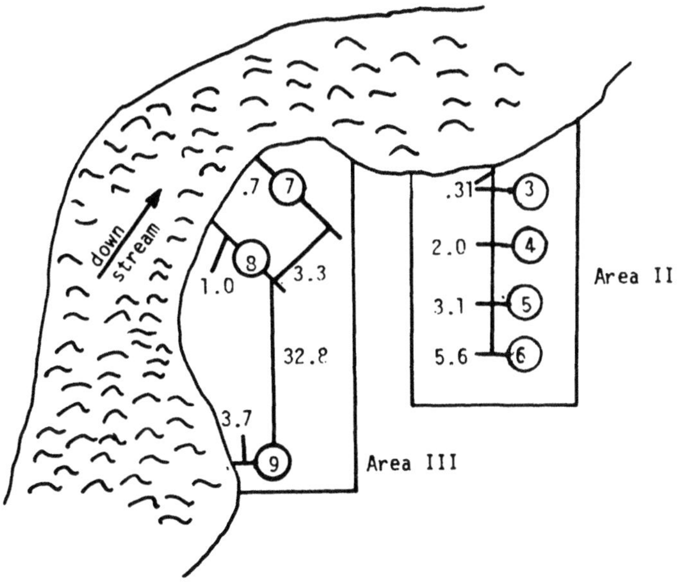

Figure 4B. Relative Position of Sites in Areas II and III, (Distances in Metres).

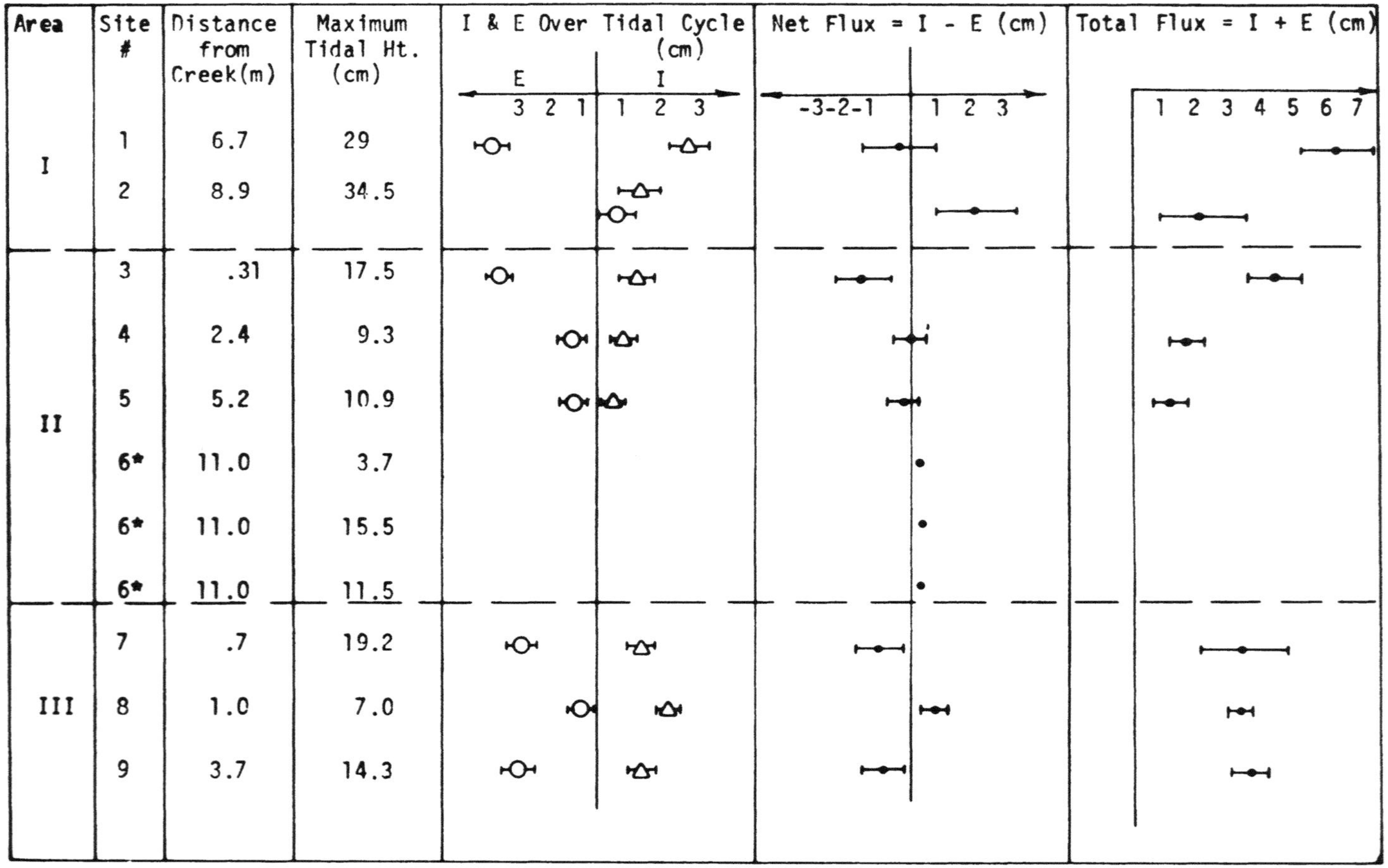

Figure 5. Flux Measurements for Three Study Areas in Sippewissett (All Runs Taken at Same Site)
Δ = Flood Tide I = Infiltration 0 = Ebb Tide E = Exfiltration * = Net Flux only obtained

Infiltration during flood tide, followed by exfiltration during ebb. While some infiltration was expected, the occurrence of exfiltration was a somewhat surprising result. Not only was exfiltration widespread, occurring at all but one of the sampling sites, but it's magnitude in many cases was nearly the same, if not greater, than the observed infiltration. Upward seepage of water into the bottom of lakes, estuaries, and coastal seas has been described by several investigators (Lee, 1977; McBride, 1975; and Cooper, 1964). It seems possible that a similar phenomenon could occur in inundated marsh flats. (2) Decrease of total flux (found by adding together infiltration and exfiltration) with distance from the creek. This conclusion is in general agreement with conventional knowledge, which defines the tidal margins as being more hydraulically active, and better drained than interior regions. Chapman (1960) mentions that areas adjacent to creeks will be better drained than more interior areas. Given the relatively higher hydraulic activity of areas adjacent to the creeks, these regions would likely have a higher total flux than more interior parts of the marsh. (3) A suggestion that a net flux of water from the sediment to the water may occur near the tidal creeks. This flux may be reversed on higher marsh.

ERROR ANALYSIS

Sources of error in the infiltrometer fall into two main categories, those caused by non-idealities in the equipment, and those inherent in the design concept. The two major equipment related errors are irreproducibility in the pump flow rate (measured to be ± 2%), and head differences between the inside and outside of the infiltration cylinder, which may cause water to leak around the confines of the cylinder. Such head differences may be due to improper balancing of the lever arms. Plots of water level vs. time for inside and outside the infiltration cylinder (Figure 6) reveal a water level difference of 5mm or less. The impact of such differences is inconclusive, and will have to wait for further analysis. It seems somewhat doubtful, however, whether such small differences would cause large errors in the observed seepage rates.

Another possible error inherent in the infiltrometer concept lies in the possibility that the portion of the infiltration cylinder submerged below the peat surface may adversely affect the flow regime, causing unnatural seepage patterns. Whether or not this poses a serious threat to the infiltrometer concept is not known at this time. There are so many factors affecting seepage that segregating out and analyzing any one phenomenon is a difficult task, made even harder by the lack of data on salt marsh hydrology.

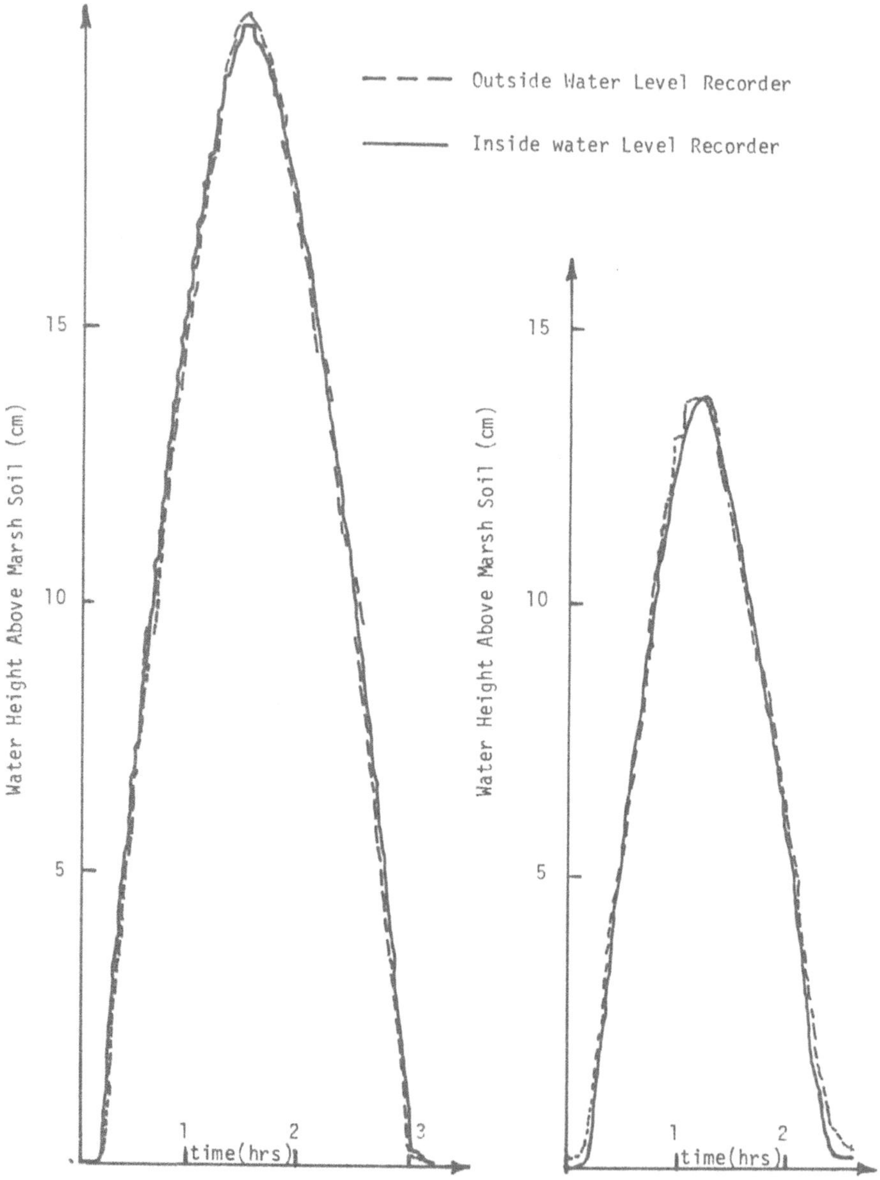

Figure 6. Water Height vs. Time for Two Different Runs

CONCLUSIONS

This study considers the design, construction, and operation of a device for measuring how much water seeps across the surface of the salt marsh soil as it is inundated by flooding tides. The device was used to measure seepage at a number of sites in Great Sippiwissett Marsh, in Cape Cod.

The infiltrometer represents a definite step forward in the attempt to understand the seepage component in the water budget of the flow over salt marshes. Initial tests have shown that the device can directly quantify the water flux across the sediment interface under relatively undisturbed conditions whereas previous approaches have either involved indirect measurement or considerable disturbance to the hydrologic regime.

REFERENCES

Cooper, H. H. 1964. Sea water in coastal aquifers. USGS Water Supply paper 1613, 64 p.

Chapman, V. J. 1960. Salt marshes and salt deserts of the world. Interscience, N. Y.

Lee, David R. 1977. A device for measuring seepage flux in lakes and estuaries. Limnol. and Oceanogr. 22: 140-147.

McBridge, M. S. 1975. The distribution of seepage within lake beds. J. Res. U. S. Geol. Surv. 3:505-512.

Rodda, J. E. 1976. Facets of Hydrology. Wiley and Sons, N. Y.

Valiela, I., J. M. Teal and E. J. Carpenter. 1978. Nutrient and particulate fluxes in a salt marsh ecosystem; tidal exchanges and inputs by precipitation and groundwater. Limnol. Oceanogr. 23: 798-812.

QUANTITATIVE ASSESSMENT OF EMERGENT SPARTINA ALTERNIFLORA BIOMASS IN TIDAL WETLANDS USING REMOTE SENSING

David S. Bartlett* and Vytautas Klemas

*Marine Environments Branch, Mail Stop 272, Marine and Applications Technology Division, NASA-Langley Research Center, Hampton, VA 23665 and College of Marine Studies, University of Delaware, Newark, Delaware, 19711

ABSTRACT

Modeling and other techniques applied to quantitative assessment of wetland energy and nutrient flux depend, in part, upon accurate data on vegetative species composition and primary production. Remote-sensing techniques have been applied to mapping of emergent wetland vegetation but not to quantitative measurement of emergent plant biomass.

A study conducted in the tidal wetlands of Delaware has shown that spectral canopy reflectance properties can be used to measure the emergent green biomass of _Spartina alterniflora_ (Salt Marsh Cord Grass) periodically throughout the peak growing season (April through September) in Delaware. The study used _in situ_ measurements of spectral reflectance in the four Landsat/MSS wavebands (4: 0.5-0.6 μm; 5: 0.6-0.7 μm; 6: 0.7-0.8 μm; and 7: 0.8-1.1 μm) for correlation with green biomass of _S. alterniflora_. Such measurements could be applied to calculations of net aboveground primary production for large areas of _S. alterniflora_ marsh in which conventional harvest techniques may be prohibitively time consuming.

INTRODUCTION

Efforts to accumulate data concerning tidal wetlands have recently come to rely more and more on remote-sensing techniques. These techniques have been extensively developed for the characterization of emergent vegetation--an important part of the wetlands ecosystem. Photographic remote sensing is now routinely used to identify wetland boundaries because of the large reduction in time and effort achieved over that required by conventional surveys. Remote sensing has not, however, been extensively utilized to assess the function of the wetlands ecosystem, primarily because the photographic methodologies commonly used do not allow quantitative relationships to be drawn between radiance received by the sensor and the characteristics of the plants from which it is reflected. Knowledge of the complex interactions between the electromagnetic radiation and the marsh plant cover is substantially lacking. Nevertheless, such knowledge is critical if quantitative evaluation of wetlands is to be extended beyond the mapping/inventory phase.

The emergent grasses and rushes of tidal wetlands bear a considerable resemblance to many terrestrial vegetation types to which rigorous remote-sensing research has been applied. Quantitative studies using field radiometers and Landsat/MSS data have generally concluded that reflectance measurements in the near-infrared spectral region (0.75 μm - 1.35 μm) and in the red region (0.6 - 0.7 μm) can be used to estimate vegetative characteristics related to biomass (Leaf Area Index, stem density, or biomass itself). Measurements in the Landsat/MSS Band 5 (red - 0.6 μm to 0.7 μm) region are inversely related to the amount of green ("live") biomass because of chlorophyll absorption of radiation in this waveband. Seevers et al. (1975) found that grass biomass estimates for rangeland management could be based on MSS Band 5 radiance measurements extracted from Landsat digital tapes. Pearson and Miller (1972) describe an inverse relationship between total grassland biomass and in situ reflectance measurements at 0.68 μm (r^2 = 0.49), but report a better, positive correlation of infrared reflectance with total biomass (r^2 = 0.71). Pearson and Miller propose that, as measurements in these two spectral regions respond with opposite trends to increasing biomass, some combination of the two bands might produce a more sensitive response than is present in either single band. The difference in reflectance at the two wavelengths and the ratio of the two reflectances were both found to correlate with biomass more highly than did reflectance in either band alone. Most impressive is the case of the infrared/red ratio which yielded a coefficient of determination (r^2) of 0.90 when linearly correlated with the amount of green biomass in the sample plots.

Several research groups and regional authorities have applied analysis of remotely sensed data to map or inventory tidal wetland resources (Anderson et al., 1973; Bankston, 1975; Carter and Anderson, 1972; Carter and Schubert, 1974; Erb, 1974; Gordon et al., 1975; Klemas et al., 1975; Bartlett et al., 1976, Pfieffer et al., 1973; Reimold et al., 1972; Stroud and Cooper, 1968; Thompson et al., 1974). Some used radiometric scanner data (Bankston, 1975; Butera, 1975; Erb, 1974; Gordon et al., 1975; Klemas et al., 1975; Anderson et al., 1973; Carter and Schubert, 1974) while most relied on aerial photography to distinguish major vegetation communities in coastal areas. Carter (1978) has written an excellent review of current remote-sensing applications in inventorying wetlands.

In a few cases, efforts have been directed not only at mapping but also collection of biomass information. Carter (1976) used measurements of the areal extent of major plant species based on interpretation of Landsat data along with typical values for primary production of each species to estimate primary production for a marsh island in Virginia. Such estimates are limited, however, in that large intraspecific variations in production, even within a small area, are known to occur (Nixon & Oviatt, 1973). Reimold et al. (1973) found that tonal variations of Spartina alterniflora recorded on aerial color infrared film could be associated with variations in emergent biomass. Increases in the relative intensity of infrared reflectance of stands having larger amounts of biomass produced "redder" image tones. The relationship observed was a somewhat qualitative one but illustrated the potential for more accurate work based on more easily quantifiable radiometric data.

A study has been carried out in Delaware with the objective of examining the interactions of wetland grass canopies with electromagnetic radiation in order to assess the potential for quantitative remote sensing of emergent biomass.

METHODS

Field measurements of the radiometric and biological characteristics of selected wetland plants were supplemented by digital modeling of canopy interactions with radiation in order to examine the relationship between vegetative parameters and the canopy. A prototype "Radiant Power Measuring Instrument (RPMI)" (Rogers et al., 1973) was used to measure both reflected radiance from the canopy and the characteristics of incident radiation for calculation of canopy reflectance. The RPMI measures radiance at spectral wavelengths identical to the four Landsat/MSS wavebands: MSS-4 (green: 0.5 μm - 0.6 μm); MSS-5 (red: 0.6 μm - 0.7 μm); MSS-6 (IR: 0.7 μm - 0.8 μm); and, MSS-7 (IR: 0.8 μm - 1.1 μm). Landsat bands were chosen because of the availability of Landsat data for comparison and the anticipation that Landsat data may be useful for repetitive evaluation of large areas of wetland at low cost.

Regression analysis was used to correlate spectral reflectance data acquired in the field by RPMI with various vegetative parameters measured simultaneously, including: canopy height, stem density, green (live) biomass, total biomass, etc. The significance of the correlations observed was further examined using the "Suits Model" (Suits, 1972) of canopy interactions with electromagnetic radiation to simulate the Spartina alterniflora canopy. Because of the importance of this species and in the interest of brevity, results will be presented primarily for S. alterniflora although field measurements of Spartina patens and Distichlis spicata were also made.

The RPMI and published algorithms (Rogers et al., 1973) were also used in attempts to account for the effects of atmospheric and solar zenith angle on radiance observed both near the vegetated surface and by the Landsat/MSS. Results of this part of the investigation are beyond the scope of this paper and will be reported in subsequent publications.

RESULTS

Field studies showed that canopy reflectance measurements in the visible spectral region (MSS Bands 4 and 5) were inversely related to the percentage, by weight, of green vegetation within the canopy. This result was particularly evident in the heavy chlorophyll absorption region of Band 5 (0.6 μm - 0.7 μm) (Fig. 1). Digital simulation confirmed this functional dependence and showed little response in the infrared bands 6 and 7 (Fig. 2). Infrared canopy reflectance appeared to be related to several measures of the amount of vegetation present in the canopy, including total and green biomass and canopy height (i.e., vertical distance from top of canopy to soil). Canopy reflectance for S. alterniflora in MSS Band 7 (0.8 μm - 1.1 μm) was best correlated with canopy height (Fig. 3). Regression relationships were consistently weaker between vegetative parameters and infrared reflectance than for visible reflectance. Digital modeling indicated a functional response of infrared reflectance to changing horizontal leaf area index (horizontally projected area of vegetation per unit area of ground (Fig. 4). Horizontal leaf area index was not measured in the field but is presumably related to the total amount of vegetation present. Except at very small values of leaf area index, visible canopy reflectance (Bands 4 and 5) is insensitive to this parameter (Fig. 4).

By ratioing infrared/red canopy reflectance, a parameter results which is proportional to green biomass of the canopy. As was the case for rangeland grasses (Pearson and Miller, 1972), regression produces high correlation ($r^2 = 0.81$) of ratioed reflectance with

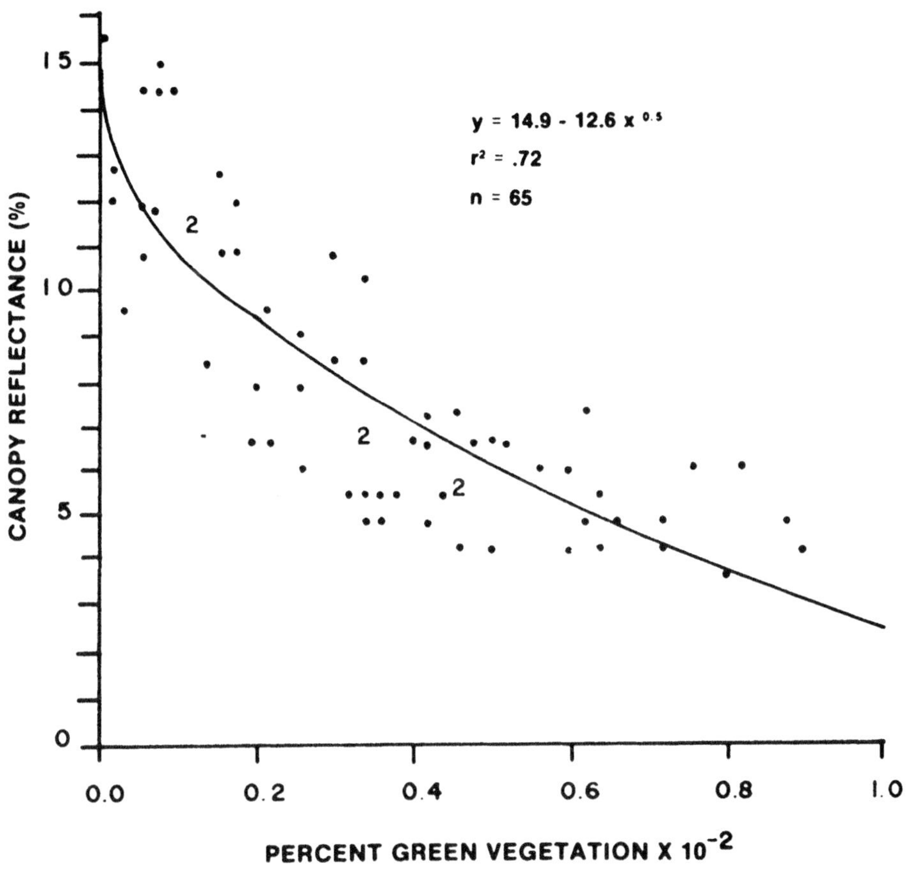

Figure 1. Plot of Band 5 (0.6 μm - 0.7 μm) canopy reflectance vs. percent green vegetation for S. alterniflora. Regression results are shown.

green biomass (Fig. 5). A significant aspect of the relationship observed for S. alterniflora is its linearity over a very wide range of green biomass values: 20-1000 g dry wt/m^2 (Fig. 5) indicating potential for use of spectral data for estimation of this parameter under a wide variety of conditions.

Spectral measurements were not as highly correlated with green biomass for the other two species tested: S. patens and D. spicata. This may result from variability in growth form (i.e., the occurrence of both vertical and lodged stands) and from the high stem densities (limiting penetration of radiation into the canopy) which

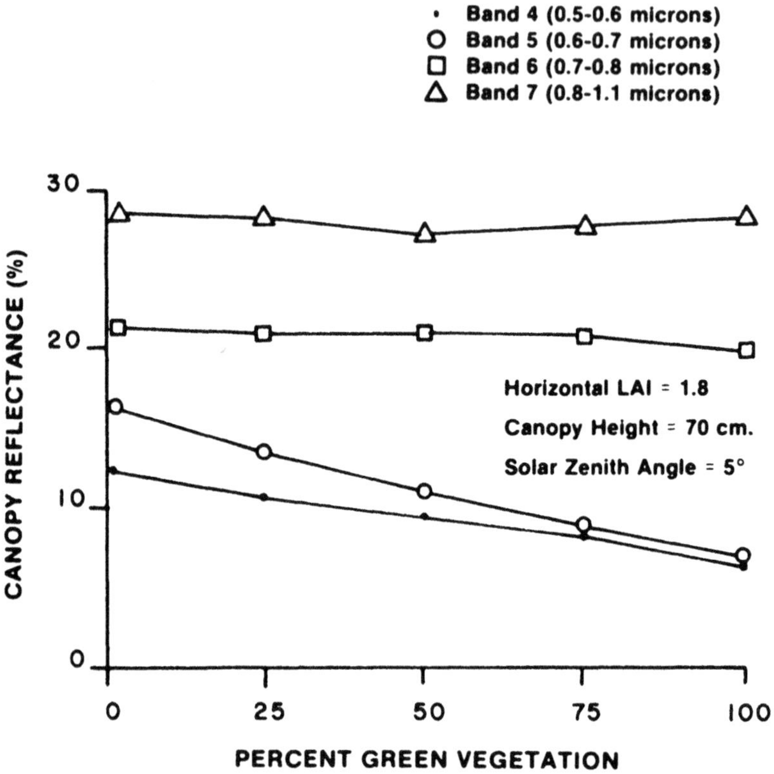

Figure 2. Simulated response of S. alterniflora canopy reflectance to changing percentage of green vegetation.

are characteristic of these species. Drake (1976) made in situ spectral measurements of S. patens, D. spicata, and Scirpus olneyi and found high correlations of red (0.66 μm - 0.71 μm) reflectance with green biomass (r^2 = 0.74 - 0.83). The range of biomass tested was restricted, however (0-350 g dry wt/m^2). It seems likely that Drake's results were produced by high intercorrelation between the percentage and the mass of green vegetation. Similar results, using visible reflectance alone have been reported in Western rangelands (Seevers et al., 1975).

CONCLUSIONS

Most efforts to evaluate characteristics of tidal wetlands and adjacent estuarine waters depend on measurements which can be difficult and time-consuming to accumulate in the field. To the

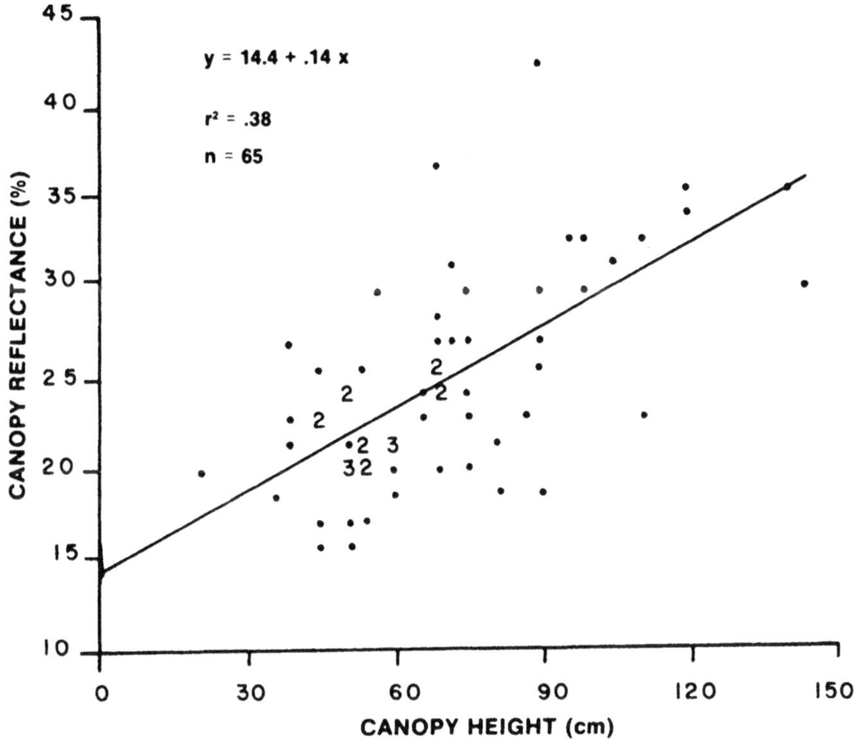

Figure 3. Plot of Band 7 (0.8 μm - 1.1 μm) canopy reflectance vs. canopy height for S. alterniflora. Regression results are shown.

extent that they can be applied, therefore, remote-sensing techniques can contribute to the available data base by providing synoptic, cost-effective information about the environment. Photographic remote sensing has established utility in delineation of wetland boundaries and areal extent and is now routinely applied to such tasks by federal and state management authorities and many research groups.

Results of this study indicate that quantitative spectral measurements made in the wavebands sensed by the Landsat/MSS contain information on the standing crop, emergent biomass of Spartina alterniflora.

The large advantages which would accrue from rapid, repetitive, cost-effective assessment of large expanses of tidal wetland should motivate continued development and extension of remote-sensing techniques beyond simple mapping tasks. Routine monitoring of biomass and productivity in wetlands using remote sensing would provide

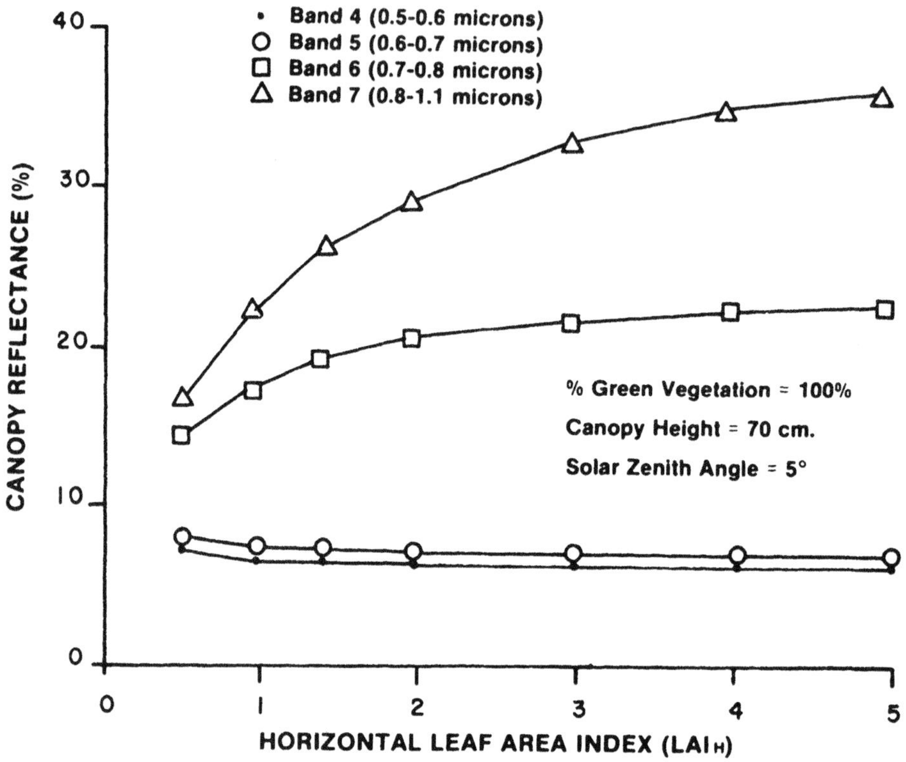

Figure 4. Simulated response of *S. alterniflora* canopy reflectance to changing horizontal leaf area index (LAI_H).

valuable management information as well as contributing data to modeling and other research efforts.

ACKNOWLEDGMENTS

The research reported is a portion of a dissertation completed by D. Bartlett in fulfillment of requirements for the doctoral degree in Marine Studies awarded by the College of Marine Studies, University of Delaware. The authors are indebted to Dr. G. H. Suits of the Environmental Research Institute of Michigan for providing his digital model of plant canopy interactions with electromagnetic radiation. The research was supported, in part, by the University of Delaware and NASA Contract NAS5-20983.

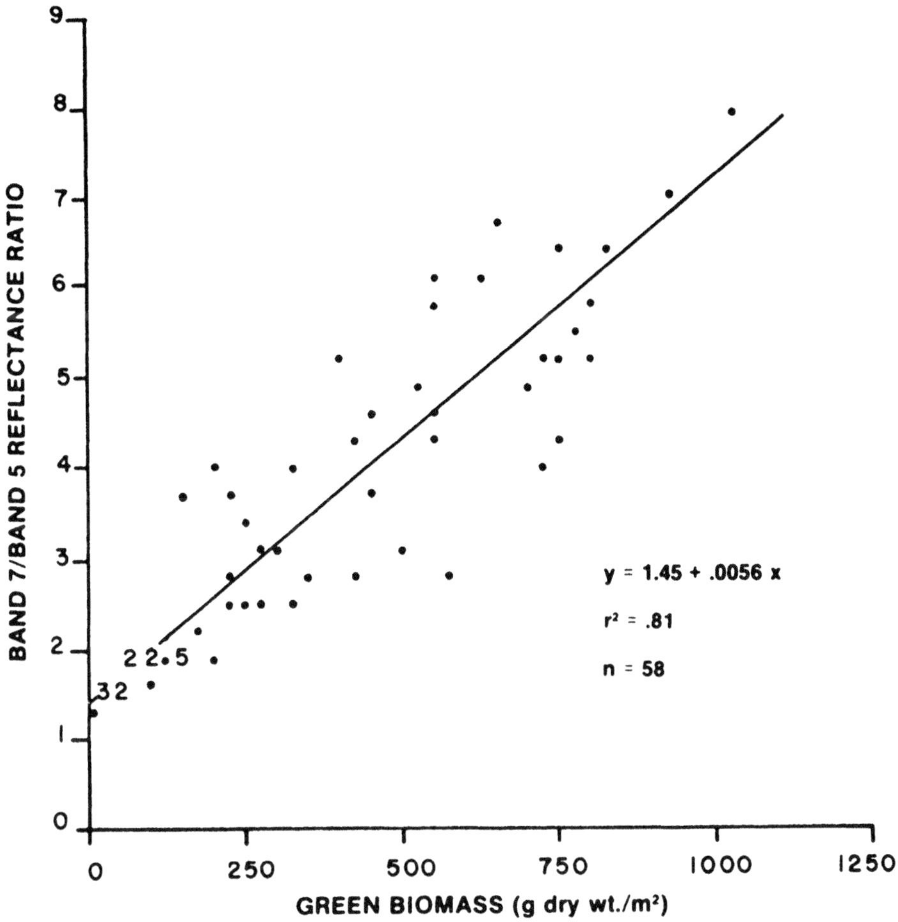

Figure 5. Plot of Band 7/Band 5 canopy reflectance ratio vs. green biomass for *S. alterniflora*. Regression results are shown.

REFERENCES

Anderson, R. R.; Carter, V.; and McGinness, J. (1973): Applications of ERTS to Coastal Wetland Ecology with Special Reference to Plant Community Mapping and Impact of Man. Proc. 3rd ERTS Symp., Washington, D. C., pp. 1225-1242.

Bankston, P. T. (1975): Remote-Sensing Applications in the State of Mississippi. Proc. NASA Earth Resources Survey Symp., Houston, TX, 284 pp.

Bartlett, D. S. (1979): Spectral Reflectance of Tidal Wetland Plant Canopies and Implications for Remote Sensing. Ph.D. Dissertation, University of Delaware, 239 pp.

Bartlett, D. S.; Klemas, V.; Crichton, O. W.; and Davis, G. R. (1976): Low-Cost Aerial Photographic Inventory of Tidal Wetlands. University of Delaware, College of Marine Studies, Center for Remote Sensing, CRS-2-76, 29 pp.

Bartlett, D. S.; Klemas, V.; Rogers, R. H.; and Shah, M. J. (1977): Variability of Wetland Reflectance and Its Effect on Automatic Categorization of Satellite Imagery. Proc. ASP and ACSM Annual Convention, Washington, D. C.

Butera, M. K. (1975): The Mapping of Marsh Vegetation Using Aircraft Multispectral Scanner Data. Proc. NASA Earth Resources Survey Symp., Houston, TX, pp. 2147-2166.

Butera, M. K. (1978): A Determination of the Optimal Time of Year for Remotely Classifying Marsh Vegetation from Landsat Multispectral Scanner Data. NASA TM-58212, 34 pp.

Carter, V. (1976): Applications of Remotely Sensed Data to Wetland Studies. Proc. 19th COSPAR Meeting, Philadelphia, PA.

Carter, V. (1978): Coastal Wetlands: Role of Remote Sensing. Proc. Coastal Zone '78, Symp. on Technical, Environmental, Socioeconomic, and Regulatory Aspects of Coastal Zone Management. Am. Soc. Civil Eng. Publ., San Francisco, CA, pp. 1261-1283.

Carter, V. P.; and Anderson, R. R. (1972): Interpretation of Wetlands Imagery Based on Spectral Reflectance Characteristics of Selected Plant Species. Proc. Am. Soc. Photogram. 38th Ann. Mtg., Washington, D. C., pp. 580-595.

Carter, V. P.; and Schubert, J. (1974): Coastal Wetlands Analysis from ERTS-MSS Digital Data and Field Spectral Measurements. Proc. 9th Internatl. Symp. on Remote Sensing of Environment, Ann Arbor, MI, pp. 1241-1259.

Drake, B. G. (1976): Seasonal Changes in Reflectance and Standing Crop Biomass in Three Salt Marsh Communities. Plant Physiology, 58, pp. 696-699.

Erb, R. B. (1974): ERTS-1 Coastal/Estuarine Analysis. NASA TMX-58118, Houston, TX, 284 pp.

Gordon, J. P.; Schroeder, R. H.; and Cartmill, R. H. (1975): South Louisiana Remote Sensing Environmental Information System. Proc. NASA Earth Resources Survey Symp., Houston, TX, Vol. II-3, pp. 217-223.

Klemas, V.; Daiber, F. C.; Bartlett, D. S.; and Rogers, R. H. (1975): Coastal Zone Classification from Satellite Imagery. Photogram. Eng. and Remote Sensing, 40:4, pp. 499-513.

Milner, C.; and Hughes, R. E. (1968): Methods for the Measurement of the Primary Production of Grassland. Internatl. Biological Programme, Blackwell Scientific Publ., London, 70 pp. (2nd printing, 1970).

Nixon, S. W.; and Oviatt, C. A. (1973): Analysis of Local Variation in the Standing Crop of Spartina alterniflora. Bot. Mar., 16, pp. 103-109.

Pearson, R. L.; and Miller, L. D. (1972): Remote Spectral Measurements as a Method for Determining Plant Cover. Tech. Rept. No. 167, U. S. Internatl. Biological Program, Colorado State University, 48 pp.

Pfeiffer, W. J.; Linthurst, R. A.; and Gallagher, J. L. (1973): Photographic Imagery and Spectral Properties of Salt Marsh Vegetation as Indicators of Canopy Characteristics. Proc. Symp. on Remote Sensing in Oceanography, Am. Soc. Photogram., Lake Buena Vista, FL, pp. 1004-1016.

Reimold, R.; Gallagher, J; and Thompson, D. (1972): Coastal Mapping with Remote Sensors. Proc. Coastal Mapping Symp., Washington, D. C., pp. 99-112.

Reimold, R. J.; Gallagher, J. L.; and Thompson, D. E. (1973): Remote Sensing of Tidal Marsh Primary Production. Photogrammetric Engineering, 39(5), pp. 477-488.

Rogers, R.; Peacocok, K.; and Shah, N. (1973): A Technique for Correcting ERTS Data for Solar and Atmospheric Effects. Third ERTS Symp., NASA SP-351, Goddard Space Flight Center, Greenbelt, MD, pp. 1787-1804.

Seevers, P. M.; Drew, J. V.; and Carlson, M. P. (1975): Estimating Vegetative Biomass from Landsat-1 Imagery for Range Management. Proc. Earth Resources Survey Symp., Houston, TX, Vol. 1-A, pp. 1-8.

Smalley, A. E. (1959): The Role of Two Invertebrate Populations, Littorina irrorata and Orchelium fiducinum in the Energy Flow of a Salt Marsh Ecosystem. Ph.D. thesis, University of Georgia, University Microfilms, Ann Arbor, MI, 126 pp.

Suits, G. H. (1972): The Calculation of the Directional Reflectance of a Vegetative Canopy. Remote Sensing of Environment, 2(1), pp. 117-125.

Thompson, D. E.; Ragsdale, J. E.; Reimold, R. J.; and Gallagher, J. L. (1974): Seasonal Aspects of Remote Sensing Coastal Resources. In: R. Shahrokhi, ed. Remote Sensing of Earth Resources, Vol. II, University of Tennessee Press, pp. 1201-1249.

Wiegert, R.; and Evans, F. C. (1964): Primary Production and the Disappearance of Dead Vegetation on an Old Field in Southeastern Michigan. Ecol. 45(1), pp. 49-63.

BETWEEN COASTAL MARSHES AND COASTAL WATERS - A REVIEW OF TWENTY YEARS OF SPECULATION AND RESEARCH ON THE ROLE OF SALT MARSHES IN ESTUARINE PRODUCTIVITY AND WATER CHEMISTRY

Scott W. Nixon

Graduate School of Oceanography, University of Rhode Island, Kingston, Rhode Island 02881 U.S.A.

TABLE OF CONTENTS

TABLE OF CONTENTS

"Certitude is not the test of certainty."
-Justice Oliver Wendell Holmes
Natural Law

INTRODUCTION

It has been almost 20 years since John Teal (1962) published his well-known paper synthesizing a variety of independent studies of production, respiration, and animal abundances in the salt marsh ecosystem of Sapelo Island, Georgia. Teal's work brought out a number of interesting points, but I think the reason the paper is most often cited is because of its last sentence. After discussing various trophic relationships in the marsh, the paper ended with the conclusion that "...the tides remove 45% of the production before the marsh consumers have a chance to use it and in so doing permit the estuaries to support an abundance of animals."

The concept that Teal outlined had a great appeal for ecologists, and the belief that large exports of organic matter from marshes support much of the secondary production of estuaries and nearshore waters quickly became a dogma of marine ecology. My impression is that this view is no longer popular, and there is a growing awareness of and frustration with the complexity of marsh-estuarine interactions. This complexity is real, but a great many studies have been carried out recently, and it cannot help but be useful to bring some of their results together.

The evolution of the idea of organic export and its extension to the fluxes of nutrients, metals, and other substances between salt marshes and estuaries is also worth exploring. It has been one of the prevailing interests of marine ecology during the past 20 years, and it reflects much of what is best and worst in ecological research. It has also led to some of the most controversial applications of ecological research to problems of coastal zone management and political decision making.

A review should be more than a collection of observations and references, however, and I have not tried to be an impartial librarian in doing this work. It is commonplace to note that hindsight makes it easier to see mistakes and problems, but it is also true that it would be a waste not to take advantage of the perspective. I have made some critical comments, indulged in some moralizing, and tried to draw together a story whenever I thought the evidence would bear it. I have also included a number of quotations in the hope that they will add life to the discussion and serve to generate a healthy perspective on the literature for those who are just beginning to work in this area.

The literature dealing with marsh-estuarine interactions is large, and I have concentrated in this review almost exclusively on questions of the exchange of carbon, nitrogen, and phosphorus, with some attention to sediments and a few of the trace metals. I have not dealt with animal migrations, fluxes of larvae, a host of exotic chemicals or a number of other possible problems. The story as chosen is rich and interesting enough.

SOME HISTORICAL PERSPECTIVE

I began this review with a quote from John Teal's paper for two reasons. It is the first formal statement in the major scientific literature about the nature of marsh-estuarine fluxes, and it is characteristic of a great deal of the writing that was to follow. It is characteristic not because it required tenuous assumptions or because it involved an indirect approach (much of our science does), but because the conclusion was pronounced with far more weight than the data warranted. A very uncertain calculation was used to support a very profound assertion about the way estuaries worked. Many of those who cite Teal's paper seem to have forgotten that his conclusion about organic export was arrived at by difference in a community energy budget. At the time it was written, it was unsupported by any direct measurement.

In 1965, 3 years after Teal's paper appeared in Ecology, Armando de la Cruz wrote an as yet unpublished Ph.D. Dissertation at the University of Georgia which apparently contained the first direct measurements of detrital flux from the Sapelo Island marsh. Some of these data appear in print in the often-cited paper by Odum and de la Cruz (1967) on detritus in the volume Estuaries (Lauff 1967). The main thrust of this paper deals with other aspects of detritus, however, and only two fluxes based on "preliminary calculations" for one neap and one spring tide were reported. I am not aware of any further publication of the Georgia flux measurements. Nevertheless, E. P. Odum applied the term "outwelling" to these marsh-estuarine fluxes and introduced the concept in his 1968 paper at the Second Sea Grant Conference in Newport, R. I., as follows:

> Most fertile zones in coastal areas capable of supporting expanded fisheries result either from the "upwelling" of nutrients from deep water or from "outwelling" of nutrients and organic detritus from shallow-water nutrient traps such as...salt marshes. The importance of [salt marshes] as "primary production pumps" that "feed" large areas of adjacent waters has only been recently recognized...

This paper contained no data on organic fluxes from the marsh, but presented a summary of ^{14}C uptake measurements indicating that phytoplankton production in the coastal water off Sapelo Island was very high. It was implied that salt marshes export nutrients which contribute significantly to sustaining this high production. The contradiction of treating marshes simultaneously as traps and sources was apparently not appreciated.

As far as I can tell, the inclusion of nutrients in the "outwelling" concept reflects a publication during the previous year of work by Larry Pomeroy and others (1967) on phosphorus in the Georgia marshes. On the basis of ^{32}P tracer studies, they concluded that "the transfer of P...from the deep sediments to the water by *Spartina* explains the high concentration of [this] element in the water of marshy estuaries." However, Pomeroy et al. neither made any marsh-estuarine flux measurements nor attempted to estimate such a flux. It might be noted in passing that the average high phosphate concentration in Georgia marsh-estuarine water was reported as about 1 μM, a value often exceeded by a factor of 2 or 3 in more northern estuaries with little or no salt marsh. Studies on nitrogen in the Georgia marshes did not become available until almost 10 years after Odum's (1968) paper appeared (Haines et al. 1976).

Adequate data on organic matter flux from the Georgia marshes have still not been published and, as Pomeroy et al. concluded in their 1976 paper on carbon flux in the marsh, "...we cannot assume that salt marshes are necessarily sources or sinks for organic materials." The current state of the evidence concerning the flux of organic matter from marshes to estuarine and offshore waters is covered in later sections of this review, but I want to emphasize here that the first measurements of the flux of organic matter from salt marshes over an annual cycle did not become available until 6 years after E. P. Odum's talk (Moore 1974). Yet, until very recently, the concept of "outwelling" was taken for granted by most ecologists. It was often taught as "gospel" in basic courses, and it formed a cornerstone of many arguments in favor of salt marsh conservation. For example, even in Rhode Island, a state not noted for an abundance of salt marsh, the legislature, in a moment of remarkable enlightenment during it's January 1969 session, responded to the concept of "outwelling" and passed a salt marsh conservation act which read in part,

> Whereas, the metabolism and katabolism of plants and animals which constitute the estuarine complex found in salt marshes furnishes the nitrates, phosphates, sugars, plankton and organic chemicals necessary for the nurture of fin fish and shellfish throughout the Narragansett Bay area and its environs...Be it resolved, that (a) any person who dumps or deposits mud, dirt, or rubbish upon, or who excavates and disturbes the ecology of intertidal salt marshes, or any part thereof...shall be fined...

There is no reason to believe that the impressions of the coastal zone management people in Rhode Island were, or are, very different from those in most other states. They have all read the ecological literature faithfully. And it is the reliability and credibility of the literature that I am objecting to, not the judgment of the management people.

"Outwelling" was a quantitative proposition, yet there were virtually no quantitative data to support it, and this situation did not go entirely unnoticed (Walker 1973).

In a recent talk at the 1979 Estuarine Research Federation Meeting at Jekyll Island, Georgia, E. P. Odum reflected on "The Status of Three Ecosystem Level Hypotheses Regarding Salt Marsh Estuaries." One of the hypotheses was "outwelling." The significant word which appears here for the first time is hypothesis. "Outwelling" was and is an important hypothesis that has only recently begun to be carefully tested. The problem is that the existence of a flux of organic matter from salt marshes to estuaries or offshore waters of sufficient magnitude to influence their production was not traditionally put forward by Teal, E. P. Odum, or most others as an hypothesis, but as a conclusion. It is not relevant that at least some of the people publishing in the field, if they had been pressed, would have admitted that "outwelling" was a hypothesis; neither is the issue of the ultimate correctness of the concept. Even in the most recent edition of the Fundamentals of Ecology textbook (Odum 1971), we find that "Another water movement that contributes to coastal fertility is what I have called outwelling..." and "In summary, it can be stated that all coastal waters capable of supporting intensive fisheries probably benefit from either (1) outwelling from shallow water "production zones" or (2) upwelling...or (3) both." The tone is not that outwelling might be an important phenomenon or that it was a good example of the kind of ecological problem that needed to be studied; rather, it seems to be taken for granted as a phenomenon, operative at varying intensity in different areas.

A relatively few researchers, however well-meaning and able, failed to make and maintain a firm distinction between what they thought was happening, or what they thought "ought" to be happening, and what they had good data to show was happening. Because they told their story so often (and at times so eloquently), because at least some of them were very well-known and respected, because the credibility of the printed scientific literature was so strong, perhaps because it sounded like such a "good" story, too many of us failed for too long to examine the concept critically. Instead, we passed it on eagerly as one of the accomplishments of marine ecological research. And we passed it on very effectively, to students, to managers, to legislatures, to funding agencies, and to each other. It occurs to me that if the early papers had clearly presented "outwelling" as an exciting hypothesis, it might not have taken so long for the data necessary to evaluate it to become available. It is reassuring that the scientific process has prevailed, and that we have begun to obtain some of the measurements necessary to evaluate the hypothesis, but it is humbling to realize how quickly and completely an idea became implanted in the literature, and in our minds, with so few data to support it.

I cannot leave this discussion of the history of the outwelling concept without adding something about the exchange of nutrients between marshes and estuaries. While the response of the ecological community to the proposition of the "outwelling" of organic matter was not particularly thoughtful, at least it did not reduce us to "doublethink," a mental process often required to assimilate the conventional wisdom on the subject of marsh-estuarine nutrient fluxes. George Orwell (1949) defined the trick as "the power of holding two contradictory beliefs in one's mind simultaneously, and accepting both of them." Let me give an example from a well-known paper on "The Value of the Tidal Marsh" (Gosselink et al. 1973):

> ...coastal marshes...all over the world export mineral and organic nutrients that support much of the production of the adjacent estuarine and coastal waters.
>
> . . . .

> When nutrient-rich effluents enter a marsh the nutrients are effectively trapped by the tidal circulation pattern, and assimilated in the productive biological system.

The idea that salt marshes are important and valuable because they "outwell" nutrients that fertilize the estuary has recently run head-on into the idea that marshes are important and valuable because they take up nutrients from the estuary and thus provide free tertiary treatment for sewage. If we bring strong statements of these conflicting themes together, the result can be perplexing, entertaining, or exasperating, depending on one's point of view. Another example:

> These results demonstrate that an undisturbed salt marsh...is a source of phosphorus for the coastal waters of the ocean...the role of *Spartina* in contributing inorganic phosphorus to the estuarine and coastal waters points out the need for cord grass to remain undisturbed in the highly productive salt marsh ecosystem.
>
> Reimold (1972)

> Detailed analysis of waste assimilation shows that marshes...have a tremendous capacity for tertiary treatment of nutrients, especially phosphorus.
>
> Gosselink, et al. (1973)

In our enthusiasm to protect the marshes, we believe them both. The Gosselink et al. (1973) paper went so far as to calculate a value of "about $2,500 per acre per year for tertiary treatment" by salt marshes, in spite of the fact that not one real marsh-estuarine nutrient uptake study was cited by the authors. The value of $2,500 appears again in a paper dealing with techniques to evaluate the benefits of road construction across a marsh in the first issue of *Coastal Zone Management* (Pope and Gosselink 1974), and I have no doubt that it has since been picked up and used by many people involved in efforts to preserve the marshes.

Again, the scientific problem is an interesting one, and I hope I can show that a considerable amount of information has recently become available that we can use to try to evaluate the probable importance of marsh-estuarine nutrient exchanges. But it does not appear from most of the earlier papers that the authors had the slightest doubt about what they thought was happening.

The situation might have been different if one of the very first papers dealing with the role of marshes in estuarine nutrient dynamics had been expanded and published in a larger journal. The Spring, 1959 issue of the Estuarine Bulletin carried a short note by F. A. Kalber, a graduate student at Delaware, titled, "A Hypothesis on the Role of Tidemarshes in Estuarine Productivity." I remember Kalber's paper from my time as an undergraduate at the University of Delaware in the early 1960's, but I cannot recall ever having seen it cited in the literature. The theme of the paper was that phytoplankton production in the Delaware Bay estuary was at certain times limited by light or some factor other than nutrients. Under these conditions, he suggested that the marshes took up nutrients which were then incorporated into plant tissue. Later in the seasonal cycle, when nutrients would have been less abundant, remineralized nutrients from the decomposition of the plants were released back into the estuary. In this way, the marshes were thought to "smooth" the estuarine nutrient cycles and thus increase estuarine production.

All of this, however, was clearly put forward as an hypothesis - it is right there, unmistakable in the title of the paper. It was also put forward 3 years before Teal's (1962)* statement about organic fluxes, 6 years before Pomeroy et al. (1965) proposed marsh sediment as "buffering" phosphate cycles, 8 years before Pomeroy et al. (1967) proposed a phosphorus flux out of the marsh in Spartina detritus, and 9 years before E. P. Odum (1968)** first published his paper on the "outwelling" of nutrients from marshes. More important than the point of historical priority, however, is the observation that during the 20 years since Kalber's note appeared, that word hypothesis was absent from the literature. We will go on later to ask whether or not the data that are available so far support Kalber's hypothesis.

The most influential early work that dealt with marsh-estuarine nutrient interactions was a short paper by Pomeroy et al. (1967) which described in general terms the results of two ^{32}P addition experiments in a marsh creek near Sapelo Island. Their paper presented two kinds of conclusions, some that could be derived directly from the results of the tracer study, and some that came out of a model similar to that developed by Teal (1962). The report of a rapid uptake of ^{32}P from the water by the marsh surface sediments is an example of the former, and the statement that "a significant part of the P is exported from the marsh in organisms and detritus" is an example of the latter. It appears that the detrital P export was arrived at by using data on the production of Spartina, along

*The substance of Teal's paper was first made available in abbreviated form as part of a report titled, "Salt Marsh Conference, March '58", distributed by the Marine Institute of the University of Georgia. At that time the nature of marsh-estuarine organic fluxes was set out in more circumspect terms.

"There is plenty of energy fixed in the salt marsh to support a large population of shrimp, fish and bottom organisms in the tidal creeks and estuaries...Data of Ragotzkie show that production of the local estuarine plankton community...is negative, indicating that most aquatic organisms must obtain their energy from some outside source, in the author's opinion the marsh."

It is interesting that in the discussion following this paper, H. T. Odum questioned the accuracy of the estimate used for bacterial decomposition in the marsh and suggested that it might be better to measure the organic export and then calculate the decomposition on the marsh by difference. The following exchange took place:

H. T. Odum: ...Why don't you concentrate on the export?

Teal: This is Ragotzkie's problem and he is working on it but he has not got it yet.

H. T. Odum: If you could get that, then you would have for the first time bacterial activity in a community.

Teal: That is right, it would be nice and it would be nice to work backwards. You have these other measurements for the macro-consumers and bacterial activity could be found by difference and you would have a better estimate of it.

**In a general article in the N. Y. Conservationist (1961), Odum touched on his later theme: "Because of the great importance of exports and imports and the diversity of production and consumption units, the entire estuary must be considered as a whole." He himself, however, cites the 1968 paper as the reference for "outwelling" in the scientific literature.

with an estimate of the P content of Spartina or detritus (it is not clear which), and the organic export calculations of Odum and de la Cruz (1967) or Teal (1962). The nature of these calculations has already been discussed. In fact, the ^{32}P was never found in Spartina and the authors made no measurements of phosphorus flux from the system. With regard to phosphorus loss through the movements of shrimp and fish, Pomeroy et al. wrote that their "very crude estimates of this loss suggest that it is small compared to the direct loss of detritus..."

Other early studies that were occasionally cited in support of the role of marshes as sources or sinks for nutrients were based solely on concentration measurements in tidal creeks (Reimold and Daiber 1970, Aurand and Daiber 1973) and cannot be used with much confidence to calculate nutrient transport from the marsh.

The paper which I quoted from earlier by Bob Reimold (1972) is also frequently cited as showing that marshes export a large amount of phosphorus to coastal waters. Reimold's study involved the addition of ^{32}P to streamside Spartina marsh sediments and looking for the rate of appearance of the labeled phosphorus in the plant tissue. He also did a number of experiments in which he showed that there was a release of ^{32}P from the plant leaves and stems when they were immersed in water. On the basis of the latter measurements, and assumptions about tidal innundation and the biomass of Spartina, Reimold estimated that the Sapelo Island marshes "could contribute 6,857 Kg P/day to the adjoining waters." While the paper provided convincing evidence that streamside Spartina actively took up phosphorus from relatively deep sediments ($\sim$1m), I will try to show by a simple argument in a later section that the flux estimate cannot be correct within orders of magnitude. Bob Reimold has long been an advocate of marsh preservation, and I admit to sharing his sympathies, but he took his numbers much further than they could go. If we had all been more skeptical and less eager to support a cause, I think it would have been clear long ago that the marshes cannot export phosphorus at anywhere near such a rate.

Because ecology often deals with nature on the scale at which we most directly perceive it, it seems to be a science especially susceptible to our feelings and prejudices. We may study abstract nutrient cycles, but we cannot divorce ourselves from the life and landscape which frame them. I suspect it is easier to remain objective about ribosomes or galaxies than it is about salt marshes or the sea. As Herman Melville put it:

> Thinking is, or ought to be, a coolness and

> a calmness; and our poor hearts throb, and
> our poor brains beat too much for that.

I suppose there is no solution. All we can do is be careful and try for the right blend of tolerance and skepticism in reading and reviewing the literature. It is not necessarily important that any one of us was right or wrong in any particular paper, and I am not suggesting that anyone intentionally misused their data. But until recently, the ecological research community as a whole has not been critical enough of the work that has been done (or not done) in studying marsh-estuarine interactions. On this point, Walker (1973) was correct when he accused us of trying too hard to use scientific data to support preconceived notions of the importance of wetland conservation.

Since the mid 1970's this situation has been changing, and the nature of the direction and magnitude of the flows of various materials between marshes and coastal waters has received a considerable amount of attention. As more people began to study the problem in more marshes, reservations about the old arguments began to appear:

> Data derived from three wetland study sites in the Chesapeake Bay show that regularly flooded tidal marshes should not be considered as sinks for available forms of the nutrients - nitrogen or phosphorus. It is concluded therefore that the utilization of marshes of any type as nutrient removal systems is questionable.
>
> Bender and Correll (1974)

My impression is that the prevailing view among those working in the field now is that chaos reigns and that we dare not make any statement at all about what marshes are importing or exporting. In fact, I said so myself in the introduction to a collection of papers devoted to "Interactions Between Tidal Wetlands and Coastal Waters" (Nixon and Odum 1976). It is even becoming popular to dismiss the question of marsh-estuarine interactions as a "non-problem" in which the flux is simply a site-specific function of marsh morphology and hydrography. According to this hypothesis, open marshes export and marshes which for any reason have restricted circulation import. This confusion and uncertainty does not yet appear to have had much of an impact on management policy, however, and I suspect that coastal zone management people will hold on to our first version of the marsh story until a really compelling case is made to the contrary. When Rhode Island's Coastal Zone Management Plan was accepted by the Office of Coastal Zone Management in NOAA in 1978, it noted the following under its "Findings" with regard to coastal wetlands:

> Coastal wetlands yield large crops of grasses that when they decay provide detritus that is an important food source both within the wetland and in coastal waters. The export of nutrients from coastal wetlands is significant in maintaining the high productivity of estuarine and coastal waters.

And a recent joint publication titled "Safeguarding Wetlands and Water Courses with 404" from the National Resources Defense Council (1978) and the National Wildlife Federation made their position clear. "The draining of wetlands destroys an invaluable resource for cleansing and safeguarding good water quality." I have a feeling that it would not be a very pleasant or rewarding task to argue otherwise, but in some cases we may have to do so. As John Tyndall observed in his work on The Forms of Water in Clouds and Rivers, Ice and Glaciers (1874), "We are here warned of the fact, which is too often forgotten, that the pleasure or comfort of a belief, or the warmth or exaltation of feeling which it produces, is no guarantee of its truth."

EVIDENCE FROM THE SEDIMENTS

Deposition

As far as I am aware, there seems to be general agreement that coastal salt marshes are sinks for suspended sediments, at least on the time scale of years to centuries. We know this because the accretion of marshes must approximately keep pace with the relative local increase in sea level, regardless of whether the apparent sea level rise is largely due to glacier melting or to land subsidence. In fact, if the few reported measurements of salt marsh accretion are examined, it appears that short-term accretion rates actually exceed the rate of sea level rise (Table 1). This imbalance in favor of marsh accretion made it possible for Shaler (1885) to find evidence in support of his classic model of salt marsh formation, even though he was unaware of sea level rise.

Most of the studies of marsh accretion rate summarized in Table 1 used markers such as brick dust (Stearns and MacCreay 1957), glitter (Harrison and Bloom 1974) or ^{137}Cs, a fallout product of

Table 1. Rates of sea level rise and salt marsh accretion

	Sea Level Rise,[1] mm y^{-1}	Marsh Accretion, mm y^{-1}
Barnstable Harbor, Cape Cod, MA[2]	3.4	18.3
Connecticut, 5 Marshes		
Spartina alterniflora zone[3]	2.5	8-10
Spartina patens zone[4]	2.5	2-5
Long Island, N. Y., Flax Pond		
Spartina alterniflora[5]	2.9	4.7-6.3
Spartina alterniflora[6]	2.9	2-4.2
Delaware, unnamed marsh[7]		
Spartina alterniflora	3.8	5.1-6.3
Distichlis spicata		
Barataria Bay, Louisiana[8]		
S. alterniflora streamside	9.2	13.5
S. alterniflora interior	9.2	7.5

[1]Hicks (1973), using nearest long-term record, 1893-1971

[2]Redfield (1972)

[3]Bloom (1967) in Richard (1978)

[4]Harrison and Bloom (1974)

[5]Armentano and Woodwell (1975)

[6]Richard (1978)

[7]Stearns and MacCreary (1957)

[8]DeLaune et al. (1978)

nuclear testing (DeLaune et al. 1978), to estimate the time-averaged marsh build up during the past 10-20 years. These techniques may overestimate accretion rates because they do not include the compaction and organic decomposition that may slow the apparent accretion rate as the labeled sediment layer is buried deeper in the marsh. However, in a very careful detailed study of the sediments of Farm River Marsh on Long Island Sound, McCaffrey (1977) used the vertical distribution of ^{210}Pb in the sediments and long-term tide gauge records to show that the rate of marsh accretion appears to have increased, along with increases in the rate of sea level rise, at least since about 1900 (Figure 1). An interesting feature of the fine structure in the record is that there did not appear to be a reduction in accretion during several short-term (5-10 y) periods of falling sea level. McCaffrey interpreted this behavior, along with vertical profiles showing a remarkable uniformity in the distribution of organic matter and a variable distribution of inorganic matter, to mean "that peat formation, rather than particle deposition, is controlling accretion." Nevertheless, on a weight basis most of the marsh soil is inorganic sediment removed from the tidal water.

The dry bulk density of marsh sediment varies considerably, but a range of 0.2-0.4 g cm^{-3} appears to be reasonable (McCaffrey 1977, DeLaune et al. 1976). This compares with values of 0.6-0.7 g cm^{-3} for New England subtidal sediments and 1.2-1.4 g cm^{-3} for upland mineral soils. According to McCaffrey, the amount of inorganic material in Farm Creek was 0.14 g cm^{-3}. If we assume a mass of 0.15 g cm^{-3} for the deposited mineral sediments and an accretion range of 5-10 mm y^{-1} (Table 1), a reasonable range of sediment trapping for marshes appears to be about 750-1500 g m^{-2} y^{-1}. The various mechanisms controlling sedimentation on marshes have been recently reviewed by Frey and Basan (1978).

It is difficult to put the sediment trapping rate in a meaningful context, a problem that will appear often throughout this review. In order to try, I will adopt the Chesapeake Bay as a test case for a number of calculations. There are several reasons for such a choice - the Chesapeake is the "Queen of Estuaries" (or so we are told in the National Estuary Study 1970), it has extensive salt marshes(416 km^2), and Don Heinle (1979) has put together a convenient summary of its physical characteristics and salt marsh inventory. I realize that this choice will tend to minimize the apparent importance of the marsh in many ways, since the ratio of marsh to open water for the Chesapeake is only about 0.04 while it approaches 2 around Sapelo Island (Teal 1962). The importance of this ratio in arriving at any conclusions regarding the potential importance of marsh-estuarine fluxes will become increasingly apparent in later sections. In the end, however, the decision of how much water is ecologically coupled to how much marsh is arbitrary. One could argue that it would be better to choose one river estuary entering

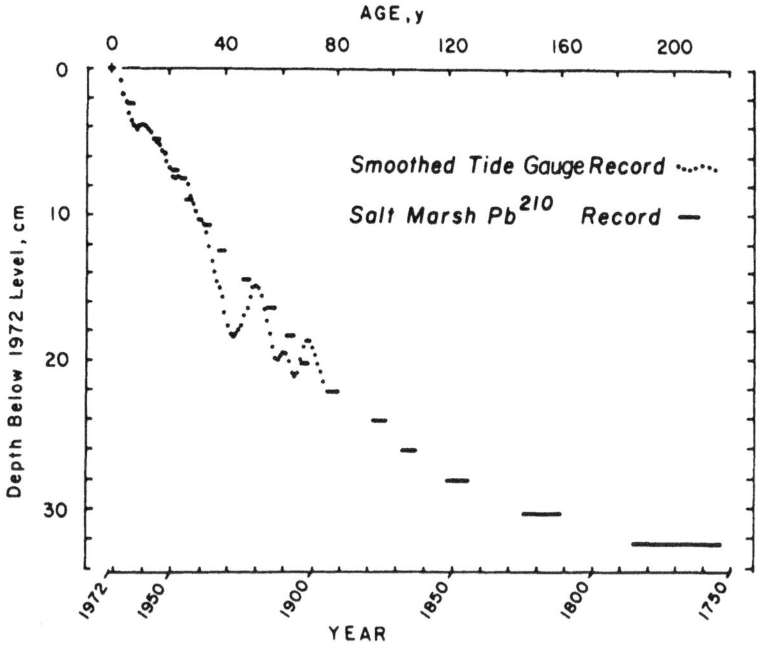

Figure 1. Variation in apparent sea level at New York City as shown by a smoothed tide gauge record and the elevation of a Spartina patens marsh in near-by New Haven, CT, calculated from the distribution of ^{210}Pb with depth in the sediment. From McCaffrey (1977).

the Chesapeake Bay or that it is more appropriate on the Georgia coast to include 10 or 20 km of the shelf water in calculating the ratio of marsh to open water (Blanton and Atkinson 1978).

If we choose the lower accretion rate for Chesapeake Bay and use Heinle's (1979) estimate of 416 km^2 of marsh, it appears that the marshes may remove over 3×10^{11} g of sediment per year. The average sediment input from the Susquehanna River, the major fresh water source, varies from 5 to 10×10^{11} g y^{-1} (Hayes 1978) and other sources of sediment may amount to an additional 8×10^{11} g y^{-1} (Schubel and Carter 1976). Thus, it seems likely that marsh deposition is an important term in the sediment budget for the Chesapeake, amounting to over 15% of the annual input. Presumably there is some lowering of water turbidity associated with this sediment removal, but it would be a tortuous exercise to attempt to estimate the magnitude of the effects.

It is likely that marshes are important in the short-term sediment dynamics of other estuaries as well. McCaffrey (1977) cites a study by Wolman (1967) in which sediment losses from watersheds with varying land use patterns are reported to range from about 40 g m^{-2} y^{-1} for forests through 80 g m^{-2} y^{-1} for urban areas to about 200 g m^{-2} y^{-1} for croplands. It is suggested that during periods of widespread construction the sediment yield might be more like 700-800 g m^{-2} y^{-1}. Thus, during periods in which there is no large construction, 1 m^2 of marsh may effectively trap the sediment lost from 4-40 m^2 of the water-shed, depending on soil type, land use, and probably a dozen other factors. However, since the current thinking is that major depositional and erosional processes are dominated by "high energy events" (big storms), it is hard to assess the role of marshes in the long-term "aging" of estuaries (Hayes (1978).

Accumulation of C, N and P in the Sediments

There is an accumulation of carbon, nitrogen, and phosphorus associated with the deposition of sediment and the accretion of marshes. But this does not mean that the marsh must necessarily be a sink for these materials with respect to the tidal waters. Other processes besides sediment deposition are involved, including rainfall, photosynthesis, respiration and decomposition, nitrogen fixation and denitrification, and fresh surface or ground water flow. If these other terms can be evaluated, the accretion rate and chemical analysis of marsh sediments can be used to give us some understanding of the probable magnitude of the net marsh-estuarine flux. The exercise is obviously more difficult for carbon and nitrogen because of the large atmospheric exchanges that take place on the marsh, but I think it is reasonable to try this approach with all three elements as a first step in getting at the question of marsh-estuarine exchanges.

Carbon

There appear to have been surprisingly few measurements of the organic carbon content of salt marsh sediments. However, if the scattering of "total" organic matter analyses which are available (measured as weight loss after ignition) are converted to carbon assuming that it represents 45% of the total, it seems that marsh sediments are much richer in organic carbon than those found offshore and perhaps those of open estuarine areas as well (Table 2). In his review of the organic content of estuarine sediments, Folger (1972) found that unpolluted areas away from marshes were characterized by organic carbon values lower than 50 mg g^{-1}. However, it is often difficult to compare the values reported in the literature

Table 2. Carbon, nitrogen and phosphorus concentrations in dried salt marsh sediments compared with those in fresh water and sub-tidal marine sediments.

Salt Marshes	C, mg g^{-1}	N, mg g^{-1}	P, µg g^{-1}
Providence River, RI (3 marshes)[1]	52	4.9	45 (31-53)
Narragansett Bay, RI (7 marshes)[1]	41	2.4	27 (10-42)
Block Island Sound, RI (2 marshes)[1]	44	3.6	20 (12-28)
North Carolina (6 marshes)[2]	16		33 (7-72)
Georgia[3]		1.5-5.0	293-431
Barataria Bay, LA[4]	80-160	5.6	20-35
Barataria Bay, LA[5]	120 (81-164)	7.2(4.6-8.8)	123 (100-148)
Lewes, DE[6]	40-240		
Fresh Water Sediments			
Chowan River, NC (3 stations)[7]			378 (46-902)
Lake Mendota, WI[8]			850 (719-961)
Pamlico River, NC (3 stations)[9]			1270 (980-1600)
Everglades Swamp, FL (5 stations)[10]			5440 (3740-7520)

[1] Nixon & Oviatt (1973), top 5 cm
[2] Broome et al. (1975), top 15 cm - a seventh marsh had P=696 µg g^{-1}
[3] N from Haines et al. (1976), P from P. R. Maye in DeLaune et al. (1976)
[4] DeLaune et al. (1976) annual range
[5] DeLaune et al. (1979) mean, range for 12 sites in one marsh
[6] Lord (1980) see Figure 2
[7] Brinson and Davis (1976)
[8] Wentz and Lee (1969)
[9] Upchurch (1972)
[10] Volk et al. (1975)

Table 2. (continued)

Marine Sediments	C, mg g^{-1}	N, mg g^{-1}	P, µg g^{-1}
Pamlico River, NC (4-10°/₀₀, 7 stations)[9]			803 (450-1610)
Pamlico River, NC (10-16°/₀₀, 11 stations)[9]			466 (20-860)
Narragansett Bay, RI (4 stations)[11]	11	1.1	325 (164-511)
Great Bay, NH[12]			285-475
Central Long Island Sound[13]	40-70		1160-1650
Eastern Long Island Sound[13]			990-1270
Gulf of Maine[12]			270-418
Matsushima Bay, Japan (3 stations)[14]	27	2.1	430
Open Ocean, Pacific[15]			1000
Open Ocean, Pacific[16]	2-3	0.9	
East Pacific Rise[15]			9000
Argentine Basin[17]	7-8	0.5-1.5	

[11] Sheith (1974)
[12] Lyons (personal communication)
[13] Lyons (1978) for P; Aller (1977) for C, top 10 cm St. NWC
[14] Okuda (1960)
[15] Froelich et al. (1977)
[16] Mueller (1977) top 10 cm
[17] Stevenson and Cheng (1972) top 10 cm

because different investigators have sampled different depth intervals in the sediment, and the distribution of organic carbon appears to vary considerably with depth in marshes (Figure 2; Valiela et al. 1976) and in subtidal sediments (Sheith 1974; Stevenson and Cheng 1972). The production of a large amount of underground root and rhizome material in marshes (Valiela et al. 1976) makes it especially difficult to interpret the sedimentary record, but on the basis of what few data we have, it appears that much of the organic carbon buried on the marsh is decomposed. This is in agreement with the measurements of high rates of respiration by salt marsh sediments (Teal and Kanwisher 1961; Hopkinson et al. 1978). It is the smaller amount of more refractory carbon that is effectively trapped in the sediment and removed from the estuarine system. The amount of carbon removed is a function of the carbon content of the deep sediment, the accretion rate, and the sediment density, but for most marshes the value appears to lie between 50 and 400 g C m^{-2} y^{-1} (Figure 3).

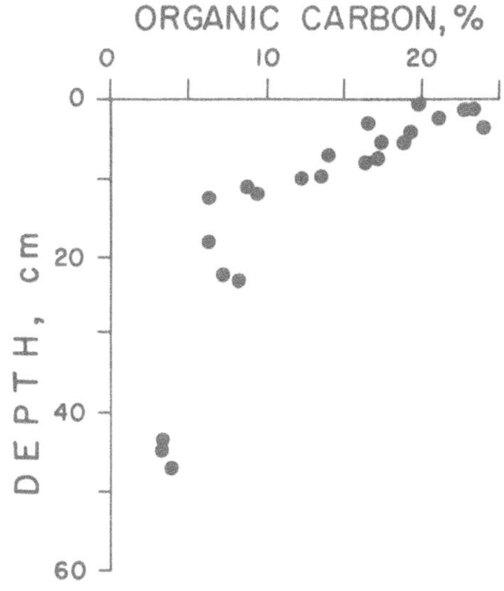

Figure 2. Distribution of organic carbon with depth in the sediment of a stunted *Spartina alterniflora* marsh in Delaware. From Lord (1980).

It is not clear, however, how much of this carbon is brought on to the marsh by tidal flooding and how much is fixed in the marsh from the atmosphere. Even if all of the organic carbon buried came from offshore, it might still be the case that photosynthetically

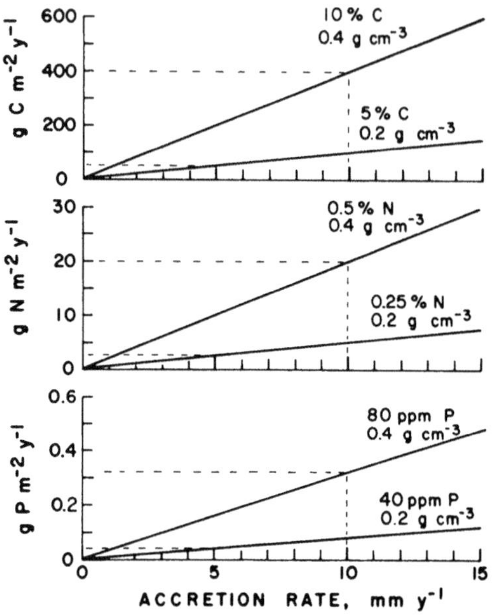

Figure 3. The accumulation of organic carbon, total nitrogen, and inorganic phosphorus in the sediments of a marsh as a function of accretion rate, sediment density, and sediment composition. On the basis of the data in Tables 1, 2 and 3, the probable range for most marshes is shown by the broken lines.

fixed carbon was exported from the marsh in greater quantity. If we assume from our earlier calculations that the annual sediment input is likely to be somewhere between 750-1500 g m^{-2}, and that the organic carbon content of this material is at least 50 mg C g^{-1}, the organic carbon input associated with sediment trapping would be 37-75 g C m^{-2} y^{-1} or more. This lies below and overlaps the lower end of the range of the amount of carbon that the literature suggests is being buried on marshes (Figure 3). While the numbers are not well enough constrained to allow us to go much further, it does seem that tidal inputs alone could account for a significant portion of the organic matter buried in some marsh sediments.

The atmospheric carbon input is a virtually unknown fraction of the annual primary production measured by harvest techniques. I am aware of only one study in which measurements of the annual net flux of carbon from the atmosphere to the marsh have been obtained. In their work at Flax Pond Marsh, Houghton and Woodwell (in press) have estimated that some 300 g C m^{-2} y^{-1} or 60% of the primary production on the marsh came from "new" atmospheric CO_2.

The rest of the plant biomass produced apparently came from CO_2 evolved from the marsh surface. The same authors report that 200 g C m^{-2} y^{-1} were buried on the Flax Pond marsh. Perhaps 50 g m^{-2} of this carbon was associated with the deposited sediment. If we assume that the marsh is "recycling" roughly the same amount of carbon each year, this means that of the 500 g C m^{-2} y^{-1} fixed on the marsh in plant biomass, 150 g would be buried, 200 g would be decomposed on the marsh, and only 150 g C m^{-2} y^{-1} or 50% of the "new" carbon input from the atmosphere would be available for export. The importance of carbon cycling within the emergent marsh itself will no doubt vary widely, but Houghton and Woodwell's work suggests that a significant fraction of the carbon fixed might be respired on the marsh and contribute to production the following year.

The flows of carbon between the atmosphere and the marsh appear large relative to the organic carbon deposited from the tidal waters on the marsh surface, but the net atmosphere-marsh exchange is very uncertain. Unfortunately, the balances between burial and atmospheric input, and between carbon associated with deposited sediment and carbon added to the sediment once it is on the marsh seem too close for us to learn much about the magnitude or direction of marsh-estuarine carbon fluxes from analyzing the marsh sediment.

Nitrogen

The nitrogen content of marsh sediments is, like carbon, higher than it is in offshore marine sediments (Table 2), but the ratio of organic carbon to nitrogen is higher (Table 3). The C/N ratio of the sediments or soils in the marsh probably reflects the fact that *Spartina* grasses have a higher C/N ratio than marine plankton, and contributes to the evidence that the carbon fixed on the marsh contributes significantly to the organic matter deposited in the sediment. The nitrogen buried in the sediments is in various organic and inorganic forms, and its abundance may decrease with depth in a manner similar to that of carbon (Haines et al. 1976). Unfortunately, I have not been able to find a detailed vertical profile of solid phase nitrogen for salt marsh sediments. However, if we assume that nitrogen values at depth in the sediments may be between 2.5 and 5 mg g^{-1} (Table 2; Haines et al. 1976), and proceed through the same exercise we followed for organic carbon, it appears that the amount of nitrogen buried on most marshes may amount to somewhere between 5 and 20 g N m^{-2} y^{-1} (Figure 3). An unknown, but appreciable fraction of this nitrogen may be brought on to the marsh along with the sediment that is deposited. If we assume that the deposited sediment has a nitrogen content of 1 mg g^{-1}, a conservative estimate, then from 2 to 4 g m^{-2} y^{-1} of the buried nitrogen may be from outside the marsh itself. This would suggest

Table 3. Atom ratios of carbon, nitrogen, and phosphorus in salt marsh and subtidal marine sediments

	C/N	C	N	P
Providence River Marshes[1]	12.4	2986	241	1
Narragansett Bay Marshes[1]	20.1	3931	195	1
Block Island Sound Marshes[1]	14.1	5631	400	1
North Carolina Marshes[2]		1209		1
Georgia Marsh[3]				
High Marsh				
0-5 cm	13.5			
10-15 cm	13.7			
25-30 cm	13.4			
Low Marsh				
0-5 cm	10.8			
10-15 cm	12.1			
25-30 cm	15.3			
Louisiana Marsh[4]	25.0	11274	451	1
Louisiana Marsh[5]	19.6	2500	128	1
Narragansett Bay Sediments[6]	11.8	92	7.8	1
Matsushima Bay Sediments[7]	15.0	161	10.7	1
Argentine Basin (∿6000 m)[8]	6.8			
Central Pacific Ocean (∿5000 m)[9]	2.6			

[1]Nixon and Oviatt (1973)

[2]Broom et al. (1975)

[3]Haines et al. (1976)

[4]DeLaune et al. (1976) median values

[5]DeLaune et al. (1979)

[6]Sheith (1974)

[7]Okuda (1960)

[8]Stevenson and Cheng (1972)

[9]Muller (1977)

that fresh water inputs or biological activities on the marsh are contributing a significant amount to the nitrogen that is being lost through burial. Since fresh surface and ground water inputs are probably not important in most Spartina marshes, and since the nitrogen input from precipitation has amounted to only 0.3 and 0.8 g N m^{-2} y^{-1} in the marshes where it has been measured (Haines 1976, Valiela et al. 1978), it is likely that biological processes on the marsh are most important.

We might first ask if it is possible for nitrogen fixation on the marsh to provide perhaps 3 to 16 g N m^{-2} y^{-1}. If the marsh cannot remove enough nitrogen from the atmosphere to account for the nitrogen buried that is in excess of that brought in with the deposited sediment, then we might conclude that the excess nitrogen is being removed from the tidal waters flooding the marsh. In which case the marsh will serve as a sink for total nitrogen with respect to offshore and estuarine waters.

There are a relatively large number of N-fixation measurements available for salt marshes, particularly during the summer (Table 4), and at least 5 groups have obtained enough measurements throughout a year to publish an estimate of the annual input of nitrogen (Table 5). David Smith (1979) has recently presented a review of all of the numerous and depressing problems with the acetylene reduction technique used in marsh N-fixation studies, and there may be much in what he says. But we must work with what is available, and I am struck by the remarkable similarity of the results, rather than by their variation. Considering the scatter that is usually found in ecological data, the consistency in the annual estimates of N-fixation by different areas of the different marshes is impressive. The estimates may all be wrong, but they are reasonably consistent in suggesting that N-fixation on salt marshes may add 5-50 g N m^{-2} y^{-1}, though the value seems more likely to be between 5-25 g N m^{-2} y^{-1} (Table 5). If the acetylene reduction measurements reported in the literature are accepted, it would seem that nitrogen fixation is of the right magnitude to account for the nitrogen buried on marshes. However, there is also a potential return of nitrogen from the marsh to the atmosphere through denitrification, and it is the value for the net exchange of nitrogen with the atmosphere that we must obtain.

Denitrification is even more difficult to measure than N-fixation, and there are very few estimates available (Table 6). The authors of the Louisiana study are not very confident of their estimate (Patrick, personal communication), and the Georgia value is based on data obtained only from December to June using a diffusion model and pore water N_2 profiles. The work reported by Kaplan et al. (1979) is the only detailed salt marsh denitrification study available, but the results from the other sites are similar.

Table 4. Approximate rates of nitrogen fixation in salt marshes during summer, mg N m^{-2} h^{-1}

	MARSH MUDS	TALL *SPARTINA ALTERNIFLORA* ZONE	STUNTED *S. ALTERNIFLORA* ZONE	HIGH MARSH *S. PATENS*, ETC.
Petpeswick Inlet, N.S.[1]	0.01-0.06	0.35-0.43	0.89-1.59	
Conrad Beach, N.S.[1]			0.78-1.00	
Sippewissett Marsh, MA[2]	0.35-0.85	1.50-3.00	1.50-3.00	0.50-1.25
Sippewissett Marsh, MA[3]	0.10-0.20 / 0.45-1.05	0.10-0.20 / 1.60-3.20	0.50-1.25 / 2.00-4.25	0-0.1 / 0.50-1.35
Flax Pond, L.I., NY[4]	0.04-0.81	0.02-0.28	0.03-0.10	
Horn Point Marsh, MD[5]	0.20			0.48
Rhode River Marsh, MD[6]				0.03
Sapelo Island, GA[7]		∿0.12-0.60	0.05-0.24	
Bank End Marsh, G.B.[8]	0.01-0.25	4.45-5.12	0.09-0.95	0.03-0.04
MEAN	0.3	1.6	1.0	0.4

[1]Patriquin and Denike (1978)
[2]Teal et al. (1979) bacterial fixation only
[3]Carpenter et al. (1978) blue-green algal fixation only. Values of 2-3 mg N m^{-2} h^{-1} were common in algal mats and pannes.
[4]Whitney et al. (1975). These measurements were made on slurries of marsh sediment using short term incubations. Values averaged 2.8 mg N m^{-2} h^{-1} for blue-green algal mats.
[5]Lipschultz (1978). Almost all of the fixation in the *S. patens* zone was bacterial while 80% of that on the mud flat was algal.
[6]Marsho et al. (1975). Marsh dominated by *Typha angustifolia*.
[7]Hanson (1977 a,b). Data for tall *S. alterniflora* zone estimated as ∿2.5x the measured rates in the stunted zone.
[8]Jones (1974). Plant zones were *Spartina anglica* for tall *S. alterniflora*, *Salicornia dolichostachya* for stunted *S. alterniflora*, and *Puccinellia maritima* for *S. patens*. Mud with algal mat ranged from 4.45-5.12 mg N m^{-2} h^{-1}.

Table 5. Annual estimates of nitrogen fixation in salt marshes, $g\ N\ m^{-2}\ y^{-1}$.

GREAT SIPPEWISSETT MARSH, CAPE COD, MA (Carpenter et al. 1978; Teal et al. 1979)	
Sand Bottom	0.6
Mud Bottom	1.7
Tall *Spartina alterniflora* Zone	9.2
Short *Spartina alterniflora* Zone	13
Spartina patens Zone	4.3
Area Weighted Mean	7.5
HORN POINT MARSH, CHOPTANK RIVER, MD (Lipschultz, 1978)	
Marsh Muds with Algae	0.7
Spartina patens Zone	2.3
Fresh Marsh *Hibiscus* Zone	0.7
SAPELO ISLAND, GEORGIA (Hanson, 1977)	
Spartina alterniflora Zone	20-50
BANK END MARSH, LANCASTER, ENGLAND (Jones, 1974)	
Mud Bottom	0.4
Mud with Algae	20
Creek Banks with Algae	32
Spartina anglica Zone	17
Salicornia Zone	2.7
Puccinellia Zone	5.0
BARATARIA BAY, LOUISIANA (Casselman 1979, DeLaune and Patrick in press)	
Mud Bottom	1.6
Tall *Spartina alterniflora* Zone	15
Short *Spartina alterniflora* Zone	5

Table 6. Annual estimates of net denitrification in salt marshes

	Denitrification, $g\ N\ m^{-2}\ y^{-1}$
Great Sippewissett, MA[1]	
Tall low marsh	12
Short low marsh	3
High marsh	4
Pannes	21
Algal mat	11
Emergent marsh average	4
Creek bottoms	22
Overall marsh average	7
Sapelo Island, GA[2]	12
Barataria Bay, LA[3]	3

[1]Kaplan et al. (1979), N_2 evolution

[2]Haines et al. (1976), pore water N_2 profiles and diffusion

[3]DeLaune and Patrick (in press), ^{15}N budget

For the purpose of argument, a value of 5-10 $g\ N\ m^{-2}\ y^{-1}$ lost from the marsh may be the best we can do. The authors considered their values to be a net N_2 exchange, since their measurements involved changes in N_2 concentrations and N-fixation was presumably proceeding along with denitrification. If this is correct, and denitrification exceeds N-fixation, the implication is that marshes are nitrogen sinks which remove dissolved and particulate nitrogen from incoming tidal waters. Once in the marsh, the nitrogen may undergo various biological transformations, but the end result would be a net loss of some 10 to 30 $g\ N\ m^{-2}\ y^{-1}$ due to the combined effects of burial and net denitrification (Figure 3, Table 6).

I am not sure how much confidence we ought to have in this kind of an analysis. It will not be easy to constrain these numbers much more, and they are all relatively close to balancing. The estimate of denitrification is badly in need of support. The data from Sippewissett Marsh may be particularly high because of large inputs of NO_3 in ground water (Teal and Valiela 1976), but the two other estimates, uncertain as they may be, do not suggest that this is a problem. The evidence may not be very compelling, but the data we have so far from the sediments and from measurements of marsh-atmosphere interactions suggest that over the long term there is a net flux of nitrogen from coastal waters into marshes.

Phosphorus

The evidence from the sediments concerning phosphorus may be more conclusive, since there are no biological exchanges with the atmosphere involved. But even for this element, the story is not simple. Virtually all of the data reported are for "available" or inorganic phosphorus extracted in various ways, and do not include phosphorus bound in organic matter. While the organic fraction is usually a small part (<15%) of the total phosphorus in subtidal sediments, it may amount to a considerably larger proportion of the total in marsh sediments. A conservative estimate based on the inorganic phosphorus content of some salt marsh sediments (Table 2) would be that between 0.05 and 0.3 g P m^{-2} y^{-1} are buried in the sediments of marshes (Figure 3). In a recent analysis based on total phosphorus measurement, however, DeLaune and Patrick (in press) estimated that 1.7 g P m^{-2} y^{-1} were being buried on a rapidly accreting marsh in Louisiana. This is considerably greater than the 0.01-0.1 g P m^{-2} y^{-1} that may fall on the marsh from atmospheric precipitation (Valiela et al. 1978; Graham and Duce 1979), and indicates that the marshes must be sinks for total phosphorus as they gain sediment from estuaries and coastal waters.

But there is more to be learned about phosphorus from the sediments. Many marine ecologists seem to be under the impression that marsh sediments are particularly rich in phosphorus. I think this situation has come about because of a comment in the paper by Pomeroy et al. (1967) that there was enough phosphorus in the top 1 m of Georgia marsh sediments to support the growth of Spartina for 500 years. Be that as it may, the evidence suggests that at least the inorganic phosphorus content of marsh sediments is very variable and often quite low compared with fresh water and offshore marine sediments (Tables 2 and 3). The few analyses of total phosphorus which are available support the conclusion that the phosphorus deficit of salt marsh sediments which is apparent in Tables 2 and 3 results at least in part from a loss of phosphorus rather than from the retention of a larger fraction of the phosphorus in organic form.

DeLaune and Patrick (in press) have obtained a detailed vertical profile of the distribution of solid phase total phosphorus in a Louisiana *Spartina* marsh which shows a marked decline with depth (Figure 4). Such a profile could result from a dramatic increase in the phosphorus content of sediments deposited during the last 7-8 years (sedimentation = 13.5 mm y^{-1}) or from a loss of phosphorus during the first 7-8 years the sediments are on the marsh. The latter seems much more likely. And as part of their studies of the chemistry of sediments in Branford Harbor, CT, Fitzgerald (1978) and his coworkers reported the results of total phosphorus analyses which show an interesting variation along a 2.8 km long transect:

Location	Sediment Depth Sampled	Total P, $\mu g\ g^{-1}$
Lower Branford River	0-45 cm	1612
Upper Branford Harbor	0-50 cm	1798
Mud Flat, Mid Harbor	16-26 cm	496
Low Marsh, Mid Harbor	0-35 cm	558
Mid-Branford Harbor	0-45 cm	930
Outer Branford Harbor	0-45 cm	1426

It would seem from these data that there is a remobilization and removal of phosphorus from the intertidal sediments.

I said earlier that we would come back to Reimold's (1972) report that *Spartina* was a "phosphorus pump," and this is a good point at which to do so. Instead of looking at ^{32}P losses from plant leaves, however, I would like to use the sediments as a longer term integration of possible plant-soil relations. If we extrapolate DeLaune and Patrick's total phosphorus curve to the surface, it looks as if sediments arriving on the Louisiana marsh have a phosphorus content of about 750 μg/g (Figure 4), and that they have declined to 500 μg/g by the time they reach a depth of 10 cm. Since it will require 7.4 years to reach this depth (DeLaune and Patrick in press), there is a loss of 33.8 μg P g^{-1} y^{-1}. The mass of sediment in the top 10 cm under 1 m^2 of streamside marsh is about 0.3 g cm^{-3} (DeLaune et al. 1976) x 10^5 cm^3 or 3 x 10^4 g. Since the loss of phosphorus amounts to approximately -33.8 μg g^{-1} y^{-1} through those 10 cm, the average flux of phosphorus from the sediments must be about 1 g P m^{-2} y^{-1}. There is some uncertainty in this calculation because I do not actually know the phosphorus content of the incoming sediment. I have also taken the sediment density from an earlier publication by the same authors working in the same area, but the data for the phosphorus profile may not be from the same core for which density data were reported. Nevertheless, DeLaune and Patrick (in press) have measured a phosphorus input to the marsh

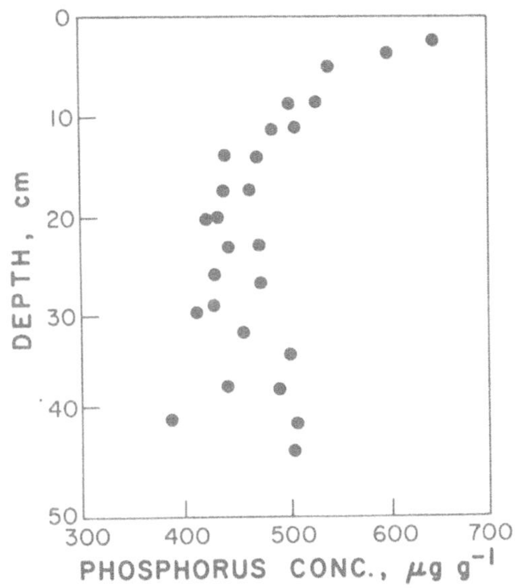

Figure 4. Distribution of solid phase total phosphorus with depth in the sediment of a Louisiana Spartina alterniflora marsh. From DeLaune and Patrick (in press).

where the vertical profile was taken of 2.3 g P m^{-2} y^{-1}, so I do not think my estimate of the phosphorus loss is likely to be too far off. Perhaps 30 to 60% of the phosphorus being deposited on the marsh is being removed by some mechanism.

The authors observed that Spartina alterniflora roots in the area of the streamside marsh were concentrated in the top 30 cm of sediment. Perhaps they were most active in the top 5 or 10 cm, where the removal of phosphorus was most rapid (Figure 4). If so, the situation in this marsh is quite different from the site in Georgia where Reimold (1972) showed a maximum uptake of ^{32}P at 1 m. In any case, the loss of about 1 g P m^{-2} y^{-1} from the marsh sediment in Louisiana is very much lower than the 222 g P m^{-2} y^{-1} that Reimold (1972) estimated was being "pumped" out of the sediment by Spartina in Georgia.

It is possible to examine the potential loss of phosphorus from marsh sediments with more general calculations. Let us assume for the purpose of illustration that the sediments at depth in marshes have 50 μg P g^{-1} and that the near-shore sediment being deposited on the marsh has 500 μg g^{-1} (Table 2). One might want to increase both these numbers, especially the value assigned to deep marsh sediments (where there may be an appreciable amount of organic phosphorus), but it is unlikely that one would want to argue for much more than a 450 μg g^{-1} difference between the source sediment and the buried sediment. Common estimates for marsh accretion seem to lie between 5 and 10 mm y^{-1} (Table 1) and for dry bulk sediment density between 0.2-0.4 g cm^{-3}. Combining the extremes gives a sediment deposition rate of 1 to 4 x 10^3 g m^{-2} y^{-1}. At steady-state, the flux of phosphorus from the marsh would then lie between 0.4 and 1.8 g P m^{-2} y^{-1}. It seems clear from this analysis that Spartina does not function as a "phosphorus pump" at anywhere near the rate reported by Reimold (1972), or, if it does, that it does not, and cannot, result in a net loss from the marsh of more than about one half of 1% of the material passing through the "pump." Even if the sediments initially contained 1000 μg P g^{-1}, and it were possible for Spartina to remove all of this phosphorus, it would require an accretion rate of 0.5 to 1 m y^{-1} just to supply the 222 g P m^{-2} y^{-1} that Reimold (1972) estimated was being "pumped" out by the grasses.

It is not clear what mechanism might be most important in removing the 0.4-1.8 g P m^{-2} y^{-1} (or less) from the marsh sediments. While Reimold's tracer study indicated that phosphorus "pumping" by Spartina must play a role, there is reason to believe that diffusion from the pore waters across the marsh surface may be more important. In his studies of nutrients in water draining from the marsh surface at low tide (when the grasses were not immersed) Gardner (1975) found a strong phosphate enrichment. But the problem is complex. Lord (1980) attempted to measure phosphate fluxes across the sediment-water interface using in situ chamber incubations on the high marsh in Delaware, but found no net flux. He attributed the lack of flux to precipitation reactions as iron and phosphate diffused together up into the oxidized surface sediment. Such a mechanism does not appear to dominate in subtidal marine sediments, however, and we have measured an annual net phosphate flux from Narragansett Bay sediments amounting to 3.7 g P m^{-2} y^{-1} (Nixon et al. in press). The surface of the sediments and the bottom water in the bay are oxidized. Of course, in offshore waters the phosphorus is continuously put back on the bottom, while water draining off the marsh removes the material from the system, at least if the flushing times of the marsh creeks are short relative to uptake or precipitation reactions. At least at certain times of the year in most marshes such conditions probably prevail. Eastman (1980) has, in fact, obtained mixing diagrams for phosphate in a Delaware marsh creek which appear to show that the salt marsh is a source of phosphate (Figure 5). It is likely, however, that at any one time or place on a marsh surface there may be a biological uptake of phosphate, a release, or some sort of precipitation. Lee (1979) has documented the metabolic patchiness of the marsh surface as reflected in nitrogen exchanges, and Clem and Lee (personal communication) have

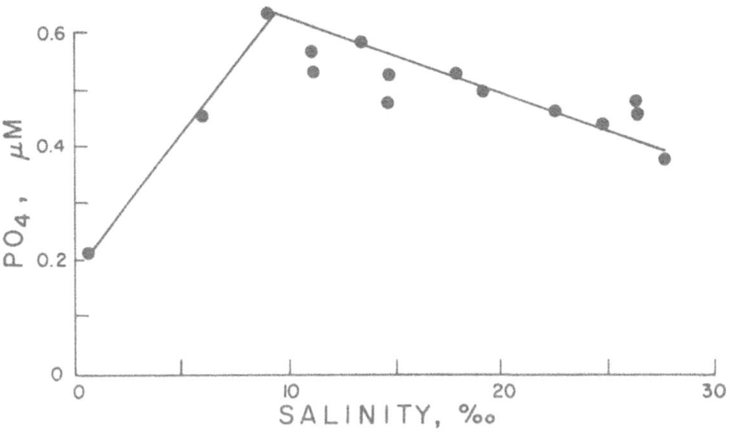

Figure 5. Mixing diagram for phosphate in a salt marsh tidal creek in Delaware. From Eastman (1980).

found similar variability in phosphate fluxes. Measurements such as those made by Gardner (1975) or in situ flume studies such as those used by Lee (1979) and Wolover et al. (1979) probably give a more integrated picture of overall marsh surface behavior than isolated chamber flux measurements. In spite of the variability shown by the short-term exchange measurements, a consideration of the sediments and the potential atmospheric input confirms that in the long run, marshes are sinks for total phosphorus with respect to the coastal waters. But on the basis of an admittedly small number of analyses, it also appears likely that an interesting transformation may take place. The remobilization and loss of some of the phosphorus from sediments buried on the marsh may, in fact, make marshes a source of reactive and organic phosphorus for adjacent waters.

Accumulation of Metals in the Sediments

Just as the accumulation of sediment on the marsh makes it a sink for phosphorus, so is it a sink for the trace metals associated with particulate material in the tidal waters (Windom 1975). But our preoccupation with tidal exchange often makes us forget that as intertidal systems, salt marshes receive inputs from the atmosphere and from upland drainage as well. In fact, this situation has been largely responsible for the few studies of trace metals which we have available. The interest of the geochemists in salt marshes appears to be relatively recent, and it has arisen in no small part because the marshes appear to be a good recording device for anthropogenic pollution inputs from the atmosphere as well as from the sea.

Lead

In 1972, Siccama and Porter published a report showing that lead concentrations increased dramatically in the sediments above about 25 cm in a salt marsh near New Haven, CT. Their interpretation of the data was that this profile reflected the increasing amount of lead in highway runoff with the increasing use of gasoline containing tetraethyl lead since about 1935. They concluded "that salt marshes are effective in removing lead from inflowing waters from urban centers and this lead is accumulating in the marsh." Two years later, another paper on this subject appeared titled, "Export of Lead from Salt Marshes" by Banus, Valiela and Teal (1974). Solid phase Pb concentrations had been determined at various depths in 45-100 cm long cores from the S. alterniflora zone in 3 salt marshes in Massachusetts, and it was again apparent that there was a dramatic increase in Pb in the more recent sediments (Figure 6). A marsh from an urban area near a "heavily travelled expressway" also showed much higher levels of excess Pb than the more rural sites. The cores were not dated, but the authors assumed an accretion rate of 0.9 mm y^{-1} and speculated that early increases in Pb began shortly after Cape Cod was settled in the mid 1600's as a result of "extensive hunting." An examination of Figure 6 and a consideration of Table 1 suggests to me, however, that the real increase in Pb is considerably more recent and that it is not necessary to implicate Pilgrim duck hunters as a significant anthropogenic input. The authors did show that Spartina on the marsh with higher Pb in the sediments also contained higher concentrations of Pb in the above-ground portions of the plants. It was this observation, coupled with Teal's old (1962) estimate that 45% of Spartina was exported from the Georgia marsh, that must have been responsible for the title of the paper, since no data showing an export of Pb from the marsh are given. Even if the 45% export figure is used, however, the loss of Pb from the marsh in grass detritus would only represent about 10% of the 11.2 mg Pb m^{-2} y^{-1} that the authors estimate is presently being input from the atmosphere. In later studies, the same authors (Banus et al. 1975) have shown that even larger additions of Pb to marshes are almost completely held in the marsh sediments and "that salt marshes act as sinks for lead..." Their rough budgets show that less than 4% of Pb inputs may be exported, even with higher lead levels in the grass of Pb-enriched marshes and assuming that 50% of the above-ground grass production is exported.

Siccama and Porter's (1972) paper helped persuade Richard McCaffrey at Yale to do a detailed study of Pb and a number of other

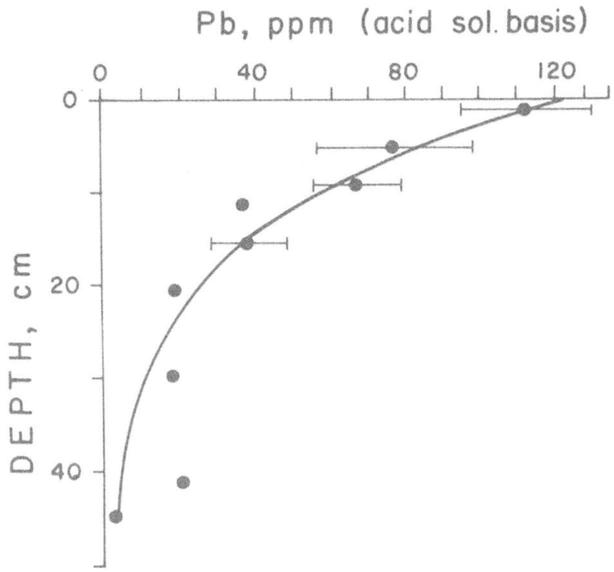

Figure 6. Distribution of solid phase lead with depth in the sediment of a Massachusetts salt marsh. Modified from Banus et al. (1974).

trace metals (Mn, Fe, Cu and Zn) in Farm Creek salt marsh, about 8 km from New Haven, CT. The report of McCaffrey's (1977) work provides an excellent analysis of the accumulation of trace metals and sediment in the S. patens marsh. Using cores "dated" with detailed ^{210}Pb profiles, he found an erratic but increasing Pb concentration in the sediments corresponding with increasing industrial activity from about the end of the Civil War, and concluded that "there is no obvious contribution due to the introduction of leaded gasoline in the 1920's." The analysis of Farm Creek marsh cores allowed McCaffrey to develop and test a model for trace metal budgets which suggested that atmospheric inputs (rather than highway runoff as in Siccama and Porter's study) probably accounted for over 90% of the Pb in the sediments (Table 7). The Pb profile at the site appeared to be supported by an input of some 90 mg Pb m^{-2} y^{-1}, a flux 8 times greater than calculated by Banus et al. (1974). However, the S. patens zone studied by McCaffrey was high marsh, exposed to the atmosphere over 90% of the time, while the S. alterniflora zone analyzed by Banus and co-workers would be submerged considerably more often.

Manganese

While marshes also appear to be sinks for total Mn, the major

Table 7. Estimated inputs of trace metals from tides and the atmosphere to the Farm River, Connecticut salt marsh (McCaffrey 1977) $\mu g\ cm^{-2}\ y^{-1}$

	Fe	Mn	Cu	Zn	Pb
Atmosphere	30	0.6	5	13	17
Tidal Input	2600	45	1	4	1
Calculated Input	2630	46	6	17	18
Measured Input	3600	<78	10	17	9

source of this metal is the sediment brought in by the tides, rather than atmospheric deposition (Table 7). The behavior of Mn once it is in the salt marsh is also dramatically different from that of Pb, as shown by several pieces of evidence. Solid phase Mn profiles found by McCaffrey (1977) in the *S. patens* marsh in CT, and by Charles Lord (1980) working in the stunted *S. alterniflora* marsh in Delaware, have shown marked depletion in Mn below the surface few cm of sediment (Figure 7). Presumably, the reduction of MnO_2 in the anoxic deeper marsh sediments liberates soluble Mn^{+2} which diffuses out of the sediment into the overlying water.

On the basis of his Mn profiles from Farm Creek marsh, McCaffrey estimated that the loss of Mn amounted to about 0.1 $g\ m^{-2}\ y^{-1}$, or about 50% of the input. While such a flux is significant in influencing the composition of the sediments, it is not necessarily an important term in the Mn budget of estuarine and near-shore waters. Numerous measurements over an annual cycle of the flux of Mn from subtidal sediments in Narragansett Bay have shown an input from the bottom to the overlying water of about 4.8 $g\ Mn\ m^{-2}\ y^{-1}$ (Carlton Hunt, Univ. Rhode Island, personal communication). It is not clear, however, how much of this benthic Mn flux is a real input of "new" Mn from the sediments (as inputs from the marsh would be) and how much is recycled Mn that has recently come out of the water column. However, as we noted earlier, McCaffrey's flux estimate was based on Mn profiles in the *S. patens* marsh. Much higher Mn fluxes were actually measured *in situ* with chambers by Lord (1980) in a stunted *S. alterniflora* marsh. In his study, the flux of Mn from the marsh surface was found to increase linearly from about zero to almost 4.4 $mg\ Mn\ m^{-2}\ h^{-1}$ as the concentration of MnO_2 in the surface sediment increased from zero to 250 $\mu g\ g^{-1}$. Interestingly, the higher flux rates were found in early spring,

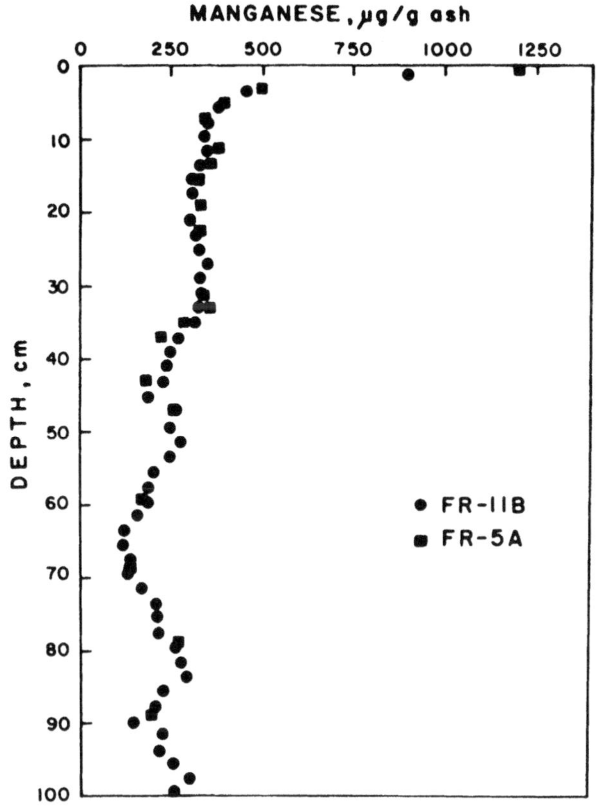

Figure 7. Distribution of solid phase manganese with depth in the sediments of two cores taken in a Spartina patens marsh near New Haven, CT. From McCaffrey (1977).

just after the thaw. Ice cover in winter apparently prevented the Mn from escaping and there was a buildup in the surface sediments (Figure 8). Mixing diagram for Mn as a function of salinity in the Delaware marsh creek were obtained by Kurt Eastman (1980) during December, and suggest that Mn is behaving conservatively in the marsh during winter (Figure 9). Lowest fluxes were found in late summer and fall after the sediments were depleted in Mn. This seasonal pattern of release is the opposite of that found in the subtidal sediments. Moreover, the highest fluxes reported by Lord (1980) are about 4 times higher than usually measured during the summer in Narragansett Bay (Hunt, personal communication).

Removal of Mn from the sediments through uptake by Spartina and the export of detritus is also possible, but difficult to assess for reasons discussed earlier. Aside from the fact that the

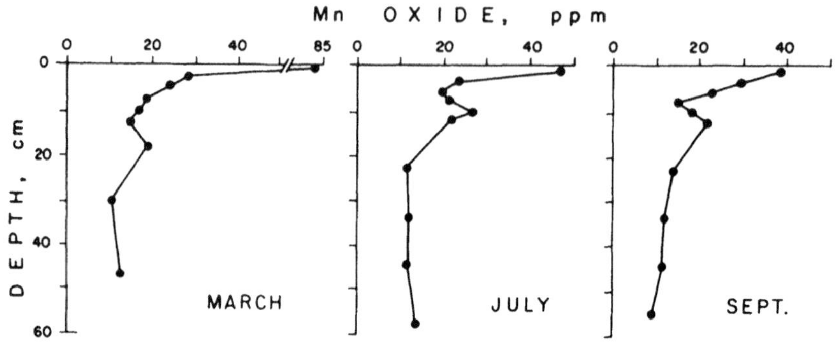

Figure 8. Distribution of manganese oxide with depth in the sediment at various times during the year in a stunted *Spartina alterniflora* marsh in Delaware. From Lord (1980).

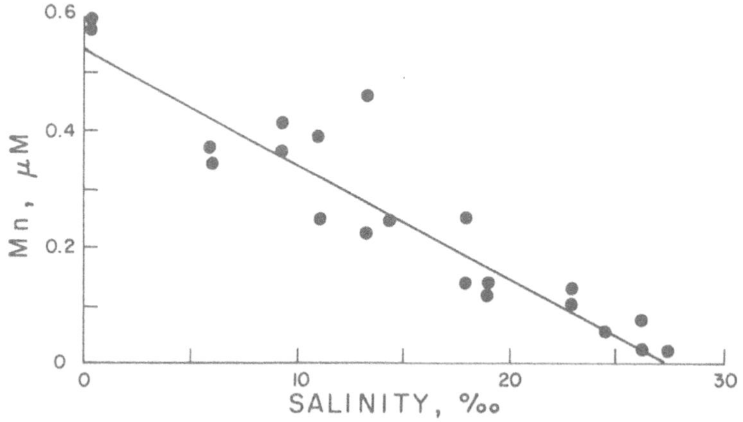

Figure 9. Mixing diagram showing apparently conservative behavior for dissolved manganese in the waters of a salt marsh tidal creek in Delaware. From Eastman (1980).

magnitude of the export of detritus is elusive, a recent study by Rice (1979) has shown very dramatically that the concentrations of metals in detritus are quite different from those in fresh *Spartina*. In contrast to all of the other metals he examined, Mn was rapidly lost from *Spartina* detritus and remained at less than 50% of the original value even after 150 days. In spite of the fact that *Spartina* could play a role in the removal of Mn from the marsh through particulate export or leaching, the magnitude of the *Spartina* pathway is probably small relative to the sediment-water flux. For example, if *Spartina* contains about 40 μg Mn g^{-1} (Rice 1979) and the above-ground production is 1000 g m^{-2} y^{-1}, the maximum Mn loss would be on the order of 0.04 g Mn m^{-2} y^{-1}. This flux would be exceeded by the sediment-water flux in less than 10 h at the maximum rate observed by Lord (1980).

Uncertainties in our knowledge of the importance of various terms in the Mn budget of coastal waters, including its uptake by the plankton, make it difficult to know if marshes really are important in "smoothing" the estuarine Mn cycle. If we go back to using the Chesapeake Bay an an example, we can put the potential marsh Mn flux in some perspective. Carpenter and coworkers (1975) have estimated that the annual Mn input to the bay from the Susquehanna River during 1965-66 was some 5×10^9 g y^{-1}. Taking the area of the Chesapeake as 11×10^3 km^2 (Heinle 1979), this is equivalent to about 0.45 g Mn m^{-2} y^{-1} input from the river. If McCaffrey's (1977) estimate is increased an order of magnitude to 1 g Mn m^{-2} y^{-1} input from the marshes, and the marsh area is 416 km^2 (Heinle 1979), the contribution to the Chesapeake from marshes would still be only about 0.04 g Mg m^{-2} y^{-1}. Of course, in areas with little river input and a large marsh to open water ratio, the balance might be quite different.

Other Metals

In the same year that Siccama and Porter's paper on lead appeared, Strom and Biggs (1972) published data on Zn, Cu, Cr and Fe in relatively long cores from the Great Marsh near Lewes, Delaware. Their analyses began about 30 cm below the surface and provided concentration measurements at intervals of over 50 cm. Unfortunately, the distribution of all of the metals was virtually uniform with depth and there is little that can be learned from this work.

More recent studies have provided data on iron (DeLaune et al. 1976, McCaffrey 1977, Lord 1980), copper (McCaffrey 1977), zinc, and cadmium (Banus et al. 1975) in near-surface marsh sediments. There must certainly be other studies available, but there are few like McCaffrey's or Lord's in which detailed and reliable vertical profiles have been obtained.

It might seem that Fe would follow the same pattern as Mn, but this appears not to be the case. The solid phase Fe profiles found by McCaffrey (1977) and Lord (1980) do not indicate a loss of iron from the marsh. The kinetics of iron oxidation are much faster than for MnO_2 formation, and it appears that Fe is effectively reprecipitated in or on the oxidized surface sediments. This observation seems to argue against earlier speculation by Williams and Murdock (1969) that the high concentration of Fe in Spartina detritus makes it an important factor in estuarine iron cycling. However, the flux calculated by Williams and Murdock amounted to an uptake of only 2.6 g Fe m^{-2} y^{-1} by Spartina detritus, while deposition may provide 18-75 g Fe m^{-2} y^{-1} assuming a sediment input of 750-1500 g m^{-2} y^{-1} and an iron concentration of 25-50 mg Fe g^{-1} (Windom 1975, McCaffrey 1977). It is not likely that the removal of a fraction of the Fe incorporated in detritus each year would be evident in the vertical Fe profile in the sediments, even if the amount of Fe in the detritus were to double over the aging process, as found by Rice (1979). In fact, the strong uptake of Fe by Spartina detritus after about the first 10 days may serve to remove Fe from tidal waters. In any case, it appears that the offshore waters and sediments (along with fresh water inputs when they are present) rather than the atmosphere are the primary sources of Fe for the marsh (Table 7).

The Cu and Zn profiles obtained by McCaffrey (1977) showed a relatively linear increase since about the mid 1800's, reflecting an anthropogenic input largely through atmospheric deposition (Table 7). There is no evidence of any appreciable mobilization of either metal from the sediments. Shallower cores analyzed by Banus et al. (1975) showed little, if any, increase in Zn above 15 cm. The latter authors concluded that the sediments were in steady-state with respect to Zn, but went on to speculate that the Zn removed by Spartina export (assumed to be 50%) must therefore be replaced in some fashion from offshore by inputs over and above sedimentation. An alternative explanation might be that there is little, if any, Zn going offshore in detritus.

Much of what little literature there is dealing with the exchange of metals between marshes and offshore waters turns on the question of the magnitude of Spartina detritus export. For the most part, analyses of the elemental composition of Spartina leaves and/or stems have been combined with production data and Teal's (1962) estimate of 45% export to arrive at an estimate of the loss of metals from marshes (Pomeroy et al. 1967, Zn; Williams and Murdoch 1967, Zn, Mn, Fe; Windom 1975, Cu, Mn, Fe, Cd, Hg). The most extensive analysis of this kind, that by Windom (1975), concluded that, with the possible exception of Hg, this was not a major pathway in estuarine trace metal cycling. The recent study by Rice (1979), as well as some of the earlier analyses by Williams and Murdoch (1967), have shown that Spartina detritus takes up some metals such as Al, Fe, Cu, and perhaps Zn, and concentrates them as much as several fold above levels found in the living plant. Thus, while Spartina may give up some metals for a short time after the leaves die (Rice 1979), the detritus may be an active net uptake site for other metals as it ages. If most of the detritus remains in the marsh, it may thus serve as a mechanism for trapping some materials rather than as a pathway for export. Overall, however, it is the bulk sediment budget which dictates that marshes are sinks for metals.

EVIDENCE FROM DIRECT FLUX MEASUREMENTS

I quoted earlier from the discussion which followed John Teal's presentation of his energy budget at the Salt Marsh Conference on Sapelo Island in 1958, and I would like to begin this section of my review by going back to that dialogue and to the comments of some other members of the audience. After the discussion had gone on for a while, A. C. Redfield asked a question of Robert Ragotzkie who, you will remember, had the problem of trying to measure the detrital export that Teal's energy budget suggested might be taking place.

Redfield: How are you going to find out how much *Spartina* goes out to sea? There should be a lot of it.

Ragotzkie: I don't know, but physically it is very difficult. The transport would be tremendous during storm tides. These would move out a lot of material in perhaps two or three months when there would be no net consumption.

Burkholder: Isn't a lot of this detritus being sloshed back and forth by the tides? It goes off the marsh and back again on the marsh. I think it is a very complicated business.

Redfield: Very complicated indeed...

I think it is safe to say that all of the efforts which have been made during the past 20 years to measure marsh-estuarine exchanges have confirmed the wisdom of these early assessments of the difficulty of the task. As far as I know, Bob Ragotzkie never did figure out how to get a reliable set of flux measurements, and no one else published an annual carbon flux measurement until 17 years later, when Settlemyre and Gardner (1975) reported on particulate organic losses from Dill Creek, S. C. In the past five years a number of papers have appeared dealing with aspects of the problem of marsh-estuarine carbon and nutrient exchanges, and there are now perhaps a half dozen or more marshes for which annual exchange estimates have been reported. But after reading the literature and trying to do some of this kind of work myself, I am skeptical that anyone has yet learned how to get a really credible set of long-term net flux measurements. Paul Burkholder and Bob Ragotzkie certainly hit on some of the major problems right away, but there are others equally daunting.

On Measuring Fluxes and Flows

As an introduction, let me set out the problem quickly for those who may not be familiar with this kind of work. The effort involves determining the exchanges of water between a marsh and some "offshore" source of tidal water and measuring the concentrations of dissolved and/or suspended material in the water entering and leaving the marsh. For a simple case, multiplying the differences in concentration between flood and ebb tides by the volume of water exchanged provides an estimate of the net flux of material. There are, of course, various ways of estimating the water exchange and the appropriate concentration of material to use in the calculation, and there are a number of ways in which one can handle the resulting data - averaging concentrations, integrating "instantaneous" fluxes, etc. I have tried to summarize very briefly the different approaches used in the better-known marsh-estuarine flux studies (Table 8), but it is necessary to read the original literature to really appreciate the tortuous efforts, assumptions, and uncertainties that lie behind such a deceptively simple number as an annual flux value. For the most part, these studies have not included a rigorous analysis of the uncertainties involved in their measurements or their calculations, and much of what we have learned about this problem has come from the efforts of two people, John Boon (1975, 1978) and B. J. Kjerfve (Kjerfve et al. 1978, Kjerfve and Proehl 1979).

Since the differences in concentration of all materials between flood and ebb tides are usually small, the dominant term in the net flux calculation is often the exchange of water. While Boon and Kjerfve's data apply strictly only to the two systems they have studies, their results document the effort needed to obtain accurate water exchange measurements and the importance of a rigorous analysis of these data to provide an estimate of the error involved in the flux calculation. As Boon (1978) pointed out, "...any inference concerning the net transport of the material during one or more tidal cycles depends on the magnitude of the net transport being at least as large as the error for either the flood or ebb portions of the total transport." Both Boon and Kjerfve have developed convincing evidence that it is not easy to measure the discharge over any given tidal cycle accurately, and it is my impression that

Table 8. Various approaches used to obtain estimates of carbon and nutrient fluxes between marshes and estuarine waters using direct measurements

LOCATION	WATER FLUX ESTIMATE	NUTRIENT FLUX ESTIMATE
Great Sippewissett Marsh, Cape Cod, MA (Valiela et al. 1978), 40 m wide channel	current meters, 2 depths every 5 m at mid-flood and mid-ebb on 7 tidal cycles. Regression analysis used to relate discharge from these data to current speed at 1 station plus tide height. One station current speeds taken monthly with chemical measurements.	water samples from 1 station collected hourly over a tidal cycle monthly for 19 months. Flux calculated from discharge and measured concentration.
Flax Pond, Long Island, NY (Hall et al. 1975, Woodwell et al. 1977), narrow tidal channel	Tidal exchange estimated from tide height and hypsographic curve.	water samples from 1 station every 4 h over a tidal cycle once each week over an annual cycle.
Canary Creek, Lewes, DE (Lotrich et al. 1979)	current meter used to obtain a mean tidal prism on 11 tidal cycles. Flood and ebb tides were assumed to be of equal volume and the fraction of total tidal exchange volume as a function of time in the tidal cycle was determined for various tide heights. This relationship was used with chemical concentrations measured at various times in the tidal cycle to obtain fluxes.	water samples collected hourly during a tidal cycle once each month over a 2 year period.

Table 8. (continued)

LOCATION	WATER FLUX ESTIMATE	NUTRIENT FLUX ESTIMATE
Gott's Marsh, Patuxent River, MD (Heinle and Flemer 1976), 6 m wide creek mouth	exchange estimated from tide height and hypsographic curve during the first year, from current meter measurements at 1-4 points during the second year. A mean flood and ebb tidal volume was calculated from 25 tidal cycles measured in various ways over the 2 year study.	water samples collected hourly for 13 h from 2 depths at one station monthly for 2 years. Average flood and average ebb concentrations were calculated for each month and multiplied by the mean flood or ebb tide volume to arrive at a mean monthly flux.
Ware Creek and Carter Creek, York River, VA (Moore 1974, Axelrad 1974), 7 m and 14 m wide creek mouths	exchange estimated from tide height and hypsographic curve and from current meter measurements at 1 station every 20 min. Current meter water fluxes adjusted so that flood and ebb tidal volumes were made equal. Measurements made at about mean tide each month for one year.	water samples collected hourly from one depth and station over a complete tidal cycle each month for one year. Fluxes were calculated using the two different water exchange methods and averaged.
Dill Creek, Charleston, SC (Settlemyre and Gardner 1975), 60 m wide creek mouth	exchange estimated from tide height and hypsographic curve.	composite flood and ebb tide water samples analyzed and weighted according to the hypsometric model for 25 tidal cycles over one year.

Table 8. (continued)

LOCATION	WATER FLUX ESTIMATE	NUTRIENT FLUX ESTIMATE
Sapelo Island, GA (Haines 1977), diked experimental marsh	"...a dense array of current meters..." Data for 8 tidal cycles over 16 months showed flood tide volume exceeded ebb by 10-20%, so accurate budgets could not be obtained.	"...frequent sampling of water parameters..." A time and volume weighted average concentration was used to arrive at a rough estimate of the flux.
Barataria Bay, LA (Happ et al. 1977)	average flushing estimates (derived from the volume and area of the system) and the tidal prism were taken from the literature.	water samples were collected monthly for one year from numerous stations in the marsh-bay area and offshore. The average annual concentration gradient was applied to a first order mixing model to estimate the annual flux.

virtually all of the direct flux studies published thus far (Valiela et al. 1978 may be an exception) suffer in this regard.

It is also troublesome that there are marked flood-ebb asymmetries in the discharge of salt marsh systems, and it appears that on any given tidal cycle a marsh is likely to show a net accumulation or loss of water (Table 9)(Boon 1975; Kjerfve and Proehl 1979). There are a number of factors in addition to the regular lunar tidal cycle which may influence the relative magnitude of flood and ebb tides, including wind, fresh water input, continental shelf waves, and short and long term variations in mean sea level caused by changes in atmospheric pressure and sea temperature (Kjerfve et al. 1978). The potential importance of seasonal changes in mean sea level is seldom appreciated in studies of marsh-estuarine interactions, but Kjerfve and his colleagues have provided a convincing analysis. By comparing the topography of the North Inlet marsh in South Carolina with careful measurements of mean sea level, they have shown that, when averaged over all of 1974-75, the top of the marsh surface was covered by tidal water 30% of the time. However, the extent of submergence varied from a monthly average of 42% of the time in October to 27% in January as mean sea level went from maximum to minimum.

Table 9. Comparison of rising and falling tidal range and ebb and flood discharge rates for three consecutive tidal cycles in the North Inlet, S. C. marsh-estuary. After Kjerfve and Proehl (1979).

	Tidal Cycle 1	Tidal Cycle 2	Tidal Cycle 3
Rising range (m)	1.81	2.41	1.64
Falling range (m)	2.05	1.99	2.27
Max. flood discharge (m^3/s)	1,392	1,888	1,218
Max. ebb discharge (m^3/s)	-1,599	-1,571	-2,135
Cumulative flood flow ($10^6 m^3$)	17.97	24.84	15.19
Cumulative ebb flow ($10^6 m^3$)	-22.11	-19.53	-30.47
Net discharge (m^3/s)	-92	118	-340

The effect of such short and long-term asymmetries on the calculation of net material fluxes from a set of single tidal cycle samples collected monthly or even weekly over a year is a depressing thought. Nevertheless, the lesson that Boon and Kjerfve have provided is as instructive as it is painful, and it may help us to understand some of the vagaries apparent in the results of the various material flux studies. With the exception of a study of Great Sippewissett marsh (Valiela et al. 1978), all of the studies which have attempted to use current meters to obtain direct measurements of tidal discharge have found an imbalance in the tidal water budget for the marshes during the study interval. Various authors have handled this problem in different ways (Table 8), but the conclusion that Boon and Kjerfve's findings reflect a general behavior of salt marsh systems seems unavoidable.

Much of the work reported in the literature, however, has been based on the use of time-averaged concentration data and exchange estimates taken from hypsographic models rather than from direct discharge measurements. None of these studies has included a good discussion of the uncertainties associated with this approach or with the assumption that a time-averaged concentration can be assigned uniformly to a tidal prism with any confidence. Moreover, Boon (1975) compared the net discharge obtained from a hypsometric model with detailed current meter measurements on 11 tidal cycles in Little Fool Creek marsh in Virginia, and found no agreement between the two methods. He concluded that "...the determination of time-varying tidal discharge by means of an area-height model, rather than by direct flow measurements, is not recommended for the calculation of net transport."

The end result is that while the past 4 or 5 years have provided us with a number of studies which have attempted to obtain direct measurements of the annual flux of material between coastal marshes and coastal waters, most, if not all, of these efforts did not attend closely enough to the problem of water exchange and did not provide us with the data needed to place any sort of confidence estimate on the results. We are left with a very large amount of information which we must admit is of very uncertain quality.

But having said all of this, I must admit that it was exciting in preparing this review to see the results of all of these studies come together. There is a remarkable agreement among the various papers with respect to the direction and magnitude of the net flux of most materials which I cannot easily reconcile with the potential errors and uncertainties in the methods used. Moreover, the general results of the direct flux studies, at least for carbon and phosphorus, are consistent with the conclusions we reached earlier on the basis of the chemical analyses of marsh sediments. Perhaps it is all fortuitous, a house built of straw. But at the very least, a considerable number of able people have made a determined effort to test a major piece of ecological dogma, and their efforts deserve our attention. There is even a nice historical symmetry here in that John Teal has been part of one of the most thorough of these efforts with his recent work in Great Sippewissett marsh.

The Direction of Marsh-Estuarine Fluxes

There are 12 tidal marshes I am aware of for which sufficient information has been reported to make an assessment of the direction of the annual net exchange of at least some forms of carbon, nitrogen, and/or phosphorus (Figure 10). The sites studied range from New England to the Gulf of Mexico and encompass a salinity range from tidal fresh water to high salinity coastal sea water. The notation of export or import or no net exchange given in Figure 10 reflects the results given in the original literature. I will hold off discussing the probable magnitude of the various fluxes and the nature of the data until later. There is at least one additional major effort presently underway to measure net exchanges between coastal marshes and coastal waters that is not included in Figure 10. The North Inlet study in South Carolina has already begun to produce exciting data (Kjerfve and Proehl 1979, Chrzanowski et al. 1979), but it is still too early to present any picture of seasonal variations or annual fluxes. Similarly, I am reluctant to extrapolate annual exchanges from the 4 detailed flux studies reported earlier from a tidal creek sub-system of the North Inlet estuary (Erkenbrecher and Stevenson 1978). It is worth pointing out, that the most consistent findings of Erkenbrecher and Stevenson's study were net exports of ATP and Chl a and imports of total suspended material in a high-marsh creek.

While there are certainly some discrepancies in the results of the various studies, it seems to me that there is general agreement that tidal marshes export dissolved and particulate organic carbon, that they export dissolved organic nitrogen and that they export dissolved phosphorus, at least as phosphate and probably in organic forms as well. I think there is also sufficient agreement to say that marshes appear to take up nitrate and nitrite. The situation with regard to ammonia and particulate nitrogen and phosphorus is not clear, and I do not find that surprising in light of what we know about nutrient dynamics in coastal marine ecosystems (Nixon in press). Ammonia is the major form of nitrogen exchanged across the sediment-water interface, it is the form excreted by marine animals, and it is the form preferentially taken up in light and dark by bacteria, phytoplankton, seaweeds, and epibenthic algae. Particulate nitrogen and phosphorus are associated with the suspended inorganic material which shows a net import, and with the detrital organic material which shows a net export. It is not unreasonable that the balance may shift in response to a great variety of factors, including sediment type and accretion rate.

The exceptions to the general behavior evident in Figure 10 are instructive as well. The import of POC by Flax Pond may be attributed to intense filter feeding by a mussel bed near the mouth of the tidal inlet (C. Hall, personal communication), and the

NET ANNUAL TIDAL FLUXES BETWEEN SALT MARSHES AND COASTAL WATERS

MARSH	‰	CHL a	DOC	POC	PN	DON	NH_4	NO_2	NO_3	PP	DOP	PO_4	ΣC	ΣN	ΣP
Great Sippewissett Cape Cod, MA	28-33			E	E	E	E	E	E			E			
Providence River Rhode Island	20-30								I						
Block Island Sound Rhode Island	30-33								I						
Flax Pond Long Island, NY	~26	I	O	I			E	I	I	I		E	I		E
Hamilton Marsh Delaware River, NJ	0						I	I	I			E			
Canary Creek Delaware Bay, DL	10-28		E	E	E	E	I	I			E		E	E	
Gott's Marsh Patuxent River, MD	0-9	I		E	E	E	E	E	E	E	E	E		E	E
Ware Creek York River, VA	0-7		E	E	E	E	E	I	I	I	E	E	E	E	I
Carter Creek York River, VA	0-12		E	E	I	E	I	I	I	I	E	E	E	E	I
Dill's Creek Charleston Harbor, SC	10-23			E								E			
Sapelo Island, GA (diked Marsh)	20-30		E	I	I	E	O	O	O						
Barataria Bay Louisiana	15-25		E	E									E		

Figure 10. A qualitative summary of the net annual export (E), input (I) or lack of exchange (O) reported in the literature for chl a, organic carbon, nitrogen and phosphorus in various tidal marshes along the Atlantic and Gulf Coasts of the U. S. Great Sippewissett from Valiela et al. (1978), Valiela and Teal (1979); Rhode Island marshes from Lee (1979); Flax Pond from Woodwell and Whitney (1977), Woodwell et al. (1977), Moll (1977), Woodwell et al. (1979); Hamilton marsh from Simpson et al. (1978); Canary Creek from Lotrich et al. (1979); Gott's marsh from Heinle and Flemer (1976); York River marshes, carbon data from Moore (1974), nutrients from Axelrad (1974); Dills Creek from Settlemyre and Gardner (1975); Sapelo Island from Haines (1977, 1979), Haines et al. (1976); Barataria Bay from Happ et al. (1977).

experimental manipulation of the Sapelo Island marsh may have altered its drainage characteristics (Haines 1977). The apparent export of NO_3 by the Sippewissett marsh was due to large inputs of ground water containing high nitrate, much of which was actually removed by the marsh (Valiela et al. 1978). The only other NO_3 export reported, that from Gott's marsh on the Patuxent River, MD. (Heinle and Flemer 1976) was from an area often exposed to very high nitrate levels (up to at least 65 μM) and influenced by sewage treatment plant inputs and runoff from cultivated uplands.

The general uptake of NO_3 by marshes seems to support their importance as sites of active denitrification as discussed earlier

(Table 6), while the consistent export of phosphate is compatible with the solid phase phosphorus profiles found in marsh sediments (Figure 4). The apparent export of organic carbon, of course, is at least qualitative support for the "outwelling" hypothesis and agrees with the analysis of potential carbon export given earlier.

A Note on Metal Fluxes

There appear to have been only two studies where the fluxes of metals between marshes and offshore waters were measured directly. As part of their work on Dill Creek, S. C., Settlemyre and Gardner (1975) found a small (probably statistically insignificant) export of zinc and iron and an import of lead over an annual cycle. Similarly, Pellenbarg and Church (1979) have recently reported a net export of copper, zinc, and iron from a Delaware marsh during three tidal cycles in spring, summer, and fall. In their review of chemical exchanges between marshes and coastal waters, Gardner and Kitchens (1978) argued that it is not likely that present analytical capabilities are sufficient to allow us to determine net trace metal exchanges of the magnitude that are likely to occur. It may be that the analysis of solid phase profiles and sediment pore waters is the most productive way to assess the activities of trace metals in coastal marshes.

The Magnitude of Marsh-Estuarine Fluxes

A knowledge of the direction of the net flux of various materials between marshes and coastal waters is instructive, but it does not tell us how significant those fluxes are likely to be. We need to know how much material is moving, and when. Some of the studies cited earlier provide answers to these questions, but the uncertainty in their individual estimates must be taken as very large. When considered as a whole, however, it seems to me that the results from all of the marshes examined so far converge remarkably well.

Organic Carbon

The net exchange of particulate organic carbon (POC) has now been determined in at least 8 different salt marshes, and in 5 of these the movement of dissolved organic carbon (DOC) was also measured. The values for POC flux range from an import of + 60 g C m^{-2} y^{-1} to an export of -300 g C m^{-2} y^{-1}, with a mean flux for all marshes of -70 ± 95 g C m^{-2} y^{-1} (Table 10). DOC was exported in all of the studies, with fluxes ranging from -8 to -140 g C m^{-2} y^{-1}. These DOC fluxes for the marsh as a whole are smaller

Table 10. Annual flux of organic carbon between salt marshes and coastal waters

	CARBON FLUX, g C m^{-2} of marsh y^{-1}		
	DOC[1]	POC	TOC
Great Sippewissett, Cape Cod, MA (Valiela et al. 1978)		-76	
Flax Pond, Long Island, NY (Woodwell et al. 1977)	-8.4	61	53
Canary Creek, Lewes, DE (Lotrich et al. 1979)	-38	-62	-100
Gott's Marsh, Patuxent River, MD (Heinle & Flemer 1976)		-7.3	
Ware Creek, York River, VA (Moore 1974)	-80	-35	-115
Carter Creek, York River, MA (Moore 1974)	-25	-116	-142
Dill Creek, Charlestown, SC (Settlemyre & Gardner 1975)		-303	
Barataria Bay, LA (Happ et al. 1977)	-140	-25	-165

[1]These DOC fluxes may be compared with Turner's (1978a) estimate of some 200-250 g C m^{-2} y^{-1} of DOC released directly by Spartina alterniflora leaves in a Georgia marsh and with laboratory core measurements reported by Pomeroy et al. (1976) amounting to some 50 g C m^{-2} y^{-1} of DOC released from the surface of a Georgia salt marsh.

than might be expected on the basis of the sum of DOC flux measurements from Spartina leaves[1] (Turner 1978a) and from the marsh surface

[1]There is some disagreement about the magnitude of DOC loss from Spartina. Values reported earlier by Gallagher et al. (1976) are much lower than those given by Turner (1978a).

(Pomeroy et al. 1976)(Table 10). They are, however, considerably higher than the DOC exports from various upland and fresh water swamp watersheds reviewed by Mulholland and Kuenzler (1979), which usually remained below -10 g C m^{-2} y^{-1}. Much of the DOC produced on the marsh may be rapidly respired or transformed into POC before it is exported (Gallagher et al. 1976), but it seems clear that these intertidal systems are less effective at retaining organic carbon than are terrestrial or fresh water communities.

On the basis of these measurements, the total flux of organic carbon from salt marshes seems likely to lie between -100 to -200 g C m^{-2} y^{-1}, or less. This is the best assessment I can make given the uncertainty in the various methods. None of the studies, for example, has adequately accounted for the export of large rafts of Spartina leaves during storms, for bed load detrital transport (Pickral and Odum 1976) or for the (probably relatively small) export of carbon in fish migration. The uptake of POC by Flax Pond has already been discussed, and the large export from Dill Creek may be due to difficulties with the volume-weighted hypsographic sampling method used in that work. In all of these studies, however, the presentation of a single annual flux value can be deceptive. If the individual net flux measurements from some of the marshes are examined as a time-series record, it is clear that we may often be integrating a very "noisy" signal that is heavily filtered by the sampling schedule that was chosen (Figure 11). It is possible that Flax Pond is a particularly "noisy" marsh, but it seems more likely to me that the "noise" is a function of the sampling frequency. Before leaving Figure 11, however, I should point out that I have used the original author's values and not my plots to obtain the annual fluxes summarized in Table 10. In some cases the discrepancy might have been quite large. For example, Woodwell et al. (1977) put a smooth polynomial regression through their data rather than emphasizing the fine scale variations as I have done.

An organic carbon export of 100-200 g m^{-2} y^{-1} is compatible with observations of the above-ground net production in salt marshes (perhaps 250-1250 g C m^{-2} y^{-1}, Keefe 1972; Nixon and Oviatt 1973), with a generous allowance for burial (Figure 3) and the consumption of a large amount of organic matter in the marsh (Teal 1962; Nixon and Oviatt 1973; Hopkinson et al. 1978; Houghton and Woodwell in press). There does not appear to be any clear correlation of export with primary production, however, and the flux data from diverse sites seem to be at least as uniform as production estimates (Turner 1976).

The potential significance of an organic carbon export of this magnitude for the secondary production of adjacent waters is, once again, a function of the relative sizes of marsh and open water systems. A lack of appreciation for this simple relationship has

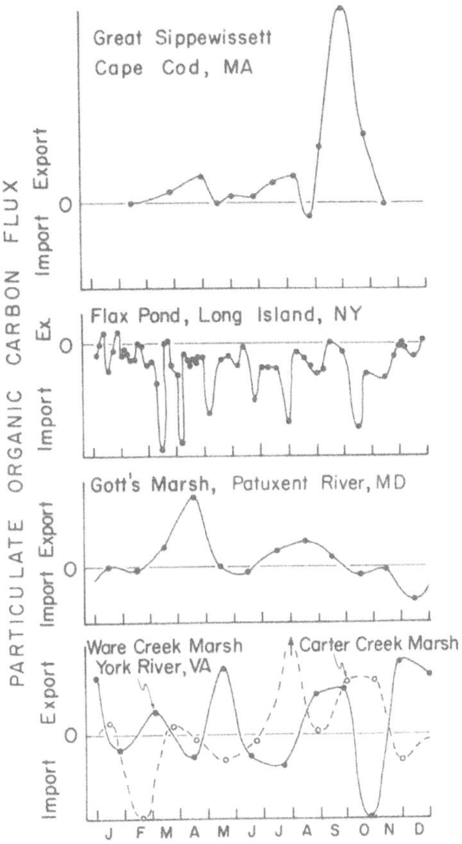

Figure 11. Time-series plots of the relative (within each data set) magnitude of the export and import of particulate organic carbon over an annual cycle reported for several marshes along the Atlantic coast of the U. S. The measurements do not include neckton, large "rafts" of Spartina, or bed load transport. In some cases the original paper did not present a plot of the data, and in others the data were displayed in different form (linear interpolations or polynomial regressions). See Figure 10 for sources.

helped confuse the discussion of marsh-estuarine interactions. It is a profound experience for a New Englander to share Sidney Lanier's vision of "a league and a league of marsh grass, waist high, broad in the blade" stretching out across Glynn County, GA. But it is also useful for those from the Gulf Coast to put themselves in the midst of a small New England pocket marsh with upland trees in sight all around.

I have tried to put these experiences in a more respectable scientific form by showing the potential organic carbon supplement from marsh exports as a function of the ratio of salt marsh area to open water and the primary production of the open water system receiving the supplement (Figure 12). Assuming an export of 100 g C m^{-2} of marsh y^{-1} (Table 10), I have simply drawn in the lines showing a marsh contribution ammounting to 10% and 50% of the open water primary production. It is no accident that the "outwelling" concept developed at Sapelo Island rather than at the laboratory on Narragansett Bay.

In keeping with the practice adopted earlier for suspended sediment, I have also attempted a more detailed calculation estimating the potential contribution of organic exports from marshes to the carbon budget for Chesapeake Bay (Table 11). While inputs from marshes appear to be important in some of the estuaries entering the Chesapeake (Heinle and Flemer 1975), it is not clear if they are of any real trophic significance in the secondary production of the open waters of the Bay itself. The net annual primary production of Chesapeake Bay appears never to have been reported, but Biggs and Flemer (1972) have used gross primary production figures from the Upper and Mid-Bay regions to derive an estimate of net phytoplankton production. Depending on whether one assumes that net production amounts to 25% or 50% of gross production, their calculations suggest an area-weighted production for the upper half to two thirds of the Bay of some 60 to 120 g C m^{-2} y^{-1}. This figure would probably increase considerably if the Lower Bay were included, since production in the less saline water appears to be severely light limited (Flemer 1970). If production in the Lower Bay is equal to that in the Middle (~100-200 g C m^{-2} y^{-1}; Flemer 1970), the average for the Bay as a whole might be some 80-160 g C m^{-2} y^{-1}. If so, the contribution from the marshes would amount to somewhere between 2 and 10% of the primary production. The upper estimates of potential marsh input (3.8-7.7 g C m^{-2} y^{-1}) are also very similar to Biggs and Flemer's (1972) measurements of 7.6 g C m^{-2} y^{-1} input to the Chesapeake from the Susquehanna River (reported as 84 x 10^3 metric tons per year).

Some may object that use of an annual flux ignores other important relationships, and that is true. An inspection of the time-series record (Figure 11) suggests that Great Sippewissett, and perhaps other marshes, especially those in the north (Nixon and Oviatt 1973), lose most of their annual export during a few months. Under these conditions, particularly if the offshore production is low at the same time, the export of organic matter from marshes may be important even where the area of marsh is relatively small.

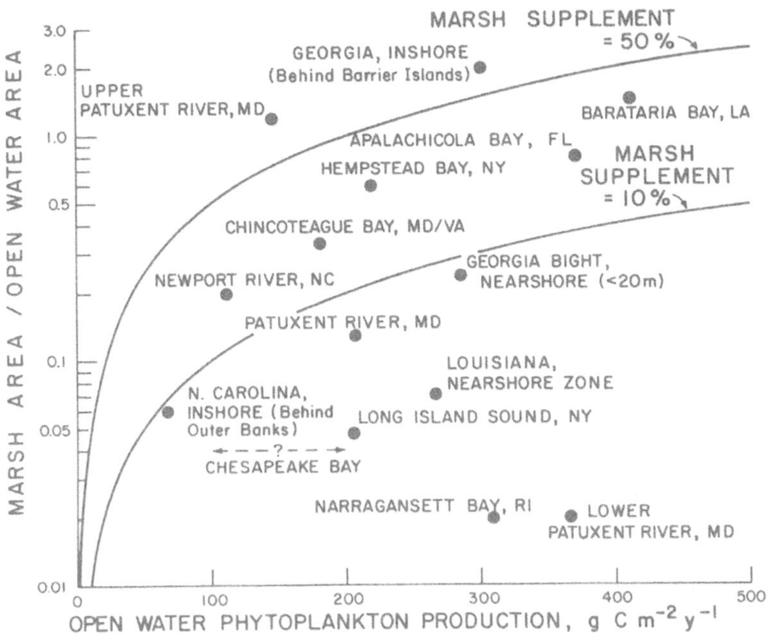

Figure 12. The ratio of vegetated marsh area to open water area in various estuarine ecosystems, and the annual phytoplankton production in the open water. The lines were drawn to show where organic carbon exports from the marshes would provide a supplement for the open waters equal to 10% and 50% of the phytoplankton production, assuming an export of 100 g C m^{-2} of marsh y^{-1} (Table 10). For Georgia and Louisiana the data were readily available to show the result of including the nearshore waters as part of an "extended estuary." The Patuxent River, MD. has been studied intensively, and I have included the upper and lower river separately, as well as the overall system, to show the dramatic variation in one estuary depending on where its boundaries are drawn. The data have been taken from a number of sources, including Teal (1962), Haines and Dunstan (1975), and Haines (1976) for Georgia; Day et al. (1973), Sklar (1976) and Day (personal communication) for Louisiana; Estabrook (1973) and the National Estuary Study (1970) for Apalachicola Bay; Udell et al. (1969) for Hempstead Bay; Boynton (1974) for Chincoteague Bay; Wolfe (1975) for the Newport River; Flemer et al. (1970) and Stross and Stottlemyer (1965) for the Patuxent River; Thayer (1971, 1974) for N. C.; Riley (1956) and the U. S. Dept. of Interior (1965) for L. I. Sound; Heinle (1969) and Biggs and Flemer (1972) for the Chesapeake Bay; and Halvorson and Gardiner (1976), Kremer and Nixon (1978) and Furnas et al. (1976) for Narragansett Bay.

Table 11. Estimate of the potential contribution of salt marsh organic carbon exports to Chesapeake Bay

Marsh Export	100 - 200 g C m^{-2} y^{-1}
Marsh Area[1]	416 km^2
Total Marsh Export	41.6×10^9 - 83.2×10^9g C y^{-1}
Bay Volume[1]	38.5×10^9 m^3
Marsh Input to Bay[1]	1.1 - 2.2 g C m^{-3} y^{-1}
	1.1 - 2.2 mg C L^{-1} y^{-1}
	3.8 - 7.7 g C m^{-2} y^{-1}

[1]Heinle (1979) mean depth 3.5 m, area = 11×10^3 km^2

Nitrogen

The quantification of nitrogen fluxes is more complicated than for carbon because there are at least 5 forms of nitrogen which are of interest, including particulate nitrogen, dissolved organic nitrogen, ammonia, nitrite and nitrate. There are at least 5 studies in which the net exchange of all of these forms has been measured directly, and a sixth is available in which only the dissolved inorganic forms were determined (Table 12). Again, I think the results for all of the forms are remarkably consistent in view of the fact that we are looking at the dynamics of a highly reactive material in six completely different locations ranging from Virginia to Cape Cod. Moreover, the characteristics of the marshes and the sampling techniques used by the different study groups were quite different (Table 8, Figure 10).

There appears to be a loss of total nitrogen from all of the marshes, but with the exception of Great Sippewissett Marsh, the net loss amounts to only a few grams of N m^{-2} y^{-1} (Table 12). This is not, however, what we expected from a consideration of burial and the net exchange of nitrogen with the atmosphere. Those data suggested that marshes were importing some 10 to 30 g N m^{-2} y^{-1} from the tidal waters. In some areas this discrepancy might be resolved by the presence of important groundwater inputs, but that will probably not help in the general case. I think it is more likely that for most marshes the net denitrification estimate I used earlier is too high. It is also possible that the tidal flux

Table 12. Annual flux of nitrogen between salt marshes and coastal waters

	NITROGEN FLUX, g N m^{-2} of marsh y^{-1}					
	PN	DON	NH_4	NO_2	NO_3	ΣN
Great Sippewissett, Cape Cod, MA (Valiela et al. 1978)	-6.7	-9.8	-4.2	-0.1	-3.8	-24.6
Flax Pond, Long Island, NY (Woodwell et al. in press)			-2.0	0	+1.0	
Canary Creek, Lewes, DE (Lotrich et al. 1979)	-2.9	-0.9	+0.7	+1.9		-1.2
Gott's Marsh, Patuxent River, MD (Heinle & Flemer 1976)	-0.3	-2.1	-0.4	0	-0.9	-3.7
Ware Creek, York River, VA (Axelrad 1974)	0	-2.3	-2.9	-0.1	+2.3	-2.8
Carter Creek, York River, VA (Axelrad 1974)	+4.6	-9.2	-0.3	0	+0.3	-4.0

measurements are all in error, but given the relative number of measurements of each process, my guess is that the denitrification number is contributing most to the problem. In spite of the rather tenuous corroboration from preliminary measurements in Georgia (Haines et al. 1976) and Louisiana (DeLaune and Patrick, in press) which I brought forward earlier, the flux measurements suggest that the net denitrification found by Kaplan et al. (1979) in Great Sippewissett Marsh may be a response to the large nitrate inputs found there in groundwater, rather than a general behavior of coastal salt marshes. The uptake of NO_3 from tidal waters by virtually all of the other areas studied (Figure 10) supports the conclusion that marshes are sites of active denitrification, but the larger export of reduced nitrogen argues that in most marshes, nitrogen fixation is a larger term in the budget. If we assume that in most systems there is no net denitrification, the nitrogen fixation measurements (problematical though they may be) suggest that it is possible for the 5 to 25 g N m^{-2} y^{-1} that might be provided by this process (Table 5) to provide for the burial of 3-16 g N m^{-2} y^{-1} calculated earlier, plus the export of some 1 to 5 g N m^{-2} y^{-1} by the tidal waters. The larger export of total nitrogen from Great Sippewissett is due to a large input of ground water in that marsh containing high concentrations of DON and NO_3. Without the groundwater, I would guess that the DON flux would be much smaller and the sign of the NO_3 flux would be reversed, making the total amount of nitrogen exported similar to that found in the other marshes.

The complex nature of the particulate nitrogen flux has already been mentioned, and the composition and biological importance of the material involved in the DON flux are virtually unknown, so it is probably not instructive to look at them much closer in this review. However, I do not think it is safe to assume that the DON is simply a refractory group of compounds of little interest. In her studies with "bell-jar" incubation chambers on marsh surfaces, Lee (1979) observed rapid uptake and release of DON at exchange rates exceeding those of the other forms of nitrogen. But the dissolved inorganic forms of nitrogen are of more general concern, and it is worth looking at a time-series display of some of the NH_4 and NO_3 flux data (Figures 13 and 14). I would not be surprised if this exercise dampens some of the enthusiasm I have tried to generate for the unambiguous nature of marsh-estuarine nitrogen exchanges. Again, while I have plotted or replotted the published data, the values shown in Table 12 are from the original author's calculations.

As we might expect, the exchange characteristics of NH_4 and NO_3 appear complex and highly variable. There is no evidence of any latitudinal gradient in moving from north to south, though some of the marshes, for example Flax Pond, show seasonal variations in their own behavior. In some cases, such as Great Sippewissett, the exchange of NH_4 seems to occur during relatively short (1 to 2 or 3

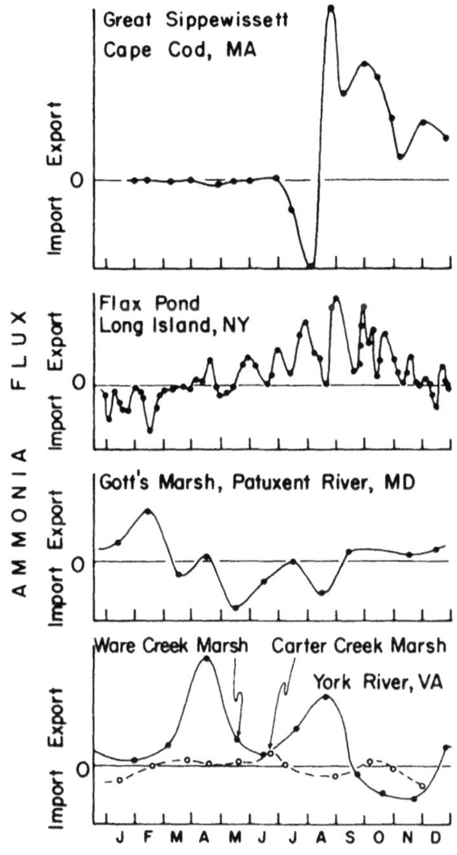

Figure 13. Time-series plots of the relative (within each data set) magnitude of the export and import of ammonia over an annual cycle reported for several marshes along the Atlantic Coast of the U. S. See Figure 10 for sources and Figure 11 for additional explanation.

month) periods, and the impact of such fluxes on surrounding waters may be considerably greater than the annual budget suggests. However, the valuable record from the intense sampling at Flax Pond makes me skeptical of reading too much into the smooth seasonal signals from other marshes.

The importance of the apparent nitrogen export from the marshes is particularly difficult to assess. In general, these systems seem to act as nitrogen transformers, importing dissolved oxidized forms and exporting dissolved and particulate nitrogen in reduced forms. The same behavior can be found in lagoons with very little intertidal

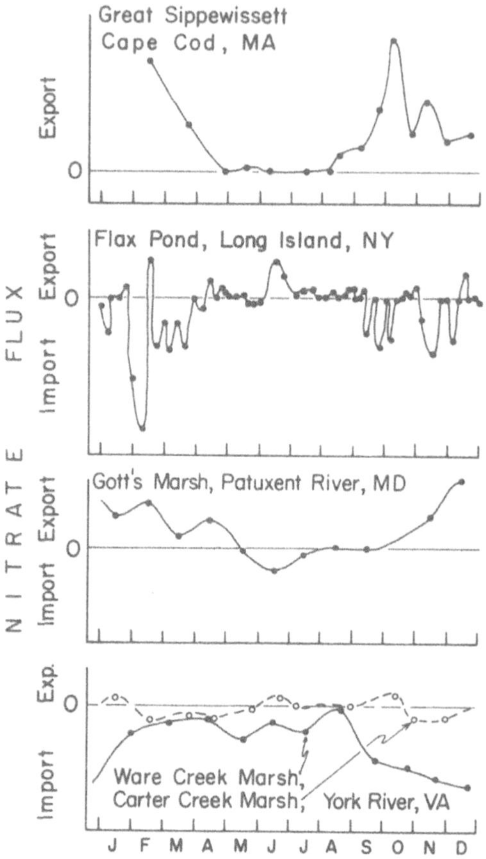

Figure 14. Time-series plots of the relative (within each data set) magnitude of the export and import of nitrate over an annual cycle reported for several marshes along the Atlantic Coast of the U. S. See Figure 10 for sources and Figure 11 for additional explanation.

marsh (Nixon and Lee, in press) and it may be a general feature of the relationship between shallow, highly productive coastal systems and the sea. It is not known how much, if any, of the exported DON is incorporated in offshore food chains, and it is this material which forms a major part of the nitrogen export (Table 12). The particulate organic nitrogen may well be consumed. If so, each gram of organic nitrogen might support the growth of 4-5 grams of animal tissue if it were completely assimilated. In fact, of course, assimilation efficiencies are considerably lower than 100%, especially for detritus, but it is still possible that the organic nitrogen supplement from marshes may enhance secondary production when the ratio of marsh area to open water is high (Figure 12). But the export of a few grams of particulate organic nitrogen from each square meter of marsh must be seen against a background production of some 15-20 g of particulate nitrogen associated with each 100 g of carbon fixed offshore.

The export of ammonia, when it occurs, will almost certainly contribute to primary production in the near-shore waters, and it may even play a role in the decomposition of some of the organic carbon that is exported (Thayer 1974). However, it is virtually impossible to derive a meaningful estimate of the primary production that might potentially be supported by the nitrogen export. Even if we knew the amount of nitrogen very accurately, we do not know how many times the nitrogen put into the open water at any particular moment might pass through the primary producers. On the basis of the limited information available, it seems that most of the primary production in coastal marine systems is supported by recycled nitrogen rather than by "new" inputs (Nixon in press). Especially during summer, the available nitrogen in near-shore waters appears to be turning over rapidly (Furnas et al. 1976), and the effect of a given amount of nitrogen on the production of the water may be multiplied many times. In some areas, however, it may be possible to compare the potential nutrient input for marshes with other terms in the estuarine nutrient budget. For example, if we use a generous estimate of 5 g N m^{-2} y^{-1} for the export of nitrogen from marshes in Chesapeake Bay (Table 12), it appears that they may provide some 0.2 g N m^{-2} y^{-1} to the open waters of the Bay (Table 13). Such an input would amount to less than 2% of the estimated nitrogen inputs from the atmosphere, freshwater runoff, and municipal and industrial loadings (Jaworski, in press).

Before leaving this discussion, I think it is important to make some additional points about marsh-estuarine nutrient dynamics, and nitrogen is a convenient and pertinent example to use. I have been developing my arguments as if there were two very distinct systems, the emergent marsh with its tall *Spartina* on one side, and the open waters with their plankton on the other, and some sort of pipe between them. That is certainly not the situation in the real world. Virginia Lee's (1979) field studies with metabolic chambers have shown that the surface of the emergent marsh is itself a complex metabolic mosaic, where not only the magnitude, but even the direction of net nitrogen fluxes may vary widely over distances of a few meters. In my own work with entrapped water masses in a marsh embayment and creek system (Nixon et al. 1976), it was evident that the sediments and detrital mats of the subtidal components of the marsh acted to conserve nitrogen rather than exchanging it with the overlying water as offshore sediments do. And Barbara Welsh (in press) has developed a nice story showing a functional coupling of

Table 13. Estimate of the potential contribution of salt marsh nitrogen exports to Chesapeake Bay compared with other inputs to the Bay

Marsh Export	5 g N m^{-2} y^{-1}
Marsh Area[1]	416 km^2
Total Marsh Export	20.8 x 10^8 g N y^{-1}
Bay Volume[1]	38.5 x 10^9 m^3
Marsh Input to Bay	0.19 g N m^{-2} y^{-1}
	(3.9 μg-at N l^{-1} y^{-1})
Other Inputs[2]	
Municipal and Industrial	2.9 g N m^{-2} y^{-1}
Upper Basin Runoff	6.5 g N m^{-2} y^{-1}
Atmosphere	0.5 g N m^{-2} y^{-1}
Total	9.9 g N m^{-2} y^{-1}

[1]Heinle (1979) mean depth 3.5 m, area = 11 x 10^3 km^2

[2]Jaworski (in press)

marshes and their adjacent mud flats, which act to trap nutrients leaving the marsh on ebb tides and then return at least some of them to the marsh on flood tides. The point I want to make with all of these studies is that the path from the emergent marsh to the open coastal water is not through a pipe, but through a complex chain of sub-systems, each of which is characterized by its own internal cycling as well as by its own inputs, outputs, transformations and storages. The flux measurements from tidal channels have occupied most of our attention, but even as the nutrients flow past our measuring points, they are entering other subsystems, exchanging with new surfaces, and mixing in broader channels on their way offshore.

Phosphorus

At least seven nutrient exchange studies have included one or more of the forms of phosphorus in their measurements (Table 14). The results are interesting, not just because they are strikingly consistent, but because they show a behavior which is quite different from that observed with nitrogen. While the oxidized inorganic forms of nitrogen (NO_2 + NO_3) were taken up by the marshes (Table 12), the flux measurements show an export of PO_4 to the offshore waters. And not only was the direction of the phosphate flux predicted from our analysis of the sediments, but the magnitude as well. On the basis of a solid phase phosphorus profile (Figure 4) and a comparison of the phosphorus content of near-shore and marsh sediments (Tables 2 and 3), it seemed that the export of remobilized phosphorus as PO_4 might amount to some 0.4-1.8 g P m^{-2} y^{-1}. The evidence from the flux measurements is consistent with this analysis (Table 14), and with the conclusion that Reimold's (1972) estimate of phosphorus "pumping" by Spartina must be over 100 times too high. It also appears that the potential phosphorus input through rainfall (0.01-0.1 g P m^{-2} y^{-1}) might make a significant contribution to the apparent PO_4 export reported for some of the marshes.

The large input of organic carbon to the sediments by the highly productive Spartina seems to create a reducing environment in which denitrifying bacteria actively remove nitrate from the incoming water. At the same time, phosphorus is remobilized in the anoxic pore waters and put into the overlying water by diffusion and by the grasses. The magnitude of the fluxes, however, is not sufficient to account for the characteristically low N/P ratio of coastal waters. The explanation for that phenomenon is more likely to lie in the subtidal sediments and in the physiology of nitrogen fixation (Nixon in press). The export of phosphate may have some influence on the fertility of coastal waters in areas with a large amount of salt marsh. But again, it is impossible to assess the magnitude of this influence without a knowledge of the turnover times of phosphorus in the near-shore systems at various times of year. For Chesapeake Bay, the role of marshes in the phosphorus dynamics of the estuary cannot be very large, since the potential input of phosphorus from the wetlands appears to amount to less than 3% of the other sources summarized by Jaworski (in press) (Table 15).

The seasonal pattern of PO_4 exchange appears to be highly variable and lacking in any consistent trend among the marshes studied (Figure 15). Since near-shore marine and estuarine waters are usually characterized by a summer phosphate maximum (Taft and

Table 14. Annual flux of phosphorus between salt marshes and coastal waters

	PP	DOP	PO_4	ΣP
Great Sippewissett, Cape Cod, MA (Valiela et al. 1978)			-0.6	
Flax Pond, Long Island, NY (Woodwell & Whitney 1977)	+1.1		-1.4	-0.3
Canary Creek, Lewes, DE (Lotrich et al. 1979)			<-0.1	
Gott's Marsh, Patuxent River, MD (Heinle & Flemer 1976)	-0.1	-0.2		-0.3
Ware Creek, York River, VA (Axelrad 1974)	+1.0	-0.2	-0.1	+0.7
Carter Creek, York River VA (Axelrad 1974)	+0.8	-0.2	-0.6	0
Dill Creek, Charlestown, SC (Settlemyre and Gardner 1975)			-6.4	

Taylor 1976; Nixon et al. in press), and at least some of the study areas show strong exports of PO_4 during summer, it does not seem likely that marshes in general are "smoothing" estuarine phosphorus cycles (Kalber 1959). It was a good hypothesis, though, and it may yet be true in some areas. I wish we had good time-series data from the marshes of Glynn.

EVIDENCE FROM NEARSHORE WATERS

So far, we have been trying to learn about the nature of marsh-estuarine interactions by analyzing marsh sediments and by measuring the composition of the tidal waters entering and leaving marshes. I think we can also learn something about the problem by looking at

Table 15. Estimate of the potential contribution of salt marsh phosphorus exports to Chesapeake Bay

Marsh Export		0.1 - 1.0 g P m^{-2} y^{-1}
Marsh Area[1]		416 km^2
Total Marsh Export		4.2 - 41.6 x 10^7 g P y^{-1}
Bay Volume[1]		38.5 x 10^9 m^3
Marsh Input to Bay		3.5 - 35 mg P m^{-2} y^{-1}
		(0.03 - 0.3 μg-at l^{-1} y^{-1})
Other Inputs[2]		
Municipal and Industrial		950 mg P m^{-2} y^{-1}
Upper Basin Runoff		340 mg P m^{-2} y^{-1}
Atmosphere		80 mg P m^{-2} y^{-1}
	Total	1370 mg P m^{-2} y^{-1}

[1]Heinle (1979) mean depth 3.5 m, area = 11 x 10^3 km^2

[2]Jaworski (in press)

the near-shore environment, at the open waters at the other end of the "pipe." In fact, this is the approach E. P. Odum took in 1968 when he introduced the "outwelling" concept.

Primary Production

A large part of E. P. Odum's (1968) argument was based on Jim Thomas's (1966) M. S. Thesis, which showed very high rates of phytoplankton production (546 g C m^{-2} y^{-1}) in nearshore waters off the Altamaha River in Georgia. Somewhat earlier, Ragotzkie (1959) had used oxygen changes in dark and light bottles to show that the plankton communities in the inshore Duplin River in the Sapelo

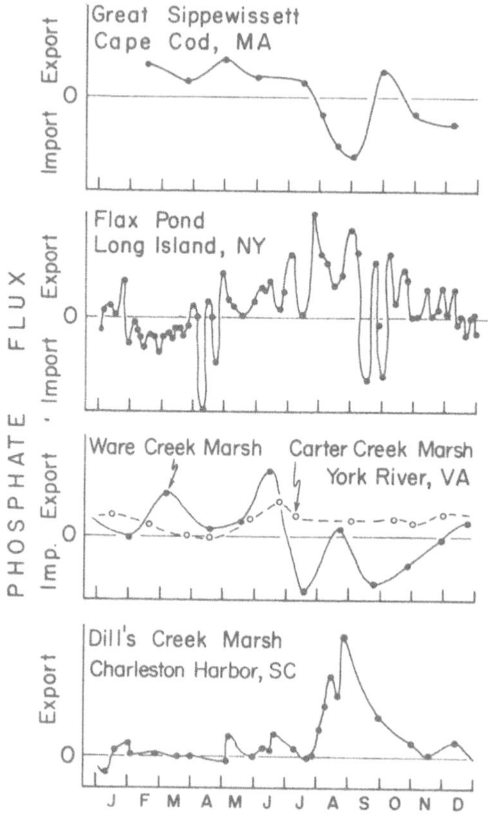

Figure 15. Time-series plots of the relative (within each data set) magnitude of the export and import of phosphate over an annual cycle reported for several marshes along the Atlantic Coast of the U. S. See Figure 10 for sources and Figure 11 for additional explanation.

Island marsh area were heterotrophic. These two observations were brought together to suggest that the high turbidity of the water in the inshore area kept phytoplankton production and nutrient uptake very low, with the result that large amounts of nutrients (presumably exported from the marshes) were carried out into the open waters. The lack of primary production in the inshore waters was compensated for by detrital organic carbon washed off the marshes and the nutrients flushed out into the less turbid nearshore (and perhaps even offshore) waters stimulated the very high phytoplankton production reported by Thomas. It was a very satisfying picture of inshore-offshore ecosystem coupling, and it continues to be developed in various forms (Turner et al. 1979).

The fact that Thomas's production measurements were so very high and taken from a location off the mouth of a large river was somewhat disturbing. But for almost 10 years there were no other comparable data for a wide area off the Georgia coast. Recently, however, intensive studies of the Georgia Bight (Haines and Dunstan 1975) have shown that the annual primary production averages only 285 $g\ C\ m^{-2}\ y^{-1}$ for nearshore waters (out to a depth of 20 m) and 130 $g\ C\ m^{-2}\ y^{-1}$ for offshore waters from 20-200 m depth. These values are not significantly higher than found in comparable areas which lack a strong potential salt marsh influence (Smayda 1973; Figure 12). It is possible that higher rates of production, such as those found by Thomas (1966), may characterize a fairly narrow band of water some 15-20 km off the coast that sits at an optimum point between declining nutrient concentrations and increasing depth of the euphotic zone (Haines 1979a). If so, this enhanced production of even a limited area now appears to be due to nutrient inputs from fresh waters rather than from the marshes (Haines 1975, 1979b).

More recent measurements of phytoplankton production using ^{14}C uptake rather than oxygen changes have also shown that there is substantial primary production in Georgia inshore estuarine waters (300 $g\ C\ m^{-2}\ y^{-1}$, Haines 1978) and even in salt marsh tidal creeks (90 $g\ C\ m^{-2}\ y^{-1}$, Turner et al. in press). Neither of these observations conflicts with earlier reports that the plankton community as a whole in tidal marsh waters is heterotrophic (Ragotskie 1959, Turner 1978a). But they do show that there is a substantial amount of organic matter being fixed (and nutrients being taken up) in the waters adjacent to the marshes. This is important to know when examining the next line of evidence from the nearshore waters. It also adds to the complexity of the passage from emergent marsh to offshore waters which I emphasized earlier. It should not surprise us any longer to learn that some marshes even appear to import phytoplankton (Moll 1977, Turner et al. in press).

Stable Carbon Isotope Ratios

I think it is a healthy sign for ecology that one of the most explicit challenges to the "outwelling" concept came from an untenured postdoctoral student working at the Sapelo Island laboratory. Beginning with a paper published in 1975, Evelyn Haines has developed the argument that organic matter exported from salt marshes does not contribute significantly to the standing crop of particulate organic carbon in the offshore, nearshore, inshore, or even tidal creek waters of Georgia (Haines and Dunstan 1975; Haines 1976; Haines 1979a). Her evidence comes from the relative amounts of

$^{13}C/^{12}C$ in the organic matter found in the waters of these areas compared with the $^{13}C/^{12}C$ ratio of Spartina grasses and marine phytoplankton (Table 16).

The use of this technique is based on the observation that plants discriminate between the two stable carbon isotopes to different degrees, depending on whether they employ the C-3 or C-4 pathways in photosynthesis. Vascular plants with C-3 pathways show $\delta^{13}C$ values of -24 to -34°/∘∘, while those (like Spartina) with a C-4 pathway show $\delta^{13}C$ of -6 to -19°/∘∘. Algae appear to have intermediate values of -12 to 23°/∘∘ (Haines 1976). As far as I know, Patrick Parker (1964) was the first to apply this information to the problem of unraveling food webs in a coastal marine ecosystem. However, the extension of this technique to the question of the importance of organic carbon from Spartina is particularly appropriate because of the large difference in the $\delta^{13}C$ for Spartina and marine phytoplankton (Table 16). Haines (1976, 1977) and Haines and Montague (1979) have also shown that the $\delta^{13}C$ of Spartina does not change during decomposition and that the stable carbon ratios of estuarine animals in Georgia do, in fact, reflect the ratios in their diets. At this point, the $\delta^{13}C$ data seem very compelling, and lead to the conclusion that most of the organic carbon in the coastal waters of Georgia is from algal production rather than from Spartina detritus. But my interpretation of the values summarized in Table 16 is that they do not indicate that there is no contribution of organic carbon from Spartina in the inshore or offshore waters. If we take the $\delta^{13}C$ of Spartina as -13°/∘∘ and phytoplankton as -21°/∘∘ (Haines 1979a) a mixture consisting of 10 to 20% marsh carbon would have a $\delta^{13}C$ of -20.2 to -19.4°/∘∘. These values fall within the range reported for suspended particulate matter in Georgia coastal waters (Table 16). In fact, the data would support a $\delta^{13}C$ for phytoplankton of -22 to -23°/∘∘ (Table 16), which would allow a mixture consisting of 30 or 40% Spartina carbon in the water. The $\delta^{13}C$ values are not well enough constrained to carry the weight of this argument much further (or perhaps even this far), but I want to make the point that the stable carbon isotope evidence is not out of line with the relative abundance of marsh carbon and phytoplankton carbon that we arrived at earlier for Georgia waters (Figure 12).

Fisheries

From the very beginning, E. P. Odum (1968) proposed that "outwelling" from salt marshes was responsible for the presence of "expanded" or "intensive" fisheries in coastal waters where upwelling was not a factor. An article titled "The Role of Tidal Marshes in the Productivity of Coastal Waters" by Armando de la Cruz (1973) expressed the prevailing view among marine ecologists that, "The

Table 16. Carbon isotope composition in marshes and coastal waters off Georgia (Haines and Dunstan 1975, Haines 1976).

	$\delta^{13}C$ ‰
MARSH	
Spartina alterniflora (live)	-12.7 to -13.6
(dead)	-12.3
Mud diatoms	-16.2 to 17.9
Organic matter in soil	-13.2
PHYTOPLANKTON	
Diatom bloom (*Skeletonema costatum*)	-22.1 to -22.7
Dinoflagellate bloom (*Kryptoperidinium* sp.)	-20.0
Green flagellate bloom	-26.3
SUSPENDED PARTICULATE MATTER	
Estuary Tidal Creeks	-19.8 to -22.8
Shelf (0-10 km offshore)	-18.0 to -24.3
Shelf (20 km offshore)	-21.0 to -23.9

$$\delta^{13}C = \left(\frac{^{13}C/^{12}C \text{ in sample}}{^{13}C/^{12}C \text{ in a standard}} - 1\right) 1000$$

obvious occurrence of extremely productive fishing and shell-fishing grounds at or near the regions of extensive tidal wetlands is indicative of the important role marshes and swamps play in the fertility of coastal waters." The implication of these, and many other papers, was that the fisheries of the Middle Atlantic, Southeastern, and Gulf coasts of the U. S. were considerably more productive than those of the Northeast, where the area of marsh is small. The mechanism for the presumed relationship between fish catch and marsh was some combination of habitat, organic carbon export, or nutrient export with an associated phytoplankton bloom.

With this in mind, it is instructive to look at the commercial landings data for the different regions of the U. S. (Table 17). While the Gulf Coast is certainly higher than the other areas, the North Atlantic catch was higher (at least in 1966) than that of the Mid or South Atlantic Regions. Even for recreational fisheries, there was no indication in a 1965 Bureau of Sport Fisheries survey that northern waters were any less productive (Table 18). It does seem to be true, however, that a much larger portion of the catch in the warmer waters is composed of estuarine-dependent species (Table 17). In fact, if the catch-per-unit area of these species is plotted as a function of the marsh/open water ratio for the various geographic areas, there is, with one marked exception, an apparently strong linear relationship between them (Figure 16). In a more detailed analysis of the commercial catch of one estuarine-dependent group, penaeid shrimp, Turner (1978b) also found higher yields from areas of the nearshore northern Gulf of Mexico associated with greater amounts of "vegetated estuary" on the adjacent Louisiana coast.

Table 17. Commercial Fish Landings (finfish and shellfish) on the East and Gulf Coasts of the United States in 1966[1]

Region	Total Landings Millions of pounds	Estuarine-dependent Landings Millions of pounds	Percent Estuarine-dependent
Northeast	611	43	7
Middle Atlantic	241	128	53
Chesapeake Bay	502	493	98
Southeast	368	350	95
Gulf of Mexico	1,196	1,149	96

[1]From U. S. Department of the Interior, Fish and Wildlife Service, National Estuary Study, Vol. 2, 1970.

Table 18. Salt Water Recreational Fishing in the United States, 1965[1]

Region	Number of Anglers	Number of Fish	Number fish/angler
	-in thousands-		
Northeast (New England and New York)	1,530	172,660	112.8
Middle Atlantic (New Jersey to Cape Hatteras)	1,375	92,126	67.0
Southeast (Cape Hatteras to Florida Keys)	1,720	190,802	111
Gulf of Mexico (Florida West Coast to Texas)	1,972	194,101	98

[1]Conducted by the Bureau of the Census for the Bureau of Sport Fisheries and Wildlife. From National Estuary Study, Vol. 2, 1970.

It does not necessarily follow, however, that the marshes themselves are the cause of these relationships. The marked deviation of Chesapeake Bay in Figure 16 may be important and instructive. Perhaps estuarine-dependent fish simply do well in shallow protected waters, the same environment that usually favors salt marsh development. Perhaps they do well in Southeastern and Gulf Coast estuaries because the input of freshwater is generally larger there. Chesapeake Bay is the only system I could find readily that fits this description, has a low marsh/open water ratio, and is large enough to merit separate fisheries statistics. And it is evident from the large catch obtained in the Chesapeake that it is not necessary to have a great deal of marsh relative to the open water to sustain a highly productive estuarine fishery. Moreover, there is a similar lesson to be learned from the total catch data for a variety of coastal systems. If we look at the total commercial catch-per-unit-area from waters with and without a potential salt

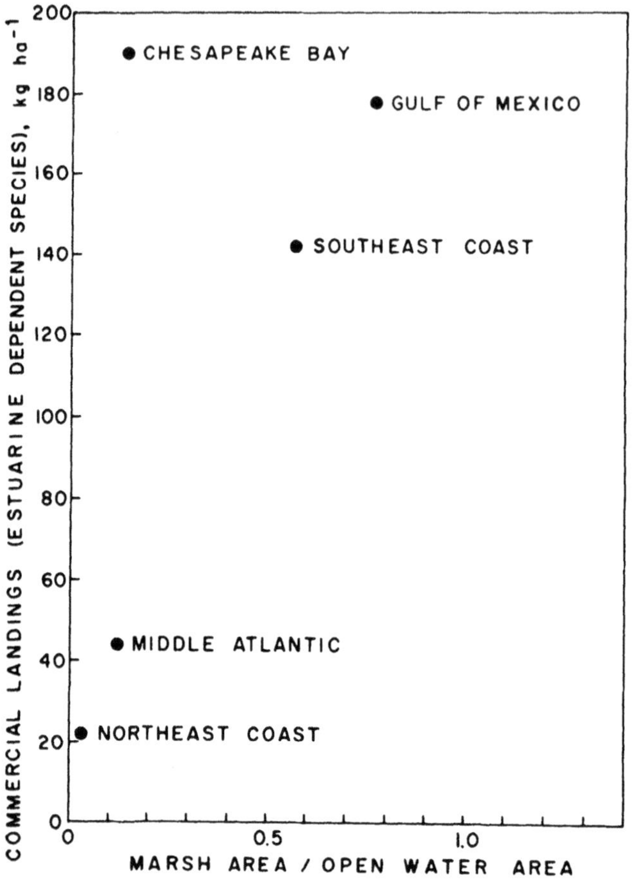

Figure 16. The commercial landings of estuarine-dependent species in the major East and Gulf Coast regions of the U. S. during 1966 (National Estuary Study 1970, Vol. 2) as a function of the ratio of marsh area to open water area in the region (Woodwell et al. 1973).

marsh influence, it is clear that the nearshore and offshore fisheries of the Northeast are at least as productive as those of the Southeastern and Gulf Coasts (Table 19). It is possible, of course, that all of this appears to be true because the fishing effort is much more intense or efficient in the North, though I have no reason to believe that is the case.

I am not sure why there is no apparent effect of marsh exports on fish production. The few data we have suggest that the primary production of the nearshore waters all along the Atlantic and Gulf Coasts is similar (200-300 g C m^{-2} y^{-1}, Figure 12), and we might

Table 19. Commercial finfish and shellfish yields (Kg ha^{-1}) from some estuarine and nearshore areas with varying amounts of saltmarsh

Location	Marsh Area / Open Water Area	Finfish	Shellfish	Shrimp	Total
Georges Bank, N. W. Atlantic[1]	0	161			161
Rhode Island, Nearshore[2]	0.01	80	31		111
Long Island Sound[3]	0.05	29	15		44
Chesapeake Bay (1962)[4]	0.04	142	12		154
Chesapeake Bay (1966)[5]	0.04	134	52		186
Apalachicola Bay, FL[5]	0.19	24	51	3	78
Barataria Bay, LA[6]	1.43	19	11	13	43
Louisiana, Nearshore[6]	0.07	46	1	4	51
Aransas Estuary, TX[5]	0.35				91
Peru Upwelling[4]	0	370			370

[1]Olsen and Saila (1976), includes foreign and U. S. fleet
[2]N.M.F.S. Area 539, W. Hahm, N.M.F.S., Woods Hole, 1975 data
[3]N.M.F.S. Area 611, W. Hahm, N.M.F.S., Woods Hole, 1975 data
[4]McHugh (1967)
[5]National Estuary Study (1970), 1966 data
[6]Day et al. (1973) and personal communication. Barataria Bay includes recreational fishing and the total nearshore catch is almost 100 Kg ha^{-1} if "trash" fish caught and discarded during shrimping are included.

expect that a carbon supplement amounting to 25-50% of the production would be apparent in the fisheries data for southern waters. The analysis summarized in Figure 12 suggests that this may be a reasonable estimate of the marsh contribution in some nearshore and inshore areas of the South. If the marshes really to export this much carbon they may still have little impact on commercial or recreational fisheries because the detrital carbon is rapidly sedimented, of little nutritive value, rapidly respired by bacteria (especially DOC), consumed by smaller fish in waters adjacent to the marshes, or a number of other reasons. Perhaps fish in warmer waters must injest more food to meet a larger respiratory demand. It is very hard to know, but at this point, I think the evidence suggests that the argument first put forward in support of the trophic importance of "outwelling" is not very convincing.

BETWEEN COASTAL MARSHES AND COASTAL WATERS - WHAT HAVE WE LEARNED?

I am reluctant to write a summary of conclusions at the end of this review. It is a long paper, everyone seems to be very busy, and the inclusion of a summary makes it tempting to flip to the end to find out how it all came out. "Are coastal salt marshes important or not?" is the question I have almost always been asked when it came out that I was working on this review. I don't think it is quite the right question. I think I know what people have in mind when they ask it, but even now, as I start to frame an answer, I have the ominous feeling that I will regret doing so. The debate over the "value" of tidal wetlands is complex and very emotional.

It is difficult to answer the question in a summary because the summary must be brief and simple, while the answer is not. I hope the body of this review conveys something of that complexity, as well as a blend of humility, frustration and satisfaction with our attempts to grapple with it. Some parts of the answer have not even been discussed in this review. I do not know if salt marshes are really important for waterfowl or mammals, or as sources of sulfur dioxide, or as storm buffers, or for a host of other possible reasons. They are important to me and to many other people who enjoy looking across the sweep and green openness of them, who like to walk out across them and observe their patterns of life and form. And these are not trivial reasons for maintaining that the marshes are important. But these sorts of things have not been the subject of this review. We have been concerned here with the influence of tidal salt marshes on estuarine and nearshore primary and secondary production and water chemistry. For much of the past twenty years it has generally been accepted that this influence was important in maintaining the great fertility of our coastal marine waters. What have we learned about this aspect of marsh-estuarine interactions? Are salt marshes important in this way?

I began this review by quoting from John Teal's (1962) paper to the effect that large amounts of organic matter exported from salt marshes made the high secondary production of estuaries possible. Many other people have said the same thing. By and large, I think we have learned that that is not true. There may be an export of organic carbon from many tidal marshes, and the export may provide a carbon supplement equivalent to a significant fraction of the open water primary production in many areas of the South. It is even possible that this carbon may contribute measurably to the standing crop of organic carbon in the water at any one time. But it does not appear to result in any greater production of finfish or shellfish than is found in other coastal areas without salt marsh organic supplements.

Marsh sediments appear to be sinks for many trace metals, including lead, copper, zinc, iron and manganese. For some metals, such as lead and copper, deposition from the atmosphere is the major input, rather than the sediments accreting on the marsh from tidal waters. There is evidence that manganese is remobilized in the anoxic pore waters of marsh sediments and exported from the marsh by diffusion across the sediment-water interface, but it is difficult to assess the importance of this flux in the estuarine manganese budget. Detritus itself appears to be an active exchange site for many metals, and the past practice of using trace metal analyses of Spartina and assumptions about the magnitude of the export of Spartina detritus to compute metal fluxes from marshes is probably not a useful way to evaluate the role of these systems in the trace metal budgets of estuaries.

On the basis of very little evidence, marshes have been widely regarded as strong terms (sources or sinks) in coastal marine nutrient cycles. The data we have available so far do not support this view. In general, marshes seem to act as nitrogen transformers, importing dissolved oxidized inorganic forms of nitrogen and exporting dissolved and particulate reduced forms. While the net exchanges are too small to influence the annual nitrogen budget of most coastal systems, it is possible that there may be a transient local importance attached to the marsh-estuarine nitrogen flux in some areas. Marshes are sinks for total phosphorus, but there appears to be a remobilization of phosphate in the sediments and a small net export of phosphate from the marsh. The magnitude of this flux is probably less than 0.5% of that suggested previously in the literature, and it is not likely to be an important term in the estuarine or nearshore phosphorus budget. It is difficult to evaluate the importance of salt marsh nutrient exports to maintaining the high primary production of coastal waters without a knowledge of the turnover times of nutrients in these systems. But it does not appear that nutrient levels or primary production are any higher in most estuaries with a large amount of marsh than they are in many areas without a strong potential salt marsh influence.

Our understanding of the interactions between coastal marshes and coastal waters is still far from complete, but it is more profound than it was twenty years ago. Anyone who has ever worked in scientific research will appreciate the large cumulative effort that so many people have had to put into this enterprise to gain that increase in understanding. It has not come easily, and my three summary paragraphs do not do justice to the difficulty of the problem or to the work that has gone into trying to resolve it. I hope the detailed review has done a better job.

But before I finish, I want to go back to the theme of the introduction to this review. I do so because I think there is a larger lesson running through the past twenty years of work than is apparent in a summary of our present knowledge. During the last few months I have had a number of vigorous discussions with various groups over my opinions about the history of ecological research on the question of marsh-estuarine interactions. In a number of these discussions, a common sentiment was expressed that it was the responsibility of the ecological community to help in the "battle" to preserve the marshes. The early efforts in this direction helped to gain time while environmental awareness developed among the public and the regulatory agencies. The momentum of the developers was so great that an atmosphere of certainty and consensus was necessary for the voice of the ecologists to be heard. The essence of the argument is that, "Yes, perhaps we overstated the case a bit, but it was important to help save the marshes. Now that is done, or at least well along, and we can go back and work on getting our science right."

I do not agree. It is a bad bargain to trade our credibility for political advantage. Science is a social enterprise, we communicate through the scientific literature, and we must do nothing to undermine the integrity of that communication. Both in sending and in receiving information, we must remain skeptical. Reading the literature on marsh-estuarine interactions convinces me that we have been too willing to trust our own preconceptions, and too eager to believe what other people are saying about their data when they agree with those preconceptions. Ecology is a young science, and we are still about the business of learning some of the basics.

I would like to end by quoting at some length from Jacob Bronowski's essay on Science and Human Values (1956). Bronowski includes a passage from the mathematician W. K. Clifford who wrote:

> ...if I let myself believe anything on insufficient evidence, there may be no great harm done by the mere belief; it may be true after all, or I may never have occasion to exhibit it in outward acts. But I cannot help doing this great wrong towards Man, that I make myself credulous. The danger to society is not merely that it should believe wrong things, though that is great enough; but that it should become credulous.

Bronowski goes on to say that:

> The fulcrum of Clifford's ethic here, and mine, is the phrase 'it may be true after all.' Others may allow this to justify their conduct; the practice of science wholly rejects it. It does not admit that the word true can have this meaning. The test of truth is the known factual evidence, and no glib expediency nor reason of state can justify the smallest self-deception in that. Our work is of a piece, in the large and in detail; so that if we silence one scruple about our means, we infect ourselves and our ends together.
>
> The scientist derives this ethic from his method, and every creative worker reaches it for himself. This is how Blake reached it from his practice as a poet and a painter.
>
> He who would do good to another
>   must do it in Minute Particulars:
> General Good is the plea of the
>   scoundrel, hypocrite & flatterer,
> For Art & Science cannot exist but
>   in minutely organized Particulars.
>
> The Minute Particulars of art and the fine-structure of science alike make the grain of conscience.

The painful shortcomings of my own work make me uncomfortable under Bronowski's uncompromising and exacting eye. We may never attend closely enough to the "Minute Particulars," and we will certainly continue to suffer from self-deception. But we can be less credulous and less comfortable.

ACKNOWLEDGMENTS

As always, a reviewer's first debt is to those who have done the research and written the literature he reviews. A number of people have also been particularly helpful in providing me with unpublished data or in helping me to find various pieces of information, including Bill Patrick, John Day, Evelyn Haines, Alice Chalmers, Walter Boynton, Don Heinle, Dave Flemer, Court Stevenson, Frank Daiber, Kurt Eastman, Charles Lord, Barbara Welsh, Carlton Hunt, Charles Hall, Elijah Swift, Virginia Lee, Jim McCarthy, John Teal, Ivan Valiela, Barry Lyons and, no doubt, a number of others. Michael Pilson, Virginia Lee and Candace Oviatt reviewed the manuscript and help to improve it a great deal. Michael Pilson also brought the wonderful quote from John Tyndall to my attention. Continuing support from the Office of Sea Grant Programs in NOAA, U. S. Dept. of Commerce has largely made it possible for me to participate in salt marsh research, and I am grateful for that opportunity.

REFERENCES

Aller, Robert C. 1977. The influence of macrobenthos on chemical diagenesis of marine sediments. Ph.D. Thesis, Yale University, New Haven, CT, 600 p.

Armentano, T. V. and G. M. Woodwell. 1975. Sedimentation rates in a Long Island marsh determined by Pb-210 dating. Limnol. Oceanogr. 20:452-456.

Aurand, D. and F. C. Daiber. 1973. Nitrate and nitrite in the surface waters of two Delaware salt marshes. Chesapeake Science 14:105-111.

Axelrad, D. M. 1974. Nutrient flux through the salt marsh ecosystem. Ph.D. Thesis, College of William and Mary, 134 p.

Banus, Mario, I. Valiela and J. M. Teal. 1974. Export of lead from salt marshes. Marine Pollution Bull. 5:6-9.

Banus, Mario, Ivan Valiela and John M. Teal. 1975. Lead, zinc, and cadmium budgets in experimentally enriched salt marsh ecosystems. Estuarine and Coastal Marine Science 3:421-430.

Bender, M. E. and D. L. Correll. 1974. The use of wetlands as nutrient removal systems. National Technical Information Service, U. S. Dept. of Commerce, 12 p.

Bender, Michael and G. Ross Heath. 1975. Marine phosphorus geochemistry. NSF Proposal, Grad. School of Oceanography, U.R.I., Kingston, RI.

Biggs, R. B. and D. A. Flemer. 1972. The flux of particulate carbon in an estuary. Marine Biology 12:11-17.

Blanton, J. O. and L. P. Atkinson. 1978. Physical transfer processes between Georgia tidal inlets and nearshore waters, pp. 515-532. In: Martin L. Wiley (ed.), Estuarine Interactions. Academic Press, N. Y.

Bloom, A. L. 1967. Coastal geomorphology of Connecticut. Final Report, O.N.R. Contract, Nonr-401(45), Task No. 388-U65, 72 p.

Boon, John, III. 1975. Tidal discharge asymmetry in a salt marsh drainage system. Limnol. Oceanogr. 20:71-80.

Boon, John, III. 1978. Suspended solids transport in a salt marsh creek - an analysis of errors, pp. 147-159. In: Bjorn Kjerfve (ed.), Estuarine Transport Processes. University of South Carolina Press, Columbia, SC.

Boynton, W. R. 1974. Phytoplankton production in Chinocoteague Bay, MD. M. S. Thesis, University of North Carolina, 142 p.

Brinson, Mark M. and Graham J. Davis. 1976. Primary productivity and mineral cycling in aquatic macrophyte communities of the Chowan River, North Carolina. Water Resources Research Institute, University of North Carolina, Report No. 120, 137 p.

Bronowski, J. 1956. Science and Human Values. Harper & Row, 119 p.

Broome, J. W., W. W. Woodhouse and E. D. Seneca. 1975. The relationship of mineral nutrients to growth of Spartina alterniflora in North Carolina. II. The effects of N, P, and Fe fertilizers. Soil Sci. Soc. Am. Proc. 39:301-307.

Carpenter, Edward J., Charlene D. VanRaalte and Ivan Valiela. 1978. Nitrogen fixation by algae in a Massachusetts salt marsh. Limnol. Oceanogr. 23:318-327.

Casselman, Maria Eugenie. 1979. Biological nitrogen fixation in a Louisiana Spartina alterniflora salt marsh. M. S. Thesis, Louisiana State University, 82 pp.

Chrzanowski, Thomas H., L. Harold Stevenson and Bjorn Kjerfve. 1979. Adenosine 5'-Triphosphate flux through the North Inlet Marsh system. Applied and Environmental Microbiology 37(5): 841-848.

Cruz, A. A. de la. 1965. A study of particulate organic detritus in a Georgia salt marsh estuarine ecosystem. Ph.D. Dissertation. University of Georgia, Athens, GA, 141 pp.

Cruz, A. A. de la. 1973. The role of tidal marshes in the productivity of coastal waters. Assoc. Southeastern Biologists Bull. 20(4); 147-156.

DeLaune, R. D., W. H. Patrick, Jr., J. M. Brannon. 1976. Nutrient transformations in Louisiana salt marsh soils. Sea Grant Publication #LSU-T-76-009, Louisiana State University, Baton Rouge, LA, 38 p.

DeLaune, R. D., W. H. Patrick, Jr., R. J. Buresh. 1978. Sedimentation rates determined by ^{137}Cs dating in a rapidly accreting salt marsh. Nature 275:532-533.

DeLaune, R. D., R. J. Buresh and W. H. Patrick, Jr. 1979. Relationship of soil properties to standing crop biomass of Spartina alterniflora in a Louisiana marsh. Est. and Coastal Mar. Sci. 8:477-487.

DeLaune, R. D. and W. H. Patrick, Jr. In press. Nitrogen and phosphorus cycling in a Gulf Coast salt marsh. Proc. Int. Estuarine Research Federation Conf., Jeckyll Island, GA, 1979.

Dunstan, W. M., H. L. Windom and G. L. McIntire. 1975. The role of Spartina alterniflora in the flow of Lead, Cadmium, and Copper through the salt-marsh ecosystem, pp. 250-256. In: Fred G. Howell, John B. Gentry and Michael H. Smith (eds.), Mineral Cycling in Southeastern Ecosystems.

Eastman, Kurt W. 1980. The mixing behavior of iron, manganese, phosphorus, and humic acid in a salt marsh creek. M. S. Thesis, College of Marine Studies, Univ. of Delaware.

Erkenbrecher, Carl W., Jr., and L. Harold Stevenson. 1978. The transport of microbial biomass and suspended material in a high-marsh creek. Can. Journal Microbiology 24:839-846.

Fitzgerald, W. F. 1978. Sedimentary geochemical and geological studies: Branford Harbor, CT. Submitted to Waterways Experiment Station, U. S. Army Corps of Engineers, Contract #DACW33-75-C-0085.

Flemer, David A. 1970. Primary production in the Chesapeake Bay. Chesapeake Science 11(2):117-129.

Folger, David W. 1972. Texture and organic carbon content of bottom sediments in some estuaries of the United States, pp. 391-408. In: Bruce W. Nelson (ed.), Environmental Framework of Coastal Plain Estuaries, The Geological Society of America, Inc., Memoir 133.

Frey, Robert W. and Paul B. Basan. 1978. Coastal salt marshes, pp. 101-169. In: Richard A. Davis, Jr. (ed.), Coastal Sedimentary Environments. Springer-Verlag, New York.

Froelich, P. N., M. L. Bender and G. R. Heath. 1977. Phosphorus accumulation rates in metalliferous sediments on the East Pacific Rise. Earth and Planetary Science Letters 34:351-359.

Gallagher, John L., William J. Pfeiffer and Lawrence R. Pomeroy. 1976. Leaching and microbial utilization of dissolved organic carbon from leaves of Spartina alterniflora. Estuarine and Coastal Marine Science 4:467-471.

Gardner, Leonard R. and Wiley Kitchens. 1978. Sediment and chemical exchanges between salt marshes and coastal waters, pp. 191-207. In: B. J. Kjerfve (ed.), Estuarine Transport Processes. University of South Carolina Press, Columbia, SC.

Gardner, Leonard R. 1975. Runoff from an intertidal marsh during tidal exposure-recession curves and chemical characteristics. Limnol. Oceanogr. 20:81-89.

Gosselink, James G., Eugene P. Odum, R. M. Pope. 1974. The value of the tidal marsh. Center for Wetland Resources, Louisiana State University, Baton Rouge, LA, 30 p.

Graham, William F. and Robert A. Duce. 1979. Atmospheric pathways of the phosphorus cycle. Geochimica et Cosmochimica Acta 43: 1195-1208.

Haines, Evelyn B. 1975. Nutrient inputs to the coastal zone: the Georgia and South Carolina shelf, pp. 303-322. In: L. E. Cronin (ed.), Estuarine Research, Vol. 1, Academic Press, N. Y., 738 p.

Haines, E. B. and W. M. Dunstan. 1975. The distribution and relation of particulate organic material and primary productivity in the Georgia Bight, 1973-1974. Estuarine and Coastal Marine Science 3:431-441.

Haines, Evelyn B. 1976. Stable carbon isotope ratios in the biota, soils and tidal water of a Georgia Salt Marsh. Est. Coastal Mar. Sci. 4:609-616.

Haines, Evelyn B. 1976. Relation between the stable carbon isotope composition of fiddler crabs, plants, and soils in a salt marsh. Limnol. Oceanogr. 21:880-883.

Haines, E. B., A. Chalmers, R. Hanson and B. Sherr. 1977. Nitrogen pools and fluxes on a Georgia salt marsh, pp. 241-254. In: M. Wiley (ed.), Estuarine Processes, Vol. 2, Academic Press, N. Y.

Haines, E. B. 1977. The origins of detritus in Georgia salt marsh estuaries. Oikos 29:254-260.

Haines, Evelyn B. 1978a. Interactions between Georgia Salt marshes and coastal waters: a changing paradigm, pp. 35-46. In: R. J. Livingston (ed.), Ecological Processes in Coastal and Marine Systems, 1979. Proceedings of the Symposium at Florida State University, April, 1978. Plenum Press.

Haines, Evelyn B. 1979b. Nitrogen pools in Georgia coastal waters. Estuaries 2:34-39.

Haines, Evelyn B. and Clay L. Montague. 1979. Food sources of estuarine invertebrates analyzed using $^{13}C/^{12}C$ ratios. Ecology 60:000-000.

Hall, C. A. S., D. Whitney, G. M. Woodwell, D. W. Juers and R. Moll. 1975. Material exchanges between the Flax Pond marsh system and Long Island Sound. Paper presented at the Third Biennial International Estuarine Research Conference, Galveston, TX.

Halvorson, William L. and William E. Gardiner. 1976. Atlas of Rhode Island salt marshes. Univ. of Rhode Island Marine Memorandum No. 44.

Hanson, Roger B. 1977. Nitrogen fixation (acetylene reduction) in a salt marsh amended with sewage sludge and organic carbon and nitrogen compounds. Applied and Environmental Microbiology 33:846-852.

Hanson, Roger B. 1977. Comparison of nitrogen fixation activity in tall and short Spartina alterniflora salt marsh soils. Applied and Environmental Microbiology 33:596-602.

Happ, Georgeann, James G. Gosselink and John W. Day, Jr. 1977. The seasonal distribution of organic carbon in a Louisiana estuary. Estuarine Coastal Marine Science 5:695-705.

Harrison, E. Z. and A. L. Bloom. 1974. The response of Connecticut salt marshes to the recent rise in sea level. Geol. Soc. Amer. Abstracts with Programs 6:35-36.

Hayes, Miles O. 1978. Impact of hurricanes on sedimentation in estuaries, bays, and lagoons, pp. 323-346. In: Martin L. Wiley (ed.), Estuarine Interactions. Academic Press, N. Y.

Heinle, D. R. and D. A. Flemer. 1976. Flows of materials between poorly flooded tidal marshes and an estuary. Mar. Biol. 35: 359-373.

Heinle, D. 1979. Characteristics of Chesapeake Bay. Presentation and mimeo handout of the Int. Symp. on Nutrient Enrichment in Estuaries. Williamsburg, VA.

Hicks, Steacy D. 1973. Trends and variability of yearly mean sea level 1893-1971. U. S. Dept. of Commerce, National Oceanic and Atmospheric Administration National Ocean Survey, NOAA Tech. Mem. NOS 12, Rockville, MD, pp. 13.

Hopkinson, Charles S., John M. Day, Jr., and B. T. Gael. 1978. Respiration studies in a Louisiana salt marsh. An. Centro Cienc. Del Mar. Y. Limnol., Univ. Nal. Auton, Mexico 5(1): 225-238.

Houghton, R. A. and G. M. Woodwell. In press. The Flax Pond ecosystem study: Exchanges of CO_2 between a salt marsh and the atmosphere. Ecology.

Jaworski, Norbert A. In press. Sources of nutrients and the scale of eutrophication problems in estuaries. International Symposium on the Effects of Nutrient Enrichment in Estuaries. Williamsburg, VA, 29-31 May 1979, Humana Press.

Jones, Keith. 1974. Nitrogen fixation in a salt marsh. J. of Ecology 62:553-565.

Kalber, Frederick A., Jr. 1959. A hypothesis on the role of tide-marshes in estuarine productivity. Estuarine Bulletin 4(1):3.

Kaplan, Warren, Ivan Valiela and John M. Teal. 1979. Denitrification in a salt marsh ecosystem. Limnol. Oceanogr. 24(4): 726-734.

Keefe, C. W. 1972. Marsh production: a summary of the literature. Contrib. Mar. Sci., Univ. of Texas 16:163-181.

Kjerfve, Bjorn, Jeffrey E. Greer and Richard L. Crout. 1978. Low-frequency response of estuarine sea level to non-local forcing, pp. 497-513. In: Martin L. Wilsy (ed.), Estuarine Interactions. Academic Press, N. Y.

Kjerfve, Bjorn and Jeffrey A. Proehl. 1979. Velocity variability in a cross-section of a well-mixed estuary. J. of Marine Research 37:409-418.

Lee, Virginia. 1979. Net nitrogen flux between the emergent marsh and tidal waters. M. S. Thesis, University of Rhode Island, 67 pp.

Lord, C. J., III. 1980. The chemistry and cycling of iron, manganese, and sulfur in salt marsh sediments. Ph.D. Dissertation, Univ. of Delaware, 177 pp.

Lotrich, Victor A., William H. Meredith, Stephen B. Weisberg, L. E. Hurd and Franklin C. Daiber. 1979. Dissolved and particulate nutrient fluxes via tidal exchange between a salt marsh and lower Delaware Bay. The Fifth Biennial International Estuarine Research Conference Abstracts, Jekyll Island, GA, Oct. 7-12, 1979.

Lyons, Barry. 1979. Early diagenesis of trace metals in nearshore Long Island Sound sediments. Ph.D. Thesis, Univ. of Connecticut.

Marsho, T. V., R. P. Burchard and R. Fleming. 1975. Nitrogen fixation in the Rhode River Estuary, Chesapeake Bay. Can. J. Microbiol. 21:1348-1356.

McCaffrey, Richard J. 1977. A record of the accumulation of sediment and trace metals in a Connecticut, U. S. A., salt marsh. Ph.D. Dissertation, Yale University, New Haven, CT, 156 p.

McHugh, J. L. 1967. Estuarine Nekton, pp. 581-620. In: George H. Lauff (ed.), Estuaries, Amer. Association for the Advancement of Science, Washington DC.

Moll, R. A. 1977. Phytoplankton in a temperate-zone salt marsh: net production and exchanges with coastal waters. Marine Biology 42:109-118.

Moore, Kenneth A. 1974. Carbon transport in two York River, Virginia marshes. M. S. Thesis, Univ. of Virginia, pp. 102.

Muller, P. J. 1977. C/N ratios in Pacific deep-sea sediments: effect of inorganic ammonium and organic nitrogen compounds sorbed by clays. Geochimica et Cosmochimica Acta 41:765-776.

Mulholland, Patrick J., and Edward J. Kuenzler. 1979. Organic carbon export from upland and forested wetland watersheds. Limnol. Oceanogr. 24:960-965.

National Resources Defense Council. 1978. Safeguarding wetlands and Watersources with 404. General distribution mimeo pamphlet.

Nixon, Scott W. and Candace A. Oviatt. 1973. Ecology of a New England salt marsh. Ecological Monographs 43(4):463-498.

Nixon, S. W., C. A. Oviatt, J. Garber, V. Lee. 1976. Diel metabolism and nutrient dynamics in a salt marsh embayment. Ecology 57(4):740-750.

Nixon, S. W. and William E. Odum. 1976. Interactions between tidal wetlands and coastal waters, pp. 217-218. In: Martin Wiley (ed.), Estuarine Processes, Vol. II, Circulation, Sediments, and Transfer of Material in the Estuary. Academic Press, N. Y.

Nixon, S. W., J. R. Kelly, B. N. Furnas and C. A. Oviatt. In press. Phosphorus regeneration and the metabolism of coastal marine bottom communities, pp. 000-000. In: K. R. Tenore and B. C. Coull (eds.), Marine Benthic Dynamics. Univ. of SC Press, Columbia, SC.

Nixon, S. W. and V. Lee. In press. The flux of carbon, nitrogen and phosphorus between coastal lagoons and offshore waters. In: P. Lassier (ed.), Coastal Lagoons: Present and Future Research, Part II UNESCO Tech. Pub. in Mar. Sci., Paris.

Nixon, S. W. In press. Remineralization and nutrient cycling in coastal marine ecosystems. International Symposium on Nutrient Enrichment in Estuaries, Williamsburg, VA, 1979, Humana Press.

O'Connor, S. G. and A. J. McErlean. 1975. The effects of power plants on productivity of the nekton, pp. 494-517. In: L. E. Cronin (ed.), Estuarine Research, Vol. 1, Academic Press, N. Y., pp. 738.

Odum, Eugene. 1961. The role of tidal marshes. New York Conservationist, June-July, p. 12.

Odum, Eugene P. and Armando A. de la Cruz. 1967. Particulate organic detritus in a Georgia salt marsh-estuarine ecosystem, pp. 383-388. In: George H. Lauff (ed.), Estuaries. American Association for the Advancement of Science, Publication No. 83.

Odum, Eugene P. 1968. A research challenge: evaluating the productivity of coastal and estuarine water, pp. 63-64. In: Proceedings of the Second Sea Grant Conference. Univ. of Rhode Island, October 1968.

Odum, Eugene P. 1971. Fundamentals of Ecology. W. B. Saunders Co., Philadelphia, PA, 574 p.

Odum, Eugene P. 1979. The status of three ecosystem-level hypotheses regarding salt marsh estuaries: tidal subsidy, outwelling and detritus-based food chains. The Fifth Biennial International Estuarine Research Conference Abstracts, Jekyll Island, GA, Oct. 7-12, 1979.

Okuda, T. 1960. Metabolic circulation of phosphorus and nitrogen in Matsushima Bay (Japan) with special reference to exchange of these elements between overlying water and sediments. Trabalhos Inst. Biol. Marit Oceanogr. 2:7-153.

Olsen, S. and Saul B. Saila. 1976. Fishing and petroleum on Georges Bank. New England Regional Commission Tech. Report 76-3, 22 p.

Olsen, Stephen B. and David K. Stevenson. 1975. Commercial marine fish and fisheries of Rhode Island. Univ. of Rhode Island Marine Technical Report 34, 117 p.

Parker, P. L. 1967. Chemical parameters, pp. 317-321. In: Pollution and Marine Ecology, Theodore A. Olson and Frederick J. Burgess (eds.), Interscience Publishers, N. Y.

Patriquin, D. G. and D. Denike. 1978. In situ acetylene reduction assays of nitrogenase activity associated with the emergent halophyte Spartina alterniflora Loisel.: methodological problems. Aquatic Botany 4:211-226.

Pellenbarg, Robert E. and Thomas M. Church. 1979. The estuarine surface microlayer and trace metal cycling in a salt marsh. Science 203:1010-1012.

Pickral, James C. and William E. Odum. 1976. Benthic detritus in saltmarsh tidal creek, pp. 280-292. In: Martin Wiley (ed.), Estuarine Processes, Vol. II, Academic Press, N. Y.

Pomeroy, L. R., R. E. Johannes, E. P. Odum and B. Roffman. 1967. The phosphorus and zinc cycles and productivity of a salt marsh, pp. 412-430. In: Symposium on Radioecology, Daniel J. Nelson and Francis C. Evans (eds.). Proceedings of the Second National Symposium held at Ann Arbor, MI, May 15-17, 1967.

Pomeroy, Lawrence R., E. E. Smith and Carol M. Grant. 1965. The exchange of phosphate between estuarine water and sediments. Limnol. Oceanogr. 10(2):167-172.

Pope, R. M. and James G. Gosselink. 1973. A tool for use in making land management decisions involving tidal marshland. Coastal Zone Management Journal 1(1):65-74.

Redfield, A. C. 1972. Development of a New England salt marsh. Ecol. Monogr. 42:201-237.

Reimold, Robert J. and Franklin C. Daiber. 1970. Dissolved phosphorus concentrations in a natural salt-marsh of Delaware. Hydrobiologia 36(3-4):361-371.

Reimold, Robert J. 1972. The movement of phosphorus through the salt marsh cord grass, Spartina alterniflora Loisel. Limnol. Oceanogr. 17(4):606-611.

Rice, Donald L. 1979. Trace element chemistry of aging marine detritus derived from coastal macrophytes. Ph.D. Thesis, Georgia Institute of Technology, 144 p.

Richard, Glenn A. 1978. Seasonal and environmental variations in sediment accretion in a Long Island salt marsh. Estuaries 1(1): 29-35.

Saila, Saul B. 1975. Some aspects of fish production and cropping in estuarine systems, pp. 473-493. In: L. E. Cronin (ed.), Estuarine Research, Vol. 1, Academic Press, N. Y., pp. 738.

Salt Marsh Conference, Marine Institute, Sapelo Island, GA, 1958, Proceedings, pp. 133.

Schubel, J. R. and Harry H. Carter. 1976. Suspended sediment budget for Chesapeake Bay, pp. 48-62. In: Martin Wiley (ed.), Estuarine Processes, Vol. II, Circulation, Sediments and Transfer of Material in the Estuary. Academic Press, N. Y.

Settlemyre, J. L. and L. R. Gardner. 1975a. Chemical and sediment budgets for a small tidal creek, Charlestown Harbor, S. C. Water Resources Research Institute Report No. 57, Clemson University, Clemson, SC.

Settlemyre, J. L. and L. R. Gardner. 1975b. A field study of chemical budgets for a small tidal creek-Charlestown Harbor, S. C. In: Marine Chemistry in the Coastal Environment, pp. 152-175. T. M. Church (ed.), ACS Symposium Series, No. 18.

Shaler, N. S. 1885. Sea-coast swamps of the Eastern United States. In: U. S. Geol. Survey 6th Annual Report, pp. 359-368.

Sheith, M-S., J. 1974. Nutrients in Narragansett Bay sediments. M. S. Thesis, Univ. of Rhode Island, Kingston, RI.

Siccama, Thomas G. and Elliot Porter. 1972. Lead in a Connecticut salt marsh. BioScience 22(4):232-234.

Simpson, Robert L., Dennis F. Whigham and Raymond Walker. 1978. Seasonal patterns of nutrient movement in a freshwater tidal marsh. pp. 243-257. In: Ralph E. Good, Dennis F. Whigham and Robert L. Simpson (eds.), Freshwater Wetlands: Ecological Processes and Management Potential, Academic Press, N. Y.

Smayda, T. J. 1973. A survey of phytoplankton dynamics in the coastal waters from Cape Hatteras to Nantucket, pp. 3-1 to 3-100. In: Saul B. Saila (ed.), Coastal and Offshore Environmental Inventory: Cape Hatteras to Nantucket Shoals, Marine Publication Series No. 2, University of Rhode Island, Kingston, RI, 02881

Smith, David W. 1979. Marsh nitrogen fixation: fact or fantasy. The Fifth Biennial International Estuarine Research Conference Abstracts, Jekyll Island, GA, Oct. 7-12, 1979.

Spinner, George P. 1969. A plan for the marine resources of the Atlantic coastal zone. American Geographical Society, pp. 80.

Stearns, L. A. and D. MacCreay. 1957. The case of the vanishing brick dust. Mosquito News 17:303-304.

Stevenson, F. J. and C.-N. Cheng. 1971. Organic geochemistry of the Argentine Basin sediments: carbon-nitrogen relationships and Quaternary correlations. Geochimica et Cosmochimica Acta 36:653-671.

Strom, Richard N. and Robert B. Biggs. 1972. Trace metals in cores from the Great Marsh, Lewes, Delaware. Sea Grant Publication #DEL-SG-12-72, Univ. of Delaware, 35 p.

Stross, Raymond G. and John R. Stottlemyer. 1965. Primary production in the Patuxent River. Chesapeake Sci. 6:125-140.

Taft, J. L. and W. R. Taylor. 1976. Phosphorus dynamics in some coastal plain estuaries. In: Estuarine Processes, Vol. 1, pp. 79-89, M. Wiley (ed.), Academic Press, N. Y.

Teal, John M. 1958. Energy flow in the salt marsh ecosystem, pp. 101-107. In: Proceedings - Salt Marsh Conference, Marine Institute, Sapelo Island, GA, March 25-28.

Teal, J. M. 1962. Energy flow in the salt marsh ecosystem of Georgia. Ecology 43:614-624.

Teal, J. M. and J. Kanwisher. 1961. Gas exchange in a Georgia salt marsh. Limnol. Oceanogr. 6(4):388-399.

Teal, J. M., I. Valiela and D. Berlo. 1979. Nitrogen fixation by rhizosphere and free-living bacteria in salt marsh sediments. Limnol. Oceanogr. 24:126-132.

Thayer, G. W. 1971. Phytoplankton production and the distribution of nutrients in a shallow unstratified estuarine system near Beaufort, N. C. Chesapeake Sci. 12:240-253.

Turner, R. E. 1976. Geographic variations in salt marsh macrophyte production: a review. Contributions in Marine Science 20:47-68.

Turner, R. E. 1978a. Community plankton respiration in a salt marsh estuary and the importance of macrophytic leachates. Limnol. Oceanogr. 23:442-451.

Turner, R. E. 1978b. Louisiana's coastal fisheries and changing environmental conditions, pp. 363-370. In: J. W. Day, D. D. Culley Jr., R. E. Turner and A. J. Mumphrey, Jr., (eds.), Proc. Third Coastal Marsh and Estuary Management Symposium. Louisiana State University, Division of Continuing Education, Baton Rouge, LA, 1979.

Turner, R. E., S. W. Woo and H. R. Jitts. 1979. Phytoplankton production in a turbid, temperate salt marsh estuary. Estuarine and Coastal Marine Science, in press.

Turner, R. E., S. W. Wood, and H. R. Hitts. 1979. Estuarine influences on a Continental Shelf plankton community. Science 206:218-220.

Tyndall, John. 1874. The forms of water in clouds and rivers, ice and glaciers. D. Appleton & Co., N. Y. 4th ed., pp. 196.

Udell, H. F., J. Zarudsky, T. E. Doheny. 1969. Productivity and nutrient values of plants growing in the salt marshes of the Town of Hempstead, Long Island. Bulletin of the Torrey Botanical Club 96:42-51.

United States Department of the Interior, Fish and Wildlife Service. 1970. National Estuary Study, Vols. 2 & 3.

Upchurch, Joseph B. 1972. Sedimentary phosphorus in the Pamlico Estuary of North Carolina. Sea Grant Publication UNC-SG-72-03, Univ. of North Carolina, 39 p.

Valiela, Ivan, John M. Teal and Norma Y. Persson. 1976. Production and dynamics of experimentally enriched salt marsh vegetation: below ground biomass. Limnol. Oceanogr. 21:245-252.

Valiela, Ivan, John M. Teal, Suzanne Volkman, Deborah Shafer and Edward J. Carpenter. 1978. Nutrient and particulate fluxes in a salt marsh ecosystem: tidal exchanges and inputs by precipitation and groundwater. Limnol. Oceanogr. 23(4):798-812.

VanRaalte, Charlene D., Ivan Valiela, Edward J. Carpenter and John M. Teal. 1974. Inhibition of nitrogen fixation in salt marshes measured by acetylene reduction. Estuarine and Coastal Marine Science 2:301-305.

Volk, B. G., S. D. Schemnitz, J. F. Gamble and J. B. Sartain. 1975. Baseline data on everglades soil-plant systems: elemental composition, biomass, and soil depth, pp. 658-672. In: Fred G. Howell, John B. Gentry and Michael H. Smith (eds.), Mineral Cycling in Southeastern Ecosystems, Proceedings of a Symposium held at Augusta, GA, May 1-3, 1974.

Walker, Richard A. 1973. Wetlands preservation and management on Chesapeake Bay: the role of science in natural resource policy. Coastal Zone Management Journal 1(1):75-101.

Welsh, Barbara L. In press. Comparative nutrient dynamics of a marsh-mudflat ecosystem. Estuarine and Coastal Marine Science.

Wentz, D. A. and G. F. Lee. 1969. Sedimentary phosphorus in lake cores-observations on depositional pattern in Lake Mendota. Environmental Science and Technology 3:754-759.

Whitney, D. E., G. M. Woodwell and R. W. Howarth. 1975. Nitrogen fixation in Flax Pond, a Long Island salt marsh. Limnol. Oceanogr. 4:640-643.

Williams, Richard B. and Marianne B. Murdoch. 1967. The potential importance of Spartina alterniflora in conveying zinc, manganese, and iron into estuarine food chains, pp. 431-439. In: Daniel J. Nelson and Francis C. Evans (eds.). Symposium on Radioecology, Proceedings of the Second National Symposium held at Ann Arbor, MI, May 15-17, 1967.

Windom, H. L., W. M. Dunstan and W. S. Gardner. 1975. River input of inorganic phosphorus and nitrogen to the southeastern salt marsh estuarine environment, pp. 309-313. In: Fred G. Howell, John B. Gentry and Michael H. Smith (eds.), Mineral Cycling in Southeastern Ecosystems, Proceedings of a Symposium held at Augusta, GA, May 1-3, 1974.

Wolaver, T., R. L. Wetzel, and K. L. Webb. 1979. Nutrient interactions between salt marsh, intertidal mudflats and estuarine waters. Abstracts of The Fifth Biennial International Estuarine Research Conference, Jekyll Island, GA, Oct. 7-12, 1979.

Wolman, M. G. 1967. A cycle of sedimentation and erosion in urban river channels. Geografisha Annaler 49A:385-395.

Woodwell, G. M., D. E. Whitney and C. A. S. Hall and R. A. Houghton. 1977. The Flax Pond ecosystem study: exchanges of carbon in water between a salt marsh and Long Island Sound. Limnol. Oceanogr. 22(5):833-838.

Woodwell, G. M. and D. E. Whitney. 1977. Flax Pond ecosystem study: exchanges of phosphorus between a salt marsh and the coastal waters of Long Island Sound. Marine Biology 41:1-6.

MODELING SALT MARSHES AND ESTUARIES: PROGRESS AND PROBLEMS

Richard G. Wiegert

Department of Zoology, University of Georgia

Athens, Georgia 30602

ABSTRACT

Explanatory models of salt marshes must be based on realistic interactions between structure (niches and flow pathways) and function (species and their ecological attributes). These interactions produce behavior (changes in rates and standing stocks of energy and nutrients). Such models may be used for (1) management, (2) prediction of perturbations or (3) development of testable hypotheses. Difficulties facing modelers involve (a) conserved flows, (b) trophic condensation, (c) time delays and (d) feedback controls. In addition hydrodynamic problems are important involving (e) sediment water exchanges, (f) matter/energy transport by tides, (g) catastrophic storms and tides and (h) spatial heterogeneity. Construction and use of a carbon flow model of a coastal Georgia salt marsh are discussed.

INTRODUCTION

What is an estuarine salt marsh?

Estuarine salt marshes of the world share the characteristic of being invaded more or less frequently by salty water. The tidal frequency and salinity vary, but must be often enough or high enough to produce a discernable effect on the floral and faunal composition.

The physiography in concert with the tidal flows produces a unique milieu and determines even the major biological processes that are found in the ecosystems. Consider a rocky intertidal by way of contrast. Here the elevation is steep, the intertidal zone is small and exposed and the physical impact of tides and waves is enormous. The same is true, to a somewhat lesser degree of a sand beach. Like these, the estuarine salt marsh is a double ecotone, sandwiched between the much larger terrestrial and open-water marine communities. Unlike the rocky and sandy coasts, which have little accumulated detritus and net primary productivity, the salt marsh and estuary have both. As a result, salt marshes are more tightly coupled to the boundary communities; ground and nutrient water flows from the land may affect the productivity. The marsh may in turn export organisms, detritus or nutrients to the estuary. The surplus of net primary production and the heavy sediment accumulation means that microorganismal degradation of energy and release of nutrients will be important regulators of the community. Incomplete mixing of tidal water in these low energy systems creates spatially heterogeneous concentrations of nutrients and energy. On high energy coasts, for example the rocky intertidal, most of the complexity and behavior of the biological community can be explained by the influence of the macrofauna and by physical disturbances (Paine 1969; Dayton 1971).

Salt marshes, in contrast, require a prior appreciation of the role of the microorganisms. The microbiological processes are so fundamental that the salt marshes with very different dominant vegetation may be similar with respect to their degradative processes. Indeed, this fundamental similarity of process in ecosystems facilitates modeling simulation and inter marsh comparisons.

Simulation models of ecosystems

A photograph of a salt marsh is a model in the sense of portraying an ecosystem. However, although the amount and quality of information conveyed by a picture may indeed be great, the picture is a static model. It cannot simulate behavior and thus cannot be used to predict. The static model has structure but no functional attributes. Addition of such functional capabilities to a static model permits interaction of structure and function to simulate behavior. The model is now dynamic.

Used in this manner, the words structure and function may appear strange to the ecologist, who is more used to seeing structure appended to the objects in the system (trees, birds, rocks and water) and function applied to the processes, such as energy flow, nutrient cycling, etc. Here structure is the boxes and arrows indicating the compartments and pathways of information and material flows in the system. Functional components are the species and abiotic parts of the system as well as the specific feedback controls operating on the flows. Functional components, constrained by the structural relationships and pathways produce system behavior in the form of flows of material and change of state in the components. This usage is more convenient here than the more traditional (and more ambiguous) usage common in the ecological literature. It follows accepted definitions in the field of systems science and is discussed more fully in Hill and Wiegert (in press).

Explanatory models must be constructed from data on functional aspects obtained independently of the behavior of the system to be modeled. The model is thus a hypothesis of how the system functions. Models derived from prior study of system behavior, as are empirical multiple regression models, can be good predictors within a defined range of behavior. But such empirical models make no pretense of explanation. They are not considered further in this paper. Dynamic models simulate; and from the simulated behavior, predictions may be made. These predictions may be used in several ways.

Use of models

There are three broadly overlapping categories of model objective. (1) The model may be asked management questions - what is the most efficient, least risky, or most profitable manner in which an ecosystem can be exploited in some way. In other words, the system is recognized as desirable of maintenance. Long term stability is sought. (2) The predicted effect of a perturbation, either acute or chronic may be wanted. In this case, the predictive simulation model may be one way in which the environmental impact of an action is explored. (3) The model may form part of a basic investigation into the operation of the ecosystem. Explanation is desired above all. The model is used to guide the research by suggesting testable hypotheses, which in turn suggest further research.

MODELING PROBLEMS

Many of the difficulties facing the modelers of salt marshes and estuaries are general, encountered in most or all modeling efforts. But certain problems are closely associated with the unique characteristics of salt marshes, particularly the tidal flows coupling the marsh with the estuary. As an example of the first group, I consider briefly questions of (a) conserved flows, (b) trophic condensation, (c) time delays and finally (d) the regulatory or control functions employed in the flux equations. Hydrodynamic aspects, all the examples in the second group; for example: (e) sediment - water, exchanges, (f) matter/energy transport by tidal movements, (g) catastrophic events and (h) the different kinds of spatial heterogeneity found in the marsh and estuary.

Conserved flows

Models simulate system behavior by computing flows of matter or energy between compartments and keeping track of the resulting changes in standing stock. Conserved units are those which do not change upon passage through the trophic network. A joule of energy in one component is equivalent to the same energy unit elsewhere in the system. Similarly with a mole of any element. However, the form in which the element or energy is found, that is the compounds and mixtures (or the free energy form) will vary widely in terms of its utility as a source of energy, as a nutrient or as a physical shelter. Therefore, each model will generally use some conserved (or easily converted) unit as a primary bookkeeping device to track flows and changes in standing stock. The form in which the conserved unit is found will be specified wherever it is important as a control factor or a resource. Other conserved units, for example nitrogen in a carbon flow model will also need to be modeled wherever the secondary element is an important nutrient. In this manner, a carbon or energy flow model can be employed even though scarcity of energy or of carbon may seldom regulate. The secondary models tracking the dynamics of the controlling nutrients can usually be considerably less comprehensive and detailed than the primary model.

Component identification

What species and which abiotic substances shall be identified as separate components of the model? All modelers face this decision made difficult by the large number of species and physical factors important in even the simplest of ecosystems. There are no absolute rules. Clearly any species, nutrient, or physical entity, must be identified separately if it is important either in processing large amounts of the conserved unit of flow, or in controlling such processing. Furthermore, when two or more species are combined as a single component in the model, they should have, whenever possible, similar trophic relationships and similar growth potentials (Wiegert 1975a). An alternative procedure is to simply ignore all but a select few of the species and physical factors in the system and construct a submodel. That all ignored parts of the system remain unaffected by the dynamics of the submodel is an implicit assumption of this approach. All predator-prey management models are of this type. Indeed, all ecosystem models are by default examples of this, since none has yet considered all the species and physical factors. The separation is rather one of degree.

Time delays

Time delays in the model should be used wherever the ecological information is sufficient to pinpoint a lag between a change in some causative agent and the resulting effect. The time delay must be used if it is long relative to the scale of changes that are being simulated in the ecosystem. Many crucial time delays are automatically incorporated into the function of the model, for example the representation of reproduction as continuous or periodic, in the case of annual species only once per year. On the other hand, behavioral time delays often have to be modeled explicitly. For example, an increase in a material resource, say a small food fish, may require time for the predators to sense it, congregate and begin feeding. Such required delays usually emerge clearly once ecological information on species interactions is obtained.

Regulatory functions

Each of the fluxes of energy and matter included in the structure of the model must have a functional notation incorporating the salient facts about how the flux is regulated. Under what conditions does the recipient ingest resources, what factors influence reproduction, mortality, feeding? Are thresholds of the resource and/or recipient important, etc? These are among the most vital questions in constructing a realistic simulation model. Without this knowledge and its proper representation in the model no advance in theory is possible. We are left with the alternative of an empirical model, useful only for prediction and requiring a prior series of perturbations.

I have discussed the problems of choosing a functional form for these equations elsewhere (Wiegert 1974, 1975b, 1979) and can comment only briefly on this problem here. The single most important mistake that the modeler can make is to pick the functional form because of some imagined analytical convenience or generality and ignore ecological reality. Thus, much of the previous work in theoretical ecology has almost no relevance to the real world because standard control functions such as the logistic and Michaelis-Menten were employed in place of more realistic forms of these linear or hyperbolic (respectively) feedback control functions. Indeed, Hill and Durham (1978) and Hill and Wiegert (in press) suggest the entire feedback control function may sometimes be replaced by a parallel regulatory network involving homeostatic regulation by changes in the fluxes to two or more ecologically related or congeneric species.

The second group of problems facing the modeler of marsh estuarine ecosystems are all related in some rather direct way to the hydrodynamics of the system.

Sediment - water exchanges

These are both difficult to study and to model. Because tidal import or export can be an important mechanism whereby materials are transported, it should be represented in the model. This in turn requires a knowledge of what is in the water at any given time. In particular the exchange of soluble nutrients between the water and sediment must be measured and also the turbulent mixing of surface sediment into the water must be known.

Tidal transport

Once the sedimentation - resuspension rates are known, studies of the hydrology of the marsh-estuary are needed to ascertain how much of the suspended/dissolved materials are exchanged with upstream or downstream water masses. Modeling this situation can be as simple as inserting a diffusional exchange coefficient into an input - output flux pathway or as complex as modeling the hydrodynamic patterns of both vertical and horizontal patches in a large estuary. This brings us to the last two modeling problems involving hydrology.

Catastrophic events

Marsh - estuarine systems are subject to different kinds of irregular perturbations, called catastrophic events in this paper. The most extreme are hurricanes, the most common are heavy rainstorms, particularly those falling on the marsh at low tide, causing both erosion and overturn due to decreases in salinity. Again, the effects of catastrophic events will be strongly conditioned by the local hydrology, but provision for them should be in every simulation.

Spatial heterogeneity

The difference in the spatial distribution of organisms and physical factors in an ecosystem makes simulation of ecosystem behavior with a single model difficult. Spatial heterogeneity in distribution of organisms is an important stabilizing aspect of ecosystems (Wiegert 1975a; Smith 1972) and must be included in models. Unfortunately, the best way to do this is to construct a separate model for each ecosystem and run these simultaneously. If the position and size of the spatially distinct units is not changing rapidly (as is true in most salt marshes) this procedure increases the complexity and cost of the model, but does not introduce conceptual difficulties as it does in the case in thermal spring communities, where the spatial heterogeneity is itself changing rapidly, (Wiegert 1975a).

MODELING STATE-OF-THE-ART: SAPELO ISLAND SALT MARSH SYSTEM

Many of the problems enumerated here have been considered and solved for specific ecosystems. A number of simulation models exhibiting a wide variety of objectives, structures, functions and behaviors are extant. Seven of these were considered in a recent review on state of the art in models of coastal and estuarine ecosystems (Wiegert, in press). I wish to conclude this paper with a more detailed discussion of the development of a salt marsh model of the *Spartina alterniflora* marshes of the Georgia coast. In this discussion, I try to show how our group of researchers used the model (1) as a continuously evolving tool with which to guide an interdisciplinary research program and (2) develop explanatory hypotheses about the operation of a large and important coastal ecosystem.

The models discussed were all based on the salt marsh and tidal creek system associated with the Duplin River, Sapelo Island, Georgia. These models have been developed and used over the past six years during which a large interdisciplinary study was conducted by faculty and staff associated with the University of Georgia Marine Institute on Sapelo. One major objective of this research was to document the movement and fate of carbon in the marsh. The use of the carbon-flow simulation model to aid this effort forms the basis of the presentation here. Because a number of researchers were involved in this modeling work (Wiegert et al. 1975; Imberger et al. in press), I shift to the use of 'we' in this concluding discussion.

Model construction

The dynamic model to be used as a research tool must first be constructed. The major steps are reviewed briefly here. The reader is referred for a more comprehensive treatment of this important activity to the book by Hall and Day (1977) and the review of simulation models by Wiegert (1975c).

(1) In our original salt marsh model we chose 14 compartments, 7 biotic and 7 abiotic (Wiegert et al. 1975). The bookkeeping unit of the model was to be carbon flow. The 14 components were:

1) CO_2 in the air
2) Algae (both benthic and phytoplankton)
3) *Spartina alterniflora* shoots
4) Grazers on *Spartina* roots

5) Standing dead *Spartina*

6) Dissolved organic carbon (DOC) in water

7) Heterotrophs in water (bacteria, invertebrates, fish)

8) Anaerobic microbes in the sediment

9) Particulate organic carbon (POC) in the water

10) CO_2 in the water

11) DOC in the sediments

12) *Spartina* roots

13) POC in the sediments

14) Aerobic microbes in the sediments

In general the transfer of carbon from any compartment to a biotic component was determined by a maximum rate of transfer multiplied by the standing stock of the recipient. This was reduced under limiting conditions by control terms that were either functions of the donor or of the recipient or both. Transfer of carbon to an abiotic component was usually represented by a specific rate times the standing stock of the donor, i.e. donor-controlled.

In this fashion a series of simulation models of the salt marsh ecosystem was constructed and revised. Three of these -- Marsh Model 1, Version 1 (MRSH1V1), MRSH1V3 and MRSH1V6 -- are discussed.

Marsh Model 1, Version 1

This model was constructed with nonlinear discontinuous feedback controls incorporating threshold responses to scarcity of resources and to crowding. There are no time delays or stochastic elements in the model. Integration uses the simple Euler method with a fixed interval of 0.1 day. Temperature and nutrient controls are implicit in the variable parameters, which were changed every three months.

The purpose of the model was:

1) to evaluate the sensitivity of the model parameters by varying them one at a time and noting the response of the components. Those parameters judged sensitive (a 5% change in any component value after 5 years) were to be studied intensively.

2) to evaluate the overall carbon balance of the marsh to ascertain if the marsh as a whole was a source or a sink for organic carbon and to examine how much this conclusion depended on the choice of model or parameter values.

The work with MRSH1V1 and the conclusions were described by Wiegert et al. (1975). Briefly we found one-fourth (21) of the parameters to be sensitive as defined. These were reused to incorporate new data before the construction of MRSH1V3.

Five-year runs with MRSH1V1 produced the following simulated carbon balances. Keeping the system closed (i.e., not permitting exchanges of CO_2 in the air above the marsh with adjacent air masses and permitting no exchanges between the marsh and estuary) resulted in large annual fluctuations in the carbon content of the air and in the standing crop of Spartina. This was paralleled by an annually fluctuating but constantly increasing transfer of organic carbon from the particulate pool to the dissolved pool. Relaxing the stricture against exchanges via wind changed CO_2 in air to a constant and caused an increase in Spartina to realistic levels (based on comparisons with standing stock data from the field). The marsh began accumulating carbon and both the DOC and POC pools increased each year. The mean annual increment was 765 gC x m^{-2} x yr^{-1}. Several simulation experiments were run in which parameter values were changed in an effort to reverse this accumulative or "carbon sink" behavior by the salt marsh system. Nothing short of clearly unreasonable increases in respiration coefficients or decreases in production rates would suffice. Clearly MRSH1V1 was pronouncing the marsh to be a sink, not a source of carbon. Equally clearly, the biological and geological data indicated only a modest increase in the sedimentary carbon and an apparent steady state for the biota! At this point, we decided to construct a version (3) of the model (MRSH1V3) that provided for transport of carbon out of the marsh via tidal flooding -- the outwelling concept initially supported for the Duplin marshes by the carbon export data of Odum and de la Cruz (1967) from Study Creek.

Marsh Model 1, Version 3

Model MRSH1V3 had two major objectives: (1) we wished to simulate a number of experimental perturbations and compare the results with similar experiments done in the field as a means of corroborating (Caswell, 1976) the model. (2) We wished to examine the effect of simulated transport of carbon via tidal flushing. We used the same basic 14-compartment model as MRSH1V1 with several parameter changes based on new data. In addition, the water-borne

DOC, POC, heterotrophs and algae were subjected to a tidal washout term that could be varied and reflected the assumed diffusive and advective transport to the estuary from the water mass covering the marsh at high tide. This tidal transport term was multiplied by the difference between the standing stock of a component in estuarine water versus that in the marsh water.

The simulation experiments with MRSH1V3 and the comparisons with field experiments are discussed in Wiegert and Wetzel (1979). The full computer program is listed in the appendix to that paper. Here we discuss only the results of simulating the effect of tidal flushing on carbon balance in the marsh. Four simulation runs of 5 years each were made. Tidal exchange was set for 0%/day, 12.5%/day, 25%/day and 50%/day. The total carbon exported from all four affected compartments (DOC, POC, heterotrophs and algae) was 0, 1025, 1974, and 1033 gC x m^{-2} x yr^{-1}, respectively. The net carbon accumulation in the marsh was 1012, 53, 26, and 17.4, respectively. The increase of 1012 gC accumulating when tidal flushing was absent (compared with the 765 gC predicted with MRSH1V1) was due to the parameter changes, particularly those based on the latest studies of Spartina productivity and respiration rates. These latest simulation results showed clearly the necessity of some mechanism whereby the excess carbon could be moved. Tidal flushing, even when represented as crudely as in MRSH1V3, could balance the carbon, the question now became one of what the real exchange rate was between water masses, i.e., we needed more information about the hydrology of the upper Duplin River and the marshes. Unfortunately, the four simulations of MRSH1V3 did not help resolve the problem about the magnitude of the exchange. If tidal flushing were the only mechanism accounting for carbon losses, then the exchange rate had to be at least 12%/day. But higher rates gave similar carbon exports and balances. Furthermore, the simulated trajectories of the four compartments involved were so close under the three export rates (12.5, 25 and 50%/day) that the field data could not be used to separate them. Thus was born the 1977 Interdisciplinary Duplin River Project (DRIP). Using salinity as a tracer and light rains as a means of producing salinity gradients, a team of scientists from the University of Georgia and a hydrologist (J. Imberger) from the University of California, Berkeley, found the true exchange rate to be 16%/day, but the difference in standing stocks of DOC, POC, heterotrophs and algae between the upper two water masses to be much less than originally assumed in MRSH1V3. Furthermore, POC was found to vary almost wholly in response to ebb or flood tide and, most importantly, heavy rain (> 6 cm/2 hours) falling on a marsh at low tide could completely overturn the upper Duplin River, the flood of fresh water eroding the marsh and carrying much material into the estuary. This increased hydrologic information plus some new parameter values were incorporated into the model to produce MRSH1V6.

Marsh Model 1, Version 6

This model had as its sole major objective the investigation of hydrologic effects on carbon transport and mass balance (Imberger et al. in press). Simulation runs with the model showed that four rainstorms per year (half falling at low tide) plus the other changes in the hydrologic structure of the model predicted a net annual accumulation of somewhat more than 200 gC x m^{-2} x yr^{-1}. Because this was still much more than the measured accumulation in the marsh, we used the model to help develop and test initially three alternative hypotheses. (1) Bedload transport could account for the additional carbon, but only if carbon transport by the shifting sediments at the mouth of the Duplin were far greater than existing measurements showed. (2) Two additional rain storms falling on the marsh at high tide could also account for the loss of the excess carbon. This hypothesis is stronger than that postulating bedload transport, but runs counter to the weather data in hand. (3) The excess carbon could be lost via motile organisms (fish, shrimp, crabs) leaving the marsh and not returning. Such trophic transfer would depend on carbon transformations originating with bacterial degradation of DOC and POC. The model was used to evaluate the reasonableness of this hypothesis by predicting the rate of predation on bacteria by detritovores that would be needed to use up > 200 gC x m^{-2} x yr^{-1}. The answer, 3% of the standing stock of bacteria per day, seemed well within the productive ability of the bacteria. Quantitative data on the migration of fish, shrimp, and crabs are nonexistent for the Sapelo marshes, although it is common knowledge that such organisms do in fact emigrate offshore. This carbon transport mechanism is not in the model, nor have respiration measurements made on the water column taken into account the CO_2 loss from the macroheterotrophs. The testing of this third hypothesis, developed from the model, is the subject of much of the current and proposed research on the Sapelo marshes.

Future models

Concurrent developments of the salt marsh carbon model to keep pace with the field research involves three distinct areas:

1) As more data accumulate on the ecological interactions of single species or specialized groups of species, i.e., sulfate oxidizers, the model must incorporate increased resolution of the biological entities. One such effort has already been made (Christian and Wetzel 1978). A larger and more detailed submodel of the aerobic heterotrophs in the salt marsh (AEROBE6) is now being written by Wetzel and Wiegert.

2) The heterogenous nature of the abiotic components now subsumed under the headings of POC and DOC needs to be recognized in the model. Similarly, the spatial heterogeneity of the marsh must be represented in the model. A second generation carbon flow model (MRSH2V1) will incorporate subdivisions of POC and DOC into labile versus refractory portions. Separate models will be used to simulate creek, creek bank and high marsh sections of the marsh ecosystem.

3) Nutrients other than carbon play a vital role in controlling various types of carbon transport and transformation. In current versions these nutrient controls are only implicit. A proposed third generation model (MRSH3V1) will merge carbon and nitrogen flows into the same model and will be employed to generate hypotheses and guide a research project on C-N interactions that is just starting at the Sapelo Marine Institute.

ACKNOWLEDGMENTS

The work on which this paper is based was supported by NSF Grant OCE75-20842 A03.

REFERENCES

Caswell, H. 1976. The validation problem. In: Systems Analysis and Simulation in Ecology. Ed. B. C. Patten. Vol. 4. Adacemic Press, N. Y. 313-325.

Christian, R. R. and R. L. Wetzel. 1978. Interaction between substrate, microbes and consumers of *Spartina* detritus in estuaries. Estuarine Interactions. Academic Press, N. Y. 93-113.

Dayton, Paul K. 1971. Competition, Disturbance, and Community Organization: The Provision and Subsequent Utilization of Space in a Rocky Intertidal Community. Ecological Monographs 41: 351-389.

Hall, C. and J. Day. 1977. Ecosystem Modeling in Theory and Practice: An Introduction with Case Histories. John Wiley and Sons, Inc.

Hill, J. and S. Durham. 1978. Input, signals and control in Ecosystems. Proc. 1978 IEEE. Int. Conf. on acoustics, Speech and Scig. Processing Tulsa, Okla. p. 391-397.

Hill, J. and R. G. Wiegert. 1980. Modeling microecosystems. In: Microcosms in Ecological Research, J. P. Giesey (Ed.), D O E Symp. Proc., Augusta, Ga. Nov. 1978.

Imberger, J., et al. (in press) The influence of water motion on the spatial and temporal variability of chemical and biological substances in a salt marsh estuary. Limnol. and Oceanogr.

Odum, E. P. and de la Cruz, A. A. 1967. Particulate organic detritus in a Georgia salt marsh-estuarine ecosystem. Estuaries (Ed. by G. H. Lauff), pp. 383-88. Am. Assoc. Adv. Sci. Publ. 83.

Paine, Robert T. 1969. The Pisaster-Tegula Interaction: Prey patches, predator food preference, and intertidal community structure. Ecology 50:6.

Smith, F. E. 1972. Spatial heterogeneity, stability, and diversity in ecosystems. Trans. Conn. Acad. Arts and Sci. 44:307-35.

Wiegert, R.G. 1974. A general mathematical representation of ecological flux processes: description and use in ecosystem models. Proc. Sixth SE Systems Symp. IEEE, Baton Rouge, LA.

Wiegert, R.G. 1975a. Simulation modeling of the algal-fly components of a thermal ecosystem: effects of spatial heterogeneity, time delays and model condensation. In Systems Analysis and Simulation in Ecology 3 (ed. B.C. Patten) Academic Press.

Wiegert, R.G. 1975b. Mathematical representation of ecological interactions. In Ecosystem Analysis and Prediction (ed. S.A. Levin) SIAM.

Wiegert, R. G. 1975c. Simulation models of ecosystems. Ann. Rev. Ecol. and Syst. 6: 311-338.

Wiegert, R. G. (In press). Modeling coastal, estuarine and marsh ecosystems: state-of-the-art: In Proc. of the Statistical Ecology Meeting, G. P. Patil (Ed.) Parma, Italy, Aug. 1978.

Wiegert, R. G. 1979. Population models: experimental tools for analysis of ecosystems, pp. 234-279. In D. Horn, G. Stairs and R. Mitchell (eds.) Analysis of Ecological Systems. Ohio State Univ. Press, Columbus, Ohio.

Wiegert, R. G., R. R. Christian, J. L. Gallagher, J. R. Hall, R. D. H. Jones, R. L. Wetzel. 1975. A preliminary ecosystem model of coastal Georgia Spartina Marsh. In Estuarine Research, Vol. 1 pp. 583-610. Academic Press, Inc. N. Y.

Wiegert, R. G. and R. L. Wetzel, 1979. Simulation experiments with a 14-compartment model of a Spartina salt marsh. In R. Dame (ed.) Marsh Estuarine Systems Simulation. Univ. S. C. Press, Columbia, S. C.

SIMULATION MODELING OF ESTUARINE ECOSYSTEMS

Robert W. Johnson

NASA Langley Research Center

Hampton, Virginia 23665

ABSTRACT

A simulation model has been developed of Galveston Bay, Texas ecosystem. Secondary productivity measured by harvestable species (such as shrimp and fish) is evaluated in terms of man-related and controllable factors, such as quantity and quality of inlet freshwater and pollutants. This simulation model used information from an existing physical parameters model as well as pertinent biological measurements obtained by conventional sampling techniques. Predicted results from the model compared favorably with those from comparable investigations. In addition, this paper will discuss remotely sensed and conventional measurements in the framework of prospective models that may be used to study estuarine processes and ecosystem productivity.

INTRODUCTION

Simulation models provide a means to relate the physical, chemical, and biological characteristics of estuarine ecosystems. This approach also can take into account the effects of man-induced changes in the natural ecosystem. In the highly productive Galveston Bay, Texas, these man-related effects include decreased freshwater inflow and discharge of pollutants. This paper has two objectives: first, to describe a computerized simulation model of the biological energy flow through an estuarine system, specifically Galveston Bay, Texas. The modeling effort studied the interrelated physical, chemical, and biological effects within a semi-enclosed ecosystem that is subject to pollution and other man-related effects.

Second, identification of forcing functions and other parameters from Galveston Bay and other models and how measurements made using remote sensing in conjunction with conventional techniques may contribute to improved ecosystem models.

GALVESTON BAY

Physical, Chemical and Biological Characteristics

The Galveston Bay ecosystem is a large estuarine embayment on the Texas coast bordering the Gulf of Mexico. The bay system is about 520 square miles in surface area and is made up of Galveston, Trinity, East, and West Bays. Average depth in the bays is about 2.4 metres (m) (8 feet) with a tidal range of 0.3-0.6 m (1-2 ft). The bays are located behind two major barrier islands with tidal exchange to the Gulf of Mexico through several restrictive passes. These passes, river locations, and dredged channels significantly affect water flow and concentrations of materials in the bays.

The system receives large quantities of freshwater inflow with organic particulates and dissolved nutrients and waste discharges from domestic and industrial sources via the Houston Ship Channel, Trinity River, and other tributaries. Due to these inputs, the bay is a highly productive commercial fishing area and, in addition, serves as a nursery area for a large portion of the total fishery product taken from the Gulf of Mexico along the Texas coast (Ward and Espey, 1971). It should be noted that waste discharges may provide energy sources to an ecosystem as well as toxic materials. The term pollution (measured as BOD loading) is used in this paper to indicate the net composite effect of waste discharges on secondary productivity in the ecosystem. Nitrogen concentrations are used here as indications of waste inputs and/or pollution (c.f., Copeland and Fruh, 1970).

Physical and chemical characteristics of the Galveston Bay system are summarized on maps of annual average isohalines (Figure 1) and mean annual concentration gradients of total nitrogen (which related to waste input and is used as a measure of pollution in this study, Figure 2). Salinity gradients are dominated by the dredged (to nominal 12.12 m) Houston Ship Channel (HSC) and the Trinity River, as shown by the saline intrusion in the HSC and low salinities in Trinity Bay. Total nitrogen concentration gradients indicate waste input sources associated with the Houston Ship Channel and the highly industrialized complex around Texas City on the western shore.

Armstrong (1973) investigated freshwater inflow quantities and waste discharges from the major tributaries of the Galveston Bay

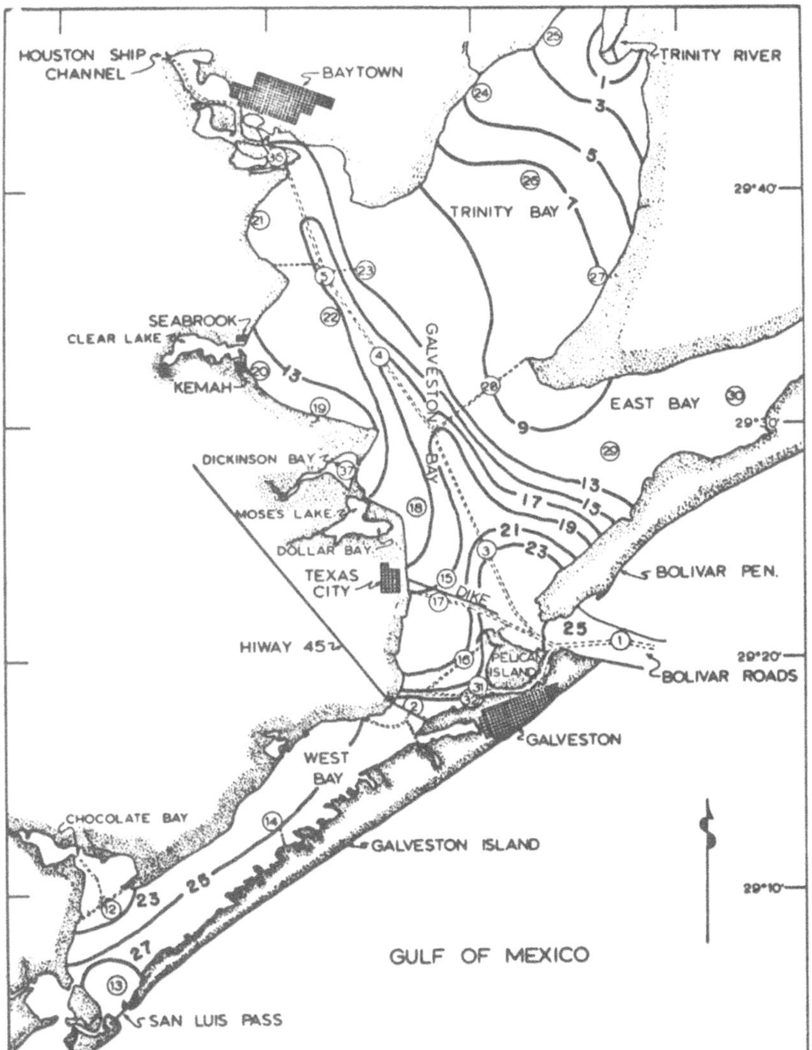

Figure 1. Isohalines for the Galveston Bay system plotted from mean annual salinity (ppt) at each station (indicated by circled numbers) (from Copeland and Fruh, 1970).

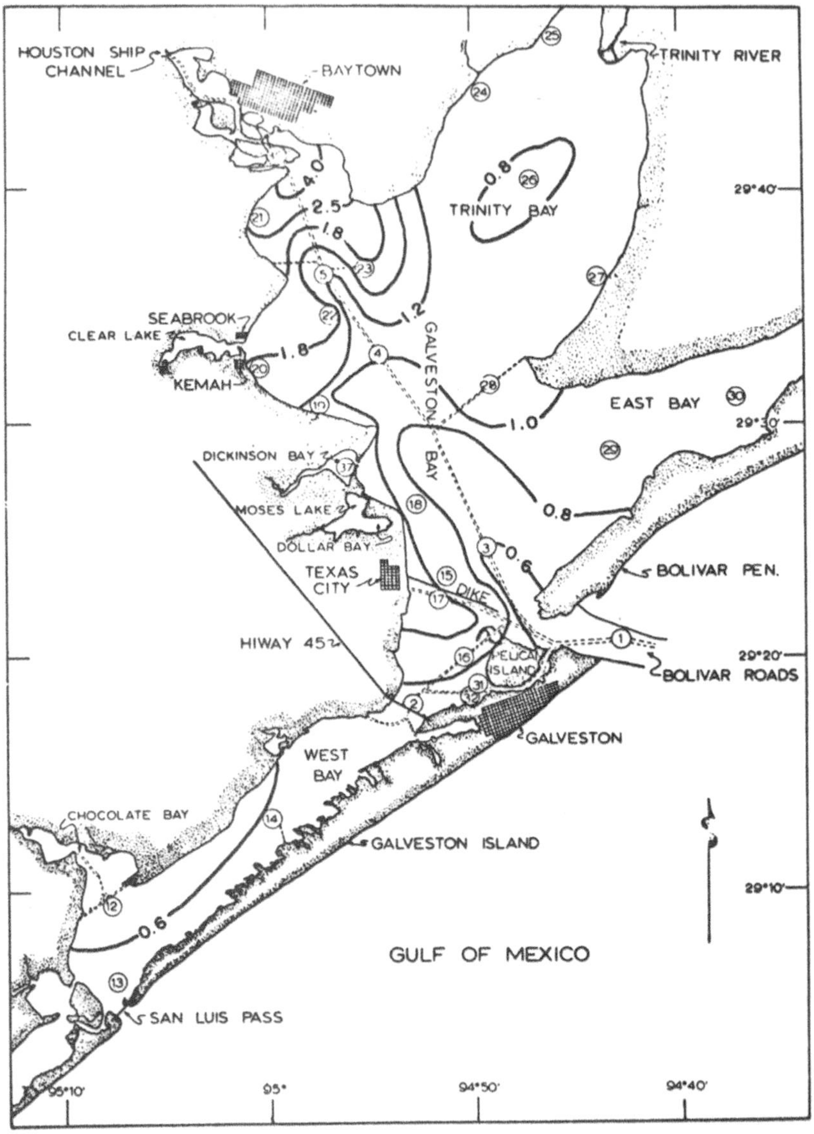

Figure 2. Mean annual concentration gradients of total nitrogen (mg/l) for the Galveston Bay system based on data taken from monthly values of the 1969 Bay Sampling Program (from Copeland and Fruh, 1970).

System, and determined source waters for the research stations indicated on Figures 1 and 2 by circled numbers. Copeland and Fruh (1970) subsequently used these same stations as locations for chemical and biological sampling in February, April, July, and October 1969, as part of a comprehensive Galveston Bay sampling program.

Galveston Bay Ecosystem

Temperate zone ecosystem characteristics are dominated by an annual seasonal cycle that is, in general, controlled by weather (Chin, 1961). Many of the biological processes in the ecosystem are related to energy sources (food) and migrating consumers with physiological adaptations that allow them to effectively compete for the available foods.

In the Galveston Bay ecosystem, freshwater flow from rivers brings in large quantities of organic particulates (detritus) and dissolved nutrients. Dissolved nutrients contribute to the growth of marsh grasses (*Spartina* spp), fixed bottom plants (turtle grass, *Thalassia testudinum*, for example), and small floating plants (phytoplankton). These materials are grazed or filtered from the water by small animals such as zooplankton, herbivores (shad, *Dorosoma cepedianum* and menhaden, *Brevoortia partronus*), and omnivores (shrimp, *Penaeus* spp and crabs, *Callinectes sapidus*). These small animals are in turn consumed by larger carnivore species (Atlantic croaker, *Micropogon undulatus*, Anchovy, *Anchoa mitchilli*, and Trout, *Cynoscion arenarius*). A generalized energy flow diagram for the Galveston Bay is shown in Figure 3. Note that each "consumer" group (with the usual exception of adult top carnivores) also is consumed by other components of the ecosystem.

One ecosystem characteristic is that certain functions are performed by one or more species, either simultaneously or at different seasons of the year. For this reason, the biological species in the Galveston Bay may be grouped based on similarity of function and feeding characteristics. In this study, consumer groups have been organized based on consuming habits and food preferences (Darnell, 1958, 1961; Odum 1971). Consumer groups and typical species in them are:

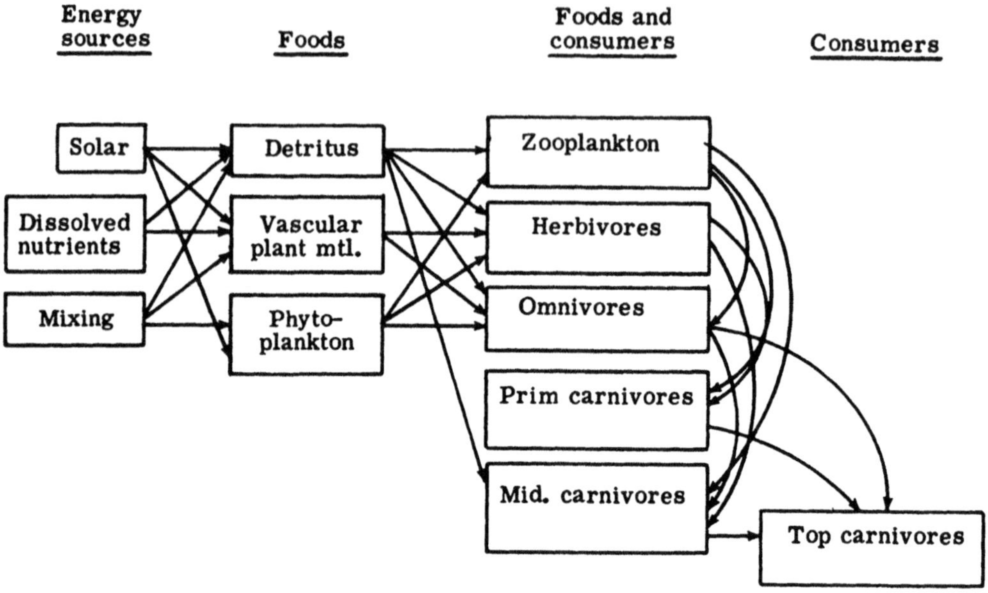

Figure 3. Biological energy flow in Galveston Bay ecosystem

Consumer Group	Typical Members
Zooplankton	
Herbivores	Menhaden
Omnivores	Shrimp
Primary Carnivores	Atlantic Croaker
Middle Carnivores	Anchovy
Top Carnivores	Trout

The above groupings are based on dominant characteristics in the first year or period of maximum rate of growth in the ecosystem. Adults do not always consume the same foods as the young of the same species; however, this is not a limitation since consumer groups and shifting of consuming habits are included in the model.

Results of biological sampling reported by Copeland and Fruh (1970) in Galveston Bay in 1969 are summarized in Figure 4. Curves

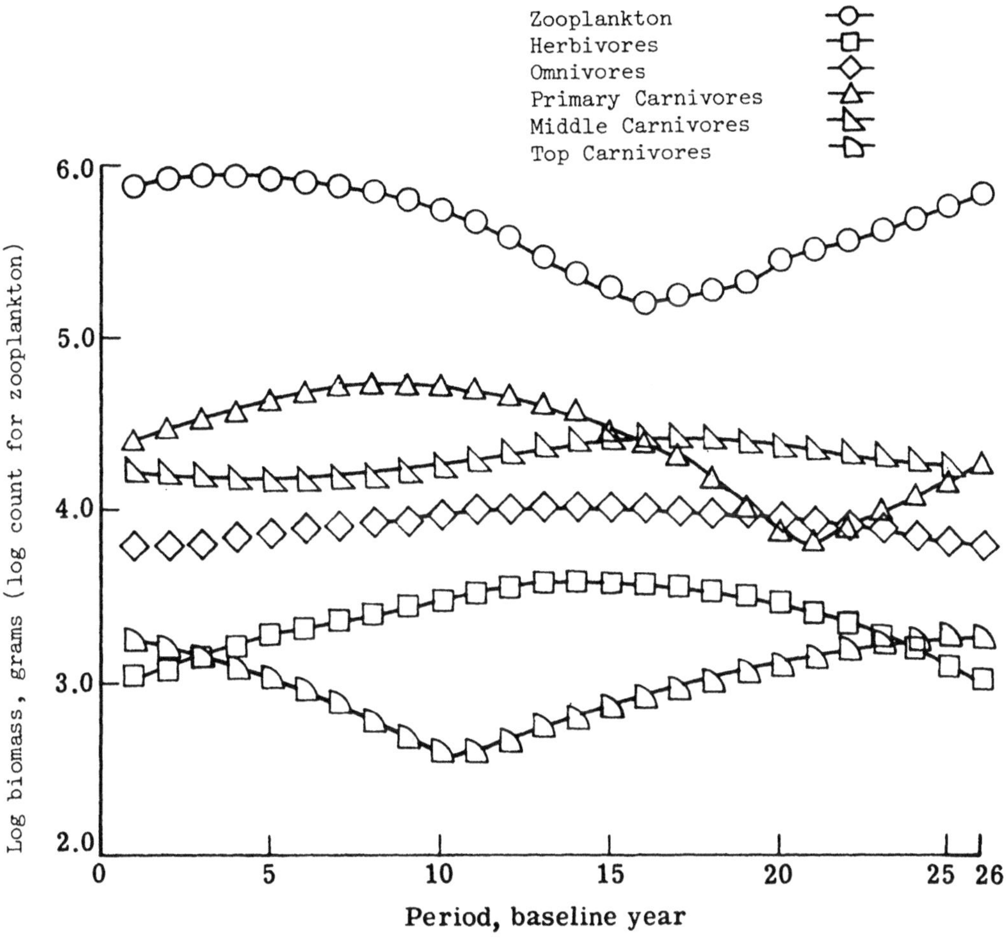

Figure 4. Measured consumer group's biomass - Galveston Bay, 1969 (from Copeland and Fruh, 1970). Periods are model increments during the year.

are shown for the zooplankton counts and other consumer groups' biomass during the year. Periods are model increments and will be discussed in the section on modeling. Biomass values are the sums from individual stations and are used as reference values in this study (i.e., rather than estimating total ecosystem biomass values).

Secondary productivity in a temperate zone estuarine system is determined primarily by immigration, growth, and emigration. Thus, the estuary meets a specific need for the species of interest during some period of their lifetime. Immigration and emigration of fish and other consumers are largely natural phenomena that represent an adaptation of particular species to the total environment in which they live and are not directly controllable or manageable by man. On the other hand, the growth phase of estuarine organisms is highly affected by food availability and by man's activities, particularly pollution associated with waste discharges and manipulation of water flows into or within the estuarine system.

Factors identified from *a priori* information as primary in the growth of estuarine species are:

1. Food and consumer densities,

2. Environmental effects, including pollution, and

3. Distribution effects, which are related to geological parameters (i.e., area of potential spawning or feeding zones).

Other parameters, such as temperature, are obviously important, but do not appear to be controlling factors in changes in Galveston Bay. In any event, it appears that their relation to growth *per se* are secondary compared to the three factors listed above.

SIMULATION MODEL OF GALVESTON BAY

A continuous simulation model format (Forrester, 1961) is used for the Galveston Bay ecosystem. Independent variables are exogenous changes in freshwater and waste discharge to the ecosystem. Dependent variables, or outputs, are biomass levels of the six identified consumer groups. Analytical and empirical relations are used to define and relate physical, chemical, and biological characteristics of the ecosystem. This section provides a functional description of the model. A detailed description, including equations and a program printout, has been reported by Johnson (1975).

The Model

In an estuarine ecosystem, as discussed previously, the dominant cycle is the seasonal calendar year. In the model, the calendar year is divided into 26 2-week periods (designated as I=1, ..., 26). These periods were short enough that rate changes within the period are insignificant.

Each of the six consumer groups is phased into the yearly cycle, but their own cycle is different from the others, as shown in consumer biomass curves developed from sampled and historical data for the first model (baseline) year, Figure 4. Each consumer group's seasonal cycle (periods designated by M) starts with M=1 defined as the period when a consumer group's biomass is at a minimum in the ecosystem (e.g., M=1 in period I=21 for primary carnivores). The biomass in the estuary at M=1 is assumed to be residual and is therefore designated as "adult" (as contrasted to this year's young or the immigration into the ecosystem). The seasonal cycle and its relationship to the calendar year is shown schematically in curve "e" of Figure 5.

Immigration starts in period M=1 and continues over 3-4 model periods (6-8 weeks). Functionally, it is taken to be a sinusoidal shaped curve (positive 180°) and is shown schematically as curve "a" of Figure 5. Immigration is considered to be a constant from year to year, independent of ecosystem variations.

Immigrated (larval and post-larval) organisms have very high growth rates decreasing with increasing organism size (Paloheimo and Dickie, 1965; Patten, 1971). A decreasing exponential function is used for the baseline year to describe the growth rate over a consumer group's year, decreasing to a value of 0 for adults at the beginning of the next year. Baseline year growth rates are shown schematically as curves "b" and "c" for immigrated and adult organisms, respectively. Note that in the baseline year all growth rate effects due to changes in exogenous variables have, by definition, values of 0. In subsequent years, year to year changes in the exogenous variables are defined in terms of growth rate change ratios, which are used to determine new values for net growth rates.

After high growth rates in the estuary, consumer organisms emigrate from the ecosystem - in this case, primarily to the Gulf of Mexico. Emigration for primary carnivores (typical species is Atlantic Croaker) is shown schematically as curve "d" in Figure 5.

Primary carnivore biomass for the baseline year is shown schematically as curve "e" in Figure 5. Biomass curves (counts for zooplankton) for the other five consumer groups were independently developed in an analogous manner. These six curves are correlated with the calendar year periods (I), which are used for period identification after the baseline year.

In the model, iterative calculations are made period by period. A period calculation consists of: (1) determining the net period growth rate, which is the product of the prior year net growth rate and current year change ratios due to consumer and food densities and exogenous variables (in the baseline year the net growth rate is taken from the exponential curve); (2) adding immigration and subtracting emigration from the biomass at the beginning of the period; and (3) multiplying the total by the net growth rate to obtain period biomass, which is also the biomass at the beginning of the next period.

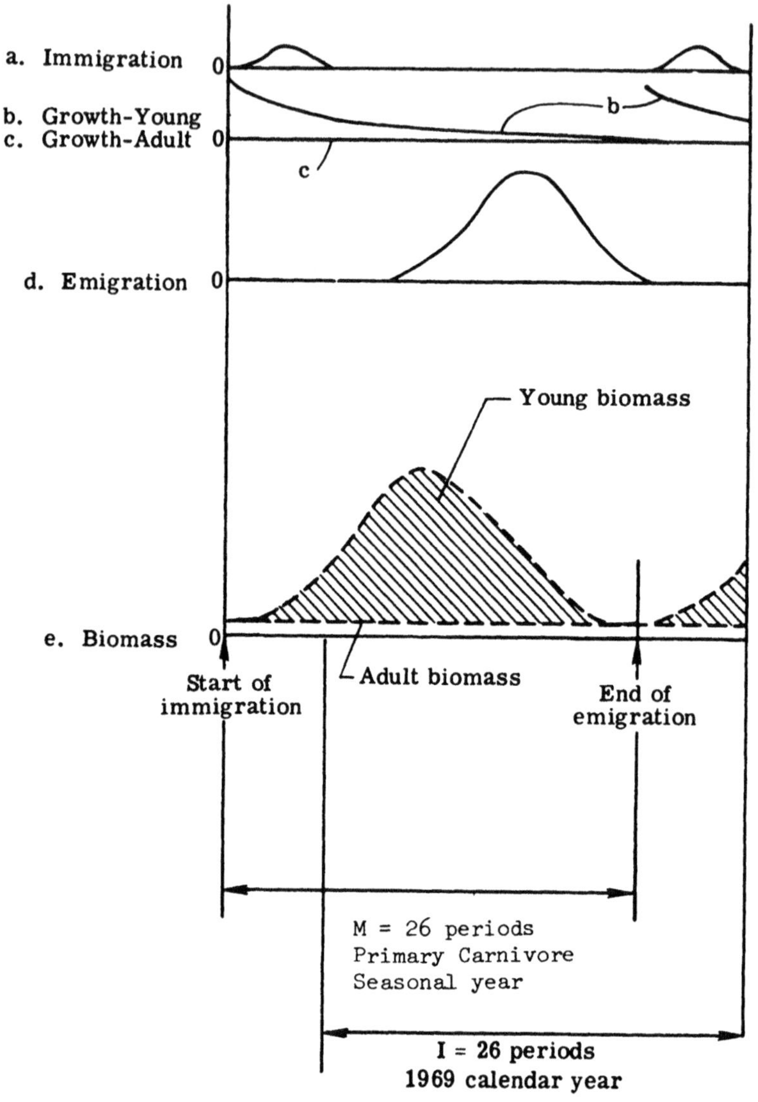

Figure 5. Development of baseline year biomass curve for primary carnivores

Model stability is aided by built-in safeguards which act as negative feedback: First, if environmental conditions remain the same, biomass curves will repeat the previous year's values, except for time-lag effects; second, the growth rate ratio is an inverse function of consumer density compared to the baseline year value (in effect, this implies an upper limit on consumer biomass in the new ecosystem); and third, model calibrations and examples are based on wide ranges of freshwater and pollution inputs that occurred over about a 20-year period.

The Baseline Year

In 1969 a comprehensive sampling program was accomplished as part of the Galveston Bay Program (Copeland and Fruh, 1970). Results of this sampling program have been used to determine numerical parameters for seasonal changes in consumer groups biomass levels for the Galveston Bay pollution effects model. For this baseline year, (year 1 of the model) all growth change ratios are set equal to 1.0. This allows the development of growth rates for the conditions that existed during the baseline year. Biomass curves for the consumer groups were calculated and empirically fitted to sampled data from 1969. The model biomass curve for the primary carnivores is shown in Figure 6. Values from the sampling program (Copeland and Fruh, 1970) are shown also.

Model Calibration

The development of numerical equations to describe changes in growth rates as a result of changes in food and consumer densities and exogenous variables is based on a number of independent investigations, each of limited scope. In the model, as in the studies, factors other than those being evaluated are held constant during that phase of model calibration. Effects on consumer group growth rates were evaluated in the following order: (1) Food and consumer densities; (2) Environmental effects; and (3) Distribution effects.

Growth rate effects due to consumer and food densities were studied by Brockson, Davis, and Warren (1970). Their field studies from three lakes investigated Sockeye salmon growth as functions of both salmon (consumer) and zooplankton (food) densities. The empirical equations in the model have a self-limiting feature based on an inverse ratio in the consumer growth function (i.e., consumer growth rates can only increase if that consumer group biomass decreases).

Environmental effects on growth, primarily due to pollution effects, has been investigated by Wohlschlag (1972). In the model, growth effects are assumed to vary linearly with the waste input parameter (BOD concentration). Increased waste input (pollution) leads to decreased growth rates.

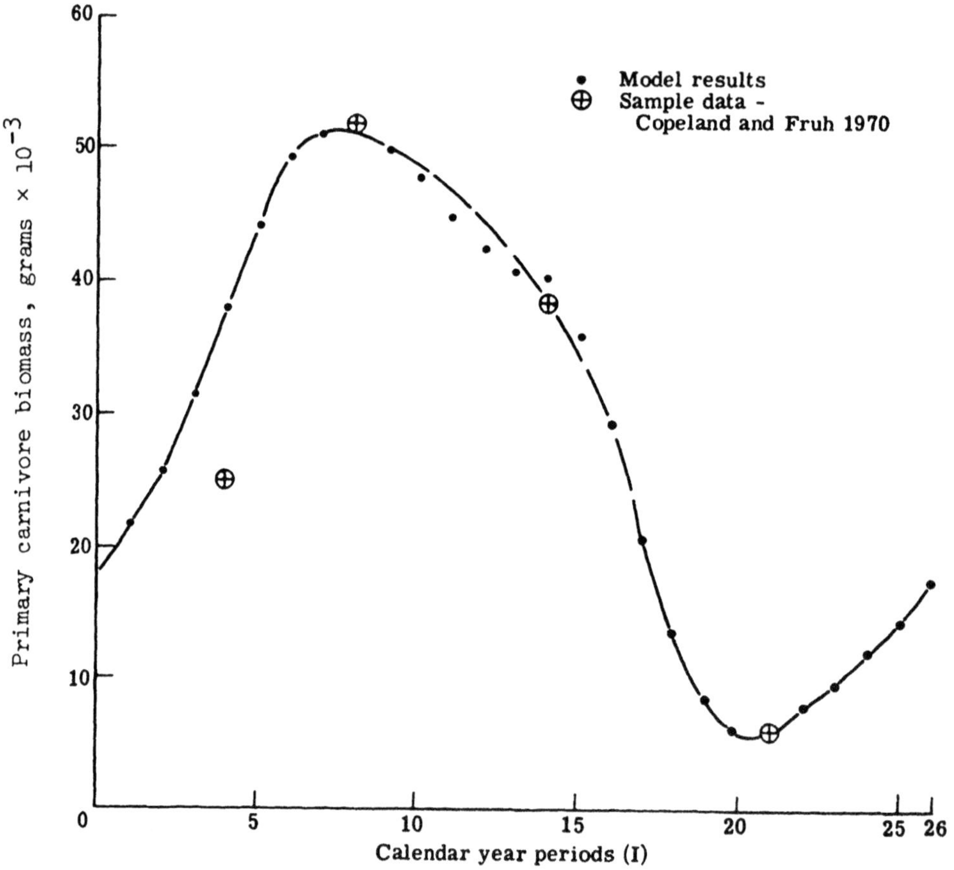

Figure 6. Model baseline year biomass curve and sample data for primary carnivores

Distribution effects in the Galveston Bay are due primarily to changes in freshwater inflow and consequent shifts of isohalines. Investigations by Copeland (1966) and Armstrong (1973) have evaluated omnivore and total biomass productivity as influenced by freshwater inflow and/or Galveston Bay water displacement rate, both of which may be directly related to salinity and pollution concentrations. Field investigations indicated a 2-year displacement (lag) of omnivore productivity to freshwater input, which is included in the model equations. This lag may be influenced by the area available for spawning and feeding in the bay for the omnivores. Both omnivores and their consumers are highly sensitive to small salinity gradients.

EVALUATION OF MANAGEMENT EFFECTS

The calibrated model developed in this study was used to study the effects of projected changes in exogenous variables over a 12-year period, specifically: (1) decrease in waste discharge to the Galveston Bay from the Houston Ship Channel (HSC), and (2) changes in freshwater inflow.

Waste Discharge

Armstrong (1973) indicated that, due to pollution control measures on the Houston Ship Channel (HSC), projected decreases in waste load from this source were:

Model Year	
1	153.2×10^6 lb BOD/yr
5	41.6×10^6 lb BOD/yr
7 (Goal)	29.1×10^6 lb BOD/yr

Effects of this pollution load reduction (Table 1) were evaluated by the model based on the following yearly waste loads (see Figure 7).

Table 1. Organic Carbon, 10^6 lb BOD/yr

Model Year	Trinity River	HSC	Other	Total
1	29.9	165.0	52.2	247.1
2	29.9	152.2	52.2	235.3
3	29.9	140.0	52.2	222.1
4	29.9	80.0	52.2	162.1
5	29.9	41.6	52.2	123.7
6	29.9	33.0	52.2	115.1
7	29.9	29.0	52.2	112.1

Freshwater inflow was unchanged as were waste discharges from other sources.

Total productivity in Galveston Bay was projected to increase by about 40 percent due to decreases of waste discharge (Figure 8). This effect is due to the overall increased basic foods (detritus

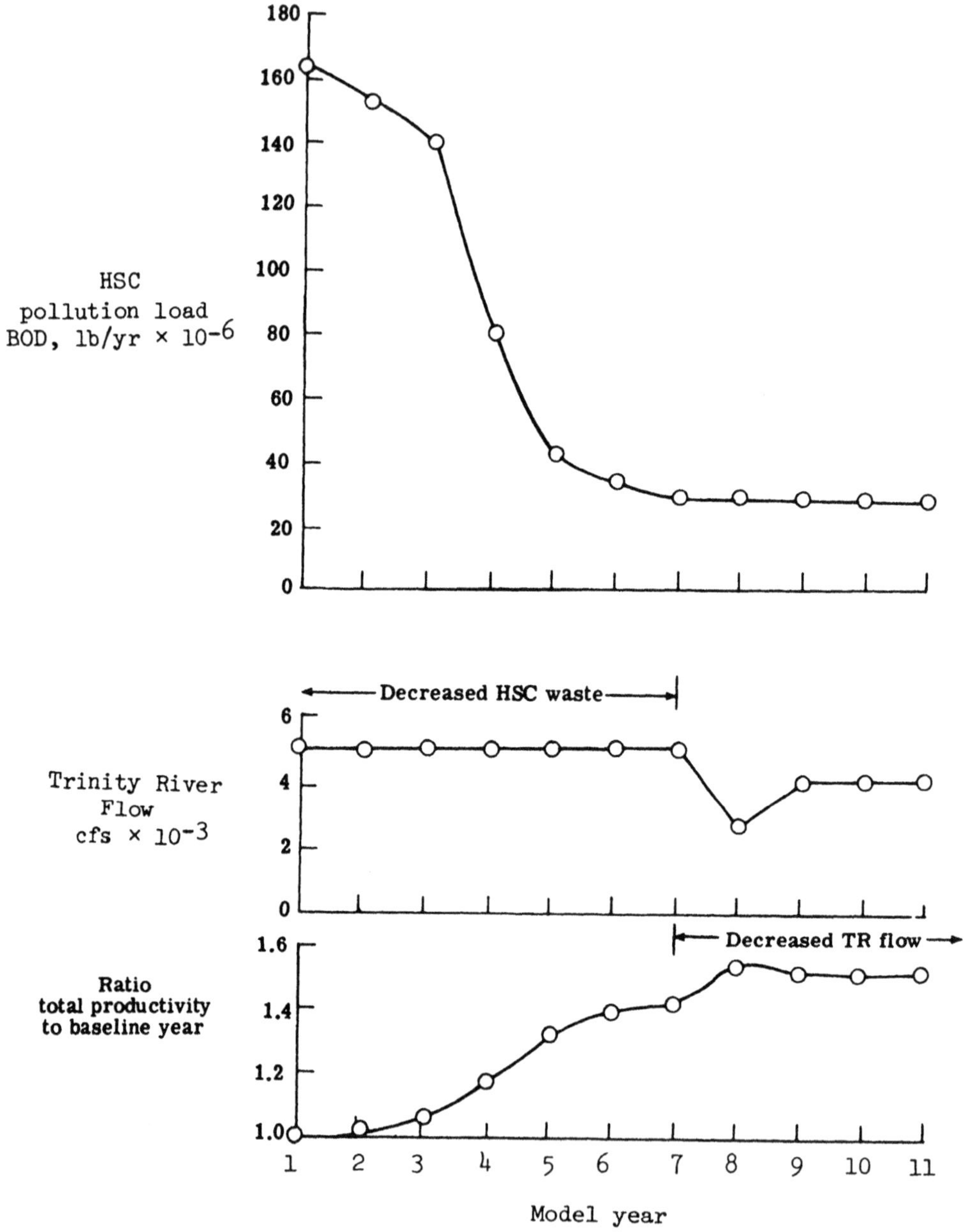

Figure 7. Effect on Galveston Bay productivity due to decreased HSC Waste Load (Model years 1-7) followed by decreased Trinity River freshwater discharge (Model years 7-11).

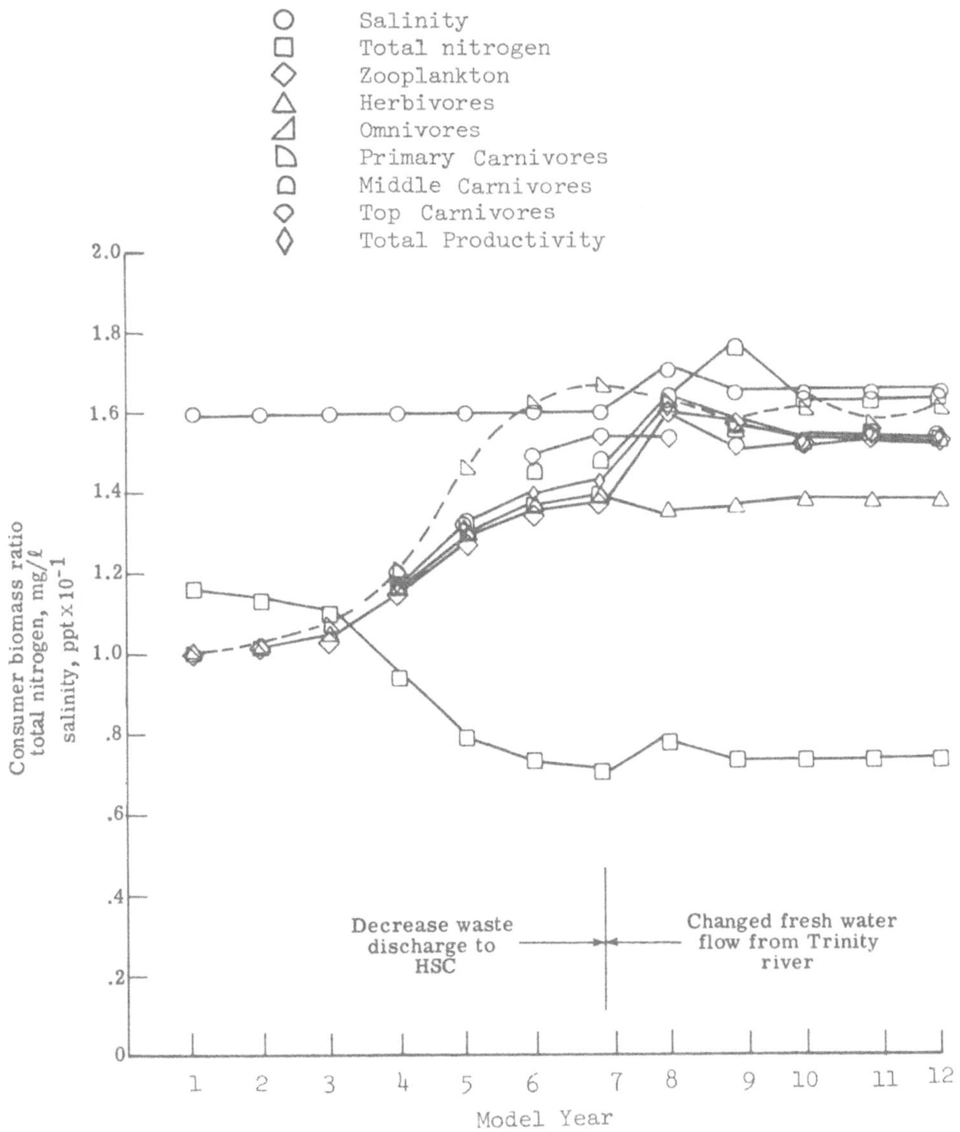

Figure 8. Productivity as result of decreased waste input and decreased freshwater inflow.

and phytoplankton) and decreased pollution toxicity effects on organism's growth. As has been discussed previously, total productivity is determined primarily by the biomass of the primary and middle carnivores; however, the other consumer groups also increase in productivity (Figure 8). Omnivores, primarily shrimp, increased relatively more than the total (67 percent over baseline year productivity) due to higher overall sensitivity to pollution effects of their own growth rates and of their food sources.

Freshwater Inflow

After the pollution load decrease from the HSC, which was assumed to occur from model years 1 through 7, a step decrease of 50 percent in freshwater discharge from the Trinity River (e.g., due to filling a reservoir) was assumed for the model year 8, followed by a restoration to 75 percent in model years 9 through 12. Freshwater inputs (Table 2) to the Galveston Bay were then (see Figure 7):

Table 2. Freshwater Inflow, 1000 CFS

Year	Trinity River	HSC	Other	Total
1- 7	5.58	10.53	2.79	18.90
8	2.79	10.53	2.79	16.11
9	4.18	10.53	2.79	17.50
10-12	4.18	10.53	2.79	17.50

Total productivity in the Galveston Bay ecosystem increased about 50 percent by year 12 above the baseline year due to the combined effects of reduced waste discharge to the Houston Ship Channel, followed by reduced Trinity River freshwater discharge. Reduced freshwater flow leads to an overall increased total productivity. The middle carnivores increased about 18 percent due to decreased freshwater and the consequent increase in salinity. The other consumer group productivities are moderately increased (about 10 percent) by the decreased flow with the exception of omnivores which were reduced 20 percent due to their sensivity to system changes (salinity and pollution concentrations). Productivity ratios (ratios of current year to baseline year) for the six consumer groups, average annual salinity, and total nitrogen concentrations are shown in Figure 7.

It is interesting to note the carryover effects in the ecosystem predicted by the model. For example, there is a strong perturbation on omnivores due to effects from prior years. The specific reason for this lag is unknown and may be due to reproduction or food effects, but has been documented from previous studies (Copeland, 1966). The significant result, however, is that short term (year to year) comparisons may be misleading where management of an ecosystem is concerned. It should be pointed out that, except as implicitly included in the calibration data, the model does not consider shifts in species composition (within the consumer groups) due to changes of environmental conditions or pollution levels.

REMOTE SENSING IN ECOSYSTEM MODELING

Review of the Galveston Bay and other ecosystem models indicates several areas where remote sensing may be applicable to new or revised models. Primary input parameters such as temperature and salinity distributions could be monitored synoptically by remote sensing and compared to the model inputs generated from physical model, weather, flow, and other information. Of particular interest would be level and/or spatial deviations that may lead to more detailed measurements. Sensitivity analyses could be effectively used to establish significant deviation ranges.

Two physical parameters that have received a good deal of attention from the remote-sensing community are suspended solids and chlorophyll a concentrations and their distribution in surface waters within the coastal zone. Suspended solids have been quantitatively mapped using remote-sensing techniques and could possibly be used in conjunction with the detritus input concentration curve. Measurement of parameters by remote sensing has been summarized by Johnson and Harriss (In Press).

Chlorophyll a concentrations and distribution over a seasonal cycle are subject to relatively wide variations due to environmental changes (sunlight, temperature, salinity, nutrients, etc.). The Galveston Bay model used a very simple approach based on limited historical data. More complete models (Kremer and Nixon, 1978; Kiefer and Enns, 1976) would appear to be more realistic and could be tested using remotely sensed data. Interestingly, chlorophyll a appears to be one of the few "analytical" (i.e., subject to several other inputs) parameters that also may be remotely sensed (Johnson, 1978) and thus may be one of the key elements to use to test model performance.

REFERENCES

Armstrong, Neal E. 1973. Personal communication. Institute of Marine Sciences, Univ. Texas, Port Aransas.

Brockson, R. W., G. E. Davis, and C. E. Warren. 1970. Analysis of Trophic Processes on the Basis of Density-dependent Functions. In Steele, J. H. (Ed) Marine Food Chains. Univ. Calif. Press, Berkeley: 468-498.

Chin, Edward. 1961. A Trawl Study of an Estuarine Nursery Area in Galveston Bay, with Particular Reference to Penaeid Shrimp. Ph.D. Dissert., Univ. Wash., Seattle.

Copeland, B. J. 1966. Effects of Decreased River Flow on Estuarine Ecology. Journal Water Pollution Control Fed. 38:1831-1839.

Copeland, B. J. and E. Gus Fruh. 1970. Ecological Studies of Galveston Bay, 1969. Final Report to Texas Water Quality Board (Galveston Bay Study Program) for Contract IAC (68-69) 408: 482 p.

Darnell, R. M. 1958. Food Habits of Fishes and Larger Invertebrates of Lake Pontchartrain, La. an Estuarine Community. Publ. Inst. Mar. Sci., Univ. Texas 5: 353-416.

Darnell, R. M. 1961. Trophic Spectrum of an Estuarine Community Based on Studies of Lake Pontchartrain, Louisiana. Ecology 42: 553-568.

Forrester, Jay W. 1961. Industrial Dynamics. MIT Press, Cambridge, Mass., 464 pp.

Johnson, R. W. 1975. A Simulation Model for Studying Effects of Pollution and Freshwater Inflow on Secondary Productivity in an Ecosystem. Ph.D. Dissertation, North Carolina State University. NASA TMX 72169.

Johnson, R. W. 1978. Mapping of Chlorophyll a Distributions in Coastal Zones. Photogrammetric Engineering and Remote Sensing, 44: 617-624.

Johnson, R. W. and R. C. Harriss, In Press. Applications of Remote Sensing of Water Quality and Biological Measurements in Coastal Waters. Photogrammetric Engineering and Remote Sensing.

Kiefer, D. A. and T. Enns. 1976. A Steady-State Model of Light-, Temperature-, and Carbon-Limited Growth of Phytoplankton. In Canale, R. P. (Ed). Modeling Biochemical Processes in Aquatic Ecosystems. Ann Arbor Science Publishers, Inc.: 319-336.

Kremer, J. N. and S. W. Nixon 1978. A Coastal Marine Ecosystem. Springer-Verlag.

Odum, W. E. 1971. Pathways of Energy Flow in a South Florida Estuary. Sea Grant Tech. Bulletin No. 7. Univ. Miami, Fla.

Palaheimo, J. E. and L. M. Dickie. 1965. Food and Growth of Fishes. I: A Growth Curve Derived from Experimental Data. J. Fish. Res. Bd. Canada 22(2):521-542.

Patten, B. C. 1971. Systems Analysis and Simulation in Ecology, Vol. I. Academic Press, NY, 607 pp.

Ward, G. H. and W. H. Espey. 1971. Case Studies. Galveston Bay. In Estuarine Modeling: An Assessment Project 16070DAV, Environmental Protection Agency, Washington, D. C.: 399-437.

Wohlschlag, D. E. 1972. Respiratory Metabolism of the Striped Mullet as an assay of low level stresses in Galveston Bay. Report on Contract IAC (72-73)-183 Texas Water Quality Board "Toxicity Studies on Galveston Bay Project", C. H. Oppenheimer, P. I.

THE MOVEMENTS OF A MARINE COPEPOD IN A TIDAL LAGOON

Ivan T. Show, Jr.

Science Applications, Inc., 1200 Prospect Street

La Jolla, California 92038

ABSTRACT

It has been hypothesized that marine copepods are able to react to water movements in an estuary in such a way as to minimize advective losses to the open ocean. This hypothesis is developed mathematically and tested by use of a stochastic model. Several processes are investigated as contributing to the minimization of advective loss.

The model treats time-varying spatial patterns of marine plankton. In the present instance, the model is used to describe the movements of Acartia tonsa (Copepoda) in a tidal lagoon on the eastern end of Galveston Island, Texas. Four distinct processes are considered: advection by currents, behavioral response to environmental variables (current velocity, temperature, and salinity fields), intraspecific aggregation, and birth-death processes. The portion of the model dealing with biological processes is a stochastic compartmental model. The biological model is driven by a three-dimensional physical dynamic model which provides numerical solutions for current velocity, temperature, and salinity fields.

The coupled physical-biological model used to simulate the distribution of A. tonsa provided numerically accurate estimates for the time histories of the physical and biological processes involved. The success of the model was probably attributable to a number of factors which involve the nature of the ecological situation modeled, the form of the parameters in the biological model, the manner in which the lagoon was compartmentalized, and the nature of the sampling data used to test the results.

Assuming then that the results of the numerical simulation were accurate by other than random chance, the most important conclusion was that A. tonsa appears to owe its spatial distribution in the lagoon to the combined effects of advection by currents and behavioral response to environmental stimuli: tides, light, temperature gradients, salinity gradients, and the population density gradients of its own species; the most important being tidal advection. It also appears that the resultant movements of the organism are sufficient to minimize losses from the lagoon to the extent that it maintains an endemic population inside the lagoon which is distinct from the population found immediately outside. Finally, it was concluded that the spatial distribution of A. tonsa is heterogenous, that the patches are of the order of 240 metres long by one or two metres deep, and that changes in density occur as a result of an increase in within-patch density rather than an increase in the number of patches.

INTRODUCTION

The purpose of this study was to investigate the processes responsible for observed spatial patterns of zooplankters in estuaries and to test certain hypotheses involving the relative importance of these processes. The study involves the development of a modeling technique designed to deal with spatial heterogeneity in marine zooplankton. The model deals specifically with the estuarine copepod, Acartia tonsa, in East Lagoon, Galveston Island, Texas.

The East Lagoon model contains components dealing with four distinct processes: current velocity fields, behavioral responses to environmental gradients, intraspecific aggregation, and birth-death processes. The biological portion of the model is driven by a dynamic physical model which provides numerical solutions for current velocity, temperature, and salinity fields.

Russell (1935) gives an excellent review of earlier attempts to correlate plankton distributions with environmental property distributions. Pierce (1939) and Russell (1939) suggested and used certain previously investigated organisms to indicate hydrographic conditions. A diagramatic method for comparing zooplankton distributions with temperature and salinity fields was developed by Bary (1963a, b, c, d; 1964).

Fisher (1943) first showed that the variances encountered in zooplankton sampling were not within the range to be expected from sampling homogeneously distributed populations. This paper established the concept of aggregation. Based on this concept, equations were developed to describe such spatial structures as fish schools and other similar aggregations (Breder 1954).

Hutchinson (1953) considered plankton populations to be highly aggregated and gave the following as potential causes of the observed spatial patterns:

(1) vectorial effects: the organisms' response to physio-chemical gradients,

(2) stochastic-vectorial effects: due to mass transport by currents,

(3) reproductive patterns,

(4) social patterns: intraspecific interactions,

(5) coactive patterns: interspecific interactions.

Cassie (1960) and Stavn (1970) reflect very much the same view.

Barnes and Marshall (1951) pointed out the difference between clumping and behavioral aggregation. They considered the former as due to external influences on the organisms and the latter as due to social behavior. They were able to show a strong correlation between vertical salinity fields and the distribution of *Calanus*, *Acartia*, and *Centropages* *nauplii* over vertical ranges of as little as ten meters. They suggested that the effective microclimate is made up of very small physical differences.

In a study of plankton populations in Wellington Harbor, New Zealand, the findings of Cassie (1959a) support those of Barnes and Marshall. Cassie presented evidence to indicate that small-scale temperature and salinity gradients could influence zooplankton spatial patterns or at least coincide with them. The within-sample standard deviation for temperature and salinity was 0.06°C and 0.05°/₀₀, respectively. Liebermann (1951) demonstrated that temperature fluctuations of 0.2°C occur commonly in the upper 60 meters of the ocean. Similar magnitude fluctuations were significantly correlated with the distribution of *Paracalanus*, resulting in a 16% drop in numbers corresponding to a 0.1°C change in temperature and a 31% drop in numbers with an 0.2°C change in temperature (Cassie 1959a). Cassie (1959b) concluded that the spatial aggregation of zooplankton can probably be demonstrated at any spatial scale and then went on to show significant aggregation on a scale of 10 meters.

Lance (1962) and Harder (1968) both showed that *Acartia* and other planktonic organisms react to salinity gradients with swimming movements which are most notable in their modification of vertical migration patterns. McNaught and Hasler (1964) showed that Daphnia reacts to extremely small changes in light intensity.

Wiebe (1970) measured zooplankton spatial patterns off Baja California. He found that patches were circular and ranged from 13.6 to 73.1 m in diameter, the diameter tending to become larger at night. He also found that the degree of aggregation depended on the density of the organism.

Fasham et al. (1974) studied the spatial patterns of ostracods and copepods. They concluded that patch size does not vary significantly between species and that dense populations result from the development of greater numbers of patch centers rather than an increase in the population density in a smaller number of patches. They further concluded that the observed aggregation is caused by behavioral responses and not by physio-chemical gradients, particularly temperature.

For the most part, the studies reviewed above deal with the distribution of zooplankton at a single instant in time. A transition to studies of time variation is represented by Ragotzkie (1953) who found a strong correlation between current velocity fields, especially divergence, and the distribution of Daphnia in a freshwater lake. Stavn (1971) demonstrated some implications of Daphnia distributions near the surface in relation to the animals' orientation to light and currents. Powell et al. (1975) found a significant correlation between phytoplankton and horizontal currents in Lake Tahoe. They hypothesized that the primary sources of spatial heterogeneity are physical mixing and the turnover rate of the population.

Of special significance here are studies which deal with plankton spatial patterns in estuaries. Ketchum (1954) considered production and net seaward transport to be equal cofactors in estuarine plankton distributions; production balances seaward transport to maintain a steady-state endemic population. The species he studied were small and very feeble swimmers: phytoplankton, a clam larva, and A. tonsa larvae.

Barlow (1955) followed the same line as Ketchum; he estimated rates of production based on losses due to net seaward transport. For Acartia, experiments showed that near-surface populations could only be maintained by recruitment from deeper water levels.

Jacobs (1968), on the other hand, appealed to the combined action of tidal dynamics and vertical migration to explain the horizontal distribution of A. tonsa in estuaries over short time periods. The influence of asymetrical marsh drainage was shown to play a significant role.

Based on an eleven day series of vertical plankton density profiles, Stickney and Knowles (1975) showed that spatial patterns of copepod nauplii in estuaries appear to be a function of dispersion and concentration caused by tidal currents. However, adults of the same species appeared to show more complexly controlled patterns. A. tonsa adults were found nearer the surface during high and rising tides regardless of the time of day.

Trinast (1975) did the first study of a coastal lagoon with no appreciable freshwater inflow. The author concluded that the loss of copepods to the ocean was minimized by a tendency to aggregate at deeper water levels during outgoing tides and at shallower levels during incoming tides.

Recently, Riley (1976) proposed a model in which vertical migration, tides, and advection were used to force the spatial and temporal distribution of zooplankton. When the resulting distributions were applied to the problem of zooplankton grazing, the joint zooplankton-phytoplankton distributions were those predicted by the animal exclusion theory of Hardy and Gunther (1935). Riley proposes that this model provides a mechanism to explain the negative correlation between zooplankton and phytoplankton density.

EAST LAGOON

East Lagoon is a long, narrow tidal lagoon on the eastern end of Galveston Island. The length of the lagoon is 2 km, its average width about 150 metres, and its average depth about two metres with a maximum depth of five metres near the landward end. The orientation of the long axis is northeast-southwest. It has no freshwater source and communicates with Bolivar Roads through seven 36-inch culverts which run through a seawall (Figure 1). Figure 2 shows the bathymetry of the lagoon.

A detailed plankton survey was carried out in conjunction with an experiment to determine the effects of copper ions on marine animals (Marvin et al. 1961). Fleminger (1959) gave the results and thereby showed that A. tonsa is the most abundant plankter in the lagoon. He also showed that A. tonsa matures during the summer in four to five weeks and that its populations are endemic to the lagoon. Fleminger conjectures that the lower densities of plankton in the lagoon (lower compared to adjacent coastal populations) are due to either marginal environmental conditions or to intense grazing by the dense populations of menhaden and shrimp larvae observed in the lagoon.

In a broad investigation of the entire biota of the lagoon, Arnold et al. (1960) found that the lagoon was the nursery ground

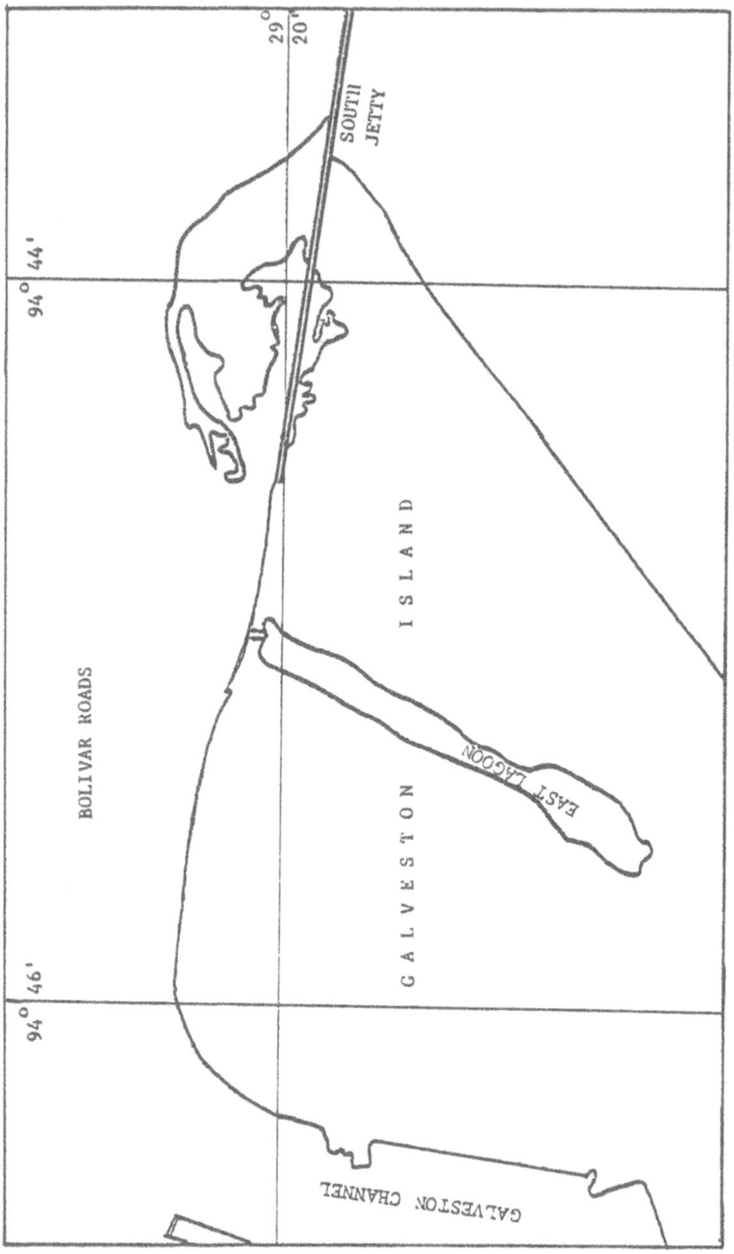

Figure 1. Geographical location of East Lagoon on Galveston Island.

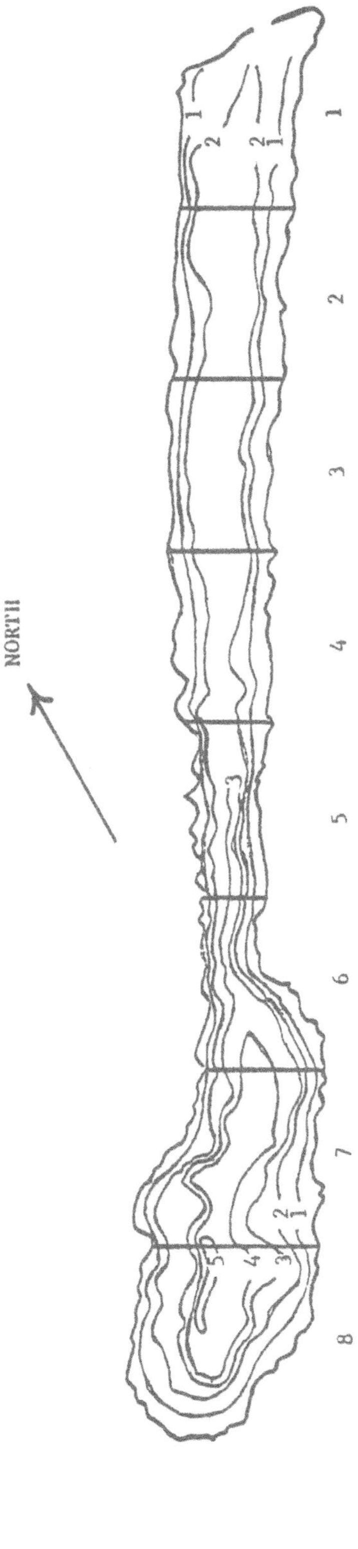

Figure 2. Bathymetry of East Lagoon. Depths are in meters. The dark vertical lines indicate the positions of compartment boundaries.

for many species of fish and invertebrates. East Lagoon represents a type of environment found commonly on the Texas coast that is critically important to the larval growth of many commercially important species (Fleminger 1959).

MODEL DESCRIPTION

The model is best understood if we consider a hypothetical body of water. This body of water is divided into a number of compartments or boxes on which there are no size or shape restrictions. However, it is convenient to form the compartments as rectangular parrallelopipeds. Figure 3 gives a three-dimensional view of such a compartment.

In Figure 3, the U's represent the x-directed transport of plankton which results from both advection and behavioral response. The V's represent the same in the y direction and the W's the same in the z direction. The variables, Z and P, at the center of the compartment are the species density and production or interaction term respectively. Figure 4 shows the grid system which results from placing a number of compartments in the positions in which they would occur in the model.

The mathematical formulation of the model is a system of "n" ordinary differential equations (ODE) where "n" is the total number of compartments in the model. In all subsequent discussion, the U, V, and W components will be referred to as the fractional transport coefficients or just transport coefficients and the variable Z will be referred to as the state variable.

Solution of the full system of "n" ODE's yields the expected values of the state variables at a particular time, "t". Thakur and Rescigno (1978) have derived, in a multi-compartmental model, approximate variances based on univariate probability distributions. They have also shown that the variances depend approximately on the amount of material entering a compartment only through the expectations of the precursor compartments. Their approximation is given by:

$$\mathrm{Var}\,[Z_n(t)] = Z_{no}e^{-h(t)}\,[1 - e^{-h(t)}] + E[Z_n'(t)] \qquad (1)$$

where Z_{no} is the initial value of the state variable in compartment "n" at time t = 0, $h(t) = \int_0^t u(s)\,ds$, and $Z_n'(t)$ is the number of

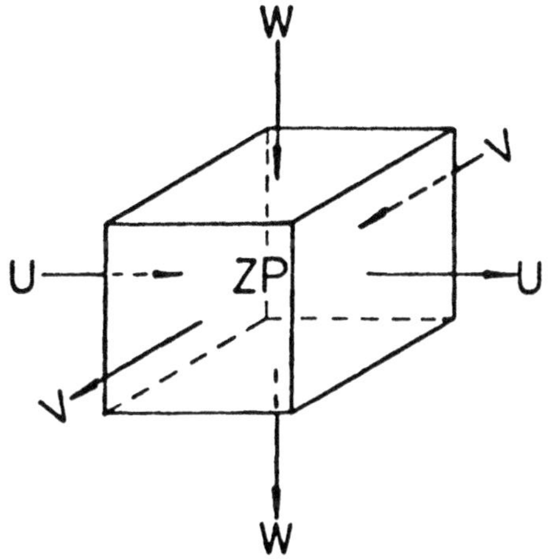

Figure 3. A single model compartment. The figure shows the relative positions of the transport coefficients, production, and zooplankton density term. U, V, and W represent transport coefficients, P represents the production term, and Z represents the zooplankton density term.

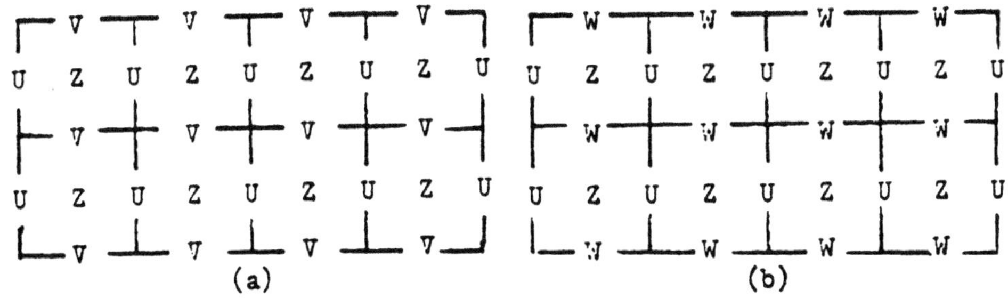

Figure 4. Plane views of the locations of model components. (a) horizontal cross-section. (b) vertical cross-section.

individuals in compartment "n" at time "t" which were not in compartment "n" at time t = 0. Here, u(s) is the fractional rate of transport of individuals out of compartment "n".

$Z_n'(t)$ is difficult to determine. However, this difficulty is overcome by considering all individuals which enter the compartment as not having been in the compartment at time t = 0 and assuming that the individuals are thoroughly mixed in the compartment. Then, given the number of individuals which are lost to the compartment in a certain period of time,

$$\text{loss to } Z_n' = Z_n'(t) \cdot \ u(t). \tag{2}$$

The physical numerical model is based on finite difference analogues of vertically integrated equations of motion, continuity, and conservation in an incompressible fluid.

The primitive equations for incompressible flow used in the model are:

$$\frac{\partial u}{\partial t} + u\frac{\partial u}{\partial t} + v\frac{\partial u}{\partial y} + w\frac{\partial u}{\partial z} - fv - \frac{1}{\rho}\left(\frac{\partial \tau_{xx}}{\partial x} + \frac{\partial \tau_{xy}}{\partial y}\right.$$

$$\left. + \frac{\partial \tau_{xz}}{\partial z}\right) + \frac{1}{\rho}\frac{\partial p}{\partial x} = 0$$

$$\frac{\partial v}{\partial t} + u\frac{\partial v}{\partial x} + v\frac{\partial v}{\partial y} + w\frac{\partial v}{\partial z} + fu - \frac{1}{\rho}\left(\frac{\partial \tau_{yx}}{\partial x} + \frac{\partial \tau_{yy}}{\partial y}\right.$$

$$\left. + \frac{\partial \tau_{yz}}{\partial z}\right) + \frac{1}{\rho}\frac{\partial p}{\partial y} = 0$$

$$\frac{\partial p}{\partial z} + \rho g = 0 \tag{3}$$

$$\frac{\partial u}{\partial x} + \frac{\partial v}{\partial y} + \frac{\partial w}{\partial z} = 0$$

$$\frac{\partial s}{\partial t} + u\frac{\partial s}{\partial x} + v\frac{\partial s}{\partial y} + w\frac{\partial s}{\partial z} - \frac{\partial}{\partial x}\left(D_x \frac{\partial s}{\partial x}\right) - \frac{\partial}{\partial y}\left(D_y \frac{\partial s}{\partial y}\right)$$

$$- \frac{\partial}{\partial z}\left(K \frac{\partial s}{\partial z}\right) = 0$$

$$\frac{\partial T}{\partial t} + u\frac{\partial T}{\partial x} + v\frac{\partial T}{\partial y} + w\frac{\partial T}{\partial z} - \frac{\partial}{\partial x}\left(D_x \frac{\partial T}{\partial x}\right) - \frac{\partial}{\partial y}\left(D_y \frac{\partial T}{\partial y}\right)$$

$$- \frac{\partial}{\partial z}\left(K \frac{\partial T}{\partial z}\right) = 0$$

where the following definitions apply:

x, y, z	Cartesian coordinates
u, v, w	velocity components
t	time
f	coriolis parameter
p	pressure
s, T	salinity °/₀₀, temperature °C
ρ	density
K	vertical eddy diffusion coefficient
τ	components of stress tensor
D_x, D_y	horizontal eddy diffusion coefficients

The model is arranged in levels of fixed thickness. If k-1/2 and k+1/2 are the upper and lower boundaries respectively of a level, then integrating over a level thickness yields the vertically integrated equations of motion which follow:

$$\frac{\partial (hu)}{\partial t} + u\frac{\partial (hu)}{\partial x} + v\frac{\partial (hu)}{\partial y} + (wu)_{k-1/2} - (wu)_{k+1/2} - fhv + \frac{h}{\rho}\frac{\partial p}{\partial x}$$

$$+ \left(\frac{\tau_{xz}}{\rho}\right)_{k+1/2} - \left(\frac{\tau_{xz}}{\rho}\right)_{k-1/2} - \frac{1}{\rho}\frac{\partial\left(A_x \frac{\partial u}{\partial x}\right)}{\partial x} - \frac{1}{\rho}\frac{\partial\left(A_y \frac{\partial u}{\partial y}\right)}{\partial y} = 0$$

$$\frac{\partial(hv)}{\partial t} + u\frac{\partial(hv)}{\partial x} + v\frac{\partial(hv)}{\partial y} + (wv)_{k-1/2} - (wv)_{k+1/2} + fhu + \frac{h}{\rho}\frac{\partial p}{\partial y}$$

$$+ \left(\frac{\tau_{yz}}{\rho}\right)_{k+1/2} - \left(\frac{\tau_{yz}}{\rho}\right)_{k-1/2} - \frac{1}{\rho}\frac{\partial\left(A_x \frac{\partial v}{\partial x}\right)}{\partial x} - \frac{1}{\rho}\frac{\partial\left(A_y \frac{\partial v}{\partial y}\right)}{\partial y} = 0$$

(4)

$$\frac{\partial(hs)}{\partial t} + u\frac{\partial(hs)}{\partial x} + v\frac{\partial(hs)}{\partial y} + (ws)_{k-1/2} - (ws)_{k+1/2} - \frac{\partial\left(hD_x \frac{\partial s}{\partial x}\right)}{\partial x}$$

$$- \frac{\partial\left(hD_y \frac{\partial s}{\partial y}\right)}{\partial y} + \left(K \frac{\partial s}{\partial z}\right)_{k+1/2} - \left(K \frac{\partial s}{\partial z}\right)_{k-1/2} = 0$$

$$\frac{\partial(hT)}{\partial t} + u\frac{\partial(hT)}{\partial x} + v\frac{\partial(hT)}{\partial y} + (wT)_{k-1/2} - (wT)_{k+1/2} - \frac{\partial\left(hD_x \frac{\partial T}{\partial x}\right)}{\partial x}$$

$$- \frac{\partial\left(hD_y \frac{\partial T}{\partial y}\right)}{\partial y} + \left(K\frac{\partial T}{\partial z}\right)_{k+1/2} - \left(K\frac{\partial T}{\partial z}\right)_{k-1/2} = 0$$

In order to compute the vertical current speeds, it is noted that

$$\frac{\partial w}{\partial z} + \frac{\partial u}{\partial x} + \frac{\partial v}{\partial y} = 0 \qquad (5)$$

which implies that

$$\int_{k+1/2}^{k-1/2} \left(\frac{\partial w}{\partial z} + \frac{\partial u}{\partial x} + \frac{\partial v}{\partial z}\right) dz = 0. \quad (6)$$

Therefore, it can be shown that

$$w_{k-1/2} = w_{k+1/2} - \left(\frac{\partial(hu)}{\partial x} + \frac{\partial(hv)}{\partial y}\right). \quad (7)$$

The boundary stresses at the air-water interface are formulated as

$$T_x^s = C \cdot (u_w^2 + v_w^2)^{1/2} u_w$$

$$T_y^s = C \cdot (u_w^2 + v_w^2)^{1/2} v_w \quad (8)$$

where C = drag coefficient, u_w = x-directed component of the wind, and v_w = y-directed component of the wind.

The boundary stress at the water-sediment interface is formulated as

$$T_x^b = R \cdot (u^2 + v^2)^{1/2} u$$

$$T_y^b = R \cdot (u^2 + v^2)^{1/2} v \quad (9)$$

where R = friction coefficient.

The vertical interfacial stress term is based on Reid (1957). The basic form is

$$\frac{Txz}{\rho} = 1_0^2 \left|\frac{\partial Q}{\partial z}\right| e^{-mRi} \frac{\partial u}{\partial x}$$

$$\frac{Tyz}{\rho} = 1_0^2 \left|\frac{\partial Q}{\partial z}\right| e^{-mRi} \frac{\partial v}{\partial y} \tag{10}$$

$$1_o = k \frac{(H-z)z}{H} \tag{11}$$

where k = 0.4 (von Karman's constant), H = total water depth, z = depth at which T is computed, $Q = (u^2 + v^2)^{1/2}$, m = 1.5, and

$$\text{Ri (Richardson number)} = \frac{g \frac{\partial \rho}{\partial z}}{\rho (\frac{\partial u}{\partial z})^2} \quad . \tag{12}$$

The finite difference analogues which correspond to the vertically integrated equations of motion for the model grid system are based on the three-dimensional model of Leendertse and Liu (1975) with some refinement of the differencing scheme which aids stability. Details may be found in Show (1977). The scheme is explicit and the time step limited by the Courant condition for surface free gravity waves. Thus:

$$\Delta t \leq \tag{13}$$

$$\min \left[\frac{\Delta x}{\sqrt{2gH_m}} , \frac{\Delta y}{\sqrt{2gH_m}} \right]$$

where H_m is the maximum water depth of the lagoon.

The boundary conditions are such that only a u, v, or w is found on a boundary. Therefore, for a closed boundary it is necessary only to set the respective variable to zero. For an open boundary in the cases considered here, the value on the boundary was specified independently of the main computations. In addition, certain terms adjacent to a closed boundary are set to zero so that no advection or diffusion of momentum, heat, or salt is allowed into the boundary. The implicit assumption is that between a closed boundary and a point adjacent to it, there are no gradients of momentum, temperature, or salinity.

In the following discussion, each of the terms which make up the transport coefficients in the biological model will be treated separately.

The advective portion of the transport coefficient is computed on the basis of the average current speed normal to the compartment face on which the coefficient is located.

$$\text{Advection} = q \;/\; d \qquad (14)$$

where q = current speed in the direction normal to the compartment face and d = length of the compartment in the direction normal to the plane of the compartment face on which the coefficient is located.

The above formulation gives a fractional transport coefficient in units of time $^{-1}$ which indicates the proportion of the volume of water in the compartment from which the transport is directed that passes through the compartment face in one unit of time. If the individuals in the compartment are assumed not to show spatial heterogeneity, then the proportion of the total number of individuals transported is equal to the proportion of the volume transported.

The response to temperature and salinity gradients is formulated as:

$$G'_{TS} = A_{max}\ \{1 - \text{EXP}[-\ \frac{dS}{dx}(\frac{dS'}{dx})^{-1}_{min} - \frac{dT}{dx}(\frac{dT'}{dx})^{-1}_{min}]\} \qquad (15)$$

where $S = |S_{obs} - S_{opt}|$, $T = |T_{obs} - T_{opt}|$, $(\frac{dS'}{dx})_{min}$ and $(\frac{dT'}{dx})_{min}$ = minimum salinity and temperature gradient to which the organism will respond. T_{opt} and S_{opt} are the optimum temperature and salinity of occurrence. A_{max} = maximum swimming speed.

Cassie (1962) showed that an organism's response to gradients in temperature and salinity can be expected to be exponential. This assumption is adopted here where the exponential part of the function represents a response to the deviation from optimum temperature and salinity, the response being weighted by the minimum gradient to which the organism might be expected to respond.

After the swimming speed in response to the gradients has been determined, the temperature and salinity portion of the transport coefficient is computed as

$$G_{TS} - G'_{TS} / d \tag{16}$$

where G_{TS} = proportion of the total number of organisms which swim through the compartment interface in a unit of time in response to the temperature and salinity gradients. G_{TS} is in units of time^{-1}.

Several authors have recently surmised that zooplankton in estuaries and coastal lagoons respond to tides (or tidal currents) by moving upward with an incoming tide and downward with an outgoing tide (see Introduction). A simple mathematical form with which to describe this response is a sine curve. Therefore, the response is taken to be of the form:

$$\frac{dx}{dt} = A \cos (kt) \tag{17}$$

where $\frac{dx}{dt}$ = resultant swimming speed of the organism, A = maximum swimming speed of the organism, t = time, and $k = 2\pi/T$. T = tidal period and t = 0 is taken as the mean tide level on a rising tide.

The computation of the tide response portion of the transport coefficient is completed by the relationship:

$$G_T = \frac{dx}{dt} / d \tag{18}$$

where G_T is the fractional transport coefficient in units of $time^{-1}$.

The response to light considered in this model is that which results in a diurnal vertical migration. Bary (1967) showed that vertically migrating organisms could be responding to the time rate of change of light intensity. If time t = 0 is taken as midnight, then a simple representation of the curve of time vs. light might be

$$I = - \sin(\omega t - \frac{\pi}{2}) . \qquad (19)$$

Therefore, the time rate of change of light intensity would be

$$\frac{dI}{dt} = - \omega \cos(\omega t - \frac{\pi}{2}) \qquad (20)$$

allowing the swimming response of the organism to be defined by

$$\frac{dx}{dt} = A \frac{dI}{dt} . \qquad (21)$$

Then, the fractional transport coefficient for the light response would be

$$G_L = A \frac{dI}{dt} / d . \qquad (22)$$

In order to derive a formulation for aggregation, two compartments must be considered. These compartments have state variables Z_1 and Z_2 and must be adjacent to one another.

Aggregation is defined in this application as the tendency for an organism to move toward an area of higher density of its own species. The basis for modeling aggregation is taken as the logistic growth model, this model being chosen because it has certain negative feedback properties and is bounded above and below.

A logistic model for the variables Z_1 and Z_2 is

$$\frac{dZ_1}{dt} = aZ_1 - bZ_1^2 \tag{23}$$

$$\frac{dZ_2}{dt} = aZ_2 - bZ_2^2$$

Now, since we are considering the response of the organism to the relative densities of the two compartments, (23) can be written as

$$\frac{dZ_1}{dt} = f(Z_1, Z_2) \cdot Z_1 - g(Z_1, Z_2) \cdot Z_1^2 \tag{24}$$

$$\frac{dZ_2}{dt} = f(Z_2, Z_1) \cdot Z_2 - g(Z_2, Z_1) \cdot Z_2^2$$

where $f(Z_i, Z_j)\cdot Z_i$ is the intrinsic rate of gain and $g(Z_i, Z_j)\cdot Z_i^2$ is the inhibition of gain due to increase in density.

It is assumed that no movement will occur if $Z_1 = Z_2$ or if either Z_1 or $Z_2 = 0$. It is also assumed that movement is never away from the compartment with highest density since diffusion is not considered. Therefore, it is required that $f(Z, Z)\cdot Z = 0$ and $f(Z_i, Z_j)$ is non-decreasing on Z_i.

If $f(Z_i, Z_j) = (Z_i - Z_j)/\Delta t(Z_i + Z_j)$, then the restrictions are satisfied. Furthermore, from a knowledge of the behavior of the logistic system, it is known that $f(Z_i, Z_j)/g(Z_i, Z_j) \leq$ the total number of individuals in compartment i. Therefore, it can be seen that f/g cannot exceed $Z_i + Z_j$. The effect of diffusion, if it were considered, would be to make f/g strictly less than $Z_i + Z_j$ and to decrease $\frac{dZ_i}{dt}$. Therefore,

$$\frac{1}{g(Z_i, Z_j)} \cdot \frac{Z_i - Z_j}{\Delta t(Z_i + Z_j)} \leq Z_i + Z_j \tag{25}$$

which implies that

$$g(Z_i, Z_j) \leq \frac{Z_i - Z_j}{\Delta t(Z_i + Z_j)^2} \quad . \tag{26}$$

Equations (24) then become approximately

$$\frac{dZ_1}{dt} = \frac{(Z_1 - Z_2)\cdot Z_1}{(Z_2 + Z_2)\Delta t} - \frac{(Z_1 - Z_2)\cdot Z_1^2}{(Z_1 + Z_2)^2\Delta t} \tag{27}$$

$$\frac{dZ_2}{dt} = \frac{(Z_2 - Z_1)\cdot Z_2}{(Z_2 + Z_1)\Delta t} - \frac{(Z_2 - Z_1)\cdot Z_2^2}{(Z_2 + Z_1)^2\Delta t} \quad .$$

The swimming response of the organism is, therefore defined by

$$\frac{dx}{dt} = \frac{AdZ}{dt} \tag{28}$$

Now, to complete the computation of the fractional transport coefficient,

$$G_A = \frac{dx}{dt} \;/\; (d\cdot Z) . \tag{29}$$

G_A is in terms of time $^{-1}$.

The computation form of K_{ij} is

$$K_{ij} = \text{advection} + \min\left(|G_{TS} + G_T + G_L + G_A|,\ |A/d|\right) \tag{30}$$

where K_{ij} represents the coefficients in equations of the form

$$\frac{dZ_i}{dt} = \sum_{j \neq i} K_{ij} \cdot Z_j - \left(\sum_{i \neq j} K_{ji} \right) Z_i \tag{31}$$

which is the mathematical formulation of the biological response portion of the model. The first coefficient of K_{ij} represents the compartment to which transport is directed; the second represents the compartment from which flow is directed. Thus a total of six fluxes, K_{ij} through each face of a typical compartment, i, are required for the solution of equation (31). It should be noted that equation (31) represents a reversible system; therefore, if $K_{ij} \neq 0$, then $K_{ji} = 0$. This restriction assumes that all individuals react in the same manner to the various stimuli. An assumption of varying response would be theoretically appealing, but difficult to apply, since some knowledge of the probability of a certain response would be required.

EAST LAGOON SIMULATION

In order to adequately simulate the physical and biological processes operating in East Lagoon, it was necessary for the computational grid of the model to resolve these processes spatially. To determine the scale of the processes involved, two sampling trips, in February and March 1976, were undertaken. During these sampling periods, only hydrographic sampling was carried out; vertical profiles of temperature, salinity, and currents were taken at intervals of 50 meters down the length of the lagoon. The procedure was repeated twice a day beginning at about 0900 and 1500 hours. Several laterally placed sets of samples were also taken. The results showed that detectable changes in current velocity, temperature, and salinity tended to occur on a scale of approximately 250 to 300 metres in the horizontal plane and 0.1 to 1.0 metre in the vertical.

During a later sampling period, 26 to 28 June 1976, both hydrographic and zooplankton sampling was carried out. Beginning at 0700 hours on 26 June, the entire lagoon was sampled for all field variables. Starting at the mouth of the lagoon, a vertical series of current velocity measurements was taken at each compartment interface and a vertical series of plankton tows and temperature and salinity measurements taken at the center of each compartment. The vertical sampling increment was one metre. The results showed _A. tonsa_ populations comparable to those described by Fleminger (1959). On the basis of these samples, it was possible to determine that a three-level model was sufficient to resolve the vertical movements of _A. tonsa_ in the lagoon.

Sampling was carried out again during September 1976. The intention then was to acquire sufficient data to simulate the sampling period and to test the results of the simulation. With this in mind, samples were taken at evenly spaced intervals of time and space throughout the entire 44-hour sampling period.

Beginning at 0000 hrs on 18 September, all physical variables were measured at each point in the model grid for the lagoon. In order to have the samples as synoptic as possible, the physical variables were measured as quickly as possible. High tide occurred at approximately 0000 hrs; therefore, the tidally driven processes probably had the smallest time rate of change at that time. The resulting data was used to provide starting conditions for the physical model. Next, beginning at 0100 hrs. a pair of plankton samples was taken from each compartment. These samples were used to provide the initial condition for the biological model. After the initial sets of samples were taken, only the vertical sets of compartments centered at 120, 960, and 1800 metres from the mouth of the lagoon were sampled. Hydrographic and zooplankton samples were taken from each of these three sets of compartments.

The first step in the numerical simulation was to derive from the physical model computed time histories for the current velocity, temperature and salinity fields. The results of the physical numerical simulation were used to drive the biological model. Table 1 gives the values of all parameters used in the physical model.

The solution technique used required complete u, v, w, T, S, and η fields from two consecutive timesteps as initial conditions with which to start the model. The simulation was begun at the real time of 0000 hrs 18 September 1976 which was at the time of a high tide. Therefore, it was assumed that the prototype water surface field was maximum and flat and that the u, v, and w current fields were zero.

Table 1. Parameters of the physical model. Parameter values used in the East Lagoon simulation

$\Delta x = 240$ m $\quad \Delta z = 1$ or 2 m $\quad \Delta t = 10$ sec

$\Delta y = 120$ m $\quad$ no. time steps = 17280

Vertical Transfer of Momentum

$$T_{xz} = [\, l_0^2 \left|\frac{\partial Q}{\partial z}\right| e^{-mRi} \,] \frac{\partial u}{\partial x} \quad m^{-2}\ sec^{-2}$$

Surface Wind Stress Coefficient

$$C_W = 3.0\ (10^{-6})\cdot$$

Bottom Frictional Stress Coefficient

$$C_R = 2.5\ (10^{-3})$$

Lateral Eddy Diffusion Coefficient (momentum)

$$A_x = A_y = 100.0\ M^2\ sec^{-1}$$

Lateral Eddy Diffusion Coefficient (heat and salt)

$$D_x = D_y = 10.0\ M^2\ sec^{-1}$$

Vertical Eddy Diffusion Coefficient (heat and salt)

$$K = [\, l_0^2 \left|\frac{\partial Q}{\partial z}\right| e^{-1.5\ Ri} \,] \quad M^2\ sec^{-1}$$

To start the model, the temperature and salinity from the sampling data values were used. Since the time increment was 10 seconds, it was assumed that these fields did not change in that time. The fields given by the exact solutions of a tidal seiche in a rectangular, uniform depth canal closed at one end were found to be sufficient for starting the model. Here:

$$\eta = \frac{a \cos(\kappa x)}{\cos(\kappa L)} \cos(\omega t)$$

$$u = \frac{a\, C_0}{h \cos(\kappa L)} \sin(\kappa x) \sin(\kappa t) \qquad (32)$$

$$w = \frac{-ak\, C\, (H-z)}{h \cos(\kappa L)} \cos(\omega x) \sin(\omega t)$$

where $\kappa = \frac{\omega}{\sqrt{gh}} = \frac{2\pi}{\lambda}$, $C_0 = \frac{gT}{2\pi}$, and $\omega = \frac{2\pi}{T}$. λ = wave length, T = period, a = amplitude, H = maximum depth of canal, L = length of canal, x = distance from open end, z = depth downward from surface, and t = time.

The placement of the current velocity variables on the boundaries dictated that the currents would have to be forced with current velocities rather than with water surface elevations. Field data fit to discrete Fourier series were used to force the circulation. The data to which the series were fit were the current velocity measurements taken near the mouth of the lagoon.

Temperature and salinity were specified at each time step at the point in the grid directly outside the computation field. Discrete Fourier series were fit to sampling data. A separate series was developed for each level at the open boundary.

The wind was set at zero for the entire simulation. At no point did the wind exceed four knots and it always blew at right angles to the long axis of the lagoon. What effect it might have had was not discernable in East Lagoon during the sampling period.

The transport coefficients in the biological model were computed in units of hr^{-1}. In order to meet the stability criterion, the time step of the biological model was 0.25 hrs. The total time simulated was 44 hrs; the simulated period began at 0000 hrs 18 September 1976.

It was assumed that the horizontal gradients of temperature, salinity, and intraspecific population density were not sufficient to elicit a behavioral response, but that vertical gradients probably were (Dr. D. R. Heinle, personal communication). Accordingly, the formulae for response to temperature and salinity gradients and for behavioral aggregation were applied only to vertical transports. In addition, the formulae for response to tidal advection and response to light were applied only to vertical transports. Advection by currents was applied to both vertical and horizontal transports. For purposes of this simulation, it was further assumed that no recruitment of A. tonsa from outside the lagoon occurred and that births exactly balanced deaths.

The first step in testing the model was to determine the accuracy of the physical numerical simulation. As an indicator of accuracy, contours of observed and computed temperatures and salinities were placed on the same figures, making direct comparison possible.

Several simulations were then run with the biological model using identical time histories of temperature, salinity, and current velocity from the physical model. In the first simulation, only advection was considered; none of the behavioral terms were included. In each of the successive simulations, behavioral terms were added one at a time. For each simulation and for each compartment for which sampling data were available, the following statistic was computed:

$$\text{Chi-square} = \sum \frac{(Z\text{comp} - Z_{\text{obs}})^2}{Z_{\text{obs}}} \quad . \qquad (33)$$

The statistic was used in two ways: as a basis of comparison between simulations and as a test of the significance of the final version of the model. In addition, time histories of the stochastic means and variances of each compartment were compared to the sampling data. The comparisons provided a basis for inferences regarding the distribution of zooplankton densities. Finally, vertical cross-sections of A. tonsa density down the long axis of the lagoon were prepared in which different computed fields were compared. These figures served to highlight the overall behavior of the physical-biological system in East Lagoon.

RESULTS OF PRELIMINARY SAMPLING IN EAST LAGOON

Preliminary sampling results were used to support certain assumptions incorporated in the biological model and to estimate parameters required by the model. The only directly testable assumption was that of no recruitment from outside the lagoon. The required biological parameters were temperature and salinity optima and the maximum swimming speed of A. tonsa.

There were two major lines of evidence to indicate that little, if any, recruitment of *A. tonsa* was taking place during the period simulated. The first was the species composition of the copepod populations sampled outside and inside the lagoon. On 18 September, samples from outside included *A. tonsa*, *Temora turbinata*, *Centropages furcatus*, *Anomalocera ornata*, *Pontella meadii*, *Paracalanus* sp., and *Oithona* spp. in significant numbers. Samples from inside the lagoon included *A. tonsa*, *T. turbinata*, and *Oithona* spp.

The second line of evidence was the length-frequency distribution of *A. tonsa* from outside compared to inside the lagoon. The mean length of the population immediately outside the lagoon exceeded that of those inside by a factor of 1.6 to 1. Since both samples were taken on a rising tide and were taken either immediately inside or immediately outside the lagoon, recruitment should have been indicated by the lack of a clear-cut difference in the two populations. However, a clear-cut difference did exist and was statistically significant.

Temperature and salinity optima were determined graphically. Figure 5 shows the plots of plankton density vs. temperature and salinity respectively. The mode of each distribution was taken as the optimum for that variable. This relationship is consistent with the temperature salinity plot given also in Figure 5.

RESULTS OF THE PHYSICAL NUMERICAL SIMULATION OF EAST LAGOON

Figures 6 through 9 show comparisons of the observed and calculated temperature and salinity fields and the calculated current velocity fields. The close correspondence between the observed and calculated fields indicates that the physical model provided a sufficient representation of the physical processes in East Lagoon during the period simulated. Therefore, it seems reasonable to assume that the physical variable fields used to drive the biological model were adequate.

Most of the transport occurred in the top two meters; there was little movement in the basin near the head of the lagoon. A small magnitude return current ran consistently counter to the main tidal flow below about three metres. *A. tonsa* was able to take advantage of this current pattern to minimize losses from the lagoon.

The lagoon remained stably stratified under calm atmospheric conditions despite the relatively small temperature and salinity gradients. During the simulation, water that was cooler and less saline than the lagoon water was forced into the lagoon with each rising tide. The introduction of this water can be seen in the tilting of the isotherms and isohalines near the mouth of the lagoon. Turbulent mixing was apparently not significant as a physical

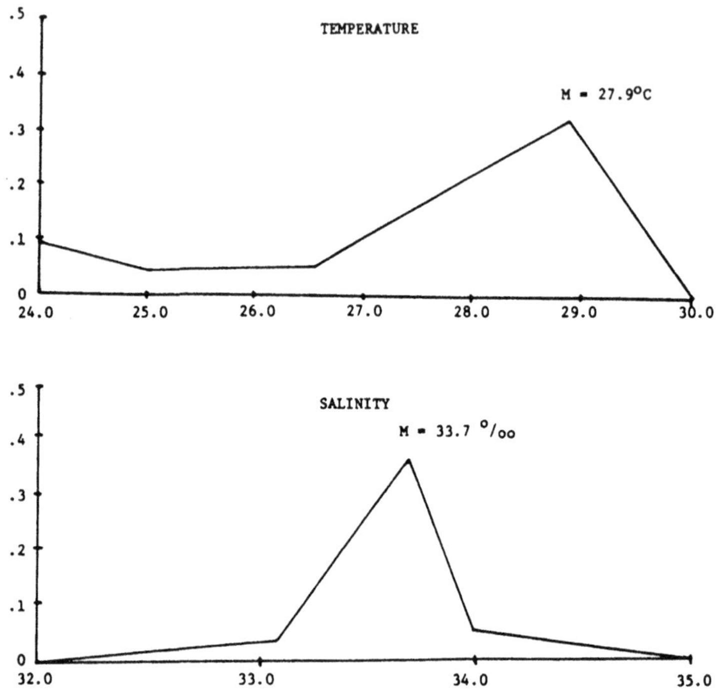

Figure 5. Optimum temperature and salinity. The graphs represent the empirical probability distributions of A. tonsa in relation to the temperatures and salinities at which it occurred. M designates the location and value of the optima.

process in the lagoon because the water introduced during each rising tide did not mix to any great extent with the lagoon water and was subsequently forced back out of the lagoon on the next falling tide. If turbulent mixing of heat and salt was minimal then it might be surmised that the non-behavioral (what Hutchinson called vectorial) mixing of lagoon and coastal populations was minimal. If this assumption is true, then the separation of lagoon and introduced water was an indication of the conditions which allowed A. tonsa to maintain an endemic lagoon population.

RESULTS OF THE BIOLOGICAL SIMULATION OF EAST LAGOON

The biological simulation was accurate in depicting the time history of A. tonsa spatial distributions in the lagoon. Statistical tests and graphic representations indicate that the model was numerically accurate and adequately portrayed the processes involved.

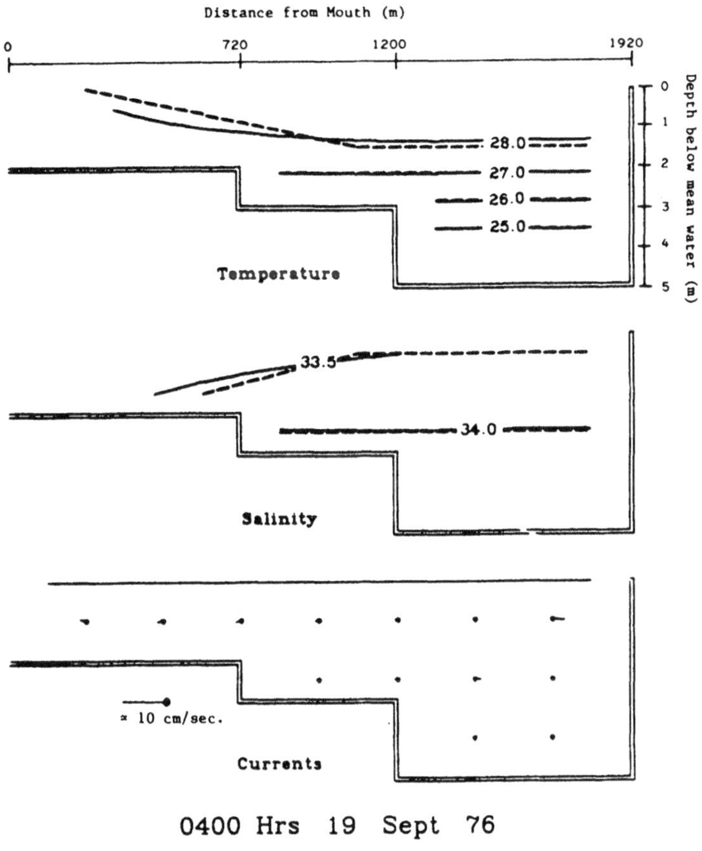

Figure 6. Comparison of physical field data with the computed results at 0400 19 September. ——— calculated fields, -------- sampling data. The currents and water surface elevations in the bottom section of the figure are from the simulation. Temperature is in units of °C and salinity in °/₀₀.

Table 2 gives the values of the χ^2 statistic. If the behavior of this statistic is taken as an indication of the sufficiency of the model, then each term was sufficient to reduce the deviation between calculated and observed zooplankton densities. In no case did the introduction of a new behavioral response term increase the χ^2 value. What is not apparent is that the tidal response term had to be included in order to achieve the consistent pattern shown. Therefore, not only does the tidal response term contribute the most toward model accuracy, but it is apparently the one crucial term without which the rest of the model cannot operate properly.

Table 2. χ^2 comparisons. Under the "compartment" column, the number is the distance in metres from the mouth of the lagoon to the center of the compartment and the letter designations are S - surface, M - middle, and B - bottom. The form of the model from which values in the body of the table result include the term at the head of the column and all the terms to the left.

Compartment	Advection	Tide	Light	T & S	Aggregation
120 S	478.54	12.97	9.43	9.41	9.22
120 B	87106.34	11.62	8.66	8.58	8.56
960 S	11459.00	11.11	7.41	7.39	7.35
960 B	26062.51	9.54	7.04	7.03	6.91
1200 S	9949.47	10.90	9.49	9.49	9.11
1200 M	9116.85	7.16	6.16	6.14	5.81
1200 B	8042.04	6.85	5.82	5.79	5.72

The advection-only form of the model represents the behavior of the population density fields in response to tidal circulation in the lagoon. The χ^2 values for this form of the model clearly indicate that it was inadequate.

Since the statistic used involved densities (which are continuous) instead of count data (which are discrete), it might deviate from a χ^2 distribution. However, if it is assumed that the statistic is distributed as a χ^2 random variable, then the results from the simulation that included all of the behavioral terms, was significant at the 95% level for all compartments tested, indicating a good fit between the calculated and observed zooplankton densities. It is perhaps significant to note, in Table 2, that χ^2 values become smaller not only as more terms are added, but also as one moves further away from the mouth of the lagoon and down from the surface. The exception to this generality is probably also significant in that it occurs near the mouth of the lagoon where advective processes dominate.

The graphic results provided by Figures 10 through 13 indicate much the same thing as the χ^2 comparisons. The advection-only and behavioral models provide quite different results.

The underlying hypothesis in this study was that zooplankton distributions result from a combination of physical processes and biological behavior. In East Lagoon, this hypothesis would imply

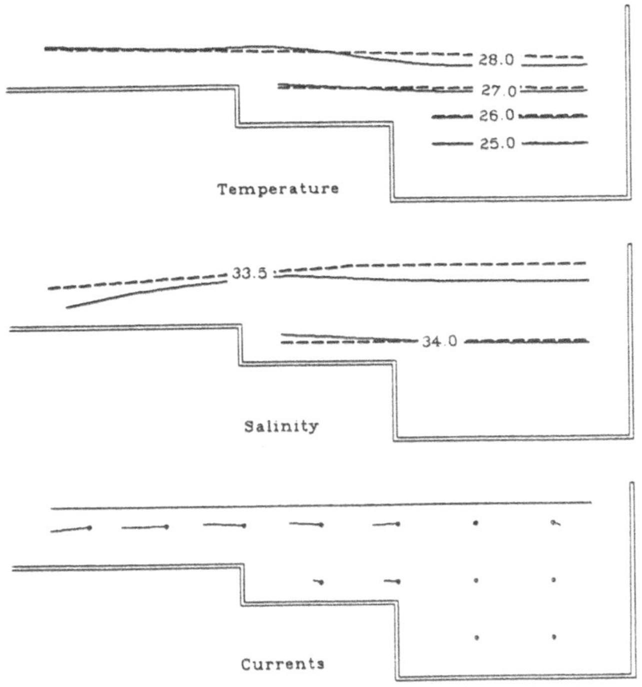

1000 Hrs 19 Sept 76

Figure 7. Comparison of physical field data with the computed results at 1000 19 September. ———— calculated fields, -------- sampling data. The currents and water surface elevations in the bottom section of the figure are from the simulation. Temperature is in units of °C and salinity in °/₀₀.

that the plankter, by responding in a particular manner to the appropriate stimuli, would be able to minimize its lateral movements, particularly those that would carry it out of the lagoon. It was also hypothesized that the behavioral movements on the part of the organisms would be primarily vertical and primarily in response to tidal movements. The density contours indicate that very little in the way of lateral movement occurred in the full behavioral model as would be predicted by the advection-only model.

The tides were closely confounded with the time rate of change of light intensity. Therefore, it was difficult to determine which of the two factors was more significant in minimizing lateral movements. However, based on the fact that the tidal response term was critical to the operation of the model, the problem became one of deciding whether the light response term contributed significantly

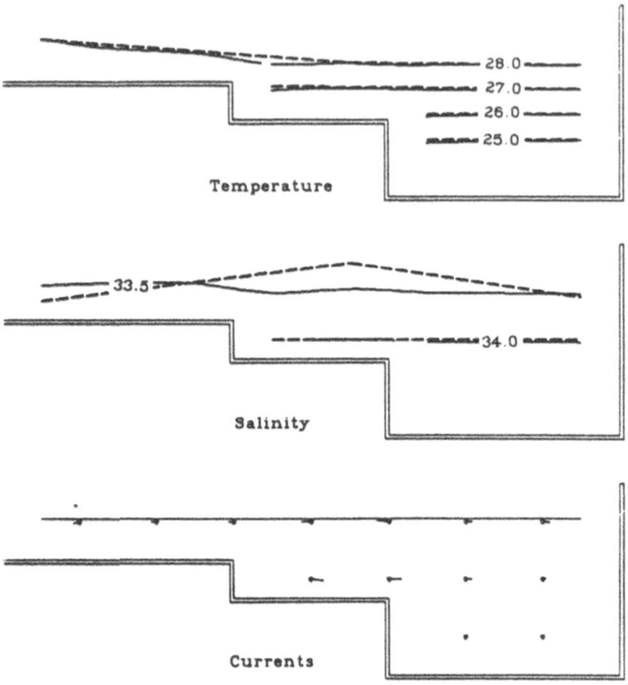

1400 Hrs 19 Sept 76

Figure 8. Comparison of physical field data with the computed results at 1400 19 September. ———— calculated fields, -------- sampling data. The currents and water surface elevations in the bottom section of the figure are from the simulation. Temperature is in units of °C and salinity in °/₀₀.

to the accuracy of the model. I believe that the light response was important since A. tonsa was shown experimentally to respond to light stimuli as were various other species investigated by Spooner (1933), Schallek (1943), and Bary (1967). I feel that light response should be smaller in magnitude than tide response as exemplified by the formulae used in this model.

The majority of the individuals in the lagoon were concentrated near the head of the lagoon. Two factors seem to be responsible for this pattern. First, the organism tended to move vertically so as to be almost always carried toward the head of the lagoon. Second, in the narrow channel portion of the lagoon near the mouth, there was little or no vertical current velocity gradient and no

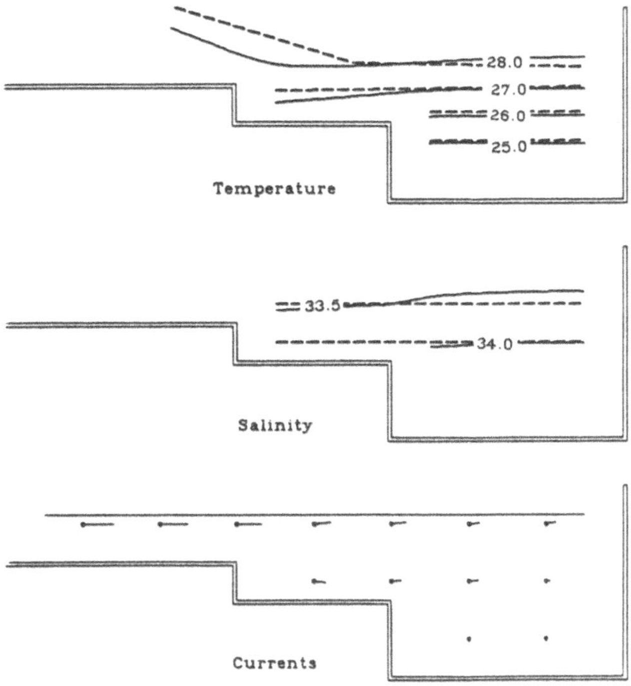

1800 Hrs 19 Sept 76

Figure 9. Comparison of physical field data with the computed results at 1800 19 September. ——— calculated fields, -------- sampling data. The current and water surface elevations in the bottom section of the figure are from the simulation. Temperature is in units of °C and salinity in °/₀₀.

return current near the bottom.

Figure 14 shows a comparison of average plankton density for the entire lagoon vs. time. The comparison is made for the advection-only form of the model, the full behavioral model, and the observed densities. Agreement is better between the full behavioral model and the observed densities. However, the most important point is that over 25% of the population was lost when the organism was assumed to be advected only while less than five percent was lost when all of the behavioral responses were assumed. Fleminger (1959) states that the time from egg to adult for A. tonsa in East Lagoon is four to five weeks. If this time span is assumed to represent population turn-over time, then even a five percent loss in a two-day period is unacceptable if an endemic population is to be

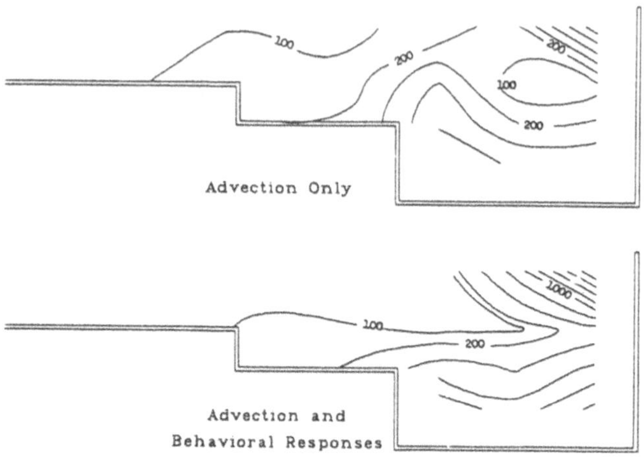

0500 Hrs 19 Sept 76

Figure 10. Comparison of the results of the advection only and behavioral forms of the biological model at 0500 19 September. The contour interval is 200 individuals/m^3 except where indicated.

maintained without significant recruitment. The above facts therefore seem to indicate that, for longer simulations, a production term would be necessary to maintain a stable population level.

Figure 15 compares means and variances of the full behavioral model with those of the observed plankton densities. Several facts can be inferred regarding the distribution of A. tonsa in the lagoon.

It appears, on the basis of both model and sample means and variances that the plankton density in a particular compartment at time "t" was distributed as a Poisson random variable. The Poisson distribution was indicated by the near equality of the sampling and calculated means and variances. In the case of the simulation results, the distribution was asymptotically Poisson since the variance approaches the mean as time becomes large. The most likely alternative would have been that the variances would have greatly exceeded the means, especially in the sampling data, indicating a heterogeneous distribution of densities.

When between compartment variances were computed from the model and from the sampling data, the variances exceeded the means by as much as a factor of ten. If the reasoning of Pielou (1969) is followed and no compelling reasons are found to reject the hypothesis, then it may be reasonably assumed that A. tonsa was aggregated, but

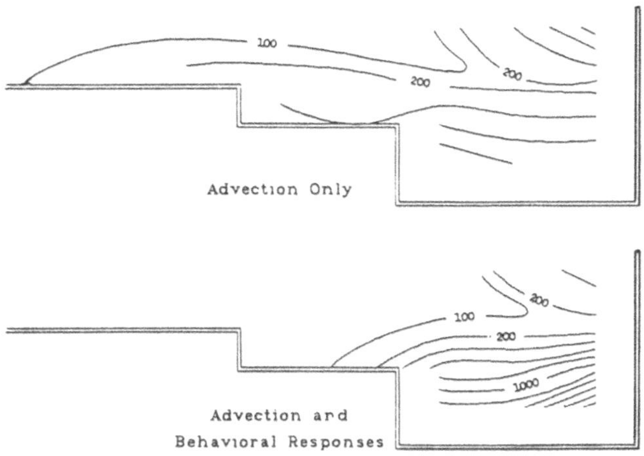

1100 Hrs 19 Sept 76

Figure 11. Comparison of the results of the advection only and behavioral forms of the biological model at 1100 19 September. The contour interval is 200 individuals/m^3 except where indicated.

on a scale larger than or equal to the compartment size used in the model. Interestingly enough, it did not matter if laterally or vertically adjacent compartments were compared, heterogeneity was still indicated. Also variances computed from sets of 50 meter long tows taken during June indicated that the variances computed from these samples exceeded the mean onoy when enough adjacent samples were used so that the total distances sampled exceeded 250 to 300 meters. This phenomenon occurred regardless of the density of the organism sampled; samples indicating densities of from 3.54 to 2378.19 individuals/m^3 all showed the same property.

The results of this study were not consistent with the conclusions of Fasham et al. (1974) that an increase in density results not from an increase in the density within patches, but from an increase in the number of patch centers. From the East Lagoon results, it appears that patch size remains fixed while increases in density result from an increase in within-patch density. The Fasham et al. study was carried out in the open sea. Perhaps differences in physical characteristics encountered in coastal lagoons and estuaries help to bring about a different mechanism by which patch formation might occur.

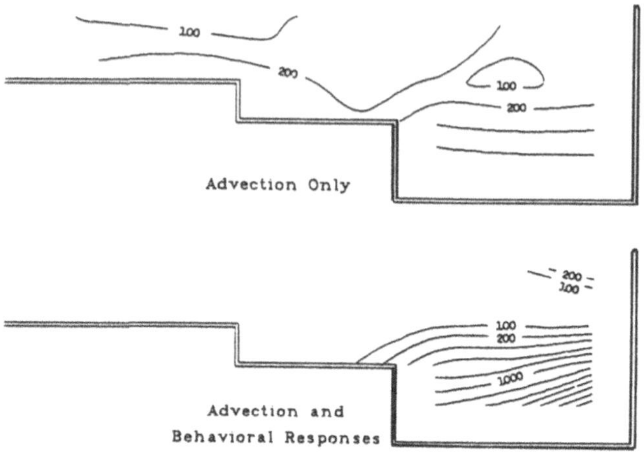

1500 Hrs 19 Sept 75

Figure 12. Comparison of the results of the advection only and behavioral forms of the biological model at 1500 19 September. The contour interval is 200 individuals/m^3 except where indicated.

CONCLUSIONS REGARDING THE EAST LAGOON SIMULATION

The coupled physical-biological model used to simulate the distribution of A. tonsa provided numerically accurate estimates for the time histories of the physical and biological processes involved. The success of the model is probably attributable to a number of factors which involve the nature of the ecological situation modeled, the selection of parameters of the biological model, the manner in which the lagoon was compartmented, and the nature of the sampling data used to test the results.

East Lagoon represents a relatively simple framework within which to study the temporal-spatial distribution of zooplankton. It is a rather small, shallow body of water and has a simple topography and simple, tidally driven circulation pattern. Also there is no fresh-water input and no significant marsh overflow. The particular period simulated also presented relatively simple conditions; in particular, atmospheric conditions were calm and stable with little or no wind and no rain. The lagoon was therefore not exposed to any strong perturbations during the period simulated.

Parameterization of the biological model was particularly significant since all parameters were based on characteristics of

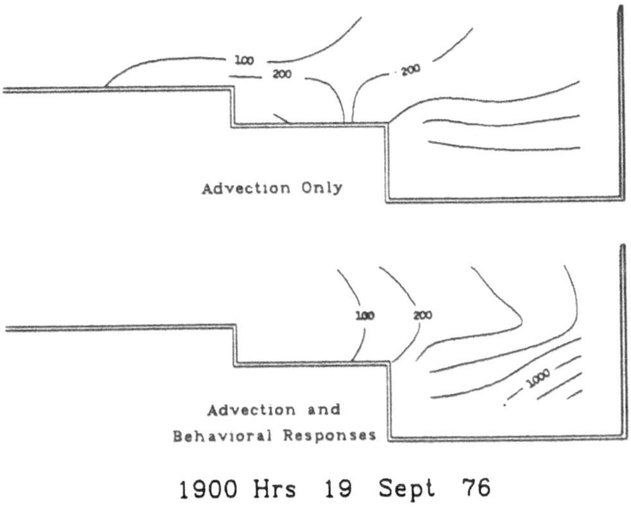

Figure 13. Comparison of the results of the advection only and behavioral forms of the biological model at 1900 19 September. The contour interval is 200 individuals/m^3 except where indicated.

the organism being modeled and were determined from data taken during the period simulated. Specifically, all of the parameters were based on the swimming speed of A. tonsa from East Lagoon, the temperature and salinity at which the organism was expected to occur most frequently on the days simulated, and the minimum temperature and salinity gradients found in East Lagoon on the days simulated.

Apparently, the vertical current velocities were not an obstacle to the vertical positioning of A. tonsa during the period simulated. The maximum observed vertical swimming speed of A. tonsa was 0.03 cm/sec while the maximum calculated vertical current speed was 0.006 cm/sec.

A major assumption of any compartmental model of present type, unless a specific form of within-compartment delay is incorporated, is that the distribution of organisms within a compartment is homogeneous. In this study, the size of the compartments was chosen to meet this criterion. The dimensions of 240 metres by one or two metres was chosen, as mentioned above, because there was at least a reasonable assurance that the distribution of A. tonsa would not show within-compartment heterogeneity at this scale.

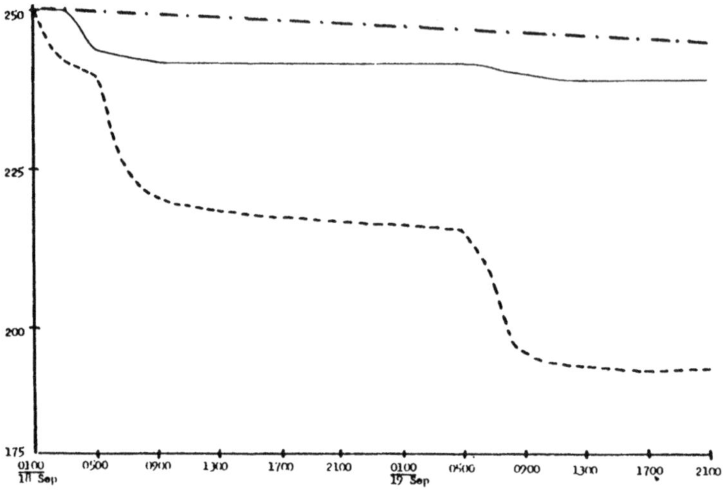

Figure 14. Comparison of average plankton densities. ———— results of the biological model with all advective and behavioral terms, -------- results of the biological model with advection only, -.-.-.-. sampling data.

Finally, no physical or biological model has ever been tested against data from a small, enclosed body of water such as East Lagoon nor have any been tested against data as temporally and spatially dense as that collected for this model (Dr. Robert E. Whitaker, personal communication). Furthermore, the apparent success of the model is at least partially attributable to the fact that the sampling program was based specifically on the requirements of supporting and testing the model.

Assuming then that the results of the numerical simulation were accurate, what can be inferred about the spatial distribution of A. tonsa in East Lagoon? The most important conclusion is that the species appears to owe its spatial distribution in the lagoon to the combined effects of advection by currents and behavioral response to environmental stimuli: tides, light, temperature gradients, salinity gradients, and the density gradients of its own species. It also appears that the resultant movement of the organism is sufficient to minimize losses from the lagoon to the extent that the species maintains an endemic population inside the lagoon which is distinct from the population of the same species found immediately outside. Finally, it may be concluded that the spatial distribution of A. tonsa was heterogeneous, that the patches were of the order of 240 metres by one or two metres in size, and

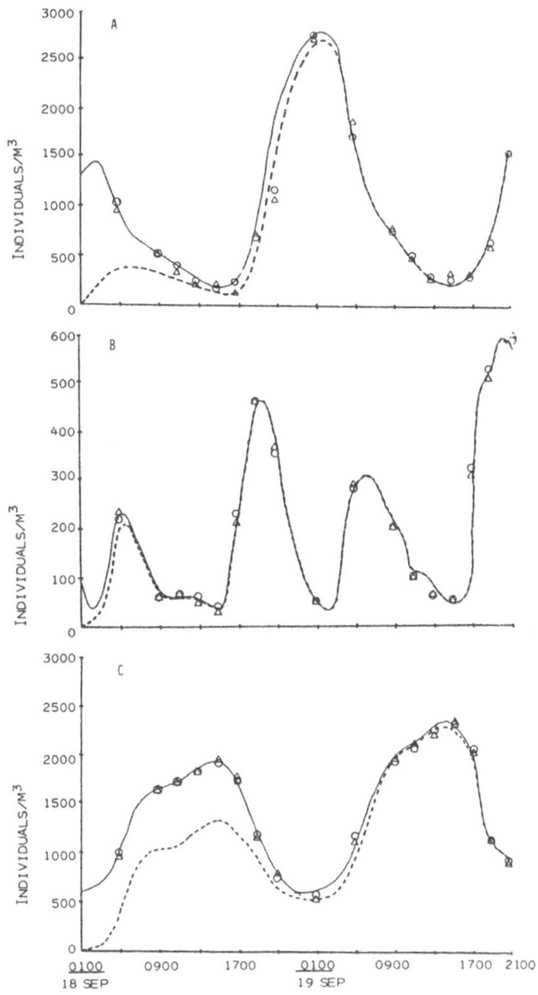

Figure 15. Behavior of means and variances. A comparison of the means and variances of the biological model and the field sampling data. ———— model means, -------- model variances, O sampling means, Δ sampling variances. All distances are measured from the mouth of the lagoon. (A) Surface compartment centered at 1800m. (B) Middle compartment centered at 1900m. (C) Bottom compartment centered at 1800m.

that changes in density occurred as a result of an increase in within-patch density rather than an increase in the number of patches.

This study represents only a first step. The possibility exists that there are other important factors effecting the spatial patterns of A. tonsa in East Lagoon, especially if the time scale were extended. Obvious possibilities include grazing, predation, and birth-death processes. The present model, of course, contributes only to a partial explanation. Further research could contribute to an understanding of processes neglected in the present investigation.

REFERENCES

Arnold, E. L., R. S. Wheeler, and K. N. Baxter. 1960. Observations on fishes and other biota of East Lagoon, Galveston Island. U. S. Fish and Wild. Serv., Spec. Sci. Rep., Fisheries #344.

Barlow, J. P. 1955. Physical and biological processes determining the distribution of zooplankton in a tidal estuary. Biol. Bull. 109: 211-225.

Barnes, H. and S. M. Marshall. 1951. On the variability of replicate plankton samples and some applications of contagious series to the statistical distribution of catches over restricted periods. Bull. Bingham Mar. Biol. Assoc. 30: 233-263.

Bary, B. M. 1963a. Distribution of Atlantic pelagic organisms in relation to surface water bodies. Roy. Soc. Canada Spec. Pubs. 5: 51-67.

Bary, B. M. 1963b. Temperature, salinity, and plankton in the eastern North Atlantic and coastal waters. J. Fish. Res. Bd. Canada. 20: 789-826.

Bary, B. M. 1963c. Temperature, salinity, and plankton in the eastern North Atlantic and coastal waters of Britain, 1957. II. The relationship between species and water bodies. J. Fish. Res. Bd. Canada. 20: 1031-1548.

Bary, B. M. 1963d. Temperature, salinity, and plankton in the eastern North Atlantic and coastal waters of Britain, 1957. III. The distribution of zooplankton in relation to water bodies. J. Fish. Res. Bd. Canada. 20: 1519-1538.

Bary, B. M. 1964. Temperature, salinity, and plankton in the eastern North Atlantic and coastal waters of Britain, 1957. IV. Its role in distribution and in selecting and using indicator species. J. Fish. Res. Bd. Canada. 21: 183-202.

Bary, B. M. 1967. Diel vertical migrations of underwater scattering, mostly in Saanich Inlet, British Columbia. Deep-Sea Res. 14: 35-52.

Breder, C. M., Jr. 1954. Equations descriptive of fish schools and other animal aggregations. Ecology 35: 361-370.

Cassie, R. M. 1959a. An experimental study of factors inducing aggregation in marine plankton. N.Z.J. Sci. 2: 239-365.

Cassie, R. M. 1959b. Microdistribution of plankton. N.Z.J. Sci. 2: 398-409.

Cassie, R. M. 1960. Factors influencing the distribution pattern of plankton in the mixing zone between oceanic and harbour waters. N.Z.J. Sci. 3: 26-50.

Cassie, R. M. 1962. Frequency distribution models in the ecology of plankton and other organisms. J. Anim. Ecol. 7: 121-130.

Fasham, J. J. R., M. V. Angel, and H. S. J. Roe. 1974. An investigation of the spatial pattern of zooplankton using the Longhurst-Hardy plankton recorder. J. Exp. Mar. Biol. Ecol. 16: 93-112.

Fisher, R. A. 1943. Relation between number of species and individuals in samples. J. Anim. Ecol. 12: 42-58.

Fleminger, A. 1959. East Lagoon zooplankton. U. S. Fish and Wildl. Ser. Circ. 62: 114-118.

Harder, W. 1968. Reactions of plankton organisms to water stratification. Limno. Oceanogr. 13: 156-168.

Hardy, A. C. and E. R. Gunther. 1935. The plankton of the South Georgia whaling grounds and adjacent waters, 1926-27. Discovery Rep. 11: 1-456.

Hutchinson, G. E. 1953. The concept of pattern in ecology. Proc. Acad. Nat. Sci. Phila. 105: 1-12.

Jacobs, J. 1968. Animal behavior and water movements as codeterminants of plankton distribution in a tidal system. Sarsia 34: 355-370.

Ketchum, B. H. 1954. Relation between circulation and planktonic populations in estuaries. Ecology 35: 191-200.

Lance, J. 1962. Effects of water of reduced salinity on the vertical migration of zooplankton. J. Mar. Biol. Assoc. U. K. 42: 131-138.

Leendertse, J. J. and S-K Liu. 1975. A three dimensional model for estuaries and coastal seas. Vol. II, Aspects of Computation. R-1764-OWRT Rand Corp., Santa Monica, Ca.

Lieberman, L. 1951. Effect of temperature inhomogeneities in the ocean on the propagation of sound. J. Acoust. Soc. Amer. 23: 563-570.

Marvin, K. T., L. M. Lansford, and R. S. Wheeler. 1961. Effects of copper on the ecology of a lagoon. Fish. Full. 61: 153-159.

McNaught, D. C. and A. D. Hasler. 1964. Rate of movement of populations of Daphnia in relation to changes in light intensity. J. Fish. Res. Bd. Canada. 21: 291-318.

Pielou, E. C. 1969. An Introduction to Mathematical Ecology. Wiley-Interscience, N. Y.

Pierce, E. L. 1939. Sagitta as an indicator of water movements in the Irish Sea. Nature 144: 784-785.

Powell, T. M., P. J. Richardson, T. M. Dillon, B. A. Agee, B. J. Dozier, D. A. Godden, and L. O. Myrup. 1975. Spatial scales of current speed and phytoplankton biomass fluctuations in Lake Tahoe. Science 189: 1088-1089.

Ragotzkie, R. A. 1953. Correlation of currents with the distribution of adult Daphnia in Lake Mendota. J. Mar. Res. 12: 157-172.

Riley, G. A. 1976. A model of plankton patchiness. Limno. Oceanogr. 21: 873-879.

Russell, F. S. 1935. On the value of certain plankton animals as indicators of water movements in the English Channel and North Sea. J. Mar. Biol. Assoc. U. K. 20: 309-332.

Russell, F. S. 1939. Hydrographical and biological conditions in the North Sea as indicated by plankton organisms. J. Cons. Int. Explor. Mer 14: 171-192.

Show, I. T. 1977. A spatial modeling approach to pelagic ecosystems. Ph.D. Dissertation. Texas A & M University, College Station, TX.

Stavn, R. H. 1971. The horizontal-vertical distribution hypothesis: Langmuir circulation and *Daphnia* distributions. Limno. Oceanogr. 16: 453-466.

Stickney, R. R. and S. C. Knowles. 1975. Summer zooplankton distribution in a Georgia Estuary. Mar. Biol. 33: 147-154.

Thakur, A. K. and A. Rescigno. 1978. On the stochastic theory of compartments. III. General, time-dependent reversible systems. Bull. Math. Biol. 40: 237-246.

Trinast, E. M. 1975. Tidal currents and *Acartia* distribution in Newport Bay, California. Estuarine and Coastal Mar. Sci. 3: 165-176.

Wiebe, P. H. 1970. Small-scale spatial distribution in oceanic zooplankton. Limno. Oceanogr. 15: 265-317.

ESTUARINE FISHERY RESOURCES AND PHYSICAL ESTUARINE MODIFICATIONS: SOME SUGGESTIONS FOR IMPACT ASSESSMENT

S. B. Saila

Graduate School of Oceanography

University of Rhode Island, Kingston, Rhode Island, 02881

ABSTRACT

A review of economically important fishery resources which are estuarine dependent in some life history stage is given and the potential value of these resources is considered. Some physical changes and consequences of estuarine modifications are categorized. Major probable fisheries impacts are described with some indication of tolerance by various life history stages.

Examples of analytical procedures designed to assess fisheries impacts of estuarine modification are presented and discussed. The advantages, disadvantages and limitations of certain procedures are pointed out.

A field experiment to determine ecological effects of an estuarine modification is outlined. It is suggested that well designed field experiments based on quantitative statistical design criteria can yield ecologically and statistically significant information about the effect of physical modifications on estuarine fisheries.

INTRODUCTION

Estuaries constitute an extremely valuable natural resource in the United States. They are a rich asset to our national economy by being portals of commerce providing ship anchoring and access as well as being biologically very productive. Some conflicts have emerged as a result of intensive demands for increased utilization by man versus the need for protecting and preserving valuable renewable natural resources. The magnitude of the estuarine resource is evident from the fact that about 85 percent of the Atlantic and Gulf coasts and about 15 percent of the Pacific coast are composed of estuaries which total about 10,000,000 hectares in area (Singer, 1969). The magnitude of the demands of man on these estuaries can be appreciated somewhat by recognizing that many large metropolitan areas of the United States have developed near estuaries, for example, New York, Los Angeles and Houston. Indeed, the coastal area (less than 10 percent of the land area of the U. S.) is occupied by approximately 30 percent of the national population.

Due to the many (sometimes conflicting) demands on the use of estuaries a need has arisen for the development of comprehensive management programs. Some of these are embraced by the Coastal Zone Management Act of 1972 (P. L. 92-583) and its amendments of 1976 (P. L. 94-370) which provide for State-Federal management, and Public Law 92-500 (October 18, 1972) entitled "Federal Water Pollution Control Act Amendment of 1972", which seeks to restore and maintain the chemical, physical and biological integrity of the Nation's waters.

The purpose of this report is to briefly describe some of the relationships between physical processes in estuaries and the fisheries resources found in them with a view toward trying to explain the probable consequences to fisheries of certain types of modifications to these estuarine processes. More specifically, the following material is considered: a) a brief review of estuarine productivity and dynamics from a fisheries point of view, b) estuarine modifications and the probable consequences on fish and invertebrate populations in estuaries, and c) some specific methodologies for assessing impacts on fisheries from certain physical modifications of estuaries.

Although much information has been summarized about estuaries in the context of estuarine management (Clark, 1977), relatively few details concerning specific assessments of possible effects of estuarine modifications are available. Most descriptions of the effects of estuarine modification are not quantitative, and little real understanding of the factors operating in the system is obtained from them.

ESTUARINE FISHERIES

According to McHugh (1976), of the total fish and shellfish landings of 3 billion kilograms in the U. S. approximately 2 billion kilograms (about 70 percent by weight) were estuarine dependent in some life history stage. He also pointed out that

Atlantic coast estuarine species increased in importance from north to south and on the Pacific coast they increased from south to north. In the Gulf of Mexico about 95 percent of the catch was considered to be estuarine. The monetary value of the fish catch seemed to be inversely related to their relative abundance; that is, the high value species are the less abundant.

A clear recognition of the significance of estuarine processes and a need for evaluating man-made changes was evident some time ago as indicated by Smith et al. (1966). Even at this time it was appreciated that not all changes in estuaries are detrimental to man's interests in fisheries, and that proper understanding of the estuarine environment would do much to provide rational answers to problems and questions concerning estuarine uses.

The total biological productivity of estuaries is usually very high, and they can support dense populations of animals, usually benthic invertebrates. Some of these invertebrates, such as oysters, clams and scallops are bivalve mollusks which are of high economic importance and they can support active commercial fisheries. These molluskan resources are of far greater significance on the Atlantic and Gulf coasts than on the Pacific coast where estuaries are less abundant. Indeed, most United States oysters and clams come from the Middle Atlantic Bight, which includes Chesapeake Bay, one of the largest and most productive estuaries in the world.

Where commercial bivalve mollusks are grown on the bottom, the average annual yield for large estuarine regions is 150 kg per hectare in the United States (Perkins, 1974). Highly sophisticated techniques of intensive bottom culture of mollusks may raise this to the order of 5000 or more kg per hectare. The potential for producing commercially important species, such as bivalves and crustaceans in estuaries is very high.

Walne (1972) has given some examples of the standing stock of commercially valuable shellfish in north European estuaries. He has estimated values of nearly 20,000 kg of dry meat per hectare for mussels, about 2000 kg per hectare of oysters and about 120 kg per hectare for cockles. A reasonably productive estuary in Britain was estimated to yield about 1000 kg live weight per hectare per year of either mussels or cockles.

With respect to finfish, the annual yields from Chesapeake Bay have been estimated to be on the order of 90 kg per hectare. Saila (1975) has summarized some of the available data on the annual yields of fish from lagoons and estuaries. The range of yields is about one order of magnitude on a global basis (50 to more than 500 kg/ha). McHugh (1966), in a review of estuarine fisheries management in the U. S., lists 39 common names of organisms (primarily fishes) identified as estuarine dependent in some life history stage. The list of scientific names is even larger, because the same common names were used for several species. It is apparent that a large part of the United States fish catch is based on species which appear to be estuarine dependent in some life history stage.

From the point of view of overall productivity estuarine waters are much higher than the open sea. Tyther (1969) has indicated that the average estuary is capable of producing organic matter at an annual rate of about 3 metric tons (dry weight) per hectare, which is 10-100 times higher than that of the open sea.

ESTUARINE DYNAMICS AND MODELS RELATED TO FISHERIES PROBLEMS

Saila (1975) has briefly described some aspects of estuarine dynamics related to fisheries. The usual definition of an estuary states that it is a semi-enclosed body of water having a free connection with the open sea and within which the sea water is measurably diluted with fresh water derived from land drainage. Officer (1976) provides a coordinated treatment of the physical oceanography of estuaries and related bodies of water with emphasis on the conceptual background, but with some application of the theory to particular estuarine problems on a global basis. This text contains the material necessary for the physical understanding of estuarine processes. Suffice it to state at this point that because circulation and mixing patterns are major aspects of estuarine behavior, hydrodynamics is fundamental to estuarine science. However, circulation and mixing are difficult to measure and analyze because they do not settle into a steady state, but instead they fluctuate in response to tides, fresh water flow variability, wind effects, irregular geometry of the estuary and density differences between fresh and salty water.

Estuarine modelling includes mathematical, statistical, hydraulic and electric analog models for simulating various processes in estuaries and for predicting the probable consequences of various forms of modifications. Ward and Espey (1971) have evaluated the use of various types of models for estuarine processes, and they provide details of the state-of-the-art of estuarine models.

From the above as well as from other information it appears that the modelling of hydrodynamic transport of a constituent in an estuary is better advanced than the modelling of estuarine kinetics. This suggests that many parameters (especially water quality parameters) behave in a highly coupled manner, and the specification of the reactions involved is inadequate. The apparent weakness of present hydrodynamic models in estuaries is their treatment of eddy diffusivities and dispersion coefficients, especially in the salinity intrusion regions. However, in relation to hydrodynamic modelling of estuaries, the modelling of estuarine biota is poorly developed indeed. Hall and Day (1977), Levin (1974) and Russell (1975) describe some aquatic ecosystem models of various levels of complexity. The summary remarks from an analysis of ecological models suggests that because of data, conceptual and computational limitations, aquatic ecosystem models must currently be viewed as useful predictors of system behavior only up to the biotic level of phytoplankton biomass. In particular, recent aquatic ecosystem models do not seem to be able to adequately deal with fish stocks (Russell, 1975, p. 8).

In summary, it appears that our knowledge of the hydrodynamics of estuaries is adequate to permit the construction of reasonable models of hydrodynamic transport processes. Models for reaction kinetics of chemical constituents (oxygen, pollutants) are less well developed but of some practical utility. Models of entire ecological systems in estuaries are the most poorly developed and have relatively little predictive value. However, it may be feasible to develop and utilize for prediction certain simulation models which combine hydrodynamic and biological models when restricted to single species. This type of study has been done for power plant assessment by Hess et al. (1975).

PHYSICAL MODIFICATIONS TO ESTUARIES AND POSSIBLE EFFECTS ON BIOTA

There are a number of physical modifications which may affect estuarine biota. For example, dams which are located somewhere in the freshwater inflow to an estuary, may impede migratory fish movements unless fishways are properly designed and installed at them. In addition these dams tend to reduce scouring and the freshwater influence from periods of high rainfall. These latter effects may be beneficial to some forms of sedentary organisms but it should be recognized that these dams also retain silt. Where the inflowing river is a major source of sediment the balance between deposition and erosion is disturbed so that the shoreline around the mouth of the estuary may recede. This has already happened in the Nile Delta since the construction of the high Aswan dam with consequent adverse effects on some biota. Conversely, a barrier or constriction of any kind across the lower reaches of a typical estuary will tend to accumulate sediment to the seaward side, because this is the main direction of its origin, and tidal scouring will be reduced. In either case fisheries resources will be modified. Indeed, in the case of damming the Nile River, significant fisheries off the mouth of the river have been adversely affected by the water control resulting from dam construction.

The lower reaches of many estuaries used as commercial waterways have channels banked by retaining walls or breakwaters. These modifications often produce hard substrates on what were previously expanses of silt or soft sediments. The impact on previously established soft bottom communities is obvious.

Dredging activity in estuaries designed to create navigation channels, turning basins, harbors, marinas, or excavation to obtain fill materials has created adverse environmental effects under some circumstances. These adverse effects include: a) increased turbidity, b) sediment build-up, c) reduced oxygen, d) disruption and removal of productive estuarine bottom and the life it contains, e) creation of stagnant deep water areas, f) disruption of estuarine circulation, and g) increased or decreased up-stream intrusion of salt and sediments.

Increased turbidity from physical modifications may alter water quality and primary production (Table 1). It appears that the effects on primary production are of very limited duration but some problems related to sorption and/or release of polluting materials from disturbed sediments remain to be further resolved. Some specialized marine habitats, such as grass communities, kelp beds (North and Schaefer, 1964) and coral reefs (Griffin, 1974) may be especially vulnerable to the effects of high concentrations of suspended particles. In addition the behavioral responses of important fish species to increased turbidity remain to be examined further. This is an important problem because some segments of the fishing industry have alleged that their catches have been adversely affected by dredging activity and the resulting turbidity (Sissenwine and Saila, 1974).

Tables 2 and 3 summarize some of the effects of turbidity on various life history stages of important invertebrates and vertebrates in estuaries. Examination of these tables indicates that:

a) Exposure to contaminated sediments decreases survival time in contrast to equal exposure to similar amounts of uncontaminated sediments.

b) Adult estuarine fishes, crustaceans and mollusks seem to survive significant increases in turbidity over short periods.

c) The effects of turbidity-producing materials on the development and growth of early life history stages (eggs and larvae) of estuarine bivalve mollusks are directly related to the concentration.

d) Turbidity affects fishes directly and indirectly. The direct effects are the greatest on the early life history stages. Some species and life history stages, such as striped bass fingerlings and certain filter-feeding species, are especially vulnerable to high concentrations of suspended materials. In general early life history stages are more vulnerable than adults, and pelagic species more vulnerable than demersal species.

e) The possibility for complex non-additive interactions among concentrations of suspended solids, dissolved oxygen and temperature which may influence survival of invertebrates and vertebrates has not been sufficiently well recognized in studies to date.

It seems clear from a review of the papers dealing with recovery and recolonization (Table 4) that these factors are very site specific. A high degree of variability in the time to recovery and the nature of the recolonization process are evident in this material. Saila (1976) and Papadakis and Saila (1976) have reviewed benthic recolonization processes and proposed models for predictive purposes. These models are basic species equilibrium mathematical models originally developed for island biogeography. They are based on constant coefficients relating to the rate of immigration and extinction of organisms and they predict the new equilibrium number of species and the time required for equilibrium to be reached.

This brief review of physical modifications to estuaries and the possible effects of these modifications on estuarine biota suggests that mathematical models for predicting the consequences of specific physical modifications on estuarine biota may be useful under some circumstances but the great variability in the estuarine systems and the variable responses of organisms to these modifications must be recognized.

In view of the above it is suggested that statistical models consisting of well designed and executed field experiments which focus experimental objectives on population and community effects may sometimes be reasonable alternatives to strictly mathematical models for predicting the outcome of certain types of modifications of the physical environment of estuaries.

Details of one statistical model will be illustrated in the next section.

Table 1. Turbidity Effects on Water Quality and Primary Production in Estuaries

WATER QUALITY INDICATOR AND ACTIVITY	EFFECTS ON CONCENTRATION	REFERENCES(S)
Metals		
dissolved heavy metals	little change	May, 1973
trace metals	no change in overlying water	Saila, et al., 1972
soluble iron, copper, lead	initial increase above ambient, followed by a decrease to below ambient; return to ambient in several days	Windom, 1972, 1973
Oxygen (DO)		
freshly dredged sediments	depressions to 50-70% of ambient, lasting 3-4 minutes, and only near the bottom	U.S. Army Engineer District, San Francisco, 1973
continuous dredging and accumulation of waste discharges	levels were 16-83 percent lower during dredging than during non-dredging periods	Brown and Clark, 1968
dredged material disposal	reduction to 0.1 ppm during disposal, rapidly returned to 7-8 ppm (normal concentrations)	U.S. Fish and Wildl. Serv., 1970
Nutrients		
total phosphorus	1,000 times greater than background, but shortlived (near dredge disposal pipe)	Biggs, 1968
total nitrogen	50 times greater than background, but shortlived (near dredge disposal pipe)	"
water quality (nutrients?)	no change at disposal sites	May, 1973
"	"	O'Neal and Sceva, 1971
"	"	Windom, 1972, 1973
Primary Production		
lab experiments	3 phytoplankton spp. (Monochripsis lutheri, Chlorella sp. and Nannochloris sp.) showed 50-90% reduction in carbon assimilation with increase in suspended matter	Sherk et al., 1974
field experiments	limited reduction in primary production near dredge channel	Ingle, 1952
	only short term effects on phytoplankton; no overall reduction in primary production	Chesapeake Biological Lab, 1970
	no significant changes in primary production due to dredging - although light is decreased, nutrients were made available	Odum and Wilson, 1962; Taylor and Saloman, 1968

field experiments (cont'd.)	gross primary production increased 3-13 X due to increased nutrients	Sabba Rao, 1973
	primary production of rooted aquatics in low energy area reduced	Kaplan *et al.*, 1974 Ingle, 1952

Table 2. Turbidity Effects on Selected Invertebrates in Estuaries

SPECIES	LIFE HISTORY STAGE	CONCENTRATIONS OF PARTICULATE	EFFECT	REFERENCES
Oyster	adult	≤ 700 ppm (lab study)	minimal	Mackin, 1961
Oyster	adult	100-700 ppm mud	no effect	Mackin, 1956
Oyster	adult	up to 4 grams/liter silt	94% reduction in filtering rate	Loosanoff and Tommers, 1948
M. edulis (mussel)	adult	1 gram/liter bentonite	no reduction in pumping rate	Chiba and Oshima, 1957
Snails:				
- *Lymnaea natalensis*	adult	190 ppm and 360 ppm suspended solids	normal egg development at both concentrations	Harrison and Farima, 1965
- *Bulinus globosus*	"	"	laid eggs in 190 ppm but not in 360 ppm	"
- *Bromphalaria pfeifferia*	"	"	high mortality levels at both concentrations	"
Eurytemora affinis	adult	250 mg/liter or more	reduced ingestion	Sherk *et al.*, 1976
Acartia tonsa	adult	50 mg/liter or more	reduced ingestion	"
Lobsters	adult	50 grams/liter (47,200 ppm turbidity) Kaolin	no mortality	Saila *et al.*, 1968
Lobsters	adult	1,600 ppm harbor sediment	no mortality	"
Venus mercenaria	eggs		development retarded by smaller concentrations of silt than by larger concentrations of clay	Davis, 1960
	larvae		larvae were more tolerant of silt than clay which clogged the digestive tract	"

oyster, mussels, barnacles	adults	death of sessile or attached forms by burial	Lunz, G. R., 1938, 1942 Wilson, 1950 Carriker, 1967 Saila, *et al.*, 1972 Rose, 1973 NAVOCEANO, 1973
Amphipod: - *Metaceradocus occidentalis*	adult	elimination of shell fragments by dredging in Newport Bay, California would destroy their habitat	Barnard and Reish, 1959
Polychaete: - *Scyphoproctus oculatus*	adult		
Oysters	adults		Lunz, 1938
Filter feeders (general)		suspended load stress affects 1) efficiency of filtering mechanisms, 2) rate of water transport 3) energy available for maintenance specific effects- 1) abrasion of gill filaments 2) clogging of gills 3) impaired respiration, feeding and excretion 4) reduced pumping rate 5) retarded egg development 6) reduced growth and survival of larvae 7) decrease in productivity	

Table 3. Turbidity Effects in Selected Vertebrates in Estuaries

SPECIES	LIFE HISTORY STAGE	CONCENTRATION OF PARTICULATE	EFFECTS	REFERENCE
Menidia menidia	adult	0.58 gram/liter Fuller's earth	24 hour LC_{10}	Sherk *et al.*, 1971, 1972, 1974
Fundulus heteroclitus	adult	24.5 grams/liter Fuller's earth	24 hour LC_{10}	"
Fundulus majalis	adults	less than 1.25 grams/liter Fuller's earth for 5 days	increased hematocrit	"
Trinectes maculatus	adults	"	increased hematocrit and increased liver glycogen depletion rate	"
Parophrys vetulus (English sole)	adults	117 grams/liter bentonite for 10 days	80% mortality	Peddicord *et al.*, 1975
Cymatogaster aggregata (shiner perch)	adults	14 grams/liter bentonite for 26 hours	almost 100% mortality	"
yellow perch, white perch striped bass, alewife	eggs	100-500 mg/liter natural, fine-grain sediment	increased hatching times (several hours), no increased mortality	Schubel and Wang, 1973
striped bass	larvae	3,411 mg/liter suspended sediment	2 day LC_{50}	Morgan *et al.*, 1973
white perch	larvae	2,679 mg/liter suspended sediment	2 day LC_{50}	"
striped bass	juvenile		21 day exposure	Peddicord and McFarlane, 1978

fish	juveniles and adults	Lab study - 1. Particles coat the respiratory epithelium or inhibit water flow between secondary lamellae. 2. Hematological changes due to reduced gas exchange. 3. Abrasion of epithelium and gill tissue. Note: bottom feeding fish were more tolerant than filter feeders; adults more tolerant than juveniles	Sherk *et al.*, 1974
fish	eggs	destruction of demersal eggs	Huet, 1965

Table 4. Recovery and Recolonization of Benthic Organisms in Disturbed Estuarine Sediments

LOCATION	SPECIES	DESCRIPTION	REFERENCES
lower Chesapeake Bay	*Nepthys incisa*	Only temporary effect on infauna in general	Harrison *et al*., 1964
	Ensis directus	Resettlement by *Nepthys* and *Ensis*	
Coos Bay, Oregon		No change in infaunal numbers seen 2 weeks later at both dredge or spoil sites - pollution resistant benthic populations were present prior to the operation.	Slotta *et al*., 1973
		Increase in diversity in spoils area - possibly due to homogenization of sediments during disposal.	
shallow bay at Goose Creek, NY		Great reduction in species, individuals, biomass in the dredged channel - no recovery within 11 months.	Kaplan *et al*., 1974, 1975
		Spp. composition changed - Increase in *Tellina agilis*, *Lyonsia hyalina*, and *Mulinia lateralis*	
		Decrease in *Clymenella torquata* and *Notomastus latericeus*	
NY Harbor		Great decrease or complete destruction of benthic communities. Pollution tolerant species (nematodes, capitellids) predominate. *Note*: continued dumping over years.	Gross, 1970(b)
Boca Ciega Bay, Florida		Lack of polychaetes, mollusks, blue crabs and pink shrimp in dredged area due to increase in silt-clay and organic C.	Taylor and Saloman (1968)

Long Island Sound	Group 1 - colonizers - arrive early, explode, high mortality - - Streblospio benedicti - Capitella capitata - Ampelisca abdita Group 2 - intermediate between 1 and 3 - Nucula annulata - Telina agilis Group 3 - may appear early, but maintain low constant densities over long periods - Nepthys incisa - Ensis directus	From 10/73-4/74, 6 X 10^5 m^3 of channel sediment was dumped in 20 meters of water. Reference station was 5.6 km NW of dump in 15 m of water. Numbers of individuals - Late fall 1974 - numbers were 2-10 X higher than reference station. Spring 1976 - numbers decreased to 4X lower than reference station.	Rhoads, D. C., P. L. Mc-Call and J. Y. Yingst, 1978
Tillamook Bay, Oregon	Crustaceans and bivalves were replaced by oligochaetes and nematodes immediately after dredging		Slotta et al., 1974
Rhode Island (lab study)	a) Nepthys incisa and Mulinia lateralis b) Streblospio benedicti c) Macoma, Yoldia, Nucula	a) could reach the surface thru 21 cm. of of sediment b) " " " " " 6 cm. c) can move horizontally from the area	Saila et al., 1972
Delaware Bay		No significant reduction in densities - maybe due to high flushing rate in area	Leathem et al., 1973
Chesapeake Bay		71% decrease in numbers immediately following disposal. No difference in numbers and diversity 1 1/2 years later. No difference in dredged channel - possibly because of sediment instability. 50% increase in biomass in deeper channels soon after dredging due to recruitment of polychaete Scolecolepides verides.	Chesapeake Biological Lab, 1970

SUGGESTED METHOD FOR ASSESSING FISHERIES IMPACTS

The purpose of this section is to briefly discuss and illustrate the use of a relatively straightforward and simple statistical model to detect the impact (if any) of physical modifications to estuaries on fisheries for vertebrates and/or invertebrates. Impact is defined as the detection, in a statistical sense, of a change in the composition of the biota.

Green (1979) has recently provided an excellent review of statistical models and methodologies for assessing environmental impacts. His review includes multivariate methods which can be used to effectively treat several variables simultaneously. The methodology suggested herein is a simpler approach to the analysis of multivariate data, which is believed to be useful for specific problems of assessing variation in community composition as related to various types of environmental perturbations. This approach for impact assessment involves formulating a null hypothesis which is the simplest one possible consistent with the evidence, with the fewest possible unknowable explanatory factors. That is, the purpose generates the question, and the null hypothesis is the simplest answer to that question, stated in a way that is testable and falsifiable.

The specific model to be described is a simplification of a method first described by Krumbein and Tukey (1956) for the simultaneous evaluation of a number of mineral or chemical constituents in rock samples. The simultaneous evaluation is accomplished for certain studies involving spatial variability by expressing several response variables as proportions or percentages, where the total adds up to 100 percent.

In order to apply statistical techniques to any experimental design and analysis it is necessary to have a clear and precise description of the experiment which identifies the measurements and the design parameters available, as well as potential sources of uncertainty and error. Such a description is provided by a model which explicitly relates the observations of interest (dependent variables) to various factors (independent variables) and error sources. An example of a practical experimental model is provided by the following equation, which decomposes a single observation (X_{ikm}) into several parts:

$$X_{ikm} = \mu + A_i + S_{ik} + T_m + AT_{im} + ST_{ikm} \tag{1}$$

where:

X_{ikm} = observation at area i, sample k and type species, m

μ = mean of all observations

A_i = effect of the ith area thought to influence X_{ikm}

S_{ik} = effect of the kth sample in area i thought to influence X_{ikm}

T_m = effect of the mth type species thought to influence X_{ikm}

AT_{im} = first order interaction term between areas and type species

ST_{ikm} = first order interaction term between samples from each area and type species within each sample

The index j is omitted in this example to permit easier comparison with the previously described model of Krumbein and Tukey.

The analysis of variance table specified by equation (1) is partitioned as follows:

Source of Variation	Degrees of Freedom	Expectation of Mean Square	
A_i	$i - 1$	$\sigma_w^2 + n_t \sigma^2_{S:A} + n_s n_t \sigma^2_A$	*
S_{ik}	$i\ (k-1)$	$\sigma_w^2 + n_t \sigma^2_{S:A}$	*
T_m	$m - 1$	$\sigma_w^2 + n_a n_s \sigma^2_T$	*
AT_{im}	$(i-1)\ (m-1)$	$\sigma_w^2 + n_s \sigma^2_{AT}$	
ST_{ikm}	$i\ (k-1)\ (m-1)$	σ_w^2	

*NOT TESTED

where n_a = number of areas, n_t = number of species, and n_s = number of samples from each area.

In the context of an actual experiment, the observation X_{ikm} would represent the transformed count (percentage of individuals per unit area) of the mth species in the kth sample located in area i. If we consider a specific experiment, say with two areas (one of which has been dredged and the other which is a control) with three samples from each area and with five types of species to be compared, then the total degrees of freedom for this experiment is (2 X 3 X 5) - 1 = 29. The total degrees of freedom can be partitioned according to the degrees of freedom column of the analysis of variance table given above as $A_i = 1$, $S_{ik} = 4$, $T_m = 4$, $AT_{im} = 4$ and $ST_{ikm} = 16$. This sum results in the 29 degrees of freedom indicated.

In the type of analysis described in this report the mean squares for main effects are of no interest, and the justification for not testing for them is given by Krumbein and Tukey (1956). Only the interaction mean squares between species and sampling levels are of interest. In the above-mentioned model interaction mean squares are calculated for areas X types of species and replicate samples within areas X types of species interactions. Using these mean square values the F ratio (variance ratio) of area X species types to replicate X species types can be used to test the hypothesis that variability in species composition between areas (dredged versus undredged) is not significantly greater than that within replicates at each area. Because temporal variability among biota is often found, it is assumed that the replicates and area compositions are made at nearly the same time.

The arrangement of the data for analysis is shown in Figure 1 using the specific example previously cited. The items in each column are first added to obtain the summation over m. The dots in Figure 1 always represent summation over the index. These summations are next combined over replicates (1 + 2 + 3) to obtain the summation over km, and these are finally combined over areas (A and B) to obtain the summation over ikm. A second condensed table is prepared by adding replicates (1 + 2 + 3) for each type species, to obtain a summation by type species, and these are combined by areas to obtain the summation over ik. By carrying the totals to the right, the summation over ikm is obtained.

The manner of computing the raw sums of squares is shown in the bottom of Figure 1. As indicated, all the original items and all the sums are squared to obtain the six numbers designated as I to III and A to C. These numbers are then used in the analysis of variance shown in Figure 2. The final sums of squares are obtained by making the subtractions indicated in the second column of Figure 2. The degrees of freedom are distributed as shown in the third column.

In conducting the analysis the first three mean squares are not used. The F test is made by taking the ratio of the last two mean squares in the last column (i.e. areas X types divided by samples within areas X types). The observed variance ratio is used

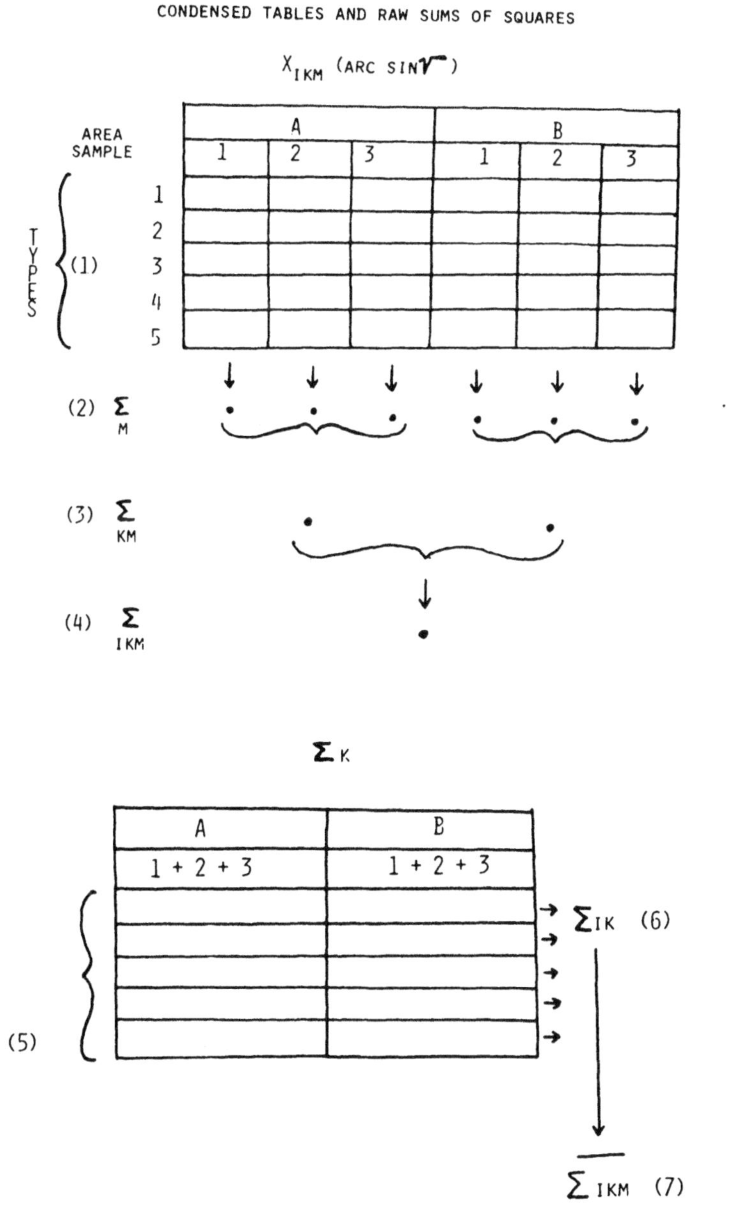

Figure 1. Schematic diagram for condensation of data and computation of mean squares for the multivariate model. The case illustrated is for two areas, three samples per area and five species types described in the text. The dots in the figure refer to sums over their respective indexes.

III: $\sum_{IK} (\sum_{M} X_{IKM})^2 / M$ SUM OF SQUARES OF ALL ITEMS (2), DIVIDED BY M.	C: $\sum_{IKM} (X_{IKM})^2$ SUM OF SQUARES OF ALL ITEMS IN (1).
II: $\sum_{I} (\sum_{KM} X_{IKM})^2 / KM$ SUM OF SQUARES OF ALL ITEMS (3) DIVIDED BY KM.	B: $\sum_{IM} (\sum_{K} X_{IKM})^2 / K$ SUM OF SQUARES OF ALL ITEMS IN (5) DIVIDED BY K.
I: $(\sum_{IKM} X_{IKM})^2 / IKM$ SQUARE OF GRAND TOTAL (7) DIVIDED BY N = IKM.	A: $\sum_{M} (\sum_{IK} X_{IKM})^2 / IK$ SUM OF SQUARES OF ALL ITEMS (6) DIVIDED BY IK.

Figure 1. (Continued)

ANALYSIS OF VARIANCE

SOURCE	SUM OF SQUARES	DEGREES OF FREEDOM	MEAN SQUARE	
BETWEEN AREAS	$II - I = SS_1$	$I - 1$	$SS_1 / I-1$	NOT USED
BETWEEN SAMPLES WITHIN AREAS	$III - II = SS_2$	$I (K-1)$	$SS_2 / I (K-1)$	
BETWEEN TYPES (SPECIES)	$A - I = SS_3$	$M - 1$	$SS_3 / (M-1)$	
AREAS X TYPES	$I + B - (II + A) = SS_4$	$(I-1) (M-1)$	$SS_4 / (I-1) (M-1)$	
SAMPLES WITHIN AREAS X TYPES	$(II + C) - (III + B) = SS_5$	$I (K-1) (M-1)$	$SS_5 / I (K-1) (M-1)$	

Figure 2. Form of multivariate analysis of variance table for the two-way nested design example. Compare with Figure 1 for the subtractions indicated in the second column.

to test the following hypothesis: The variation in over-all species composition between areas (treated versus untreated) is not significantly greater than the variation in species composition between samples in a given area.

An extension to the model described herein (from a two level to a three level nested sampling plan) is given by Krumbein and Tukey (1956). Further extensions to include additional factors are also possible, but are not considered.

CONCLUSIONS

(1) The effects of physical modifications to estuaries may induce several kinds of ecological response by organisms depending on a variety of factors.

(2) The hydrodynamic and reaction kinetic models of estuaries developed to date are of considerably higher predictive value than present ecosystem models of estuaries.

(3) Statistical models are recommended as interim solutions to assess impacts from estuarine modifications. It seems possible to design field experiments which yield ecologically and statistically significant information about the community effects of estuarine modifications.

(4) An example of a statistical model designed specifically for estuarine impact assessment is provided and explained.

REFERENCES

Barnard, J. L. and D. J. Reish. 1959. Ecology of amphipoda-polychaeta of Newport Bay, California. Allan Hancock Foundation Occasional Paper No. 21.

Biggs, R. B. 1968. Environmental effects of overboard spoil disposal. Journal of Sanitary Engineering Division, American Society Civil Engineers 94: 477-487.

Brown, C. L. and R. Clark. 1968. Observations on dredging and dissolved oxygen in a tidal waterway. Water Resources Res. 4(6): 1381-1384.

Carriker, M. R. 1967. Ecology of estuarine benthic invertebrates: a perspective. pp. 442-487 In G. H. Lauff (ed.), Estuaries. American Association Advancement of Science Publication No. 83. Washington, D. C.

Chesapeake Biological Laboratory. 1970. Gross physical and biological effects of overboard spoil disposal in upper Chesapeake Bay. NRI Special Report No. 3, University of Maryland, Solomons, Maryland. 66p.

Chiba, K. and Y. Oshima. 1957. Effect of suspended particles on pumping and feeding of marine bivalves, especially the Japanese little neck clam (in Japanese, English summary). Bulletin Japanese Society Scientific Fisheries 23: 348-354.

Clark, J. 1977. Coastal Ecosystem Management. J. Wiley & Sons, New York. 927 pp.

Davis, H. C. 1960. Effects of turbidity producing materials in sea water on eggs and larvae of the clam (Venus (Mercenaria) mercenaria). Biological Bulletin 118(1): 48-54.

Green, R. H. 1979. Sampling design and statistical methods for environmental biologists. J. Wiley & Sons, New York. 257 pp.

Griffin, G. E. 1974. Dredging in the Florida Keys, case history of a typical dredge-fill project in the northern Florida Keys - effects on water clarity, sedimentation rates, and biota. Publ. No. 33, Harbor Branch Foundation, Ft. Pierce, Florida.

Gross, M. G. 1970. Analyses of dredged wastes, fly ash and waste chemicals - New York metropolitan region. State University of New York, Marine Science Research Center, Technical Report No. 7. 34 p. (Abstract only).

Hall, C. A. S. and J. W. Day. 1977. Ecosystem modelling in theory and practice. John Wiley & Sons. 684 pp.

Harrison, A. D. and T. D. W. Farina. 1965. A naturally turbid water with deleterious effects on the egg capsules of planorbid snails. Annals Tropical Medical Parasitology 59: 327-330.

Harrison, W., M. P. Lynch and A. G. Altschaeffl. 1964. Sediments of lower Chesapeake Bay, with emphasis on mass properties. Journal Sedimentology and Petrology 34(4): 727-755.

Hess, K. W., M. P. Sissenwine and S. B. Saila. 1975. Simulating the impact of the entrainment of winter flounder larvae. pp. 2-29 In S. B. Saila (ed.), Fisheries and Energy Production. D. C. Heath and Co., Lexington, Mass.

Huet, M. 1965. Water quality criteria for fish life. pp. 160-167 In C. Tarzwell (ed.), Biological Problems in Water Pollution. U. S. Public Health Service Publication No. 999-WP-2S.

Ingle, R. M. 1952. Studies on the effect of dredging operations upon fish and shellfish. Tech. Ser. No. 5, Florida State Board of Conservation, St. Petersburg, Florida. 26 p.

Kaplan, E. H., J. R. Welker and M. G. Kraus. 1974. Some effects of dredging on populations of macrobenthic organisms. U. S. National Marine Fisheries Service, Fishery Bulletin 72(2): 445-480.

Kaplan, E. H., R. R. Walker, M. G. Kraus and S. McCourt. 1975. Some factors affecting the colonization of a dredged channel. Marine Biology 32: 193-204.

Krumbein, W. C. and J. W. Tukey. 1956. Multivariate analysis of minerologic, lithologic, and chemical composition of rock bodies. Journal of Sedimentology and Petrology 26: 322-337.

Leathem, W., P. Kinner, D. Mawer, R. Biggs and W. Treasure. 1973. Effect of spoil disposal on benthic invertebrates. Marine Pollution Bulletin 4(8): 122-125.

Levin, S. A. (ed.) 1974. Ecosystem analysis and prediction. SIAMS-SIMS Conference Proceedings, Alta, Utah. 337 pp.

Loosanoff, V. L. and F. D. Tommers. 1948. Effect of suspended silt and other substances on rate of feeding of oysters. Science 197: 69-70.

Lunz, G. R. 1938. Oyster culture with references to dredging operations in South Carolina and the effects of flooding of the Santee River in April 1936 on oysters in the Cape Romain area of South Carolina (Part II). March, 1952. U. S. Army Engineering District, Charleston, C.E., Charleston, S.C. 135 p.

Lunz, G. R. 1942. Investigation of the effects of dredging on oyster leases in Duval County, Florida. In Handbook of Oyster Survey, Intracoastal Waterway Cumberland Sound to St. Johns River. Special Report U. S. Army Corps of Engineers, Jacksonville, Florida.

Mackin, J. G. 1956. Studies on the effect of suspensions of mud in sea water on oysters. Report No. 19, Texas A & M Research Foundation Project 23, College Station, Texas.

Mackin, J. G. 1961. Canal dredging and silting in Louisianna bays. Publ. Institute of Marine Science, University of Texas 7: 262-314.

May, E. G. 1973. Environmental effects of hydraulic dredging in estuaries. Alabama Marine Resources Bulletin No. 9: 1-85.

McHugh, J. L. 1966. Management of estuarine fisheries. In A Symposium on Estuaries. American Fisheries Society Special Publication No. 3, pp. 133-154.

McHugh, J. L. 1976. Estuarine fisheries: are they doomed? In M. Wiley (ed.), Estuarine Processes, Vol. 2, pp. 15-25.

Morgan, R. P. II, V. J. Rasin and L. A. Noe. 1973. Effects of suspended sediments on the development of eggs and larvae of striped bass and white perch. Ref. No. 73-110. University of Maryland Natural Resources Institute, College Park, Maryland.

NAVOCEANO. 1973. Preliminary report - environmental investigation of a dredge spoil disposal site near New London, Connecticut. Naval Oceanography Division, Washington, D. C., NAVOCEANO Technical Note No. 7300-3-73. 87 p.

North, W. J. and M. B. Schaefer. 1964. An investigation of the effects of discharged waste on kelp. Publication No. 26, Resources Agency of California, State Water Quality Control Board, Sacramento, California.

Officer, C. B. 1976. Physical oceanography of estuaries (and associated coastal waters). J. Wiley & Sons, New York, 465 pp.

Odum, H. T. and R. F. Wilson. 1962. Further studies on reaeration and metabolism of Texas bays, 1958-1960. Publ. Institute of Marine Science, University of Texas 8: 23-55.

O'Neal, G. and J. Sceva. 1971. The effects of dredging on water quality in the Northwest. Environmental Protection Agency, Region X, Seattle, Washington.

Peddicord, R. K., V. A. MacFarland, D. P. Belfiori and T. F. Byrd. 1975. Dredge disposal study, San Francisco Bay and estuary; Appendix G, Physical impact, effects of suspended solids on San Francisco Bay organisms. U. S. Army Engineer District, San Francisco, C. E., San Francisco, California.

Papadakis, J. S. and S. B. Saila. 1976. Benthic colonization processes - a review and a proposed new model. pp. 377-382. In Middle Atlantic Continental Shelf and the New York Bight. ASLO Special Symposium, Vol. 2, M. Grant Gross (ed.).

Pritchard, D. W. 1952. The physical hydrography of estuaries and some applications to biological problems. Trans. 16th North Amer. Wildlife Conf.: 368-376.

Rhoads, D. C., P. L. McCall, and J. Y. Yingst. 1978. Disturbance and production on the estuarine seafloor. American Scientist 66(5): 577-586.

Rose, C. D. 1973. Mortality of market-sized oysters (Crassostrea virginica) in the vicinity of a dredging operation. Chesapeake Science 14(2): 135-138.

Russell, C. S. (ed.) 1975. Ecological modeling in a resource management framework. Resources for the Future, Inc. RFF Working Paper QE-1, 394 pp.

Rhyther, J. H. 1969. The potential of the estuary for shellfish production. Proc. National Shellfisheries Assoc. 59: 18-22.

Perkins, E. J. 1974. The biology of estuaries and coastal waters. Academic Press, London, 678 pp.

Sabba Rao, D. V. 1975. Effects of environmental perturbations on short-term phytoplankton production off Lawson's Bay, a tropical coastal embayment. Hydrobiologia 43(1&2): 77-91.

Saila, S. B. 1975. Some aspects of fish production and cropping in estuarine systems. pp. 473-493 In Estuarine Research, Vol. 1, L. E. Cronin (ed.).

Saila, S. B. 1976. Sedimentation and food resources: animal-sediment relationships. pp. 479-492 In Marine Sediment Transport and Environmental Management, D. J. Stanley and D. J. P. Swift (eds.), John Wiley & Sons.

Saila, S. B., S. D. Pratt and T. T. Polgar. 1972. Dredge spoil disposal in Rhode Island Sound. Univ. Rhode Island, Marine Technical Report No. 2, 48 p.

Saila, S. B., T. T. Polgar and B. A. Rogers. 1968. Results of studies related to dredged sediment dumping in Rhode Island Sound. Proc. Ann. Northeastern Reg. Antipollution Conf. 22-24 July 1968. pp. 71-80.

Schubel, J. R. and J. C. S. Wang. 1973. The effects of suspended sediments on the hatching success of Perca flavescens (yellow perch), Morone americana (white perch), Morone saxatilis (striped bass), and Alosa pseudoharengus (alewife) eggs. Special Report 30, Chesapeake Bay Institute, Johns Hopkins University, Baltimore, Maryland.

Sherk, J. A., Jr. 1971. The effects of suspended and deposited sediments on estuarine organisms: literature summary and research needs. Chesapeake Biological Laboratory, Solomons, Maryland, Contribution No. 443. 73 pp.

Sherk, J. A., Jr., J. M. O'Conner and D. A. Neumann. 1972. Effects of suspended and deposited sediments on estuarine organisms, Phase II, Ref. No. 72-9E. Natural Resources Institute, University of Maryland Chesapeake Biological Laboratory, Solomons, Maryland.

Sherk, J. A., J. M. O'Conner, D. A. Neumann, R. D. Prince and K. V. Wood. 1974. Effects of suspended and deposited sediments on estuarine organisms. Phase II. Final Report No. 74-20, University of Maryland, Natural Resources Institute, Prince Frederick, Maryland. 259 pp.

Sherk, J. A., Jr., J. M. O'Conner and D. A. Neumann. 1976. Effects of suspended solids on selected estuarine plankton. Miscellaneous Report No. 76-1, U. S. Army Coastal Engineering Research Center, CE, Fort Belvoir, Virginia.

Singer, S. F. 1969. Federal interest in estuarine zone builds. Environmental Science and Technology 3(2): 124-131.

Sissenwine, M. P. and S. B. Saila. 1974. Rhode Island Sound dredge spoil disposal and trends in the floating trap fishery. Trans. Amer. Fish. Soc. 103(3): 498-505.

Slotta, L. S., D. R. Hancock, K. J. Williamson, and C. K. Sollitt. 1974. Effects of shoal removal by propellor wash, December 1973, Tillamook Bay, Oregon. Oregon State University, Corvallis, Oregon. 155 pp.

Slotta, L. S., C. K. Sollitt, D. A. Bella, D. R. Hancock, J. E. McCauley and R. Parr. 1973. Effects of hopper dredging and in channel spoiling in Coos Bay, Oregon. Oregon State University, Corvallis, Oregon. 141 p.

Taylor, J. L. and C. H. Saloman. 1968. Some effects of hydraulic dredging and coastal development in Boca Ciago Bay, Florida. U. S. Fish Wildlife Service, Fishery Bulletin 67(2): 205-241.

Smith, R. R., A. H. Swartz and W. H. Massmass. 1966. A symposium on estuarine fisheries. American Fisheries Society Special Publication No. 3: 154 p.

U. S. Army Engineer District, San Francisco, CE. 1973. Effects of dredged materials on dissolved oxygen in receiving water. Prepared on contract No. DACW07-73-C-0051, San Francisco, California.

U. S. Fish and Wildlife Service. 1970. Effects on fish resources of dredging and spoil disposal in San Francisco and San Pablo Bays, California. Unnumbered Special Report, Nov. 1970, Washington, D. C.

Walne, P. R. 1972. The importance of estuaries to commercial fisheries. The Estuarine Environment, R. S. R. Barnes and J. Greed (eds.), Applied Science, London pp. 107-118.

Ward, G. H., Jr. and W. H. Espey, Jr. (eds.) . 1971. Estuarine modelling: an assessment. Water Pollution Control Research Series, Rept. No. 16070 DZY. Stock No. 5501-0129. U. S. Government Printing Office, Washington, D. C.

Wilson, W. 1950. The effects of sedimentation due to dredging operations on oysters in Copano Bay, Texas. Annual Report, Marine Laboratory, Texas Game, Fish, and Oyster Commission, 1948-1949.

Windom, H. L. 1972. Environmental aspects of dredging in estuaries. Journal Waterways Harbor Coastal Engineering Division, American Society Civil Engineers 98: 475-487.

Windom, H. L. 1973. Water quality aspects of dredging and dredge spoil disposal in estuarine environments. Skidaway Institute of Oceanography, Savannah, Georgia.

COMBINED FIELD-LABORATORY METHOD FOR CHRONIC IMPACT DETECTION IN MARINE ORGANISMS AND ITS APPLICATION TO DREDGED MATERIAL DISPOSAL

Willis E. Pequegnat[1], Roger R. Fay[1] and T. A. Wastler[2]

[1]TerEco Corporation, College Station, Texas

[2]Chief, Marine Protection Branch, USEPA, Washington, D. C.

INTRODUCTION

General

One of the difficult problems facing scientists who are concerned with the environmental effects of the disposal of dredged material and industrial wastes into the aquatic environment is determining whether or not given waste components elicit chronic deteriorative responses in important species of organisms (Pequegnat et al., 1978a). The full importance of such low-level, nonlethal effects is not known, but it is suspected that repeated exposures may result in ecosystem changes equally as important as those caused by more easily determinable acute effects. Such considerations are particularly important to the aquatic environment, where dumped pollutants may be quickly diluted to legal nonlethal concentrations, but may still bring forth cumulative chronic response patterns.

Bioaccumulation and Biomagnification

Equally difficult to monitor in the field are the related phenomena of metabolic bioaccumulation and trophic level biomagnification that may develop in response to low levels of pollutants found in the water (Pequegnat, 1978b). Initially the organism takes up and stores the dilute pollutant in its tissues (e.g., liver and/or fat) until it reaches a level above that in the environment. The point of concern here is whether or not the organism can subsequently rid itself of the pollutant. Such depuration is relatively easy to observe in the laboratory, but up to now has been difficult to demonstrate in the field.

Trophic level biomagnification involves a marked jump in tissue concentration of a pollutant at each step in a food chain from, say, aquatic plants, to herbivore, to first carnivore, etc., perhaps eventually to man. The chief concern here is than an acceptable level of a pollutant in the water may attain sufficiently high concentrations by food-chain transfer that it has deleterious effects on the top carnivores or renders food organisms at this trophic level unfit for human consumption (Pequegnat et al., 1979a).

From the above considerations, it is evident that chronic responses, bioaccumulation, and trophic level augmentation are related in that they could develop from subacute concentrations of a pollutant or contaminant in the water or sediments.

Problems with Detection of Early Chronic Responses

Part of the problem in discerning these subtle chronic responses is that they take time to develop and to reach detectable levels. When such work has been attempted in the field, it has been impossible to demonstrate in a mobile environment that the organisms sampled for testing were actually previously exposed to waste substances. On the other hand, when the work is done in the laboratory, there is always some doubt that the findings can be extrapolated to the natural environment. Laboratory bioassay analyses of dredged material have increased substantially since appearance of the EPA 1977 Regulations and Criteria in the Federal Register, Vol. 42, No. 7, and the 1977 joint issuance by EPA and CE of the "Green Manual" on bioassays (EPA/CE, 1977). Until very recently, most such tests were handled as static bioassays, a technique that has many shortcomings. Today, some laboratories have shifted to flow-through systems. Unfortunately, such systems must be more elaborate and are thus more costly. Yet doubts still remain as to the transferability of the results to the field. It seems to us that well-devised field bioassays coupled with laboratory analyses could provide more realistic information on the impacts of the disposal of dredged material.

STUDY OBJECTIVES

The principal objective of the present study has been to develop a method for conducting meaningful bioassays in the field of the impacts of dredged material and various industrial wastes. The usual gauge of the severity of the impact is the percentage mortality measured against a standard unit of time, e.g., 50 percent mortality in 96 hours of exposure. Measurements of this type can be conducted in the field, but a high percentage of mortality occurring in a relatively short time reflects acute more than chronic impacts. In our view chronic effects cannot be gauged by mortality unless tests are carried out for such inordinately long periods of time (e.g., one month) that costs become a serious factor. The gauge used here is metabolic enzymes that will signal that the organism is under stress. Since enzyme levels cannot be measured in the field, the method that TerEco has developed is properly called a field/laboratory technique.

DEVELOPMENT OF A FIELD ASSAY OF CHRONIC EFFECTS

The Biotal Ocean Monitor

In October 1974 TerEco Corporation was retained by Shell Chemical Company to monitor certain potential chemical changes in the surface water that might result from stack emissions released from the incinerator vessel VULCANUS during the burning of organo-chlorine wastes in the northern Gulf of Mexico (TerEco Corporation, 1974; Wastler et al., 1975). The principal components of the incineration gases were hydrogen chloride and unburned organo-chlorine compounds. No significant effects from the hydrogen chloride were found, but there was an unanswered question whether the unburned organochlorine compounds were producing significant impacts. Although incineration of the organics was 99.95% efficient, the small percentage of unburned material was a substantial amount when 16 thousand or more metric tons were consumed during the course of a complete burn cycle. During the monitoring of a second burn in December 1974, TerEco scientists, with support from the Marine Protection Branch of the Environmental Protection Agency (TerEco Corporation, 1975), drew up designs for devices that could make it possible to study either accute or chronic responses of selected marine organisms to water borne pollutants. These devices, called Biotal Ocean Monitors (BOMs), are suitable for investigations of the low-level responses mentioned above. Laboratory procedures are required to prove whether or not chronic responses are developing in field-exposed organisms. In order to reduce the field time needed for appearance of these responses, TerEco is attempting to develop techniques for following metabolic enzyme responses of the exposed organisms. It must be emphasized that the methodology needs more research before one can say that it is a useful technique.

MATERIALS AND METHODS

Materials -- The BOMs were designed with the following characteristics:

Prior to launching, the various compartments of the BOM are supplied with appropriate test organisms. And, as mentioned earlier, the BOM is equipped to trap water column and/or bottom dwelling species.

Ordinarily the pelagic BOMs are permitted to drift with the currents, much as the planktonic and other weak swimmers would do (Figure 3). This ensures that the organisms in the BOM are not only exposed to the contaminated water for longer periods but also any natural dilution of the toxic materials will also affect the organisms in the BOM (dye tests have demonstrated ease of exchange of water through the mesh). In some instances, however, it may be preferable to anchor the BOM in position. This can be done very easily with a simple bridle, a Danforth anchor, and a length of cable sufficient to reach the bottom and also provide reasonable slack. The benthic BOMs are, of course, lowered to the bottom where they remain for the prescribed 4 or more days (Figure 4). In order to prevent tampering it is desirable that no evidence of the benthic BOM be observed on the surface. This can be achieved by means of a sub-surface float short-strung by a loop to a time-release device which when set off releases the looped cable allowing the float to pop to the surface.

Expendable Pelagic Units

At present TerEco is designing much smaller pelagic BOMs that would be used to determine the effects of oil spills on the marine ecosystem. It is anticipated that they will be readily portable and could be deployed from small boats or aircraft including helicopters. The mesh cylinder carrying the test organisms will float beneath the surface of a slick where it will contact the soluble fractions of the petroleum.

Use of the Benthic Units

The B-BOMs were designed primarily to monitor the impacts of dredged material on the aquatic environment. But they can be used to study the effects of any high-density waste that drops to the bottom. Some knowledge of the current profile of the water column is required in order to deploy the units where the waste might be expected to reach the bottom.

(1) Ability to retain live organisms that ranged in size from diatoms to fishes

(2) The BOM would be constructed of a flow-through mesh that would permit ambient water carrying deleterious substances to circulate freely into and out of their interior where the experimental organisms were housed.

(3) Equipped with traps to catch indigenous species at the monitoring site and hold them apart from the laboratory bred species

(4) Designed to be (a) set adrift with housed specimens, moving with whatever currents are transporting organisms in ambient waters, or (b) anchored to retain its position at a desired location, or (c) lowered to the sea floor to remain in place for varying intervals of time prior to retrieval

To meet these specifications it was necessary to design two types of monitors. The first was constructed to hold pelagic organisms. Accordingly, it was called the Pelagic Biotal Ocean Monitor (P-BOM). The second unit was designed to hold benthic organisms and is called the B-BOM.

Pelagic Units

The structure of the P-BOMs is shown in Figure 1. Three sizes are presently in use. The cylinders or bags may be 1.25, 3, or 6 m in diameter and 3.7, 7.4 or 14.4 m in length, respectively. The bags are made of monofilament nylon screen and the mesh size varies from 0.4 to 2.2 mm, depending on the organisms to be retained.

The 6-meter P-BOMs are designed so that four or more smaller mesh cylinders can be hung inside from the supporting truss system. These can be of different size mesh so that one can hold phytoplankton, another zooplankton, a third small fish, and the fourth might be loaded with benthic organisms, such as the common mussel *Mytilus edulis*. Larger fishes could then be placed in the master cylinder.

These units are equipped with blinker lights and radio beacons, permitting detection up to 25 miles from the monitoring vessel. Samples can be removed from the cylinders at any time in order to establish serial analyses of possible biological impacts of materials introduced into the water column.

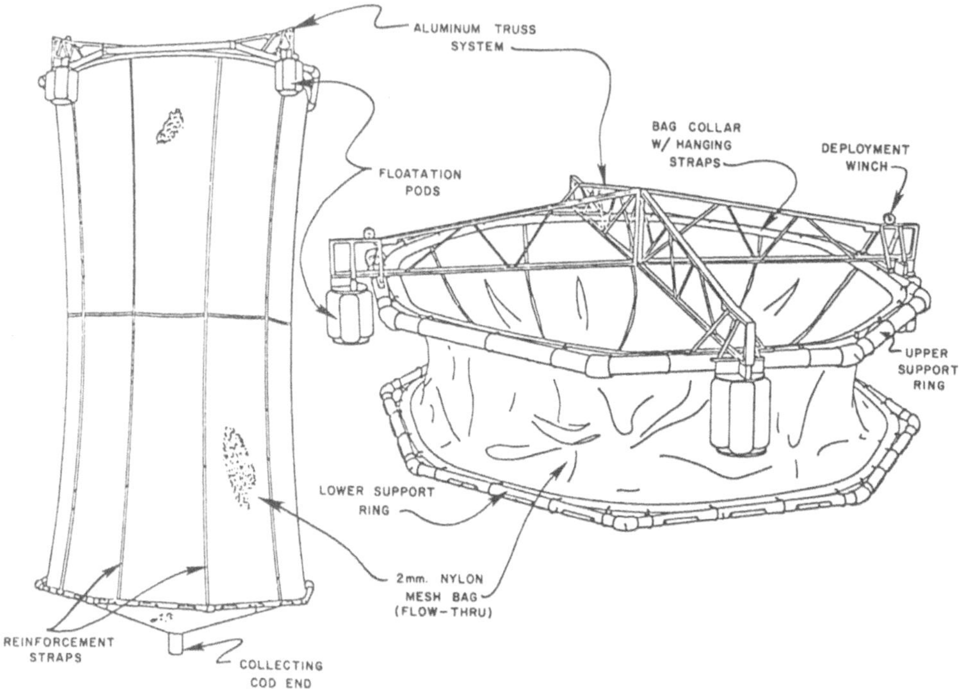

Figure 1. Pelagic Biotal Ocean Monitor - side view fully extended and partially collapsed showing support and flotation system.

Benthic Units

The structure of the Benthic Biotal Ocean Monitors (B-BOMs) is shown in Figure 2. These units are about 2 meters tall and 1.2 m across at the base. The stainless steel base weighs about 159 kilos and anchors the BOM securely to the bottom. The upper part is made of 2 mm monofilament nylon mesh and is shaped like a closed cone in order to reduce the possibility of bottom currents tipping it over. Both the base and the cone are compartmented with trays and pockets which permit various types of organisms to be retained and protected from fishes that will swim freely within the monitor. Trap cones are provided in both the base portion, to capture indigenous bottom organisms, and in the upper net for the capture of free swimming fish.

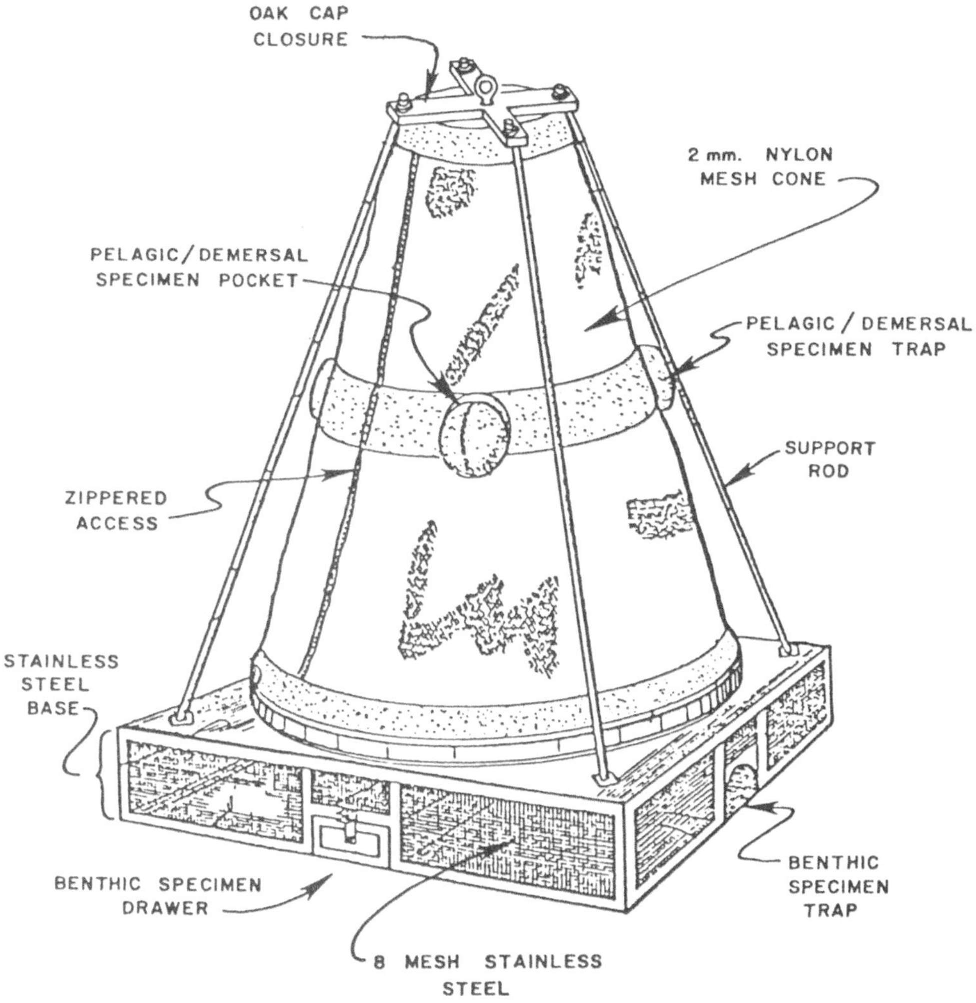

Figure 2. Benthic Biotal Ocean Monitor

Field Methods

In practice the BOMs are used in pairs, one member of which is considered the experimental and other the control. The experimental is placed in a position where the organisms it houses will be exposed to the waste material, whether it be sewage sludge, acid waste, stack emissions dissolved in the water column, or dredged material.

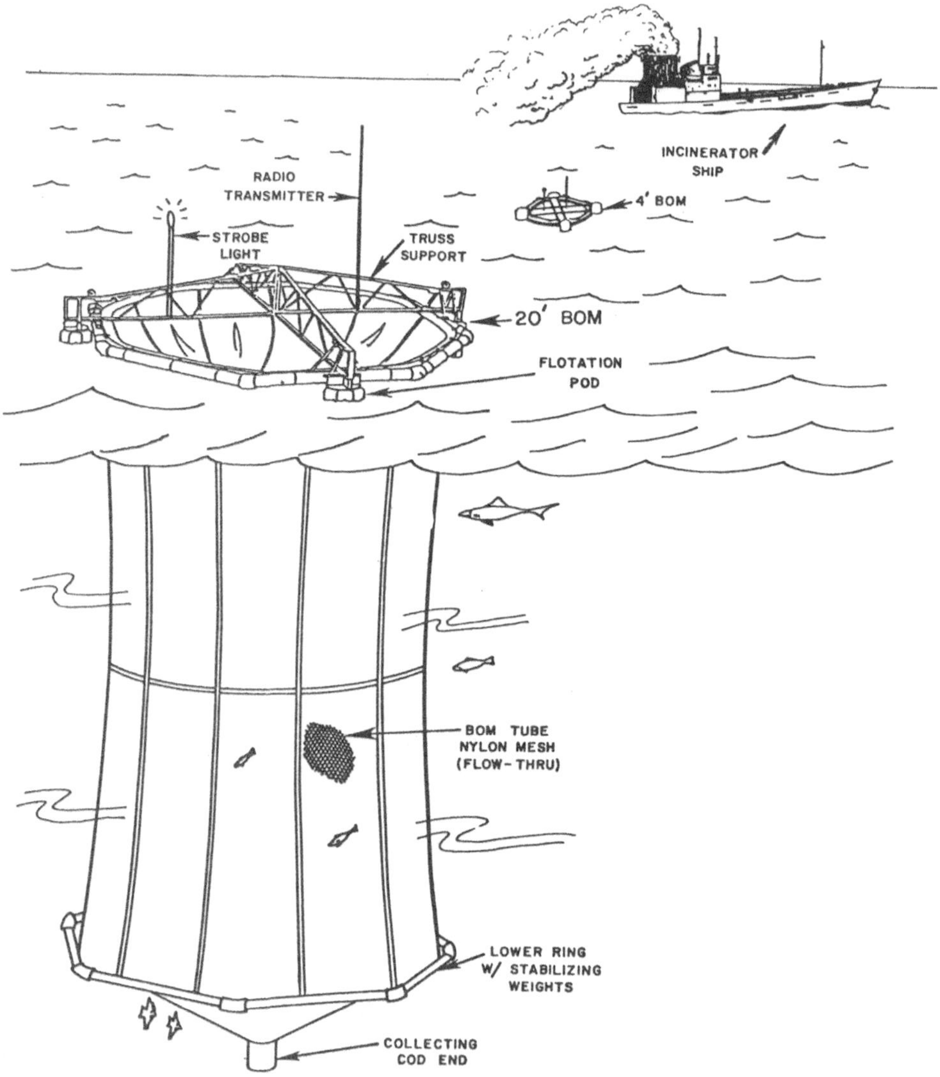

Figure 3. Pelagic Biotal Ocean Monitors in use during the ocean incineration of organohalogen wastes.

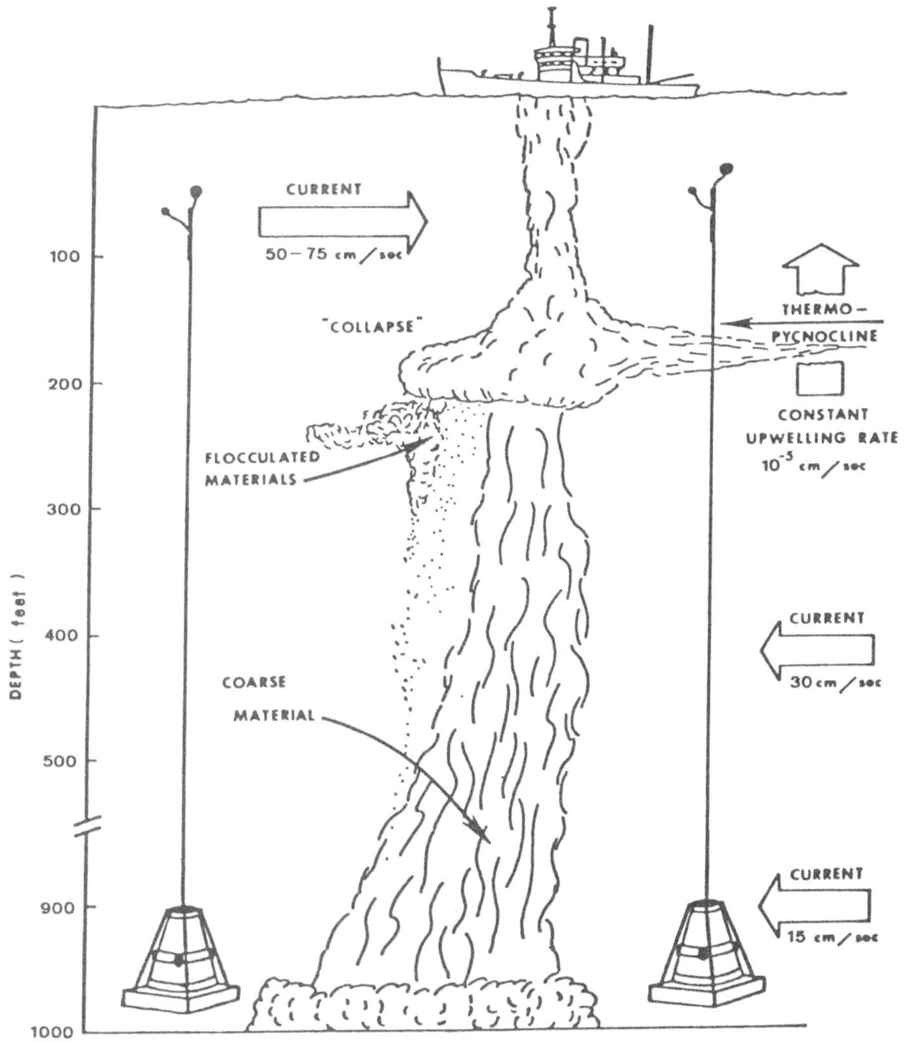

Figure 4. Benthic Biotal Ocean Monitors deployed beneath a hopper dredge shown releasing its load of dredged material.

LABORATORY METHODS

Initial Work with Metabolic Enzymes

Living cells maintain themselves by extracting energy from their food and transforming it into energy to drive biosynthetic reactions and other energy-requiring reactions characteristic of the many types of cells of the body (Atkinson, 1969). All of these activities are mediated through enzymes that are quite functionally specific. It is well known that changes occur at various points in the enzyme systems when an organism is under environmental stress (Bend and Hook, 1977). The initial response may be either an increase or decrease of enzyme levels in cells, say, of the liver (Chambers and Yarbrough, 1976). Eventually there may be a failure of the cell function catalyzed by the particular enzyme under study. The crux of TerEco's approach to the detection of chronic impacts is that prior to complete failure of a given organ or organ system the organism as a whole will exhibit a chronic nonlethal response. Thus, the disposal of toxic chemical wastes in the aquatic environment, which is usually periodic, may well serve as the stimulus for chronic stress responses. Between disposals the organism may be able to rid itself of sufficient toxic material to exhibit only nonlethal responses. Even so, its subnormal response is signalling that if it is subjected to an intensification of the disposed materials, through either increasing amounts at a time or shorter periods between dumps, acute and lethal responses may well be generated.

TerEco is not yet satisfied that the enzymes it has selected for testing are the best or most appropriate. Several have been abandoned, among them acetylcholinesterase and xanthine oxidase. At present we are using ATPase, which is found in the cell mitochondria and responds to excess biphenyls in the environment; catalase, which is dissolved in the cytosol and responds to excesses of toxic metals; and cytochrome P-420 plus P-450, which respond to metals, organochlorines, and to cyclic and long-chain hydrocarbons. Calabrese et al. (1974) and MacInnes et al. (1977) have studied the responses of such liver enzymes as aspartate aminotransferase and glucose-6-phosphate in the cunner (*Tautogolabrus adspersus*) to increase in cadmium (as cadmium chloride) in the environment. They found long-term (30-60 days) exposures of the cunner to 0.1 ppm-Cd caused increased mortality (kidney failure), lowered transaminase activity and elevated glucose-6-phosphate dehydrogenase.

More recently TerEco investigators have been studying the applicability of the adenylate energy charge system to this problem. The advantages of this technique are that a complicated set of enzyme reactions are reduced to a single parameter that relates all control mechanisms to the energy level of the cell, and that value is expressed as the following ratio.

$$\text{Energy Charge EC} = \frac{\overset{(ATP)}{\text{Adenosintriphosphate}} + 1/2\ \overset{(ADP)}{\text{Adenosindiphosphate}}}{\text{ATP} + \text{ADP} + \text{Adenosinmonophosphate (AMP)}}$$

The energy charge of a healthy cell approximates 0.85. Only at and above this level can growth and reproduction occur. Viability is maintained between 0.8 and 0.5, but death occurs below 0.5. Chlorinated hydrocarbons and some metals act as inhibitors of the electron transport system enzymes with a consequent lowering of the energy charge ratio. Some of these systems have been tested in the field.

TESTING RESULTS

A Third Monitoring of Ocean Incineration of Organochloride Wastes

During the period from March 6 to 14, 1977, TerEco personnel monitored a third incineration of organochlorine wastes by the M/T VULCANUS. The burn took place at the now designated ocean incineration site some 165 statute miles south of Galveston, Texas in the northern Gulf of Mexico.

The experimental and control P-BOMs were loaded with fish (*Fundulus grandis*), phytoplankton, sea-urchin eggs, and the crustacean *Mysidopsis bahia*. The experimentals were exposed to the water touchdown of stack emissions of the VULCANUS and permitted to drift. The control units were deployed well upwind and upcurrent of the VULCANUS. After an interval of seven days one hundred twenty fish from the exposed and control BOMs were collected, their livers removed and frozen, and sent to the laboratory. Quantitative analyses were then run for three enzymes, viz., ATPase, which responds to PCBs, metal-sensitive catalase, and cytochrome P-450, which responds to organochlorine compounds. As might have been anticipated from the assay of the waste there was no significant change in the ATPase (Table 1) and only a slight depression in catalase activity (Table 1) in the fish livers from the exposed compared with the control BOMs; however, the P-450 showed a significant response (Table 1), possibly to unburned organochlorines. It is important to note that exposed fish which were returned live and acclimated for a few days in laboratory aquaria before being tested had depurated and showed control levels of all three enzymes. This may be what occurs in the field, at least with pelagic organisms

Table 1. Metabolic Enzyme Changes in Fundulus Grandis Exposed in P-BOMs to Water Contaminated with Stack Emissions from Burning of Organochlorine Wastes in Northwestern Gulf by M/T VULCANUS

	Catalase O.D. 240 nm/min/mg Protein		ATPase O.D. 340 nm/min/mg Protein		P-450 nmoles/g liver	
	Control	Exposed	Control	Exposed	Control	Exposed
No. of fish	50	70	50	70	50	70
Mean	27.8	23.1	3.25	3.08	13.6	36.7
Standard deviation	6.0	4.2	0.76	0.61	6.3	10.4
Exposed fish after 3 days' depuration	26.2		3.21		14.3	

exposed to wastes of ocean incineration. What happens to bottom dwelling organisms as the result of the disposal of contaminated dredged material or sewage sludge may well be quite different.

Monitoring of Sludge Disposal at the Philadelphia Dumpsite

During the period from July 26 to August 12, 1977, TerEco Corporation personnel monitored the dumping of sewage sludge at the Philadelphia dumpsite some 45 miles southeast of Cape May, New Jersey (TerEco Corporation, 1977). Studies were made of the trace metal levels and enzyme activity in several groups of organisms, viz., (1) laboratory animals (*Fundulus grandis*, *Mytilus edulis*, and *Mysidopsis bahia*) that were transported to the site and placed in both the P-BOMs and B-BOMs, (2) those species that were trapped in the B-BOMs, and (3) those species that were collected by use of 3-meter beam trawl. As before, tests were run on organisms exposed and captured in both control and exposed areas. With the exception of *Fundulus* it was found that the invertebrate organisms exposed in the P-BOMs in the water column did not acquire significantly increased burdens of trace metals even though they were in BOMs deployed in the waste plume behind the sludge barge as it was dumping.

Enzyme analyses were run on the *Fundulus* that were transported to the site. The only significant difference between the controls and exposed fishes was found in catalase. This is compatible with the above finding that *Fundulus* did indeed acquire a burden of metals. This indicates also that the particular sludge involved was low in organochlorines and biphenyls (Lear and Pasch, 1975).

Monitoring of Dredged Material Disposal in the New York Bight

During July and August of 1978, TerEco personnel carried out a test monitoring of the disposal of dredged material at the Mud Dumpsite in the New York Bight Apex. Trace metal analyses were carried out on sediments, inside and outside the dumpsite and on various species of organisms exposed in P-BOMs and B-BOMs deployed at the dumpsite. In addition, metal analyses were done on indigenous species either trapped in the B-BOMs or trawled in and around the dumpsite. While laboratory species exposed in P-BOMs at the dumpsite for seven days did not exhibit significantly increased metal burdens, those exposed in B-BOMs did, somewhat in proportion to their concentrations in the surficial sediments (Table 2).

Enzyme analyses for catalase and cytochrome P-450 were carried out using liver tissue from the *Fundulus grandis* exposed in P-BOMs and B-BOMs at the Mud Dumpsite. Little catalase response (to metals) was noted but slight increases in P-450 in organisms exposed in

Table 2. Metals in Sediments Taken from Control and Stations Inside and Downcurrent of Mud Dumpsite Compared with Organisms held in P-BOM and B-BOM Inside Site and with *Cancer irroratus* Trapped in B-BOM and Trawled Inside and Outside Site

1 $NHNO_3$ Leaching Acid. New York Bight, 1978

MEAN VALUES OF STATIONS (n = no. of replicates)	Hg	Cd	Cu	Cr	Fe	Mn	Ni	Pb	Zn
SEDIMENTS									
6 stations inside site	.12	.60	21.5	25.9	3183	83	2.6	32.3	55.3
3 stations downcurrent	.20	.89	32.3	51.0	7403	81	5.3	42.3	96.0
Control station	.16	.25	12.0	17.0	3140	36	1.2	22.0	31.0
ORGANISMS IN P-BOMs									
Mercenaria mercenaria (n=5)	.06	.39	9.5	.29	97	22	3.0	.98	106
Mytilus edulis (n=5)	.07	4.1	6.3	1.10	122	20	2.1	1.40	202
ORGANISMS IN B-BOMs									
Mercenaria mercenaria (n=5)	.24	.66	13.0	3.30	308	14	4.5	6.0	104
Mytilus edulis (n=5)	.24	7.00	10.0	2.20	322	9	3.7	7.0	309
Fundulus grandis (n=5)	.36	.40	5.8	.07	175	71	.3	1.2	152
CANCER IRRORATUS									
Trawled inside site (n=5)	.08	.02	13.1	.56	484	79	1.1	4.1	68
Trapped & held in B-BOM (n=5)	.09	.02	19.0	.84	455	86	1.1	7.2	67
Trawled at control station (n=5)	.08	.03	17.0	.50	1221	48	1.0	5.5	118

P-BOMs apparently were attributable to hydrocarbon contamination in surface water from ship traffic (Table 3). No significant increase in P-450 assays occurred in those fish exposed on the bottom. The adenylate energy charge system (E.C.) was employed for the first time in a monitoring exercise. The grass shrimp, Palaemonetes pugio were exposed in both the P-BOMs and B-BOMs for seven days (Table 3). Values obtained for the shrimp in the P-BOMs were normal (0.87), but the values in the B-BOMs were depressed, averaging only 0.77. This indicates that the animals were under stress on the bottom.

DISCUSSION

The above results suggest that the combining of the Biotal Ocean Monitors with appropriate laboratory analyses may provide an advanced technique for bioassaying acute and, especially, chronic effects of certain pollutants in the marine environment. One merit of this approach is the fact that the BOMs can and will capture indigenous species and expose them alongside laboratory raised species to whatever type of waste is being monitored at the time. One criticism of more conventional bioassays is that the usual laboratory species have frequently been selected for their general tolerance of environmental perturbations, so that they are not good indicators of the responses of wild species. It is our experience that many indigenous species, especially from the open ocean, do so poorly in the laboratory that it is unlikely that they can be used for meaningful bioassays. SCUBA observations confirm that wild species captured by the B-BOMs appear not to be disturbed by confinement in their natural milieu.

At this time, satisfactory chronic bioassay monitoring of ocean incineration of such organochlorine waste components as 1, 2, 3, trichloropropane, 1, 2, dichloroethane, and 1, 1, 2 trichloroethane can be achieved by liver assay of cytochrome P-450. The wastes of this type that TerEco has monitored thus far are relatively low in metals (Wastler et al., 1975). Where metals may be the chief concern, as in the case of some sewage sludges, industrial wastes, and certain dredged material, our indicator would be catalase. If polychlorinated biphenyls are known to be involved, then one should determine the levels of ATPase. While we are as yet uncertain as to its applicability, preliminary indications are that the adenylate energy charge ratio may be a good indicator of an organism's general condition while responding to a variety of impacts.

ACKNOWLEDGMENTS

Assistance of the Marine Protection Branch of the U. S. Environmental Protection Agency in developing this program is gratefully

Table 3. Means of Catalase Values Cytochrome P-450 from Liver Tissue of Fundulus grandis and Adenylate Energy Charge of Palaemonetes pugio Exposed at New York Bight Mud Dumpsite in July-August 1978

	Catalase	
Site and BOM Type 45 Fundulus/BOM	Bergmeyer Units* per mg liver protein	S. D.
P-BOM	1.30	± .006
B-BOM	1.60	± .002
	Cytochrome P-450	
30 Fundulus/BOM	nmole P-450 per mg liver protein ($x\ 10^{-2}$)	S. D.
P-BOM	6.90	1.25
B-BOM	6.00	0.13
	Energy Charge	
25 Palaemonetes/BOM Body wt. < 200 mg	Mean ATP nM/mg Body wt.	Mean Energy Charge
P-BOM	.583 ± .11	.87 ± .04
B-BOM	.636 ± .32	.77 ± .01

*Bergmeyer Unit = Amount of catalase required to decompose 1 g of hydrogen peroxide in 1 minute.

acknowledged (Contract: EPA 68-01-2893, T. A. Wastler, Project Officer).

REFERENCES

Atkinson, D. E. 1969. Regulation of enzyme function. Amn. Rev. Microbiol., 23:47-68.

Bend, J. R. and G. E. R. Hook. 1977. Hepatic and extrahepatic mixed-function oxidases. In: (S. R. Geiger, ed.,) Handbook of Physiology, Sect. 9: Reactions to Environmental Agents, Waverly Press, Inc., Baltimore 21202. pp. 419-440.

Calabrese, A., R. S. Collier, and J. E. Miller. 1974. Physiological response of the cunner, *Tautogolabrus adspersus*, to cadmium. NOAA Tech. Rpt. NMFS SSRF-681, Seattle, Washington.

Chambers, J. E. and J. D. Yarbrough. 1976. Xenobiotic biotransformation systems in fishes. Comp. Biochem. Physiol., Vol. 55C. pp. 77-84.

Environmental Protection Agency/Corps of Engineers. 1977. Ecological evaluation of proposed discharge of dredged material into ocean waters. U. S. Army Engineers Waterways Experiment Station, CE, Vicksburg, Miss.

Lear, D. W. and G. G. Pasch. 1975. Effects of ocean disposed activities on mid-continental shelf environment of Delaware and Maryland. EPA 903/9-75-015. U. S. Environmental Protection Agency, Region III. 28 pp.

MacInnes, J. R., F. P. Thurberg, R. A. Greig, and E. Gould. 1977. Long-term cadmium stress in the cunner, *Tautogolabrus adspersus*. Fishery Bull., vol. 75:199-203.

Pequegnat, W. E. 1978b. Combined field-laboratory approaches to detecting impacts of waste materials on aquatic organisms. Procs. 3rd Annual Conf. on Treatment and Disposal of Industrial Wastewaters and Residues. Houston, Texas.

Pequegnat, W. E., B. M. James, E. A. Kennedy, A. D. Fredericks, and R. R. Fay. 1978a. Development and application of a biotal ocean monitor system to studies of the impacts of ocean dumping. Tech. Rept. to Environmental Protection Agency, Washington, D. C. Contracts 68-01-2893 and 68-01-4797.

Pequegnat, W. E., B. M. James, E. A. Kennedy, A. D. Fredericks, R. R. Fay, and F. G. Hubbard. 1979. Application of the Biotal Ocean Monitor System to a preliminary study of the impacts of ocean dumping of dredged material in the New York Bight. Tech. Rept. to U. S. Army Engineer Waterways Experiment Station. Vicksburg, Miss. Contract No. 68-01-4797.

TerEco Corporation. 1974. A field monitoring study of the effects of organic chloride waste incineration on the marine environment in the northern Gulf of Mexico. Report on the first Research Burn to Shell Chemical Company, Houston, Texas. Contract No. 74-202-10.

TerEco Corporation. 1975. Sea-level monitoring of the incineration of organic chloride waste by M/T VULCANUS in the northern Gulf of Mexico: Shell waste burn No. 2. Report to the Marine Protection Agency, Washington, D. C. Contract No. 68-01-2829.

TerEco Corporation. 1977. A report on the Philadelphia dumpsite and Shell incineration monitorings. Report to the Marine Protection Branch, the Environmental Protection Agency, Washington, D. C.

Wastler, T. A., C. K. Offutt, C. K. Fitsimmons, and P. E. DesRosier. 1975. Disposal of organochlorine wastes at sea. EPA-43019-75-014, Division of Oil and Special Materials Control, Office of Water and Hazardous Materials, U. S. Environmental Protection Agency, Washington, D. C. 20460.

INDEX

GPSR Compliance

The European Union's (EU) General Product Safety Regulation (GPSR) is a set of rules that requires consumer products to be safe and our obligations to ensure this.

If you have any concerns about our products, you can contact us on ProductSafety@springernature.com

In case Publisher is established outside the EU, the EU authorized representative is:

Springer Nature Customer Service Center GmbH
Europaplatz 3
69115 Heidelberg, Germany

Batch number: 10855257

Printed by Printforce, the Netherlands